CAMBRIDGE
UNIVERSITY PRESS

University Printing House, Cambridge CB2 8BS, United Kingdom

Cambridge University Press is part of the University of Cambridge.

It furthers the University's mission by disseminating knowledge in the pursuit of education, learning and research at the highest international levels of excellence.

www.cambridge.org
Information on this title: www.cambridge.org/9781107424722

© Y. Kuramoto and Y. Kato 2009

First published 2009
First paperback edition 2014

A catalogue record for this publication is available from the British Library

Library of Congress Cataloguing in Publication data
Kuramoto, Y. (Yoshio)
Dynamics of one-dimensional quantum systems : inverse-square interaction models / Yoshio Kuramoto and Yusuke Kato.
p. cm.
ISBN 978-0-521-81598-7 (hardback)
1. Electronic structure–Mathematical models. 2. Matrix inversion. 3. Many-body problem.
I. Kato, Y. (Yusuke), 1929– II. Title.
QC176.8.E4K867 2010
530.4'11015118–dc22

2009023529

ISBN 978-0-521-81598-7 Hardback
ISBN 978-1-107-42472-2 Paperback

DYNAMICS OF ONE-DIMENSIONAL QUANTUM SYSTEMS

One-dimensional quantum systems show fascinating properties beyond the scope of the mean-field approximation. However, the complicated mathematics involved is a high barrier to non-specialists. Written for graduate students and researchers new to the field, this book is a self-contained account of how to derive a quasi-particle picture from the exact solution of models with inverse-square interparticle interactions.

The book provides readers with an intuitive understanding of exact dynamical properties in terms of exotic quasi-particles that are neither bosons nor fermions. Powerful concepts, such as the Yangian symmetry in the Sutherland model and its lattice versions, are explained. A self-contained account of non-symmetric and symmetric Jack polynomials is also given. Derivations of dynamics are made easier, and are more concise than in the original papers, so readers can learn the physics of one-dimensional quantum systems through the simplest model.

YOSHIO KURAMOTO is a Professor of Physics at Tohoku University, Japan. He is an expert on strongly correlated electron systems, and has authored and co-authored several books and many papers in related research fields. He is a member of the Physical Society of Japan, and has served as one of the editors of the Society's journal.

YUSUKE KATO is an Associate Professor in the Department of Basic Science at the University of Tokyo. His working fields are physics of condensed matter, correlated electron systems in one dimension, integrable systems, superconductivity and Bose–Einstein condensation.

Dynamics of One-Dimensional Quantum Systems

Inverse-Square Interaction Models

YOSHIO KURAMOTO

Tohoku University

YUSUKE KATO

University of Tokyo

Contents

Contents

Preface

This book is concerned primarily with the exact dynamical properties of one-dimensional quantum systems. As a crucial property of exactly soluble models, we assume that the interaction decays as the inverse square of the distance. The family of these models is called the inverse-square interaction $(1/r^2)$ models. In the one-dimensional continuum space, the model is often referred to as the Calogero–Sutherland model. In the one-dimensional lattice, on the other hand, the first $1/r^2$ models appeared as a spin model, which is now called the Haldane–Shastry model. Soon after the discovery of the Haldane–Shastry model, it was recognized that the imposition of supersymmetry allows the model to acquire the charge degrees of freedom, while keeping the exactly soluble nature. The resultant one-dimensional electron model is called the supersymmetric t–J model. Various generalizations of these models have been proposed.

Recent experimental progress in quasi-one-dimensional electron systems, especially by neutron scattering and photoemission spectroscopy, has enhanced the theoretical motivation for exploring the dynamics over a wide frequency and momentum range. The $1/r^2$ models are ideally suited to meet this situation, since the model allows derivation of exact dynamical information most easily and transparently. In spite of the special appearance of the $1/r^2$ models, the intuition thus obtained contributes greatly to understanding low-dimensional physics in general. This kind of approach to dynamics is complementary to another powerful approach using the bosonization and conformal field theory. The latter is especially suitable to asymptotics of correlation functions at long spatial and temporal distances.

The literature relevant to the $1/r^2$ models is vast and scattered. Moreover, many papers include a difficult-looking mathematical set-up. This situation may cause newcomers to see a barrier too high to jump over before enjoying the rich and beautiful ingredients of the $1/r^2$ models. For several years, the authors have realized the necessity of a comprehensive treatise. This book is intended to be accessible to non-specialists who are interested in strongly correlated quantum systems. It explains the wonderfully beautiful physics and related mathematics in a self-contained manner, without assuming special knowledge on theories in one dimension. In order to make a coherent discussion, we have included many results that are newly derived for this book, in addition to summarizing what has been reported in the literature. We hope that this book is useful not only to experts already working in the field, but also to graduate students and researchers trying to delve into the fascinating physics in low dimensions.

We are grateful to our collaborators, former students, and scientific colleagues who have worked in this area and helped our understanding of the subject, especially to M. Arikawa, N. Kawakami, T. Kimura, R. Nakai, O. Narayan, Y. Saiga, B. S. Shastry, B. Sutherland, T. Yamamoto, H. Yokoyama, and J. Zittartz. Our deepest thanks go to M. Arikawa, who carefully read the first version of the manuscript and made many useful suggestions.

1

Introduction

1.1 Motivation

Interactions in many-body systems bring about collective phenomena such as superconductivity and magnetism. In many cases, simple mean-field theory provides a basic understanding of these phenomena. In fermion systems in one dimension, however, neither the mean-field theory nor perturbation theory works if it starts from the non-interacting fermions. This is because the interaction effects in one dimension are much stronger than those in higher dimensions. Intuitively speaking, two particles cannot avoid collision in a single-way track in contrast with two and three dimensions. Thus the interaction effects appear in a drastic way in one dimension.

Another aspect in one dimension, which overcompensates the difficulty of perturbation and mean-field theories, is that a complete account of interaction effects is possible under certain conditions. The class of systems satisfying such conditions is referred to as exactly solvable. Soon after the establishment of quantum mechanics, Bethe solved exactly the Heisenberg spin model in one dimension [28]. The basic idea of the solution is now called the Bethe ansatz. Since then, theoretical physics in one dimension has developed into a magnificent edifice, including sophisticated mathematical techniques. In many cases, the eigenfunctions derived by the Bethe ansatz consist of plane waves that are defined stepwise for each spatial configuration of particles. Since the coefficients of plane waves depend on the configuration, the property of the wave function cannot be made explicit without detailed knowledge of these coefficients. We mention some of the recent monographs on the Bethe ansatz and its extensions [54, 118, 179]. A comprehensive account on exactly solvable models has recently been given by Sutherland [178].

The models solved by the Bethe ansatz are characterized by short-range interactions such as on-site repulsion or the next-nearest-neighbor exchange interaction. On the other hand, it was found by Calogero that another class of models also permits exact solution [34, 35]. The models have repulsive interactions decaying as the inverse square of the interparticle distance r. In order to prevent the blow-up of particles toward infinite distance, an attractive harmonic potential can be added to the system. Alternatively, one takes the periodic boundary condition with the system length L, and employs superposition of the $1/r^2$ potential as

$$\sum_{n=-\infty}^{\infty} \frac{1}{(r+nL)^2} = \left(\frac{\pi/L}{\sin \pi r/L}\right)^2. \tag{1.1}$$

Then by construction the system does not blow up, while keeping the translational invariance. This model was proposed by Sutherland [172, 174], and hence is called the Sutherland model. If one refers to both models simultaneously, it seems appropriate to call them the Calogero–Sutherland models. Some years later, Moser analyzed the classic mechanical version of these models mathematically [135], and his name is sometimes added in referring to the models.

The $1/r^2$ models have much simpler mathematical (algebraic) structure, compared to the conventional integrable models solved by the Bethe ansatz. This simplicity enables us to derive explicitly the exact expressions of dynamical correlation functions such as the Green function, the density–density correlation function, and the spin–spin correlation function. The resultant expressions are remarkably simple, but still keep nontrivial features inherent to interacting particle systems. Further, the mathematical tools used in the derivation are far from complicated. Thus, the $1/r^2$ models provide comprehensible examples for studying dynamics of interacting particles.

In contrast with the Fermi liquid in three dimensions, the one-dimensional fermions behave as the Tomonaga–Luttinger liquid in the limit of long time and long distance. Here the conformal field theory (CFT) describes nicely the asymptotics of correlation functions. According to the CFT, characterization of the interaction parameters can be done through analysis of the finite-size correction of the ground state energy. Since the $1/r^2$ models allow for calculation of the finite-size correction much more easily than the Bethe-solvable models, the $1/r^2$ models serve as an instructive example to visualize how the CFT works in the Tomonaga–Luttinger liquid. The importance of the $1/r^2$ models does not, however, lie only in the mathematical structure. Through the study of the $1/r^2$ models, one can also learn about the dynamics of the correlated electrons in real systems. For example, the neutron scattering intensity of $S = 1/2$ antiferromagnetic spin chain reveals a similarity to the spectral function of the spin correlation function of the $1/r^2$ exchange interaction model, which is called the Haldane–Shastry model [77, 161]. A related model with charge degrees of freedom is still exactly solvable provided a supersymmetry is imposed [119]. The spin–charge separation of one-dimensional electrons can then be explicitly seen in the spectral weight of the Green function of the supersymmetric t–J model.

1.2 One-dimensional interaction as a disguise

As we shall explain in detail, the wave function in the ground state of the $1/r^2$ models can be derived explicitly as the product of two-body wave functions. This feature is quite in contrast with cases solved by the Bethe ansatz. The special feature of the $1/r^2$ interaction already appears in most elementary quantum mechanics. Let us consider a free particle with mass $m = 1/2$ in the three-dimensional space. The Hamiltonian is given by

$$H = -\frac{\partial^2}{\partial x^2} - \frac{\partial^2}{\partial y^2} - \frac{\partial^2}{\partial z^2} = -\frac{1}{r^2}\frac{\partial}{\partial r}r^2\frac{\partial}{\partial r} + \frac{l^2}{r^2}, \qquad (1.2)$$

where $r^2 = x^2 + y^2 + z^2$, and l is the angular momentum operator. We take the units $\hbar = 1$ throughout the book. In the polar coordinates, there appears a fictitious potential leading to the centrifugal force. Namely, the free motion in higher dimensions generates a fictitious potential if the radial motion alone is extracted [146]. Conversely, the potential $l(l + 1)/r^2$ in the radial coordinate is a disguise of free motion in higher dimensions. The form of the radial kinetic energy in (1.2) is interpreted as coming from the metric of the one-dimensional space. Pursuing this idea in many-body systems, one gains a perspective that interactions in exactly solvable models are a disguise of some kind of free motion in another space [146]. Alternatively, a matrix model has been constructed where the coordinates of N particles are regarded as eigenvalues of an $N \times N$ matrix. The transformation matrix for diagonalization appears as the $1/r^2$ potential [151].

In the early stage of the Tomonaga–Luttinger theory, all low-energy excitations are regarded as bosons. Actually, the statistics of excitations need not be restricted to bosons. In some cases, the interaction among bosons is absorbed into a new statistics describing exclusion of available one-body states. This idea applies to many interacting systems approximately, and to the $1/r^2$ models exactly. The exclusion includes fermions and bosons as special cases. Generally, however, the statistics is fractional. In order to account for the resultant quasi-particles obeying fractional exclusion statistics, concepts such as the Yangian symmetry and the supersymmetry turn out to be useful. These new concepts make it much easier to understand exact dynamics (and also thermodynamics) intuitively. Our key strategy in this respect is to rely on the picture of quasi-particles obeying fractional exclusion statistics. In terms of these exotic quasi-particles, the dynamics of one-dimensional systems can be understood intuitively.

In the last decades, intensive study of the $1/r^2$ models has brought about deep intuition into the structure of the excitation spectrum in one-dimensional systems in general. The most remarkable observation is that elementary excitations behave as free particles subject to certain statistical constraints. As a result, these particles obey the statistics of neither fermions nor bosons. In other words, the exchange of two excitations leads to a scattering phase shift which is independent of their momenta, but which is neither π (antisymmetric) nor 0 (symmetric).

The situation may become clearer if we make an analogy to the Fermi liquid theory. The excitations in the Bethe-soluble models have a phase shift that does depend on their momenta. Therefore, certain parameters are necessary to characterize the momentum dependence. These parameters are analogous to Landau parameters that describe interactions between the quasi-particles in the Fermi liquid. In this analogy, the excitations in the $1/r^2$ models do not need the analogue of the Landau parameters, and are comparable to free fermions except for the statistics. Just as the understanding of metals in general has been much facilitated by the free-electron model, the dynamics in one dimension should be much better understood by reference to "free" models, i.e., the $1/r^2$ models.

1.3 Two-body problem with $1/r^2$ interaction

We demonstrate the peculiar features of the $1/r^2$ model by taking the simplest example. Let us consider the two-body problem with Hamiltonian

$$H_2 = -\frac{\partial^2}{\partial x_1^2} - \frac{\partial^2}{\partial x_2^2} + g \left[\frac{\pi/L}{\sin \pi(x_1 - x_2)/L} \right]^2. \tag{1.3}$$

For the moment we assume that the two particles are distinguishable, and do not care about the symmetry of the wave function. If the distance $|x_1 - x_2|$ is much smaller than L, the interaction reduces to $g/(x_1 - x_2)^2$. The center of gravity $X = (x_1 + x_2)/2$ has free motion with wave number Q. In terms of X and the relative coordinate $x = x_1 - x_2$, the wave function is factorized into the form $\psi_g(x_1, x_2) = \phi_g(x) \exp(iQX)$, where $\phi_g(x)$ is an eigenfunction of a one-body Hamiltonian H_1 given by

$$H_1(x) = H_2 - \frac{1}{2}Q^2 = -2\frac{\partial^2}{\partial x^2} + g\left(\frac{\pi/L}{\sin \pi x/L}\right)^2. \tag{1.4}$$

Instead of solving (1.4) in the standard way, we discuss alternative ideas which are useful in generalizing to the many-body problem. Let us first examine the wave function $\phi_g(x)$ for $|x| \ll L$ where the potential in H_1 tends to g/x^2. Then $H_1(x)$ has the scaling property

$$H_1(ax) = a^{-2}H_1(x).$$

An eigenfunction should also have the scaling property for $x \sim 0$

$$\phi_g(ax) = a^\lambda \phi_g(x), \tag{1.5}$$

with certain number λ. The only solution with property (1.5) is the power-law function $\phi_g(x) = x^\lambda$. Upon differentiation twice, we obtain $\lambda(\lambda - 1)\phi_g(x)/x^2$. By taking $\lambda(\lambda - 1) = g/2$, the kinetic term cancels the potential term. Then $\phi_g(x)$ turns out to be the eigenfunction of H_1. Since we have $\lambda = (1 \pm \sqrt{1 + 2g})/2$, only the case of $g \geq -1/2$ is meaningful. Otherwise, the attractive potential causes the system to collapse as in the classical system, and the ground state cannot be defined. This situation has already been discussed by Landau and Lifshitz [122] and by Sutherland [172]. In the following we only consider the case $g > 0$, and take the positive λ as the relevant solution. We can extend the range of x so as to be consistent with the periodic boundary condition, simply by replacing x^α by $|\sin \pi x/L|^\alpha$.

It is possible to derive all the eigenvalues and eigenfunctions by using the factorization method [89], which has been refined under the name of "supersymmetric quantum mechanics" [192]. We introduce a variable $\eta \equiv \pi x/L$ and rewrite (1.4) as

$$H_1 = 2\left(\frac{L}{\pi}\right)^2 [p_\eta^2 + W_\lambda(\eta)^2 + W_\lambda'(\eta) + \lambda^2] \equiv 2\left(\frac{L}{\pi}\right)^2 \mathcal{H}_\lambda, \tag{1.6}$$

where $p_\eta = -i\partial/\partial\eta$ and $W_\lambda(\eta) = \lambda \cot \eta$. Then $\mathcal{H}_\lambda$ takes a factorized form

$$\mathcal{H}_\lambda = (p_\eta - iW_\lambda)(p_\eta + iW_\lambda) + \lambda^2 \equiv A_\lambda^\dagger A_\lambda + \lambda^2. \tag{1.7}$$

An eigenfunction of $\mathcal{H}_\lambda$ is given by

$$\phi_\lambda(\eta) = \sin^\lambda \eta = \exp[U_\lambda(\eta)], \tag{1.8}$$

where we have introduced $U_\lambda(\eta) = \lambda \ln \sin \eta$. This gives $U_\lambda'(\eta) = W_\lambda(\eta)$, and it is evident that $A_\lambda \phi_\lambda(\eta) = 0$. Since $A_\lambda^\dagger A_\lambda$ is a non-negative operator, there are no states with lower energy. Hence, ϕ_λ gives the ground state of $\mathcal{H}_\lambda$ with energy λ^2.

We note the property

$$A_\lambda A_\lambda^\dagger = p_\eta^2 + W_\lambda(\eta)^2 - W_\lambda'(\eta) = p_\eta^2 + \frac{\lambda(\lambda+1)}{\sin^2 \eta} - \lambda^2$$
$$= A_{\lambda+1}^\dagger A_{\lambda+1} - \lambda^2, \tag{1.9}$$

which corresponds to the shift $\lambda \to \lambda + 1$ in $\mathcal{H}_\lambda$ together with subtracting the constant term λ^2. Combination of (1.7) and (1.9) makes it possible to derive all the excited states. Let us take the ground state $\phi_{\lambda+1}$ of $H_{\lambda+1}$ with the eigenvalue $(\lambda + 1)^2$. Namely, we have

$$A_\lambda A_\lambda^\dagger \phi_{\lambda+1} = [(\lambda+1)^2 - \lambda^2]\phi_{\lambda+1}. \tag{1.10}$$

Applying $A_\lambda^\dagger$ from the left, we obtain

$$A_\lambda^\dagger A_\lambda A_\lambda^\dagger \phi_{\lambda+1} = [(\lambda+1)^2 - \lambda^2]A_\lambda^\dagger \phi_{\lambda+1}. \tag{1.11}$$

Thus the state $A_\lambda^\dagger \phi_{\lambda+1}$ proves to be an excited state of $A_\lambda^\dagger A_\lambda$.

We now explain briefly the idea of the supersymmetric quantum mechanics. We may treat the pair $A_\lambda A_\lambda^\dagger$ and $A_\lambda^\dagger A_\lambda$ as components of a 2×2 matrix:

$$\mathcal{H}_{\text{pair}} = \begin{pmatrix} A_\lambda^\dagger A_\lambda & 0 \\ 0 & A_\lambda A_\lambda^\dagger \end{pmatrix} = QQ^\dagger + Q^\dagger Q \equiv \{Q, Q^\dagger\}, \tag{1.12}$$

where

$$Q = \begin{pmatrix} 0 & 0 \\ A_\lambda & 0 \end{pmatrix}, \quad Q^\dagger = \begin{pmatrix} 0 & A_\lambda^\dagger \\ 0 & 0 \end{pmatrix}. \tag{1.13}$$

The space of the 2×2 matrix can be regarded as a pseudo-spin spanned by the Pauli matrices. Here Q and $Q^\dagger$ have an analogy with spin-flips $s_\pm = s_x \pm i s_y$. Alternatively, we may include the pseudo-fermion operators $f, f^\dagger$ by the identification

$$\frac{1}{2}(1 - \sigma_z) = f^\dagger f. \tag{1.14}$$

Then the operators $Q, Q^\dagger$ in (1.12) are written as

$$Q = f^\dagger A_\lambda, \quad Q^\dagger = A_\lambda^\dagger f. \tag{1.15}$$

It is obvious that $Q^2 = (Q^\dagger)^2 = 0$ and

$$[\mathcal{H}_{\text{pair}}, Q] = [\mathcal{H}_{\text{pair}}, Q^\dagger] = 0. \tag{1.16}$$

The last equality means that $\mathcal{H}_{\text{pair}}$ is invariant against the pseudo-spin rotation, and the conserved quantity Q is called the supercharge. In this framework, the degeneracy demonstrated by (1.11) is interpreted as a consequence of the supersymmetry. The use of $Q, Q^{\dagger}$ motivates us to refer to the factorization method as "supersymmetric quantum mechanics".

The operators A_{λ} and $A_{\lambda}^{\dagger}$ have the commutation rule

$$[A_{\lambda}, A_{\lambda}^{\dagger}] = -2W_{\lambda}. \tag{1.17}$$

In a special case of $W_{\lambda} = -x/2$, the commutation rule reduces to that of bosonic creation and annihilation operators. Hence, A_{λ} and $A_{\lambda}^{\dagger}$ can be regarded as a generalization of bosonic operators.

Now we iterate the procedure of increasing λ by unity to obtain all excited states. Let us use the fact $A_{\lambda}^{\dagger}A_{\lambda} = A_{\lambda-1}A_{\lambda-1}^{\dagger} + (\lambda - 1)^2 - \lambda^2$ as derived from (1.9). After this substitution in (1.11), we multiply $A_{\lambda-1}^{\dagger}$ from the left to obtain

$$A_{\lambda-1}^{\dagger}A_{\lambda-1}A_{\lambda-1}^{\dagger}A_{\lambda}^{\dagger}\phi_{\lambda+1} = [(\lambda+1)^2 - (\lambda-1)^2]A_{\lambda-1}^{\dagger}A_{\lambda}^{\dagger}\phi_{\lambda+1}, \tag{1.18}$$

which shows that $A_{\lambda-1}^{\dagger}A_{\lambda}^{\dagger}\phi_{\lambda+1}$ is an excited state of $A_{\lambda-1}^{\dagger}A_{\lambda-1}$. This process can be iterated. The wave function $\phi_{n+1;\lambda+1} \equiv A_{\lambda-n}^{\dagger}\cdots A_{\lambda-1}^{\dagger}A_{\lambda}^{\dagger}\phi_{\lambda+1}$ with $n \geq 0$ satisfies the equation

$$A_{\lambda-n}^{\dagger}A_{\lambda-n}\phi_{n+1;\lambda+1} = [(\lambda+1)^2 - (\lambda-n)^2]\phi_{n+1;\lambda+1}. \tag{1.19}$$

Equivalently we obtain for $m \geq 1$

$$\mathcal{H}_{\lambda}\phi_{m;\lambda} = (\lambda+m)^2\phi_{m;\lambda}. \tag{1.20}$$

We identify $\phi_{0;\lambda}$ as ϕ_{λ} to include the case of $m = 0$ in the above. In this way we can derive all the excited states of $\mathcal{H}_{\lambda}$ starting from the ground state of $H_{\lambda'}$ with appropriate $\lambda' > \lambda$. Figure 1.1 shows the situation where the ordinate κ gives the energy as κ^2.

Conversely, starting from a free state $\phi_{m;0} = \sin m\eta$ at $\lambda = 0$, we can construct the eigenfunctions of $\mathcal{H}_{\lambda}$ as $\phi_{m;\lambda} = A_{\lambda-1}\cdots A_1 A_0 \phi_{m+\lambda;0}$. Figure 1.1 also shows this inverse direction of construction. Note that the spectrum of $\mathcal{H}_{\lambda}$ above the ground-state energy λ^2 is the same as that of the free system. To derive $\phi_{n;\lambda+n}$ explicitly, we use the relation $A_{\lambda}^{\dagger} = \exp(-U_{\lambda})p_{\eta}\exp(U_{\lambda})$ and obtain

$$\begin{aligned}
\phi_{n;\lambda+n} &= \exp(-U_{\lambda+1})p_{\eta}\exp(U_{\lambda+1} - U_{\lambda+2})\cdots p_{\eta}\exp(U_{\lambda+n})\phi_{\lambda+n} \\
&= \exp(-U_{\lambda})[\exp(-U_1)p_{\eta}]^n \exp(U_{\lambda+n})\phi_{\lambda+n} \\
&= \phi_{\lambda} \times (1-y^2)^{-\lambda}\left(-\mathrm{i}\frac{\mathrm{d}}{\mathrm{d}y}\right)^n (1-y^2)^{\lambda+n},
\end{aligned} \tag{1.21}$$

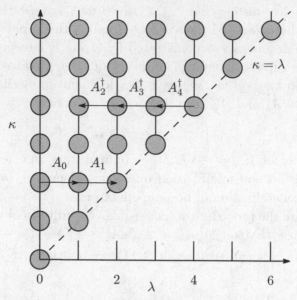

Fig. 1.1. The spectrum of the one-body Sutherland model $\mathcal{H}_\lambda$. The creation operator $A_\lambda^\dagger$ generates an excited state for $H_{\lambda-1}$ from an eigenstate of $\mathcal{H}_\lambda$. These two states are degenerate, and their energy is given by κ^2. The annihilation operator A_λ generates an eigenstate of $H_{\lambda+1}$ from that of $\mathcal{H}_\lambda$.

where $y = \cos\eta$. The last expression includes, apart from the normalization factor, the Rodrigues formula for the nth-order Gegenbauer polynomial $C_n^{\lambda+1/2}(y)$. Namely, we obtain

$$\phi_{n;\lambda+n}(\eta) = C_n^{\lambda+1/2}(\cos\eta)\phi_\lambda(\eta). \tag{1.22}$$

The generating function of Gegenbauer polynomials is given by

$$(1 - 2yt + t^2)^{-\lambda-1/2} = \sum_{n=0}^{\infty} C_n^{\lambda+1/2}(y)t^n. \tag{1.23}$$

In the special case $\lambda = 0$, it is reduced to the Legendre polynomial $C_n^{1/2}(y) = P_n(y)$. Because of the similarity to the Legendre polynomial, $C_n^{\lambda+1/2}(y)$ is also called the ultraspherical polynomial.

Let us come back to the two-body system. For each particle j, we introduce the complex coordinate $z_j = \exp(2\pi i x_j/L)$ which specifies a point on the unit circle. The wave function of two particles is written as

$$\psi_{n;\lambda+n}(z_1, z_2) = \exp(iQX)\phi_{n;\lambda+n}(\eta), \tag{1.24}$$

where $\eta = \pi(x_1 - x_2)/L$. We assert that $\psi_{n;\lambda+n}(z_1, z_2)$ is a homogeneous polynomial of z_1 and z_2 times an integer (or half-integer) power of $z_1 z_2$.

The sum q of powers of z_1 and z_2 in $\psi_{n;\lambda+n}(z_1, z_2)$ is related to the total momentum Q as $q = LQ/(2\pi)$. To prove the assertion, we note the relations

$$2\cos\eta = (z_1/z_2)^{1/2} + (z_2/z_1)^{1/2} = (z_1 z_2)^{-1/2}(z_1 + z_2), \tag{1.25}$$

$$2\mathrm{i}\sin\eta = (z_1/z_2)^{1/2} - (z_2/z_1)^{1/2} = (z_1 z_2)^{-1/2}(z_1 - z_2), \tag{1.26}$$

$$\exp(\mathrm{i}QX) = (z_1 z_2)^{q/2}. \tag{1.27}$$

Since $\phi_{n;\lambda+n}(\eta)$ is a polynomial of $\cos\eta$ and $\sin\eta$, it is a polynomial of z_1 and z_2, times an integer or half-integer power of $z_1 z_2$. Thus the assertion is proved. The homogeneous polynomial of z_1 and z_2, which originates from a Gegenbauer polynomial of $\cos\eta$, corresponds to a special case of the Jack polynomial. The latter is defined for arbitrary number n of complex variables $z_1, z_2, \ldots, z_n$, as will be discussed in detail later.

We now proceed to the case of two identical (indistinguishable) particles. If the particles are bosons, we should take symmetric (even) wave functions. If the particles are spinless fermions, on the other hand, we take antisymmetric (odd) wave functions. For example,

$$\phi_\lambda(x) = |\sin\eta|^{\lambda-1}\sin\eta \tag{1.28}$$

describes the ground state of two fermions for H_2. If fermions have spin $1/2$, the spatial part of the wave functions is either symmetric (spin singlet) or antisymmetric (spin triplet). In the case of bosons, the even–odd property becomes the opposite to that of fermions; exchange of spatial and spin coordinates at the same time gives the same wave function as that before the exchange. In order to discuss the case with internal degrees of freedom, we consider a generalized model as given by

$$H_{2;K} = -\frac{\partial^2}{\partial x_1^2} - \frac{\partial^2}{\partial x_2^2} + \left(\frac{\pi}{L}\right)^2 \frac{2\lambda(\lambda - K_{12})}{\sin^2 \pi(x_1 - x_2)/L}, \tag{1.29}$$

where we have introduced the coordinate exchange operator K_{12}. As a complementary factor, we also introduce the spin permutation operator $P_{12} = 2\boldsymbol{S}_1 \cdot \boldsymbol{S}_2 + 1/2$. They act on a two-body wave function with spin coordinates σ_1, σ_2 as

$$K_{12}\psi(x_1, x_2; \sigma_1, \sigma_2) = \psi(x_2, x_1; \sigma_1, \sigma_2), \tag{1.30}$$

$$P_{12}\psi(x_1, x_2; \sigma_1, \sigma_2) = \psi(x_1, x_2; \sigma_2, \sigma_1), \tag{1.31}$$

$$P_{12}K_{12}\psi(x_1, x_2; \sigma_1, \sigma_2) = \psi(x_2, x_1; \sigma_2, \sigma_1) = \pm\psi(x_1, x_2; \sigma_1, \sigma_2). \tag{1.32}$$

Namely, we have $K_{12}P_{12} = \pm 1$ depending on whether the particles are bosons or fermions.

For symmetric spatial wave functions with $K_{12} = 1$, we know from (1.4) and (1.20) that the spectrum is given by

$$E_n = \frac{1}{2}Q^2 + 2\left(\frac{\pi}{L}\right)^2 (n+\lambda)^2 = k_+^2 + k_-^2, \tag{1.33}$$

where $k_\pm = [Q/2 \pm \pi(n+\lambda)/L]^2$ with $n \geq 0$. Note that E_n can be written as if it consists of the kinetic energy of free particles with momenta $k_\pm$. The interaction effect appears only in the restriction $k_+ - k_- \geq 2\pi\lambda/L$, which becomes the same as the Pauli exclusion principle in the case of $\lambda = 1$. For the antisymmetric (odd) wave function with $K_{12} = -1$, we have $\lambda(\lambda + 1)$ in (1.29), and accordingly replace $\lambda \to \lambda + 1$ in (1.33). This is alternatively interpreted as taking the excitation level n one step higher. The odd-function ground state in particular is given by E_1, where $n = 1$ is the smallest degree of antisymmetric polynomials of z_1 and z_2 with $z_j = \exp(2i\pi x_j/L)$. If there are spin degrees of freedom for a pair of fermions, the ground state energy becomes E_0 for the spin singlet, and E_1 for the triplet.

We have thus found that different symmetries of the wave functions appear only as a shift of energy levels. In particular, the difference between the singlet and triplet states in each ground state is given by

$$E_g(S = 1) - E_g(S = 0) = \pm(2\pi/L)^2 (\lambda + 1/2), \tag{1.34}$$

where the plus sign is for fermions and the minus sign for bosons. The signs mean the antiferromagnetic interaction for fermions and the ferromagnetic interaction for bosons. With use of $K_{12}P_{12} = \pm 1$ for identical particles, we obtain the Hamiltonian equivalent to $H_{2;K}$ as

$$H_{2;P} = -\frac{\partial^2}{\partial x_1^2} - \frac{\partial^2}{\partial x_2^2} + \left(\frac{L}{\pi}\right)^2 \frac{2\lambda(\lambda \pm P_{12})}{\sin^2 \pi(x_1 - x_2)/L}, \tag{1.35}$$

where the plus sign is for fermions and the minus sign for bosons. By construction, this model has both the charge and the spin degrees of freedom. One can extract only the spin degrees of freedom by taking a limiting procedure as explained next.

1.4 Freezing spatial motion

Let us consider the limiting case $\lambda \gg 1$ in $H_{2;K}$. Accordingly $\phi_\lambda(x)$ tends to a delta function peaked at $x = L/2$. This limit for a large number of particles is relevant to the spin chain and the supersymmetric (SUSY) t–J model since the particles should crystallize to avoid the repulsion, leaving only the center of mass motion. The excitation spectrum of two bosons with $Q = 0$ is given by

$$E_n - E_0 = 2 \left(\pi/L \right)^2 \left[(n + \lambda)^2 - \lambda^2 \right] = \left(2\pi/L \right)^2 \left(n\lambda + n^2/2 \right). \quad (1.36)$$

Thus the relative motion in the leading order of λ is linear in n. In the spinless case, this motion corresponds to a small vibration around the equilibrium distance $x = L/2$, analogous to phonons in the many-body system. If one removes this part, one is left with dynamics of internal degrees of freedom only. For example, the spin exchange is described by

$$\frac{1}{\lambda} H_{2;S} = \left(\frac{\pi}{L} \right)^2 \frac{2(P_{12} - 1)}{\sin^2 \pi (x_1 - x_2)/L} = \left(\frac{2\pi}{L} \right)^2 \left(S_1 \cdot S_2 - \frac{1}{4} \right), \quad (1.37)$$

where $|x_1 - x_2| = L/2$. This example shows that the spin chain with the $1/r^2$ exchange interaction is closely related to the continuum model with SU(2) internal degrees of freedom.

In the case of many-body problem with N particles, it is easier to analyze $H_{N;K}$ than $H_{N;P}$ since the energy spectrum can be derived by generalization of the $N = 2$ case. For example, the grand partition function $Z_{N;K}$ of $H_{N;K}$ can be derived for any internal symmetry. The grand partition function Z_N of the spinless case of $H_{N;K}$ can also be derived easily. Then the grand partition function $Z_{N;S}$ of the spin chain, or its generalizations, can be derived as the limit $Z_{N;S} = \lim_{\lambda \to \infty} Z_{N;K}/Z_N$. The degrees of freedom for the spatial motion are projected out in this limiting procedure. We shall see later that the spin chain has the Yangian symmetry related to traceless 2×2 matrices sl_2, while general 2×2 matrices gl_2 are relevant with the spatial motion. The Yangian symmetry is explained in detail later in this book. If one fully exploits the sl_2 Yangian symmetry, there is an alternative way to derive all the spectrum of the spin chain from knowledge of the highest-weight states of the sl_2 Yangian.

1.5 From spin permutation to graded permutation

The spin permutation operator is written as

$$P_{ij} = 2S_i \cdot S_j + \frac{1}{2} n_i n_j = \sum_{\sigma\sigma'} X_i^{\sigma\sigma'} X_j^{\sigma'\sigma}, \quad (1.38)$$

where n_i is the electron number operator, which is always unity in spin systems. We have introduced the X-operators defined by $X^{\alpha\beta} = |\alpha\rangle\langle\beta|$, which describe transition from one state β to another state α at a given site. The spin-flip is described by

$$S_i^- = X_i^{\downarrow\uparrow}, \quad S_i^+ = X_i^{\uparrow\downarrow}. \tag{1.39}$$

On the other hand, the hopping term is described by

$$\mathcal{P}c_{i\sigma}\mathcal{P} = X_i^{0\sigma}, \quad \mathcal{P}c_{i\sigma}^\dagger\mathcal{P} = X_i^{\sigma 0}, \tag{1.40}$$

with $\mathcal{P}$ being the projection operator excluding the doubly occupied site. We can combine the spin permutation and the hopping into a single operator:

$$\tilde{P}_{ij} = P_{ij} - X_i^{00}X_j^{00} - \sum_\sigma (X_i^{\sigma 0}X_j^{0\sigma} + X_i^{0\sigma}X_j^{0\sigma})$$

$$= \sum_{\alpha\beta} p(\beta)X_i^{\alpha\beta}X_j^{\beta\alpha}, \tag{1.41}$$

which is called the *graded* permutation operator because of the minus signs involved. In the second line, we have introduced the sign factor $p(\beta) = -1$ for $\beta = 0$ and $p(\beta) = 1$ otherwise. Note that $\tilde{P}_{ij}$ includes X_i^{00} representing projection to the vacant state at site i.

In order to see the property of the graded permutation in the simplest manner, let us consider a two-site system:

$$H_{\text{SUSY}} = t\tilde{P}_{12}. \tag{1.42}$$

Figure 1.2 shows the energy levels of the two-site system with the number of holes n_h varying from 0 to 2. The singlet–triplet splitting is the same as the bonding–antibonding splitting in this model. Moreover, the hole attraction contained in (1.41) makes the two-hole state have the same energy as the bonding state.

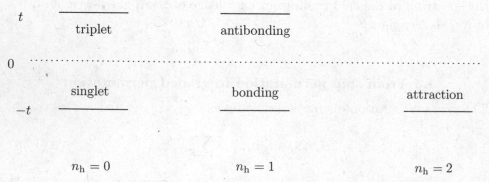

Fig. 1.2. Energy levels of the graded permutation operator for two sites. The number of holes in the system is specified by n_h.

The degeneracy means that $\mathcal{H}_{SUSY}$ is invariant not only under the global SU(2) operation, but also under the global supersymmetry operation which is generated by

$$Q_\sigma = X_1^{\sigma 0} + X_2^{\sigma 0} \tag{1.43}$$

and its Hermitian conjugate $Q_\sigma^\dagger$. These operators play a similar role as the operators introduced by (1.15), and are properly called the supercharge. The degeneracy in $\mathcal{H}_{SUSY}$ originates from the commuting property

$$[\tilde{P}_{12}, X_1^{\alpha\beta} + X_2^{\alpha\beta}] = 0 \tag{1.44}$$

for any combination of $\alpha, \beta = \uparrow, \downarrow, 0$. In Chapter 6 we discuss the supersymmetric t–J model, which is the N-site version of (1.42).

1.6 Variants of $1/r^2$ systems

The original $1/r^2$ model proposed by Calogero [34, 35] takes the form

$$H_C = -\sum_{i=1}^{N} \frac{\partial^2}{\partial x_i^2} + \sum_{i>j} \left[2(x_i - x_j)^2 + \frac{g}{(x_i - x_j)^2} \right], \tag{1.45}$$

where the units of the spatial coordinate and the energy are taken so as to adjust the strength for the harmonic potential. The wave function tends to that of the Sutherland model as the distance $x_i - x_j$ goes to zero. However, the spectra of both models have an interesting difference as well as similarity.

Let us first consider the simplest case of $N = 2$. By introducing the relative coordinate $x = x_1 - x_2$ and the center of gravity $X = (x_1 + x_2)/2$, one can write the wave function in the product form $\psi_g(x_1, x_2) = \phi_g(x) \exp(iQX)$ as in the Sutherland model. The total momentum is given by Q. The relative motion is described by

$$H_1(x) = H_C(N = 2) - \frac{1}{2}Q^2 = -2\frac{\partial^2}{\partial x^2} + 2x^2 + \frac{g}{x^2}, \tag{1.46}$$

which is to be compared with (1.4). One can use the same factorization technique to derive all energy levels and eigenfunctions. Namely, we put $g = 2\lambda(\lambda - 1)$ and write $H_1 \equiv 2H_\lambda$ as

$$H_\lambda = p^2 + x^2 + \frac{\lambda(\lambda - 1)}{x^2}, \tag{1.47}$$

with $p = -i\partial/\partial x$. In analogy with Section 1.3, we introduce the operator

$$B_\lambda = p - ix + i\frac{\lambda}{x} \equiv p + iV_\lambda(x) \tag{1.48}$$

and its Hermitian conjugate $B_\lambda^\dagger = p - iV_\lambda$. Then H_λ can also be written as

$$H_\lambda = B_\lambda^\dagger B_\lambda + 2\lambda + 1. \tag{1.49}$$

The ground state of H_λ is given by the (unnormalized) function

$$\phi_\lambda(x) = |x|^\lambda \exp\left(-\frac{1}{2}x^2\right) \equiv \exp[U_\lambda(x)], \tag{1.50}$$

where $U_\lambda(x)' = V_\lambda(x)$. Since we have $B_\lambda\phi_\lambda = 0$, the ground-state energy is given by $2\lambda + 1$. By using the property $H_{\lambda+1} = B_\lambda B_\lambda^\dagger + 2\lambda - 1$ we obtain

$$B_\lambda B_\lambda^\dagger \phi_{\lambda+1} = 4\phi_{\lambda+1}. \tag{1.51}$$

Application of $B_\lambda^\dagger$ from the left on both sides of (1.51) shows that $B_\lambda^\dagger\phi_{\lambda+1}$ gives the first excited state of H_λ with eigenvalue $2\lambda + 5$. In this way we obtain the ladder of energy levels which looks like those in Fig. 1.1. Note that in the Calogero model the ordinate is proportional to energy. The nth excited state has the eigenvalue $E_{\lambda;n} = 2\lambda + 1 + 4n$, which increases by a constant amount of four when n increases by one.

The explicit form of the wave function $\phi_{\lambda;n}(x)$ is given in terms of $U_\lambda = \ln\phi_\lambda$ by

$$\begin{aligned}
\phi_{\lambda;n}(x) &= \exp(-U_{\lambda+1})p\exp(U_{\lambda+1} - U_{\lambda+2})\cdots p\exp(U_{\lambda+n})\phi_{\lambda+n} \\
&= \exp(-U_\lambda)\left(\frac{-i}{x}\frac{\partial}{\partial x}\right)^n \exp(U_{\lambda+n})\phi_{\lambda+n} \\
&\propto \phi_\lambda(x)L_n^{(\lambda)}(x^2),
\end{aligned} \tag{1.52}$$

where we have used the Rodrigues formula for the associated Laguerre polynomial

$$L_n^{(\lambda)}(y) = \frac{1}{n!}y^{-\lambda}e^y\frac{d^n}{dy^n}y^{n+\lambda}e^{-y}. \tag{1.53}$$

In the N-body case, the original Calogero model becomes awkward to solve after separating the center of gravity. Although the harmonic potential can be rewritten in the same form in terms of the relative coordinates known as the Jacobi coordinates [34], the $1/r^2$ repulsion has a complicated dependence on relative coordinates. To circumvent the difficulty, Sutherland modified the model as

$$H_{CS} = \sum_{i=1}^N \left(p_i^2 + x_i^2\right) + \sum_{i>j}\frac{\lambda(\lambda - 1)}{(x_i - x_j)^2}, \tag{1.54}$$

where the mutual harmonic attraction is replaced by an external harmonic potential. This model is called the Calogero–Sutherland model in the narrow sense. Note that the H_{CS} in the one-body case $N = 1$ describes the relative motion of H_{C} with $N = 2$. The Calogero–Sutherland model has a simple spectrum for any N, which derives from the one-body case, and also allows the freezing limit as λ goes to infinity [58, 149]. The frozen positions of particles are not equally distant, but correspond to zeros of the Nth-order Hermite polynomial [150]. Because of these nice features, the model has been actively investigated in the literature [88, 95, 190].

The most general expression of the $1/r^2$-type potential is given by

$$\wp(x) = \frac{1}{x^2} + \sum_{m^2+n^2 \neq 0} \left[\frac{1}{(x - nL - iml)^2} - \frac{1}{(nL + iml)^2} \right], \quad (1.55)$$

where n, m are integers. The function $\wp(x)$ is called the Weierstrass elliptic function. In the complex plane $x \to z$, $\wp(z)$ is doubly periodic with a period L along the real axis, and another period l along the imaginary axis. The potential has the limiting forms

$$\wp(x) \to \begin{cases} x^{-2} & (L, l \to \infty), \\ \left(\dfrac{L}{\pi}\right)^2 \sin^{-2}\left(\dfrac{\pi x}{L}\right) & (l \to 0), \\ \left(\dfrac{l}{\pi}\right)^2 \sinh^{-2}\left(\dfrac{\pi x}{l}\right) & (L \to 0). \end{cases} \quad (1.56)$$

The first and second limits have already been discussed above. The third limit may be interpreted as analytic continuation $x \to ix$ in the second limit. Let us consider the N-body system with spin $1/2$:

$$H_{\mathrm{I}} = \sum_{i=1}^{N} p_i^2 + \sum_{i>j} \lambda(\lambda - P_{ij})\wp(x_i - x_j), \quad (1.57)$$

where P_{ij} is the spin exchange operator. This model is known to be integrable [146, 178], which means that one can identify conserved quantities up to the total number of degrees of freedom. However, the wave functions are very complicated in the general case of L and l. If one takes the limit $\lambda \to \infty$, the nth particle $(n = 1, \ldots, N)$ crystallizes at $x_n = nL/N$ as described in Section 1.4. After taking the third limit in (1.56), one is still left with the free parameter l. In the case of small enough l, the exchange interaction remains significant only for nearest-neighbor sites. In the opposite limit of large l, the exchange interaction decays as $1/x^2$. The latter limit describes the Haldane–Shastry model. Thus the sinh-type interaction interpolates the nearest-neighbor Heisenberg model and the $1/r^2$ exchange model. The exact solution of the sinh-type spin model has been discussed by Inozemtsev

[90,91]. Note that the model of (1.57) can be defined for any kind of internal symmetries, including the $SU(2, 1)$ supersymmetry.

Another direction of generalizing the Calogero–Sutherland model has been proposed by Ruijsenaars and Schneider [152]. They constructed the relativistic energy and momentum tensor by introducing an additional parameter which plays a role of the light velocity. The relativistic version of the model has a clean algebraic structure and can be solved exactly [153]. In the limit of infinite light velocity, the Ruijsenaars–Schneider model is reduced to the Sutherland model.

1.7 Contents of the book

As we have seen, there are a considerable variety of models even within the $1/r^2$ interaction family. In the rest of this book, we concentrate our attention on Sutherland models and their lattice cousins. They have translational symmetry, and the wave functions are much simpler than the most general periodic case with the potential $\wp(x)$, or the relativistic generalization [152]. The results obtained for the periodic lattice models can be compared with real one-dimensional systems most straightforwardly, even though the interaction is not exactly the same.

We adopt in this book an approach to handle many-body wave functions in the coordinate representation. The merit of this first quantization approach is that it is straightforward and valid for any size of system. The key development in this approach owes substantially to mathematicians who have found a number of useful properties of multi-variable orthogonal polynomials in recent decades. This book, written by physicists, emphasizes intuitive pictures as far as possible, instead of pursuing mathematical rigor.

We organize the contents of this book into two parts: in the first part, we discuss physical features of $1/r^2$ models, relegating the mathematical details to the second part. We proceed in the first part from the simplest continuum model to more complicated cases. The most complicated, yet interesting, system to understand is that of lattice electrons with both spin and charge degrees of freedom. In order to follow the route culminating in the dynamics of lattice electrons, one must have solutions for a continuum model with arbitrary interaction strength, and with internal degrees of freedom. A great advantage of considering continuum models is that we can apply powerful analytical techniques developed mainly by mathematicians. One can appreciate the situation by considering the usefulness of differential equations as compared with difference equations, which are relevant to lattice models by naive approaches. Some sections such as Section 2.8 with an asterisk (*) contain somewhat advanced technical aspects, and can be omitted during a first reading.

Chapter 2 deals with the single-component Sutherland model. Beginning from the derivation of eigenvalues and eigenfunctions, we proceed to thermodynamics with the idea of fractional exclusion statistics. Then in terms of Jack polynomials, which describe arbitrary excited states, we derive exact dynamical correlation functions in a self-contained manner. We also discuss the physical implication of the results in terms of quasi-particles. We demonstrate how the CFT reproduces the asymptotic behavior of the dynamical correlation functions of the Sutherland model. As prerequisites we give brief accounts of the finite-size scaling theory of the CFT.

Chapter 3 introduces internal degrees of freedom in the Sutherland model. The multi-component Sutherland model generalizes the discussion in Chapter 2. We give a self-contained account of how to derive thermodynamics and dynamics exactly. As a key mathematical technique, non-symmetric Jack polynomials are introduced and used extensively.

In Chapter 4, we turn to the simplest lattice systems. Namely, we deal with the $1/r^2$ Heisenberg model with spin 1/2 which is called the Haldane–Shastry model. We first discuss the ground-state wave function, and the resultant static spin correlation functions. Then we proceed to energy spectra, elementary excitations called spinons, thermodynamics, and finally the dynamical spin correlation functions. We show that the spinon picture gives a complete interpretation of the thermodynamics and spin dynamics.

Chapter 5 generalizes the spin 1/2 Haldane–Shastry model to the $SU(K)$ chain with K species of internal degrees of freedom. The subtlety of statistical parameters appears in the case of $K \geq 3$, which is explained both for thermodynamics and dynamics.

Chapter 6 discusses the $1/r^2$ t–J model, which includes the charge degrees of freedom, in addition to spins. We begin with the ground-state wave function and the static correlation functions of spin and charge. Then we proceed to the spectra of elementary excitations: spinons, holons, and their antiparticles, and discuss their statistics. On this basis, we derive thermodynamics including the specific heat and the magnetic susceptibility. Finally, we derive exact results on dynamical correlation functions and interpret the results in terms of the quasi-particle picture.

In the second part beginning with Chapter 7, we give a self-contained account of mathematics for $1/r^2$ systems. We begin with the Jack polynomial and its various generalizations. We intend that the reader will be able to understand the basic properties of various kinds of Jack polynomials without recourse to other literature.

Then Chapter 8 explains the Yangian symmetry restricted to the simplest case of SU(2) internal symmetry. The fundamental relation, called the Yang–Baxter relation, appears naturally as a result of the identity of quantum particles, together with the spatial dependence of eigenfunctions in the Sutherland model. Combining the charge degrees of freedom with SU(2) spins, the Yangian is usually referred to as $Y(gl_2)$. We proceed in Chapter 9 to discuss the Yangians for the general case of an SU(K) spin chain, and the supersymmetric t–J model. We provide an intuitive explanation of the beautiful algebraic property from a physicist's point of view.

Chapter 10 explains a generalization of the Jack polynomial proposed by Uglov. The generalization utilizes a special case of parameters in related polynomials called Macdonald symmetric polynomials. Uglov's theory provides an elegant mathematical setting to derive the dynamics of the multi-component Sutherland model. Finally, in the Afterword, we give a brief outlook and mention remaining problems as well as alternative theoretical approaches that are not covered in this book.

Since we use a sizable number of mathematical symbols, a list is provided near the end of the book in (loose) lexicographic order. It turns out to be difficult to avoid completely using the same notation for different meanings, such as P_μ for a momentum in one case, and for a Macdonald symmetric polynomial in another case. Since these cases never occur within a single chapter, we hope that the reader will not be confused.

Part I

Physical properties

2

Single-component Sutherland model

In the present chapter, we discuss the Sutherland model for particles without internal degrees of freedom such as spin. We call this model the single-component Sutherland model. For two particles, the eigenenergies and eigenstates of the Sutherland model (1.3) have been obtained explicitly in the previous chapter. The most striking feature of the Sutherland model is that one can derive not only the energy spectrum but also the dynamics for the many-particle case with an exact account of interaction effects. Thus, the Sutherland model provides an ideal framework to study a one-dimensional quantum liquid in detail.

In Section 2.1, we derive the eigenenergies of eigenstates. In Section 2.2, we present different but equivalent physical pictures for the energy spectrum. Namely, the energy spectrum is naturally regarded as that of interacting bosons or fermions. The same spectrum can also be interpreted as that of free particles obeying nontrivial quantum statistics, i.e., free anyons in one dimension. The exclusion statistics proposed by Haldane will be explained on this occasion. Spectrum and statistics of elementary excitations are derived in Section 2.3. In Section 2.4, we discuss thermodynamic properties, which can be rewritten as those of free anyons. In Section 2.5, we identify the eigenfunctions with Jack symmetric polynomials, and discuss their basic properties. In Section 2.6, we consider dynamical correlation functions such as Green's functions and the density correlation function. These quantities are derived with the use of Jack polynomials, and are naturally interpreted in terms of elementary excitations with fractional charge.

The Sutherland model is the simplest model to realize the Tomonaga–Luttinger liquid. In Section 2.8, long-distance and long-time asymptotic behaviors of dynamical correlation functions are reproduced by the theory of the Tomonaga–Luttinger liquid.

2.1 Preliminary approach

2.1.1 Jastrow-type wave functions

The Sutherland model for N particles is given by [172]

$$H = -\sum_{i=1}^{N} \frac{\partial^2}{\partial x_i^2} + 2\left(\frac{\pi}{L}\right)^2 \sum_{1 \leq i < j \leq N} \frac{\lambda(\lambda - 1)}{\sin^2\left[\pi(x_i - x_j)/L\right]}. \qquad (2.1)$$

The variables $x = (x_1, x_2, \ldots, x_N)$ represent the spatial coordinates of particles moving along a circle of perimeter L. The coupling parameter λ is taken to be in the range $[1/2, \infty)$ without loss of generality, according to the argument in Section 1.3. We regard particles as bosons without internal degrees of freedom. Wave functions are subject to the periodic boundary condition

$$\Psi(x_1, \ldots, x_i + L, \ldots, x_N) = \Psi(x_1, \ldots, x_i, \ldots, x_N). \qquad (2.2)$$

Furthermore, the identity of particles requires the following condition:

$$\Psi(x_1, \ldots, x_i, \ldots, x_j, \ldots, x_N) = \Psi(x_1, \ldots, x_j, \ldots, x_i, \ldots, x_N), \qquad (2.3)$$

which is called the Fock condition. When the coordinates of two particles x_i and x_j are close, the problem essentially reduces to a two-body problem. By the argument in Section 1.3, the wave function has the asymptotic form

$$|x_i - x_j| \to 0, \quad \Psi \to |x_i - x_j|^{\lambda}. \qquad (2.4)$$

We consider an N-particle wave function

$$\Psi_0^{B}(x_1, \ldots, x_N) = \prod_{1 \leq i < j \leq N} \left| \sin\left[\pi(x_i - x_j)/L\right] \right|^{\lambda}, \qquad (2.5)$$

which satisfies both the property (2.4) and the periodic boundary condition (2.2). We will show that (2.5) is an eigenfunction of the Hamiltonian (2.1), which was originally found by Sutherland [173]. Such a wave function that takes a product form of two-body factors is called a Jastrow-type wave function in general.

First we note that the derivative of Ψ_0^{B} with respect to x_i is given by

$$\frac{\partial}{\partial x_i} \Psi_0^{B} = \sum_{j(\neq i)} \frac{\lambda \pi}{L} \cot \frac{\pi(x_i - x_j)}{L} \Psi_0^{B}. \qquad (2.6)$$

Further differentiation with respect to x_i and summing over i gives

$$\left(\Psi_0^B\right)^{-1} \sum_{i=1}^{N} \frac{\partial^2}{\partial x_i^2} \Psi_0^B$$

$$= \sum_{i=1}^{N} \left[\sum_{j(\neq i)} \frac{\lambda\pi}{L} \cot \frac{\pi(x_i - x_j)}{L} \right]^2 - \sum_{i\neq j} \frac{\lambda \left(\pi/L\right)^2}{\sin^2 \left[\pi\left(x_i - x_j\right)/L\right]}.$$

$$(2.7)$$

Apart from the multiplicative factor $(\lambda\pi/L)^2$, the first term on the right-hand side (RHS) in (2.7) is the sum of two-site terms

$$\sum_{i\neq j} \cot^2 \frac{\pi(x_i - x_j)}{L}$$

$$(2.8)$$

and three-site terms

$$\sum_{i} \sum_{j(\neq i)} \sum_{k(\neq i,j)} \cot\left[\pi\left(x_i - x_j\right)/L\right] \cot\left[\pi\left(x_i - x_k\right)/L\right].$$

$$(2.9)$$

The two-site terms (2.8) are rewritten as

$$-N(N-1) + \sum_{i\neq j} \sin^{-2} \frac{\pi(x_i - x_j)}{L}.$$

$$(2.10)$$

The three-site terms (2.9) can be shown to be a constant $N(N-1)(N-2)/3$ with use of the identity

$$\cot(ij)\cot(ik) + \cot(ji)\cot(jk) + \cot(ki)\cot(kj) = -1,$$

with $\cot(ij) = \cot\left[\pi\left(x_i - x_j\right)/L\right]$. Combining these results, the RHS of (2.7) turns into

$$E_{0,N} - \sum_{i\neq j} \frac{\lambda(\lambda-1)\left(\pi/L\right)^2}{\sin^2\left[\pi\left(x_i - x_j\right)/L\right]},$$

$$(2.11)$$

with a constant defined by

$$E_{0,N} = \left(\pi\lambda/L\right)^2 N(N^2 - 1)/3.$$

$$(2.12)$$

The second term on the RHS of (2.11) is the minus of the interaction term in the Hamiltonian (2.1). As a result, we obtain

$$H\Psi_0^B = E_{0,N}\Psi_0^B,$$

$$(2.13)$$

which shows that the Jastrow wave function is the eigenfunction with the eigenenergy $E_{0,N}$.

Let us remark on the statistics. When we take an antisymmetric wave function

$$\Psi(x_1, \ldots, x_i, \ldots, x_j, \ldots, x_N) = -\Psi(x_1, \ldots, x_j, \ldots, x_i, \ldots, x_N) \qquad (2.14)$$

instead of (2.3), we also obtain the eigenfunctions of the Jastrow form. Namely, another Jastrow-type wave function

$$\Psi_0^F = \Psi_0^B \prod_{i<j} \left(\sin\left[\pi(x_i - x_j)/L\right] / |\sin\left[\pi(x_i - x_j)/L\right]| \right) \qquad (2.15)$$

which also satisfies

$$H\Psi_0^F = E_{0,N}\Psi_0^F. \qquad (2.16)$$

The wave function (2.15) satisfies the periodic boundary condition (2.2) for odd N, and the antiperiodic boundary condition

$$\Psi(x_1, \ldots, x_i + L, \ldots, x_N) = -\Psi(x_1, \ldots, x_i, \ldots, x_N) \qquad (2.17)$$

for even N. We further remark that the Jastrow wave functions Ψ_0^B and Ψ_0^F are not only an eigenstate but also the ground state, as will be shown at the end of Section 2.1.2.

For later convenience, we introduce the complex coordinate $z_i = \exp(i2\pi x_i/L)$. Then the following wave function:

$$\Psi_{0,N} = \prod_{1 \leq i < j \leq N} (z_i - z_j)^\lambda \prod_{i=1}^\lambda z_i^{-(N-1)\lambda/2} \qquad (2.18)$$

reduces to

$$\Psi_{0,N} = \begin{cases} (2i)^{N(N-1)\lambda/2}\Psi_0^B, & \text{for even integer } \lambda, \\ (2i)^{N(N-1)\lambda/2}\Psi_0^F, & \text{for odd integer } \lambda. \end{cases} \qquad (2.19)$$

2.1.2 Triangular matrix for Hamiltonian

In Section 1.3, the wave functions of excited states for a two-particle system have been given by the ground-state wave function multiplied by the Gegenbauer polynomials. For an N-particle system, we seek eigenfunctions in the form

$$\Psi = \Psi_0^B \Phi. \qquad (2.20)$$

Here $\Phi = \Phi(x_1, \ldots, x_N)$ is symmetric with respect to the interchange between x_i and x_j. First we consider the eigenvalue problem of Φ. Using

(2.20), the left-hand side (LHS) of $H\Psi = E\Psi$ becomes

$$H\Psi_0^B \Phi = \Phi \underbrace{H\Psi_0^B}_{=E_{0,N}\Psi_0^B} - \Psi_0^B \sum_{i=1}^{N} \frac{\partial^2 \Phi}{\partial x_i^2} - 2 \sum_i \underbrace{\frac{\partial \Psi_0^B}{\partial x_i}}_{(2.6)} \frac{\partial \Phi}{\partial x_i}. \qquad (2.21)$$

The eigenvalue problem for Φ is written as

$$\left(-\sum_i \frac{\partial^2}{\partial x_i^2} + \sum_{i \neq j} \frac{2\pi\lambda}{L} \cot \frac{\pi(x_i - x_j)}{L} \frac{\partial}{\partial x_i} \right) \Phi = (E - E_{0,N})\,\Phi. \qquad (2.22)$$

Since Φ satisfies the periodic boundary condition

$$\Phi(x_1, \ldots, x_i + L, \ldots, x_N) = \Phi(x_1, \ldots, x_i, \ldots, x_N), \qquad (2.23)$$

it is convenient to regard Φ as a function of variables $(z_1, \ldots, z_N)$, where $z_i = \exp[\mathrm{i}2\pi x_i/L]$. Using

$$\frac{\partial}{\partial x_i} = \frac{\mathrm{i}2\pi z_i}{L} \frac{\partial}{\partial z_i}, \quad \cot \frac{\pi(x_i - x_j)}{L} = \mathrm{i}\frac{z_i + z_j}{z_i - z_j}, \qquad (2.24)$$

(2.22) becomes

$$\mathcal{H}\Phi = (\mathcal{H}^{(1)} + \lambda\mathcal{H}^{(2)})\Phi = \mathcal{E}\Phi, \qquad (2.25)$$

with $\mathcal{E} = [L/(2\pi)]^2 (E - E_{0,N})$ and

$$\mathcal{H}^{(1)} = \sum_i \left(z_i \frac{\partial}{\partial z_i} \right)^2, \quad \mathcal{H}^{(2)} = \sum_{i<j} \left(\frac{z_i + z_j}{z_i - z_j} \right) \left(z_i \frac{\partial}{\partial z_i} - z_j \frac{\partial}{\partial z_j} \right). \qquad (2.26)$$

In the following, we set a basis for Φ and calculate the matrix elements of $\mathcal{H}$. We define a bosonic wave function ϕ_η^B:

$$\phi_\eta^B = \sum_{p \in S_N} z_{p(1)}^{\eta_1} z_{p(2)}^{\eta_2} \cdots z_{p(N)}^{\eta_N} \qquad (2.27)$$

with a set of integers $\eta = (\eta_1, \ldots, \eta_N)$. In (2.27), $p = (p(1), \ldots, p(N))$ is an element of the symmetric group S_N of order N. The set of ϕ_κ^B forms a basis for Φ, with $\kappa = (\kappa_1, \ldots, \kappa_N)$ satisfying

$$\kappa_1 \geq \kappa_2 \geq \cdots \geq \kappa_N. \qquad (2.28)$$

The action of $\mathcal{H}^{(1)}$ on ϕ_κ^B gives $(\sum_i \kappa_i^2)\phi_\kappa^B$. In order to consider the action of $\mathcal{H}^{(2)}$, we consider the two-particle case. When $\kappa_1 = \kappa_2$, the function ϕ_κ^B

is an eigenfunction with eigenvalue zero. When $\kappa_1 > \kappa_2$, the action of $\mathcal{H}^{(2)}$ on $\Phi_\kappa^B = z_1^{\kappa_1} z_2^{\kappa_2} + z_1^{\kappa_2} z_2^{\kappa_1}$ leads to

$$
\begin{aligned}
&\left(\frac{z_1 + z_2}{z_1 - z_2}\right)\left(z_1 \frac{\partial}{\partial z_1} - z_2 \frac{\partial}{\partial z_2}\right)(z_1^{\kappa_1} z_2^{\kappa_2} + z_1^{\kappa_2} z_2^{\kappa_1}) \\
&= (\kappa_1 - \kappa_2)\left(\frac{z_1 + z_2}{z_1 - z_2}\right)(z_1^{\kappa_1} z_2^{\kappa_2} - z_1^{\kappa_2} z_2^{\kappa_1}) \\
&= (\kappa_1 - \kappa_2)(z_1 + z_2)\left(z_1^{\kappa_1 - 1} z_2^{\kappa_2} + \cdots + z_1^{\kappa_2} z_2^{\kappa_1 - 1}\right) \\
&= (\kappa_1 - \kappa_2)\left(z_1^{\kappa_1} z_2^{\kappa_2} + 2z_1^{\kappa_1 - 1} z_2^{\kappa_2 + 1} + \cdots + 2z_1^{\kappa_2 + 1} z_2^{\kappa_1 - 1} + z_1^{\kappa_2} z_2^{\kappa_1}\right).
\end{aligned}
\tag{2.29}
$$

These results are rearranged as

$$
\mathcal{H}^{(2)} \phi_\kappa^B = (\kappa_1 - \kappa_2)\phi_\kappa^B + 2(\kappa_1 - \kappa_2)\sum_{l=1}^{(\kappa_1 - \kappa_2 - 1)/2} \phi_{(\kappa_1 - l, \kappa_2 + l)}^B,
\tag{2.30}
$$

when $\kappa_1 - \kappa_2$ is odd, or

$$
\begin{aligned}
\mathcal{H}^{(2)} \phi_\kappa^B &= (\kappa_1 - \kappa_2)\phi_\kappa^B + 2(\kappa_1 - \kappa_2)\sum_{l=1}^{(\kappa_1 - \kappa_2)/2 - 1} \phi_{(\kappa_1 - l, \kappa_2 + l)}^B \\
&\quad + (\kappa_1 - \kappa_2)\phi_{((\kappa_1 - \kappa_2)/2, (\kappa_1 - \kappa_2)/2)}^B,
\end{aligned}
\tag{2.31}
$$

when $\kappa_1 - \kappa_2$ is even. For example, with $\kappa = (4, 0), (3, 1), (2, 2)$, the relation (2.31) becomes

$$
\begin{aligned}
\mathcal{H}^{(2)} \phi_{(4,0)}^B &= 4\phi_{(4,0)}^B + 8\phi_{(3,1)}^B + 4\phi_{(2,2)}^B, \\
\mathcal{H}^{(2)} \phi_{(3,1)}^B &= 2\phi_{(4,0)}^B + 2\phi_{(2,2)}^B, \\
\mathcal{H}^{(2)} \phi_{(2,2)}^B &= 0.
\end{aligned}
$$

From these results on the two-particle system, we notice the following properties. First, $\mathcal{H}^{(2)} \phi_\kappa^B$ is the sum of a finite number of ϕ_μ^B, where μ is a set of two integers (μ_1, μ_2). At first glance, it seems that the action of $\mathcal{H}^{(2)}$ on ϕ_κ^B yields a pole at $z_1 - z_2$. However, the part $(z_1 \partial/\partial z_1 - z_2 \partial/\partial z_2)\phi_\kappa^B$ has the factor $(z_1 - z_2)$, which cancels the pole. As a result, $\mathcal{H}^{(2)} \phi_\kappa^B$ is spanned by a finite number of basis functions. Second, $\mathcal{H}^{(2)} \phi_\kappa^B$ does not contain ϕ_μ^B when $\mathcal{H}^{(2)} \phi_\mu^B$ contains ϕ_κ^B. By using this property, we can define an ordering of the basis function. These properties hold also in a general N-particle case.

Now we turn to the N-particle system, and consider the action of $\mathcal{H}^{(2)}$ on ϕ_κ^{B}. First we note that the operator

$$\mathcal{H}_{ij}^{(2)} = \left(\frac{z_i + z_j}{z_i - z_j} \right) \left(z_i \frac{\partial}{\partial z_i} - z_j \frac{\partial}{\partial z_j} \right)$$

is symmetric with respect to the exchange $z_i \leftrightarrow z_j$ and hence, the operator $\mathcal{H}^{(2)} = \sum_{ij} \mathcal{H}_{ij}^{(2)}$ is rewritten as

$$\mathcal{H}^{(2)} = \sum_{ij} \mathcal{H}_{p(i)p(j)}^{(2)} \tag{2.32}$$

for any permutation $p \in S_N$. Second we note that in the definition of ϕ_κ^{B}, the summation of p runs over all elements of S_N, and hence ϕ_κ^{B} is rewritten as

$$\phi_\kappa^{\mathrm{B}} = \sum_{p \in S_N} \prod_{k=1}^{N} z_{p(k)}^{\kappa_k} = \sum_{p \in S_N} \prod_{k=1}^{N} z_{p'(k)}^{\kappa_k} \tag{2.33}$$

with a permutation p' satisfying

$$(p'(1), \ldots, p'(N)) = (p(1), \ldots, \overset{i}{p(j)}, \ldots, \overset{j}{p(i)}, \ldots, p(N)). \tag{2.34}$$

From (2.33), the function ϕ_κ^{B} is further rewritten as

$$\phi_\kappa^{\mathrm{B}} = \sum_{p \in S_N} \frac{1}{2} \left(\prod_{k=1}^{N} z_{p(k)}^{\kappa_k} + \prod_{k=1}^{N} z_{p'(k)}^{\kappa_k} \right)$$

$$= \frac{1}{2} \sum_{p \in S_N} \left(z_{p(i)}^{\kappa_i} z_{p(j)}^{\kappa_j} + z_{p(i)}^{\kappa_j} z_{p(j)}^{\kappa_i} \right) \left(\prod_{k \neq (i,j)} z_{p(k)}^{\kappa_k} \right). \tag{2.35}$$

With (2.32) and (2.35), we obtain

$$\mathcal{H}^{(2)} \phi_\kappa^{\mathrm{B}} = \frac{1}{2} \sum_{p \in S_N} \sum_{i<j} \left(\prod_{k \neq (i,j)} z_{p(k)}^{\kappa_k} \right) \mathcal{H}_{p(i)p(j)}^{(2)} \left(z_{p(i)}^{\kappa_i} z_{p(j)}^{\kappa_j} + z_{p(i)}^{\kappa_j} z_{p(j)}^{\kappa_i} \right). \tag{2.36}$$

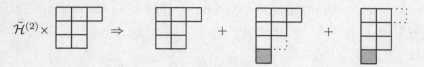

Fig. 2.1. Graphical representation of (2.39). The action of $\mathcal{H}^{(2)}$ on ϕ_κ^{B} generates ϕ_μ^{B} only if μ is obtained by moving a square (squares) from a row j to another row k. The dotted squares and shaded ones represent removed and added squares, respectively.

The evaluation of $\mathcal{H}^{(2)}_{p(i)p(j)}\left(z_{p(i)}^{\kappa_i} z_{p(j)}^{\kappa_j} + z_{p(i)}^{\kappa_j} z_{p(j)}^{\kappa_i}\right)$ has been done in the two-particle case. From (2.29) and (2.36), we obtain

$$
\begin{aligned}
\mathcal{H}^{(2)}\phi_\kappa^{\mathrm{B}} &= \sum_{i<j}|\kappa_i - \kappa_j|\phi_\kappa^{\mathrm{B}} + \sum_{i<j}|\kappa_i - \kappa_j|\sum_{l=1}^{\kappa_i-\kappa_j-1}\phi_{(\kappa_1,\dots,\kappa_i-l,\dots,\kappa_j+l,\dots,\kappa_N)}^{\mathrm{B}} \\
&= \sum_{i<j}|\kappa_i - \kappa_j|\phi_\kappa^{\mathrm{B}} + \sum_{1\le i<j\le N}\sum_{l=1}^{[(\kappa_i-\kappa_j)/2]}V(l,\kappa_i,\kappa_j)\phi_{\kappa_1,\dots,\kappa_i-l,\dots,\kappa_j+l,\dots,\kappa_N}^{\mathrm{B}}.
\end{aligned}
\tag{2.37}
$$

Here $[\cdot]$ denotes the Gauss's symbol defined as

$$
[x] = n, \quad \text{if } n \le x < n+1,
$$

with an integer n. The function $V(l,\kappa_j,\kappa_k)$ is given by

$$
V(l,\kappa_j,\kappa_k) = \begin{cases} \kappa_j - \kappa_k, & \text{for } l = (\kappa_j-\kappa_k)/2, \\ 2(\kappa_j - \kappa_k), & \text{otherwise.} \end{cases}
\tag{2.38}
$$

The rightmost part of (2.37) reduces to (2.30) or (2.31) when $N = 2$.

As an example, we consider the case where $N = 4$ and $\kappa = (3,2,2,0)$. We then obtain

$$
\mathcal{H}^{(2)}\phi_{3,2,2,0}^{\mathrm{B}} = 11\phi_{3,2,2,0}^{\mathrm{B}} + 2\phi_{3,2,1,1}^{\mathrm{B}} + 3\phi_{2,2,2,1}^{\mathrm{B}}.
\tag{2.39}
$$

Figure 2.1 illustrates (2.39). The action of $\mathcal{H}^{(2)}$ on ϕ_κ^{B} generates ϕ_μ only if μ is obtained from κ by replacement of a pair $(\kappa_i,\kappa_j) \to (\kappa_i - l, \kappa_j + l)$, where $\kappa_i > \kappa_j$ and l is an integer satisfying $1 \le l \le [(\kappa_i - \kappa_j)/2]$. The procedure to produce μ from κ is called "squeezing" in [172]; squeezing $(\kappa_i,\kappa_j) \to (\kappa_i - l, \kappa_j + l)$ makes the width of the diagram of the pair narrower. Figure 2.2 shows an example.

If μ is produced by squeezing κ, the converse is impossible; κ is not produced by squeezing μ. It follows that $\langle\kappa|\mathcal{H}^{(2)}|\mu\rangle = 0$ if $\langle\mu|\mathcal{H}^{(2)}|\kappa\rangle \ne 0$. Here $|\kappa\rangle$ denotes the state vector whose wave function is ϕ_κ^{B}. Thus we introduce an ordering of the basis between κ and μ satisfying $|\kappa| = |\mu|$, where

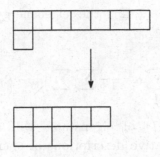

Fig. 2.2. Squeezing a pair $(\kappa_1, \kappa_2) = (7, 1)$ to $(\mu_1, \mu_2) = (5, 3)$.

$|\kappa| = \sum_{i=1}^{N} \kappa_i$. We define the ordering $\kappa > \mu$ if $|\kappa| = |\mu|$ and the first non-vanishing $\kappa_i - \mu_i$ is positive. For example, we have $(3, 2, 2, 0) > (3, 2, 1, 1) > (2, 2, 1, 1)$. We then find that $\langle \mu | \mathcal{H}^{(2)} | \kappa \rangle$ is nonzero only if $\kappa \geq \mu$. The matrices of $\mathcal{H}^{(2)}$ and hence $\mathcal{H}$ in this ordered basis are triangular.

For example, with $N = 4$ and $\kappa = (3, 2, 2, 0), (3, 2, 1, 1), (2, 2, 2, 1)$, the matrix elements are given by

$$\mathcal{H} \begin{pmatrix} \phi_{3,2,2,0}^{\mathrm{B}} \\ \phi_{3,2,1,1}^{\mathrm{B}} \\ \phi_{2,2,2,1}^{\mathrm{B}} \end{pmatrix} = \begin{pmatrix} 17 + 9\lambda & 4\lambda & 6\lambda \\ 0 & 15 + 7\lambda & 4\lambda \\ 0 & 0 & 13 + 3\lambda \end{pmatrix} \begin{pmatrix} \phi_{3,2,2,0}^{\mathrm{B}} \\ \phi_{3,2,1,1}^{\mathrm{B}} \\ \phi_{2,2,2,1}^{\mathrm{B}} \end{pmatrix}. \quad (2.40)$$

In (2.40), the function $\phi_{2,2,2,1}^{\mathrm{B}}$ is obviously an eigenfunction with eigenenergy $13 + 3\lambda$. The other eigenfunctions are given in the form of

$$\Phi_{3,2,1,1}^{\mathrm{B}} = \phi_{3,2,1,1}^{\mathrm{B}} + c\phi_{2,2,2,1}^{\mathrm{B}},$$
$$\Phi_{3,2,2,0}^{\mathrm{B}} = \phi_{3,2,2,0}^{\mathrm{B}} + c'\phi_{3,2,1,1}^{\mathrm{B}} + c''\phi_{2,2,2,1}^{\mathrm{B}}.$$

The eigenenergies of $\Phi_{3,2,1,1}^{\mathrm{B}}$ and $\Phi_{3,2,2,0}^{\mathrm{B}}$ are, respectively, given by $17 + 9\lambda$ and $15 + 7\lambda$ from the diagonal elements of the matrix in (2.40).

Generally, all the eigenfunctions of $\mathcal{H}$ can be written in the form

$$\Phi_{\kappa}^{\mathrm{B}} = \phi_{\kappa}^{\mathrm{B}} + \sum_{\mu(<\kappa)} a_{\mu} \phi_{\mu}^{\mathrm{B}}. \quad (2.41)$$

The eigenenergy $\mathcal{E}[\kappa]$ of $\Phi_{\kappa}^{\mathrm{B}}$ is given by the diagonal element of $\mathcal{H}$ in the basis $\{\phi_{\kappa}^{\mathrm{B}}\}$ as

$$\mathcal{E}[\kappa] = \sum_{i=1}^{N} \kappa_i^2 + \lambda \sum_{1 \leq i < j \leq N} (\kappa_i - \kappa_j)$$

$$= \sum_{i=1}^{N} \kappa_i^2 + \frac{\lambda}{2} \sum_{i} \sum_{j} |\kappa_i - \kappa_j|. \quad (2.42)$$

With use of (2.25) and (2.42), the eigenenergy $E[\kappa]$ for the original Hamiltonian H is given by

$$E[\kappa] = \left(\frac{2\pi}{L}\right)^2 \left[\sum_i \kappa_i^2 + \frac{\lambda}{2}\sum_i\sum_j |\kappa_i - \kappa_j|\right] + E_{0,N}. \qquad (2.43)$$

The wave functions (2.41) of eigenstates will again be discussed in Section 2.5, where alternative descriptions of energy spectra and elementary excitations are given.

From (2.42), we immediately see that the Jastrow wave function Ψ_0^B or Ψ_0^F is the wave function of the ground state. For an N-particle system, $\kappa = (0,\ldots,0)$ yields $E[\kappa] = 0$, which is the minimum of (2.42) for all the states.

2.1.3 Ordering of basis functions

The explicit expression for the wave function of the eigenstate (2.41) is obtained by diagonalization of a triangular matrix with finite dimension. The linear space is spanned by ϕ_κ^B, and only those ϕ_μ^B with μ which are obtained by multiple squeezing of κ. We then label each basis function by $|j\rangle$ with $j = 1, 2, \ldots$ according to the ordering defined in the previous subsection. When $\kappa = (3, 2, 2, 0)$, we put

$$|1\rangle = \phi_{3,2,2,0}^B, \quad |2\rangle = \phi_{3,2,1,1}^B, \quad |3\rangle = \phi_{2,2,2,1}^B.$$

In this basis, the matrix elements of $\mathcal{H}$ are written

$$\langle l|\mathcal{H}|j\rangle = \epsilon_l \delta_{lj} + I_{lj}, \qquad (2.44)$$

with

$$I_{lj} = \begin{cases} 0, & l \leq j, \\ \text{non-negative}, & \text{otherwise}. \end{cases} \qquad (2.45)$$

When we write the eigenfunction Φ_κ^B as $\sum_j c_j |j\rangle$, the eigenvalue equation is given by

$$\sum_j (\delta_{lj}\epsilon_j + I_{lj})c_j = \epsilon_1 c_l. \qquad (2.46)$$

It follows that for $l \geq 2$

$$c_l = \frac{\sum_{1\leq j<l} I_{lj}c_j}{\epsilon_1 - \epsilon_l}. \qquad (2.47)$$

We will show that $\epsilon_1 - \epsilon_l$ in the denominator is positive at the end of this subsection. With the use of (2.47), we can obtain c_l recursively starting from c_1. We take $c_1 = 1$ and define a matrix M_{lj} by

$$M_{11} = 1, \quad M_{lj} = \frac{I_{lj}}{\epsilon_1 - \epsilon_l}, \quad \text{for } (jl) \neq (11). \tag{2.48}$$

When $\kappa = (3, 2, 2, 0)$, (2.47) becomes

$$c_2 = M_{21}, \quad c_3 = (M^2)_{31}, \ldots, c_l = (M^{l-1})_{l1}, \tag{2.49}$$

with

$$M = \begin{pmatrix} 1 & 0 & 0 \\ \frac{2\lambda}{1+\lambda} & 0 & 0 \\ \frac{3\lambda}{2+3\lambda} & \frac{2\lambda}{2+3\lambda} & 0 \end{pmatrix} \tag{2.50}$$

and

$$M^2 = \begin{pmatrix} 1 & 0 & 0 \\ \frac{2\lambda}{1+\lambda} & 0 & 0 \\ \frac{\lambda(3+7\lambda)}{(1+\lambda)(2+3\lambda)} & 0 & 0 \end{pmatrix}. \tag{2.51}$$

From these results, we obtain

$$\Phi^{\mathrm{B}}_{3,2,2,0} = \phi^{\mathrm{B}}_{3,2,2,0} + \frac{2\lambda}{1+\lambda}\phi^{\mathrm{B}}_{3,2,1,1} + \frac{\lambda(3+7\lambda)}{(1+\lambda)(2+3\lambda)}\phi^{\mathrm{B}}_{2,2,2,1}. \tag{2.52}$$

In this way, eigenstates are obtained by calculation of the power of the finite-dimensional triangular matrix defined in (2.48). In Section 2.5, the properties of the resulting function Φ^{B}_κ will be identified as a symmetric Jack polynomial.

We now show that $E[\kappa] > E[\mu]$ if μ is obtained by a squeezing or multiple squeezing of κ. It suffices to show that

$$E[\kappa] > E[\mu] \tag{2.53}$$

for

$$\mu_i = \kappa_i - 1, \quad \mu_j = \kappa_j + 1 \tag{2.54}$$

with a pair $(i < j)$ and

$$\mu_k = \kappa_k \tag{2.55}$$

for $k \neq i, j$. Any squeezing of κ can be generated by this elementary squeezing (2.54). We can assume without loss of generality that μ is ordered non-increasingly:

$$\mu_{i-1} \geq \mu_i \geq \mu_{i+1} \geq \cdots \geq \mu_{j-1} \geq \mu_j \geq \mu_{j+1} \geq \cdots .$$

The contribution from the kinetic energy in (2.53) is given by

$$\sum_i (\kappa_i^2 - \mu_i^2) = \kappa_i^2 + \kappa_j^2 - [(\kappa_i - 1)^2 + (\kappa_j + 1)^2]$$

$$= 2(\kappa_i - \kappa_j - 1) \geq 0. \tag{2.56}$$

In the expression

$$\sum_{k<l} (|\kappa_k - \kappa_l| - |\mu_k - \mu_l|) \tag{2.57}$$

from the interaction energy, the contribution with $k \leq i - 1$ or $l \geq j + 1$ vanishes. Thus (2.57) becomes

$$(\kappa_i - \kappa_j) - (\mu_i - \mu_j) + \sum_{k=i+1}^{j-1} [(\kappa_i - \kappa_k) + (\kappa_k - \kappa_j) - (\mu_i - \kappa_k) - (\kappa_k - \mu_j)]$$

$$= 2(j - i) > 0. \tag{2.58}$$

The inequalities (2.56) and (2.58) prove (2.53).

2.2 Descriptions of energy spectrum

The expression (2.42) for the eigenvalue can be rewritten in different ways. In this section we present three equivalent descriptions.

2.2.1 Interacting boson description

We introduce the momentum distribution function by

$$\nu^{\mathrm{B}}(\kappa) = \sum_{i=1}^N \delta_{\kappa, \kappa_i}, \tag{2.59}$$

which takes arbitrary non-negative integers. Then (2.42) becomes

$$\mathcal{E}[\kappa] = \sum_{\kappa=\infty}^{\infty} \kappa^2 \nu^{\mathrm{B}}(\kappa) + \frac{\lambda}{2} \sum_{\kappa=-\infty}^{\infty} \sum_{\kappa'=-\infty}^{\infty} \nu^{\mathrm{B}}(\kappa) \nu^{\mathrm{B}}(\kappa') |\kappa - \kappa'|. \tag{2.60}$$

This form is useful to construct the partition function or thermodynamic potential of the Sutherland model, as discussed in the following section.

Another expression for (2.42) is available with use of the relation

$$\sum_{i<j}(\kappa_i - \kappa_j) = \sum_{i=1}^{N}(N+1-2i)\,\kappa_i. \tag{2.61}$$

Namely we obtain

$$\mathcal{E}[\kappa] = \sum_{i=1}^{N}\left[\kappa_i^2 + \lambda(N+1-2i)\kappa_i\right] \tag{2.62}$$

$$= \sum_{i=1}^{N}\tilde{\kappa}_i^2 - \sum_{i=1}^{N}\tilde{\kappa}_{i,0}^2 \tag{2.63}$$

in the form of the energy difference of free particles. Here we have introduced the rapidity

$$\tilde{\kappa}_i = \kappa_i + \frac{\lambda}{2}\left(N+1-2i\right), \quad \tilde{\kappa}_{i,0} = \frac{\lambda}{2}\left(N+1-2i\right), \tag{2.64}$$

which is interpreted as a generalized momentum including the interaction effect. The eigenenergy $E[\kappa]$ (2.43) is written in a surprisingly simple form:

$$E[\kappa] = \sum_{i=1}^{N}\left(2\pi\tilde{\kappa}_i/L\right)^2. \tag{2.65}$$

The relation (2.64) between $\{\tilde{\kappa}_i\}$ and $\{\kappa_i\}$ is rewritten as

$$\tilde{\kappa}_i = \kappa_i + \frac{\lambda}{2}\sum_{j(\neq i)}\mathrm{sgn}(\tilde{\kappa}_i - \tilde{\kappa}_j), \tag{2.66}$$

using the relation

$$\sum_{j(\neq i)}\mathrm{sgn}(\tilde{\kappa}_i - \tilde{\kappa}_j) = \sum_{j=1}^{i-1}(-1) + \sum_{j=i+1}^{N}(+1) = N - 2i + 1. \tag{2.67}$$

Then (2.66) is written alternatively as

$$\tilde{\kappa}_i = \kappa_i + \frac{1}{2\pi}\sum_{j(\neq i)}\theta^{\mathrm{B}}(\tilde{\kappa}_i - \tilde{\kappa}_j), \tag{2.68}$$

which is analogous to a Bethe ansatz equation. The scattering phase shift

$$\theta^{\mathrm{B}}(\tilde{\kappa}) = \pi\lambda\,\mathrm{sgn}(\tilde{\kappa})$$

is relevant to the two-particle scattering problem with the interaction in (2.1). Hence the dispersion relation of elementary excitations and the thermodynamic potential of the Sutherland model can be calculated with use of the method developed in the Bethe ansatz theory.

2.2.2 Interacting fermion description

We can rewrite (2.68) as

$$\tilde{\kappa}_i = I_i + \frac{1}{2\pi}\sum_{j(\neq i)}\theta^F(\tilde{\kappa}_i - \tilde{\kappa}_j) \tag{2.69}$$

in terms of the scattering phase shift

$$\theta^F(\tilde{\kappa}) = \pi(\lambda - 1)\mathrm{sgn}(\tilde{\kappa})$$

relevant to the two (spinless) fermions. A set $\{I_1, I_2 \ldots\}$ satisfying $I_1 > I_2 \cdots$ of fermionic quantum numbers is given by

$$I_i = \kappa_i + \frac{1}{2}\left(N + 1 - 2i\right), \tag{2.70}$$

which satisfies

$$I_i - I_{i+1} = \kappa_i - \kappa_{i+1} + 1 \geq 1. \tag{2.71}$$

By substituting (2.70) into (2.62), we can rewrite the energy as that of interacting fermions:

$$\begin{aligned}
\mathcal{E}[\kappa] &= \sum_{i=1}^{N}\left[I_i^2 + (\lambda - 1)(N + 1 - 2i)I_i\right] \\
&= \sum_{i=1}^{N}I_i^2 + \frac{\lambda - 1}{2}\sum_i\sum_j|I_i - I_j| \\
&= \sum_I I^2\nu^F(I) + \frac{\lambda - 1}{2}\sum_I\sum_{I'}|I - I'|\nu^F(I)\nu^F(I').
\end{aligned} \tag{2.72}$$

In the last equality, we have introduced the fermionic momentum distribution function $\nu^F(I)$, which is either 0 or 1. Note that the parameter $\lambda - 1$ appears in (2.72), in contrast with λ in (2.60).

2.2.3 Exclusion statistics

We have seen that the energy spectrum can be described as that of interacting bosons or fermions. Alternatively, the energy spectrum can be regarded as that of free particles. The price for eliminating the interaction effect is paid by the nontrivial exclusion rule for the rapidity $\{\tilde{\kappa}_1, \tilde{\kappa}_2, \ldots, \tilde{\kappa}_N\}$. Namely, the restriction is given by

$$\tilde{\kappa}_i - \tilde{\kappa}_{i+1} = \kappa_i - \kappa_{i+1} + \lambda \geq \lambda. \tag{2.73}$$

This exclusion rule (2.73) can be regarded as an example of exclusion statistics [79], which we will explain now.

Consider a many-particle system where the dimension G of the one-particle Hilbert space is proportional to the volume of the system. Models defined on a lattice are examples of these systems. Another example is given by a Landau level of two-dimensional electrons in a magnetic field; the degeneracy of each Landau level is proportional to the area of the system. Here the dimension G is meant to be the number of available one-particle states *in the absence of other particles* under the fixed boundary condition. If a particle occupies a one-particle state, the number of available one-particle states D for another particle depends on the statistics. Namely, if particles are fermions, the occupied state is not available owing to the Pauli exclusion, and hence $D = G - 1$. If particles are bosons, on the other hand, the occupied state can accommodate another particle, and hence $D = G$ for bosons. Thus when $N - 1$ particles are present, the number of available states for the Nth particle is given by

$$D = \begin{cases} G, & \text{for bosons,} \\ G - N + 1, & \text{for fermions.} \end{cases} \tag{2.74}$$

Using D and N, the number of microscopic states for an N-particle system is written as

$$W = \frac{(D + N - 1)!}{N!\,(D - 1)!}, \tag{2.75}$$

for both bosons and fermions. Then (2.75) can also be regarded as the definition of D for N-particle systems. Through the N-dependence of D, the statistical parameter g is defined by

$$\Delta D = -g\Delta N. \tag{2.76}$$

In this definition, $g = 0$ for bosons and $g = 1$ for fermions.

Beyond the conventional (i.e., Bose or Fermi) statistics, we consider the generalized statistics for any positive value of g. Correlated electron systems provide examples of elementary excitations obeying such statistics; spinon excitations in an antiferromagnetic spin chain obey the statistics $g = 1/2$, which will be discussed in Chapter 4. Quasi-particles in the quantum Hall effect is another example. The statistical parameter g describes the strength of exclusion. Thus, these new statistics are termed exclusion statistics by Haldane [79].

In the case where particles have internal degrees of freedom such as spin or color, we can easily generalize the above framework. Let N_α be the number of particles with spin or color α, and D_α be the number of available one-particle states for α species in the presence of N_β particles with $\beta (\neq \alpha)$, in addition to $N_\alpha - 1$ particles with α. The number of microscopic states with N_α fixed for all α is given by

$$W = \prod_\alpha \frac{(D_\alpha + N_\alpha - 1)!}{N_\alpha!\,(D_\alpha - 1)!}. \tag{2.77}$$

The statistics is described by a matrix $g_{\alpha\beta}$ defined by

$$\Delta D_\alpha = -\sum_\beta g_{\alpha\beta} \Delta N_\beta. \tag{2.78}$$

The exclusion statistics can be generalized to that in the continuum model, such as free Bose gas, free Fermi gas, and the Sutherland model. Let us consider a d-dimensional free Bose or Fermi gas in the system with volume V with periodic boundary condition. Divide the k-space (momentum space) into regions with finite volume. Each region is labeled by α and has the volume $\tilde{V}_\alpha$. In the k-space, there is a one-particle state per cell with volume $(2\pi)^d/V$. Hence in each region α, the number of one-particle states is $V/\tilde{V}_\alpha (2\pi)^d$, which is proportional to the volume of the system V. Thus, the exclusion statistics can be defined in each region α. In the case of the Sutherland model, the k-space is replaced by the rapidity space in the above argument. We then find that the relation (2.73) is indeed an example of exclusion statistics.

2.3 Elementary excitations

The low-energy excitation spectra in interacting systems can often be understood in terms of a kind of nearly free particle even when the interaction effect is strong. These particles are called *elementary excitations* in condensed matter physics. The Landau theory of Fermi liquids is the most successful example. However, calculation of the spectra is sometimes out of reach in the perturbation theory. Elementary excitations in the fractional quantum Hall state are the best known example. Thus, exactly solvable models such as the Sutherland model are very useful in understanding the concept of elementary excitations.

The excitation energy $\Delta E[\kappa]$ and momentum Q_κ of a state specified by κ are given by

$$\Delta E = \left(\frac{2\pi}{L}\right)^2 \sum_{i=1}^{N} \left[\kappa_i^2 + \lambda(N + 1 - 2i)\kappa_i\right] \tag{2.79}$$

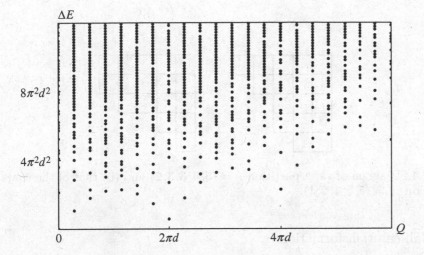

Fig. 2.3. Low-energy excitation spectrum of the Sutherland model with $\lambda = 2$ and $N = 7$. The momentum and the energy are scaled by the density $d = N/L$ and d^2, respectively.

and

$$Q_\kappa = \left(\frac{2\pi}{L}\right) \sum_{i=1}^{N} \kappa_i, \tag{2.80}$$

where (2.62) has been used. The density of the particles is given by $d = N/L$. Figure 2.3 shows the energy spectrum in the case of $\lambda = 2$ and $N = 7$. There is a similarity to the spectrum of free fermions in one dimension. This observation suggests that the energy spectrum can be constructed from particle-like and hole-like excitations, which we pursue in the present section.

2.3.1 Partitions

A combination of momenta κ_i specifying an eigenstate forms a set

$$\mathcal{L}_N^+ = \{\kappa | \kappa = (\kappa_1, \dots, \kappa_N) \in \mathbf{Z}; \kappa_1 \geq \cdots \geq \kappa_N\}, \tag{2.81}$$

where $\mathbf{Z}$ is the set of integers. The subset $\Lambda_N^+ \subset \mathcal{L}_N^+$ with non-negative momentum is defined by

$$\Lambda_N^+ = \{\mu | \mu = (\mu_1, \dots, \mu_N) \in \mathbf{Z}_{\geq 0}; \mu_1 \geq \cdots \geq \mu_N \geq 0\}, \tag{2.82}$$

where $\mathbf{Z}_{\geq 0}$ is the set of non-negative integers. Any element in Λ_N^+ is called a partition. The eigenstate specified by $\kappa \in \mathcal{L}_N^+$ is related to a partition μ

(a)   (b)

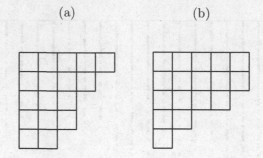

Fig. 2.4. Diagram of (a) a partition $\mu = (5, 4, 3, 3, 2)$ and (b) that of the conjugate partition $\mu' = (5, 5, 4, 2, 1)$.

via Galilean transformation by

$$\mu = (\mu_1, \ldots, \mu_N) = (\kappa_1 + n, \ldots, \kappa_N + n) \in \Lambda_N^+, \quad n \in \mathbf{Z}. \tag{2.83}$$

The partition $\mu \in \Lambda_N^+$ is easier to study than $\kappa \in \mathcal{L}_N^+$ because partitions are graphically described by Young diagrams.

Let us introduce some notation for partitions and related Young diagrams. We call the nonzero μ_i the parts of the partition $\mu \in \Lambda_N^+$, and call the number of parts length $l(\mu)$. We define the weight of μ as

$$|\mu| = \sum_{i=1}^{N} \mu_i.$$

The Young diagram of a partition μ consists of squares, the coordinates of which are (i, j) with $1 \leq i \leq l(\mu)$ and $1 \leq j \leq \mu_i$. The set of those (i, j) is denoted by $D(\mu)$. In drawing diagrams, the first coordinate i (the row index) increases as one goes downwards and the second coordinate j (the column index) increases from left to right. In Fig. 2.4(a), we show the diagram of the partition $\mu = (5, 4, 3, 3, 2)$, where $5(= \mu_1)$ squares are in the first row, $4(= \mu_2)$ squares are in the second row, and so on. The conjugate partition of a partition μ is obtained by interchanging the rows and columns of the diagram of μ. In Fig. 2.4(b), we show the diagram of the conjugate partition $\mu' = (5, 5, 4, 2, 1)$ of the partition $\mu = (5, 4, 3, 3, 2)$.

For square s specified by the coordinate (i, j), we define arm length $a(s)$, leg length $l(s)$, arm colength $a'(s)$, and leg colength $l'(s)$ by

$$a(s) = \mu_i - j, \quad a'(s) = j - 1, \quad l(s) = \mu_j' - i, \quad l'(s) = i - 1, \tag{2.84}$$

as shown in Fig. 2.5.

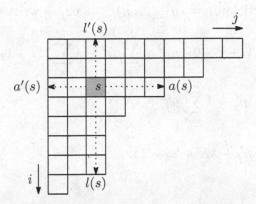

Fig. 2.5. An example is shown for the arm length $a(s) = 3$, leg length $l(s) = 4$, arm colength $a'(s) = 2$, and leg colength $l'(s) = 2$ for the shaded square $s = (3,3)$.

Using the notation introduced above, we derive some formulae. Let us start with the following relation:

$$\sum_{s \in D(\mu)} a'(s) = \sum_{s \in D(\mu')} l'(s). \tag{2.85}$$

This relation holds because the columns and rows are interchanged in $D(\mu)$ and $D(\mu')$. Both sides are written as

$$\text{LHS} = \sum_{i=1}^{l(\mu)} \sum_{j=1}^{\mu_i} (j - 1) = \frac{1}{2} \sum_{i=1}^{l(\mu)} \mu_i(\mu_i - 1), \tag{2.86}$$

$$\text{RHS} = \sum_{j=1}^{\mu_1} \sum_{i=1}^{\mu'_j} (j - 1) = \sum_{j=1}^{\mu_1} \mu'_j(j - 1), \tag{2.87}$$

respectively. Equating (2.86) and (2.87) and using an obvious relation,

$$\sum_{i=1}^{l(\mu)} \mu_i = \sum_{j=1}^{\mu_1} \mu'_j, \tag{2.88}$$

we obtain

$$\sum_{i=1}^{l(\mu)} \mu_i^2 = \sum_{j=1}^{\mu_1} (2j - 1)\mu'_j. \tag{2.89}$$

(2.88) and (2.89) are useful if the energy is to be written in terms of $\{\mu'_j\}$.

Namely for a partition $\mu = (\mu_1, \ldots, \mu_N)$, (2.62) is written as

$$
\sum_{i=1}^{N} \left[\mu_i^2 + \lambda \left(N + 1 - 2i \right) \mu_i \right] = \sum_{i=1}^{N} \mu_i^2 - \lambda \sum_{i=1}^{N} \left(2i - 1 \right) \mu_i + \lambda N \sum_{i=1}^{N} \mu_i
$$

$$
= \sum_{j=1}^{\mu_1} (2j-1)\mu_j' - \lambda \sum_{j=1}^{\mu_1} \mu_j'^2 + \lambda N \sum_{j=1}^{\mu_1} \mu_j'
$$

$$
= \sum_{j=1}^{\mu_1} \mu_j' \left(-\lambda \mu_j' + \lambda N + 2j - 1 \right). \tag{2.90}
$$

2.3.2 Quasi-particles

With the periodic boundary condition (2.2), we take N odd so that the ground state is non-degenerate in both bosonic and fermionic systems. The N-particle ground state is described by $\kappa_i = 0$ for all i. Then we have the rapidities for the ground state:

$$
\tilde{\kappa}_{i,0} = \lambda \left(\frac{N+1}{2} - i \right). \tag{2.91}
$$

Namely, the rapidities are subject to the exclusion rule. Hence the distribution of $\tilde{\kappa}_i$ is expressed in a way analogous to fermionic quantum numbers.

Figure 2.6(a) shows the rapidity distribution in the ground state for $\lambda = 2$ and $N = 7$. The sequence element "1" stands for occupied rapidities, and "0" for unoccupied. In the ground state (a), the rapidity distribution is given by $\{6, 4, 2, 0, -2, -4, -6\}$. The dashed lines represent the pseudo-Fermi points, which correspond to the Fermi surface in one dimension, of this pseudo-Fermi sea. These representations are called "motif" and frequently used in the literature [73, 78, 125] relating to the Sutherland model. In the following, we consider the one-particle addition spectrum for positive integer λ.

When the original particles are bosons and λ is an even integer, both $\{\kappa_i\}$ and $\{\tilde{\kappa}_i\}$ are integers. In this case, an $(N+1)$-particle state is possible with

$$
\tilde{\kappa}_1 \geq \tilde{\kappa}_{1,0} + \lambda, \quad \tilde{\kappa}_{i+1} = \tilde{\kappa}_{i,0}, \quad i \in [1, N], \tag{2.92}
$$

or equivalently,

$$
\kappa_1 \geq \kappa_2 = \cdots = \kappa_{N+1} = \lambda/2. \tag{2.93}
$$

Note that κ_i for $i > 1$ is shifted by $\lambda/2$ in order to make the rapidity $\{\tilde{\kappa}\}$ remain the same. An example of (2.92) is shown in Fig. 2.6(b). The state (2.92) consists of the Fermi sea for an N-particle system and an additional particle. We call this particle outside the Fermi sea a *quasi-particle*.

(a) $\cdots$ 0 0 0 $\vdots$ 1 0 1 0 1 0 1 0 1 0 1 0 1 $\vdots$ 0 0 0 0 0 $\cdots$

(b) $\cdots$ 0 0 0 $\vdots$ 1 0 1 0 1 0 1 0 1 0 1 0 1 $\vdots$ 0 0 0 0 1 $\cdots$

(c) $\cdots$ 0 0 0 $\vdots$ 1 0 0 1 0 1 0 1 0 1 0 0 1 $\vdots$ 0 0 0 0 0 $\cdots$

Fig. 2.6. Motif representation of eigenstates for $\lambda = 2$ and $N = 7$: (a) the ground state; (b) a one-particle addition state; (c) a one-particle removal state. In the sequence, "1" and "0" represent, respectively, occupied and empty rapidities. The dotted lines represent the pseudo-Fermi point. In (c), λ consecutive zeros correspond to a quasi-hole with charge reduced to $1/\lambda$ compared to that of the original particles.

The excitation energy ΔE and momentum Q of the state (2.92) are given by

$$\Delta E = \left(\frac{2\pi}{L}\right)^2 \tilde{\kappa}_1^2, \quad Q = \left(\frac{2\pi}{L}\right) \tilde{\kappa}_1, \tag{2.94}$$

with $\tilde{\kappa}_1 \geq \lambda N/2$. In addition to the dimensionless rapidity $\tilde{\kappa}_1$, we introduce another rapidity

$$p = 2\pi \tilde{\kappa}_1/L, \tag{2.95}$$

with the dimension of physical momentum. Taking the thermodynamic limit $N \to \infty$ with $N/L = d$ fixed, we obtain

$$\Delta E = \epsilon_{\mathrm{p}}(p) + \mu, \quad Q = p, \tag{2.96}$$

with $p \geq \pi \lambda d$, and

$$\epsilon_{\mathrm{p}}(p) = p^2 - \pi^2 \lambda^2 d^2. \tag{2.97}$$

The chemical potential μ is given by

$$\mu = \lim_{N \to \infty} \frac{\partial E_{0,N}}{\partial N} = \pi^2 \lambda^2 d^2. \tag{2.98}$$

The ground-state energy $E_{0,N}$ has been derived in (2.12). The quasi-particle in this state is right-moving.

We consider another $(N + 1)$-particle state with the rapidities

$$\tilde{\kappa}_{N+1} \leq \tilde{\kappa}_{N,0} - \lambda, \quad \tilde{\kappa}_i = \tilde{\kappa}_{i,0}, \quad i \in [1, N]. \tag{2.99}$$

We have to shift κ_i by $-\lambda/2$ in order to satisfy (2.99). Namely, we have

$$\kappa_{N+1} < \kappa_N = \cdots = \kappa_1 = -\lambda/2. \qquad (2.100)$$

In the thermodynamic limit, the energy spectrum of (2.99) tends to

$$\Delta E = \epsilon_{\mathrm{p}}(p) + \mu, \quad Q = p, \quad p \le -\pi\lambda d, \qquad (2.101)$$

which represents a left-moving quasi-particle.

Quasi-particles obey the generalized statistics with the exclusion parameter λ. This can be understood by considering an $(N+2)$-particle state with rapidities

$$\tilde{\kappa}_1 > \tilde{\kappa}_2 > \tilde{\kappa}_{1,0} + \lambda, \quad \tilde{\kappa}_i = \tilde{\kappa}_{i-2,0}, \quad i \in [3, N+2]. \qquad (2.102)$$

In this state, the rapidities $\tilde{\kappa}_1$ and $\tilde{\kappa}_2$ of quasi-particle states are subject to the exclusion rule

$$\tilde{\kappa}_1 \ge \tilde{\kappa}_2 + \lambda. \qquad (2.103)$$

Let us now consider the one-particle addition spectrum of bosons for odd λ. Under the periodic boundary condition with N odd, $\tilde{\kappa}_i$ for $(N+1)$-particle states is a half odd integer but $\tilde{\kappa}_{i,0}$ for the N-particle ground state is an integer. As a result, the state given by (2.92) actually does not exist. Alternatively, we consider a state with

$$\tilde{\kappa}_1 \ge \tilde{\kappa}_{1,0} + \lambda, \quad \tilde{\kappa}_{i+1} = \tilde{\kappa}_{i,0} - \frac{\lambda}{2}, \quad i \in [2, N]. \qquad (2.104)$$

In the state (2.104), the rapidities in the condensate of the Fermi sea are shifted, which is called the backflow effect. Correspondingly, the spectrum of (2.104) in the thermodynamic limit is given by

$$\Delta E = \epsilon_{\mathrm{p}}(p) + \mu, \quad Q = p - \pi\lambda d, \quad p \ge \pi\lambda d.$$

We note that the momentum in this spectrum is shifted compared to (2.96). Thus the one-particle addition spectrum of the bosonic system for odd λ is more complicated than that for even λ. As shown in Section 2.6, an exact solution of the Green function of the Sutherland model for bosons has been obtained only when λ is even.

We next consider the case where the original particles are fermions; the one-particle addition spectrum again depends on whether λ is even or odd. The N-particle ground state is given by Ψ_0^{F}, and the spectrum is given by (2.91). When λ is odd, κ_i for $(N+1)$-particle states is a half odd integer in order that the total wave function Ψ satisfies the periodic boundary condition. Rapidities $\tilde{\kappa}_i = \kappa_i + \lambda(N + 2 - 2i)/2$ for $(N+1)$-particle states are thus integers. By taking $\{\kappa_i\}$ as

$$\kappa_1 > \kappa_i = \frac{\lambda}{2}, \quad i \in [2, N], \tag{2.105}$$

we obtain the state (2.92) with one quasi-particle and no backflow.

When the particles are fermions and λ is an even integer, Ψ_0^F for $(N + 1)$-particle states satisfies the periodic boundary condition. Hence κ_i's for $(N + 1)$-particle states are integers and rapidities are half odd integers. In this case, one quasi-particle cannot be created without the momentum shift of the condensate in the Fermi sea. When particles are fermions, an exact solution of the Green function is available only for an odd integer λ.

2.3.3 Quasi-holes

We calculate the spectrum of one-particle removal states from the N-particle ground state. The particle number N is taken to be odd so that the ground state is non-degenerate under the periodic boundary condition. We focus on the model for bosons with even integer λ, or fermions with odd λ for simplicity.

When particles are bosons and λ is even, both κ_i and $\tilde{\kappa}_i$ are integers in $(N - 1)$-particle states. The $(N - 1)$-particle states with the pseudo-Fermi points fixed have rapidity distribution satisfying the condition

$$\tilde{\kappa}_1 = \tilde{\kappa}_{1,0} = \frac{\lambda(N - 1)}{2}, \ldots, \tilde{\kappa}_{N-1} = \tilde{\kappa}_{N,0}. \tag{2.106}$$

An example of the state with (2.106) is given in Fig. 2.6(c) for $\lambda = 2$ and $N = 7$. When a particle is removed from the Fermi sea for $\lambda = 2$, two "00" are created. This means that the states satisfying (2.106) are a two-parameter family parameterized by two quantum numbers.

For general even integer λ, the states satisfying (2.106) are a λ-parameter family, as shown below. From (2.106), we obtain the largest (κ_1) and the smallest (κ_{N-1}) momenta as

$$\kappa_1 = \tilde{\kappa}_1 - \lambda(N-3)/2 = \lambda/2, \quad \kappa_{N-1} = \tilde{\kappa}_{N-1} - \lambda(1-N)/2 = -\lambda/2. \tag{2.107}$$

The momentum distribution

$$\left(\kappa_1 + \frac{\lambda}{2}, \ldots, \kappa_{N-1} + \frac{\lambda}{2}\right) = (\mu_1, \ldots, \mu_{N-1}),$$

which is Galilean shifted by $\lambda/2$ from (2.107), is described by Young diagrams with λ columns. Each row has $\mu_i = \kappa_i + \lambda/2$ squares. Obviously such Young diagrams are equivalently parameterized by the set $(\mu'_1, \ldots, \mu'_\lambda)$

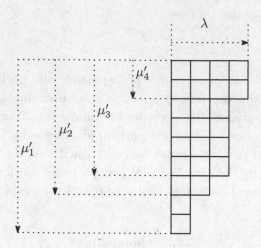

Fig. 2.7. Young diagram of a one-particle removal state for bosons with $\lambda = 4$.

of lengths of each column. Figure 2.7 shows an example for $\lambda = 4$. The excitation energy of the $(N-1)$-particle state is given by

$$
\begin{aligned}
\mathcal{E} &= \sum_{i=1}^{N-1}\left[\kappa_i + \frac{\lambda(N-2i)}{2}\right]^2 - \sum_{i=1}^{N}\left[\frac{\lambda(N+1-2i)}{2}\right]^2 \\
&= \sum_{i=1}^{N-1}\left[\mu_i + \frac{\lambda(N-1-2i)}{2}\right]^2 - \sum_{i=1}^{N}\left[\frac{\lambda(N+1-2i)}{2}\right]^2 \\
&= \sum_{i=1}^{N-1}\left[\mu_i^2 + \lambda(N-1-2i)\mu_i\right] - \left(\frac{\lambda(N-1)}{2}\right)^2.
\end{aligned}
\tag{2.108}
$$

Using (2.90) in (2.108), we obtain

$$
\mathcal{E} + \frac{1}{4}\lambda^2(N-1)^2 = \sum_{j=1}^{\lambda} \mu_j'\left[-\lambda\mu_j' + \lambda(N-2) + 2j - 1\right].
\tag{2.109}
$$

Let us introduce (dimensionless) rapidities by

$$
\dot{\mu}_j = \mu_j' - \frac{N-1}{2} + \frac{\lambda+1-2j}{2\lambda}, \quad \dot{\mu}_{j,0} = \frac{\lambda+1-2j}{2\lambda},
\tag{2.110}
$$

and write (2.109) as

$$
\mathcal{E} + \frac{1}{4}\lambda^2(N-1)^2 = -\lambda\sum_{j=1}^{\lambda}\left[\dot{\mu}_j^2 - \dot{\mu}_{j,0}^2\right].
\tag{2.111}
$$

Using (2.110), the momentum of the quasi-hole state is expressed as

$$Q = \frac{2\pi}{L}\sum_{i=1}^{N-1}\kappa_i = \frac{2\pi}{L}\sum_{i=1}^{N-1}\left(\mu_i - \frac{\lambda}{2}\right)$$

$$= \frac{2\pi}{L}\sum_{j=1}^{\lambda}\mu_j' - \frac{\pi}{L}(N-1)\lambda = \frac{2\pi}{L}\sum_{j=1}^{\lambda}\dot{\mu}_j. \qquad (2.112)$$

The spectrum in the thermodynamic limit is then given by

$$\Delta E + \mu = \lambda\sum_{j=1}^{\lambda}\left[\pi^2 d^2 - p_j'^2\right], \quad Q = \sum_{j=1}^{\lambda}p_j', \qquad (2.113)$$

where we have introduced the quasi-hole rapidity $p_j' = (2\pi/L)\dot{\mu}_j$ with $|p_j'| \le \pi d$.

We have thus found that a one-particle removal state is interpreted as an excited state with the number λ of quasi-holes. The exclusion rule is given by

$$\dot{\mu}_j \ge \dot{\mu}_{j+1} + \frac{1}{\lambda}. \qquad (2.114)$$

It is clear that these quasi-holes obey the generalized statistics with a parameter $1/\lambda$. Furthermore, (2.73) and (2.114) suggest that the Sutherland model with a parameter λ is related to another with $1/\lambda$ via the particle–hole transformation. This property is called duality. We note that in the case of odd λ, (2.113) is valid as the removal spectrum for the system with N fermions.

2.3.4 Neutral excitations

Now we consider the spectrum of excited states where the number N of particles remains the same as that in the ground state. In this case, rapidities $\{\tilde{\kappa}_i\}$ of excited states are integers for both bosons and fermions with integer λ. As a result, one quasi-particle can be excited, keeping the pseudo-Fermi points fixed (i.e., without shift of the Fermi sea). The set of rapidities in this type of excited state has the form

$$\tilde{\kappa}_1 > \tilde{\kappa}_2 + \lambda, \quad \tilde{\kappa}_2 \le \tilde{\kappa}_{1,0}, \quad \tilde{\kappa}_N \ge \tilde{\kappa}_{N,0}, \qquad (2.115)$$

which is expressed in terms of momenta κ_i as

$$\kappa_1 \ge \lambda, \quad \kappa_2 \le \lambda, \quad \kappa_N \ge 0. \qquad (2.116)$$

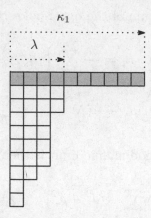

Fig. 2.8. Young diagram of the states with a right-moving quasi-particle and λ quasi-holes.

Figure 2.8 shows an example of a Young diagram for partition κ satisfying (2.116). This diagram consists of a quasi-particle, represented by shaded squares, and quasi-holes, represented by unshaded squares. One quasi-particle excitation accompanies λ quasi-hole excitations. This is interpreted as the charge conservation

$$(\text{number of quasi-particles}) \times \underbrace{e_p}_{1} + (\text{number of quasi-holes}) \times \underbrace{e_h}_{-1/\lambda} = \Delta N,$$

(2.117)

where e_p and e_h denote the charge of a quasi-particle and a quasi-hole, respectively. $\Delta N (= 0$ in the present case) is the difference of the particle numbers between excited states and the ground state. Excitation of one quasi-particle and λ quasi-holes is the minimum set of neutral excitation.

The excitation energy

$$\Delta E = \left(\frac{2\pi}{L}\right)^2 \sum_{i=1}^{N} \left[\kappa_i^2 + \lambda(N+1-2i)\kappa_i\right]$$

(2.118)

is decomposed into two parts. First, the component $i = 1$ is the energy of the quasi-particle:

$$\left(\frac{2\pi}{L}\right)^2 \left[\kappa_1^2 + \lambda(N-1)\kappa_1\right] = p_1^2 - p_{1,0}^2,$$

(2.119)

with

$$p_1 = \frac{2\pi}{L}\left(\kappa_1 + \frac{\lambda(N-1)}{2}\right), \quad p_{1,0} = \frac{2\pi}{L} \cdot \frac{\lambda(N-1)}{2}.$$

(2.120)

The other terms with $i \in [2, N]$ are rearranged as quasi-hole excitations:

$$\left(\frac{2\pi}{L}\right)^2 \sum_{i=2}^{N} \left[\kappa_i^2 + \lambda(N + 1 - 2i)\kappa_i\right] = \left(\frac{2\pi}{L}\right)^2 \sum_{i=1}^{N-1} \left[\mu_i^2 + \lambda(N - 1 - 2i)\mu_i\right]$$

$$= -\lambda \sum_{j=1}^{\lambda} \left[p'^2_j - p'^2_{j,0}\right], \tag{2.121}$$

with $\mu_i = k_{i+1}$ for $i \in [1, N-1]$ and

$$p'_j = \frac{2\pi}{L}\left(\mu'_j - \frac{N-1}{2}\right) + p'_{j,0}, \quad p'_{j,0} = \frac{\pi}{L} \cdot \frac{\lambda + 1 - 2j}{\lambda}. \tag{2.122}$$

Definition of μ'_j is shown in Fig. 2.7.

In summary, the excitation energy (2.118) is represented in terms of elementary excitations by

$$\Delta E = p_1^2 - p_{1,0}^2 - \lambda \sum_{j=1}^{\lambda} \left[p'^2_j - p'^2_{j,0}\right], \tag{2.123}$$

$$p_1 \geq \frac{\pi\lambda(N+1)}{L}, \quad |p'_j| \leq \frac{\pi}{L}\left(N - \frac{1}{\lambda}\right).$$

The corresponding momentum is given by

$$Q = p_1 + \sum_{j=1}^{\lambda} p'_j. \tag{2.124}$$

The excitation spectrum with one left-moving quasi-particle and λ quasi-holes can be obtained in a similar way.

Figure 2.9 shows an example with $\lambda = 2$ and $N = 7$. There is a similarity to the particle–hole excitation spectrum of free electrons in one dimension. In the present case, however, the lower threshold consists of two arcs, instead of one for free electrons. We note that multiple excitations of quasi-particles and λ times as many quasi-holes are necessary in order to reproduce the spectrum in Fig. 2.3.

2.4 Thermodynamics

In Section 2.2, we have interpreted the energy spectrum of the Sutherland model as either interacting bosons, interacting fermions, or free particles obeying exclusion statistics. Correspondingly, the thermodynamic potential can be derived in different ways. We shall give each description in this section, together with another description in terms of elementary excitations. Furthermore, thermodynamics of free anyons are discussed without taking the Sutherland model.

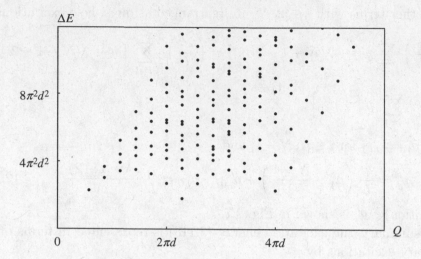

Fig. 2.9. Excitation spectrum with one quasi-particle and $\lambda(=2)$ quasi-holes in the case of $N = 7$ and the density $n = N/L$.

2.4.1 Interacting boson picture

For the Hamiltonian (2.1), the eigenenergy (2.43) is given relative to the ground-state energy $E_{0,N}$ by

$$\Delta E[\nu] = \left(\frac{2\pi}{L}\right)^2 \left[\sum_\kappa \kappa^2 \nu(\kappa) + \frac{\lambda}{2} \sum_\kappa \sum_{\kappa'} |\kappa - \kappa'| \nu(\kappa)\nu(\kappa')\right]. \quad (2.125)$$

We derive the thermodynamic potential

$$\Omega = -T \ln Z, \quad Z = \mathrm{Tr}\, e^{-\beta(E-\mu N)}, \quad (2.126)$$

following the method of [177]. Here μ is the chemical potential and $\beta = 1/T$.

First we take $\{\kappa_i\}_i^N$ as an index of one-particle states. We divide the set of one-particle states into many subsets which consist of a macroscopic number of κ_i (see Fig. 2.10). We denote the subsets by α and a representative value of one-particle state in α by κ_α. We take G_α as the number of κ_i in α so that the condition

$$1 \ll G_\alpha \ll N \quad (2.127)$$

is satisfied. For a given eigenstate specified by $\{\nu(\kappa)\}_{\kappa=\infty}^\infty$, we introduce the average of the distribution function

$$\bar{\nu}_\alpha = \frac{N_\alpha}{G_\alpha}, \quad N_\alpha = \sum_{\kappa \in \alpha} \nu(\kappa). \quad (2.128)$$

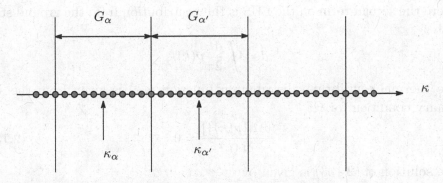

Fig. 2.10. Division of a set of one-particle states into many subsets α, α',... consisting of a macroscopic number $G_\alpha, G_{\alpha'},\ldots$ of one-particle states.

In the thermodynamic limit, only the extensive part $\bar{E}$ ($\propto L$) is important in (2.125). The energy $\bar{E}$ depends on $\{\nu(\kappa)\}$ only through N_α and is expressed relative to $E_{0,N}$ as

$$\Delta\bar{E} = \left(\frac{2\pi}{L}\right)^2 \left[\sum_\alpha \kappa_\alpha^2 N_\alpha + \frac{\lambda}{2}\sum_\alpha\sum_{\alpha'} |\kappa_\alpha - \kappa_{\alpha'}| N_\alpha N_{\alpha'}\right]. \qquad (2.129)$$

The number of ways for $N_\alpha (= \bar{\nu}_\alpha G_\alpha)$ bosons to be distributed among G_α one-particle states is given by

$$W_\alpha = \frac{(G_\alpha + N_\alpha - 1)!}{(G_\alpha - 1)! N_\alpha!}. \qquad (2.130)$$

Using the Stirling formula, the extensive part S of the entropy is given by

$$S = \sum_\alpha \left[(\bar{\nu}_\alpha + 1)\ln(\bar{\nu}_\alpha + 1) - \bar{\nu}_\alpha \ln\bar{\nu}_\alpha\right] G_\alpha. \qquad (2.131)$$

From (2.129) and (2.131), the thermodynamic potential Ω can be obtained.

In the thermodynamic limit, $\bar{\nu}_\alpha$ can be regarded as a function of the continuous variable $2\pi\kappa_\alpha/L$. When we rewrite $2\pi\kappa_\alpha/L \to k$, $2\pi G_\alpha/L \to dk$, and $\bar{\nu}_\alpha \to \nu(k)$, the expression for Ω/L becomes

$$\Omega[\nu(k)]/L - \pi^2\lambda^2 d^3/3$$
$$= \frac{1}{2\pi}\int_{-\infty}^{\infty} dk(k^2 - \mu)\nu(k) + \frac{\lambda}{4\pi}\int_{-\infty}^{\infty} dk \int_{-\infty}^{\infty} dk' |k - k'|\nu(k)\nu(k')$$
$$- \frac{T}{2\pi}\int_{-\infty}^{\infty} dk\left[(\nu(k) + 1)\ln(\nu(k) + 1) - \nu(k)\ln\nu(k)\right], \qquad (2.132)$$

where the second term on the LHS is the contribution from the ground state with

$$d = \int \frac{\mathrm{d}k}{2\pi} \nu(k).$$

The thermal equilibrium distribution function $\nu(k)$ is given from the stationary condition for Ω:

$$\frac{\delta\Omega[\{\nu(k)\}]}{\delta\nu(k)} = 0. \tag{2.133}$$

The solution of (2.133) is given by

$$\nu(k) = \{\exp\left[(\tilde{\epsilon}(k) - \mu)/T\right] - 1\}^{-1}, \tag{2.134}$$

$$\tilde{\epsilon}(k) = k^2 + \lambda \int_{-\infty}^{\infty} \mathrm{d}k' |k - k'| \nu(k'). \tag{2.135}$$

With use of these two equations and (2.132), the thermodynamic potential and other thermodynamic quantities are obtained. Although (2.134) is of the same form as the free boson distribution function, $\tilde{\epsilon}(k)$ is obtained by solving the self-consistent integral equation (2.135). Namely, $\tilde{\epsilon}(k)$ includes interaction effects, and is analogous to the dressed energy or renormalized energy in the thermodynamic Bethe ansatz theory [118].

Calculation of thermodynamic quantities is simplified with the use of rapidity (2.66), as shown in the next subsection.

2.4.2 Free anyon picture

Multiplying (2.66) by $2\pi/L$ and setting $p_i = 2\pi\tilde{\kappa}_i/L$ and $k_i = 2\pi\kappa_i/L$, we obtain

$$p_i = k_i + \frac{\pi\lambda}{L} \sum_{j(\neq i)} \mathrm{sgn}(p_i - p_j). \tag{2.136}$$

Henceforth, we regard $\mathrm{sgn}(0)$ as 0. Introducing the rapidity distribution function

$$\rho(p) = \frac{2\pi}{L} \sum_i \delta(p - p_i), \tag{2.137}$$

we write (2.136) as

$$p_i = k_i + \frac{\lambda}{2} \int \mathrm{d}p' \mathrm{sgn}(p_i - p')\rho(p'), \tag{2.138}$$

which defines a continuous function $p(k)$. Taking the derivative of both sides of (2.138) with respect to p, we obtain

$$1 = \frac{\mathrm{d}k}{\mathrm{d}p} + \lambda\rho(p) = \frac{\rho(p)}{\nu(k)} + \lambda\rho(p), \tag{2.139}$$

where we have used the property

$$2\pi \mathrm{d}N/L = \rho(p)\mathrm{d}p = \nu(k)\mathrm{d}k. \qquad (2.140)$$

Using (2.139), we obtain the relation between the distribution functions:

$$\nu(k) = \frac{\rho(p)}{1 - \lambda\rho(p)}, \quad \rho(p) = \frac{\nu(k)}{1 + \lambda\nu(k)}. \qquad (2.141)$$

The energy, including the ground-state contribution, is then written in a compact form as

$$E = \sum_i p_i^2 = L \int \frac{\mathrm{d}p}{2\pi} p^2 \rho(p). \qquad (2.142)$$

The entropy S given by (2.131) is rewritten as

$$S = L \int_{-\infty}^{\infty} \frac{\mathrm{d}p}{2\pi} \left[(\rho + \rho^*)\ln(\rho + \rho^*) - \rho\ln\rho - \rho^*\ln\rho^*\right], \qquad (2.143)$$

where ρ^* is the hole distribution given by

$$\rho^*(p) = 1 - \lambda\rho(p) = \frac{1}{1 + \lambda\nu(k)}. \qquad (2.144)$$

The expression (2.143) is equivalent to $\ln W$ in (2.75) and (2.76) in the thermodynamic limit with $\rho^* = D/G$, $\rho = N/G$, and $g = \lambda$. Thus (2.142) and (2.143) describe thermodynamic quantities of the Sutherland model in terms of free anyons.

2.4.3 Exclusion statistics and duality

In this subsection, we discuss the thermodynamics of free particles obeying exclusion statistics in general. The dispersion relation is given by $\epsilon(p)$ and the statistical parameter g. The thermodynamic potential is given by

$$\Omega/L = (E - TS - \mu N)/L$$
$$= \int \frac{\mathrm{d}p}{2\pi}(\epsilon - \mu)\rho - T \int_{-\infty}^{\infty} \frac{\mathrm{d}p}{2\pi} \left[(\rho + \rho^*)\ln(\rho + \rho^*) - \rho\ln\rho - \rho^*\ln\rho^*\right], \qquad (2.145)$$

with $\rho^* = 1 - g\rho$. We call p the rapidity also in the present case. In the particular case of the Sutherland model, we have $\epsilon = p^2$ and $g = \lambda$. The distribution function $\rho(p)$ is obtained from the stationary condition

$$\delta\Omega/\delta\rho(p) = 0, \qquad (2.146)$$

which gives

$$\ln\left(1+w\right)+g\ln\left(1+w^{-1}\right)=\frac{\epsilon-\mu}{T}. \tag{2.147}$$

In terms of the quantity

$$w=\rho^*/\rho, \tag{2.148}$$

(2.147) is written alternatively as

$$\exp\left[\left(\epsilon-\mu\right)/T\right]=w^g\left(1+w\right)^{1-g}. \tag{2.149}$$

We further obtain

$$\rho=1/(w+g),\quad\rho^*=w/(w+g). \tag{2.150}$$

Then the thermodynamic potential takes a simple form:

$$\Omega/L=-T\int\frac{\mathrm{d}p}{2\pi}\ln\left[1+w^{-1}\right]. \tag{2.151}$$

In the present case, (2.149) and (2.151) determine thermodynamic quantities. For some special values of g, we can explicitly obtain the analytic form of distribution functions. Let us consider these cases in the following.

Free bosons:
When $g=0$, the solution of (2.149) is given by

$$w=\exp\left[\left(\epsilon-\mu\right)/T\right]-1, \tag{2.152}$$

which gives

$$\rho=\frac{1}{\exp\left[\left(\epsilon-\mu\right)/T\right]-1},\quad\rho^*=1 \tag{2.153}$$

and

$$\Omega=T\int\frac{\mathrm{d}p}{2\pi}\ln\left[1-\exp[-(\epsilon-\mu)/T]\right]. \tag{2.154}$$

Thus the thermodynamics of free bosons is reproduced.

Free fermions:
When $g=1$, the solution of (2.149) is given by

$$w=\exp\left[\left(\epsilon-\mu\right)/T\right], \tag{2.155}$$

$$\rho=\frac{1}{\exp\left[\left(\epsilon-\mu\right)/T\right]+1},\quad\rho^*=\frac{1}{\exp\left[-\left(\epsilon-\mu\right)/T\right]+1}, \tag{2.156}$$

and

$$\Omega=-T\int\frac{\mathrm{d}p}{2\pi}\ln\left[1+\exp[-(\epsilon-\mu)/T]\right]. \tag{2.157}$$

Thus the thermodynamics of free fermions is also reproduced.

Free semions and ultrafermions:

When $g = 1/2$, we obtain

$$w = \frac{1}{2} \left[-1 + \sqrt{1 + 4e^{2(\epsilon-\mu)/T}} \right], \tag{2.158}$$

$$\rho = \frac{2}{\sqrt{1 + 4e^{2(\epsilon-\mu)/T}}}, \quad \rho^* = 1 - \frac{1}{\sqrt{1 + 4e^{2(\epsilon-\mu)/T}}}. \tag{2.159}$$

These particles are called semions. For $g = 2$, on the other hand, we obtain

$$w = \frac{1}{2} \left[e^{(\epsilon-\mu)/T} + \sqrt{e^{2(\epsilon-\mu)/T} + 4e^{(\epsilon-\mu)/T}} \right], \tag{2.160}$$

$$\rho = \frac{1}{2} \left(1 - \frac{1}{\sqrt{4e^{-(\epsilon-\mu)/T} + 1}} \right), \quad \rho^* = \frac{1}{\sqrt{4e^{-(\epsilon-\mu)/T} + 1}}. \tag{2.161}$$

Since the exclusion is stronger than free fermions, these particles may be called ultrafermions. The exclusion statistics is often called the fractional exclusion statistics for a case where g is a fractional number between 0 and 1.

Figure 2.11 shows the distribution functions ρ for $g = 1/2$ and $g = 2$ as functions of one-particle energy ϵ at several temperatures. At zero temperature, the solution of (2.149) is given by

$$w = \begin{cases} 0, & \epsilon < \mu, \\ \infty, & \epsilon > \mu, \end{cases} \tag{2.162}$$

which gives the particle distribution function

$$\rho = \begin{cases} 1/g, & \epsilon < \mu, \\ 0, & \epsilon > \mu. \end{cases} \tag{2.163}$$

We see from Fig. 2.11 that the profiles of ρ are similar to those of fermions. This applies to arbitrary positive real g. Owing to this property, it is possible to perform the Sommerfeld-type expansion at low temperatures [121]. Then the specific heat tends to T in the low-temperature limit. Another example will be given in Section 4.10.3.

The relation between (2.161) and (2.159) is an example of a duality between ρ^* for g, and ρ for $1/g$. The duality relation is written in the form

$$\rho^* \left(\frac{\epsilon - \mu}{T}; g \right) = \frac{1}{g} \rho \left(\frac{\mu - \epsilon}{gT}; \frac{1}{g} \right), \tag{2.164}$$

which can be derived from (2.148), (2.149), and (2.150).

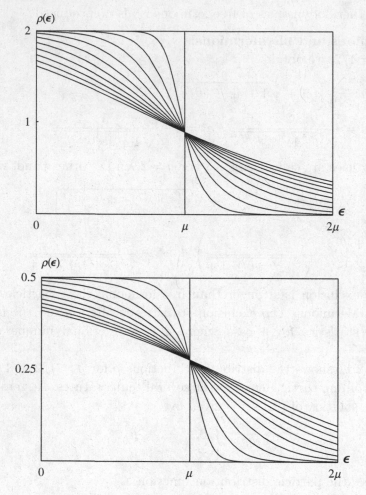

Fig. 2.11. Distribution function for $g = 1/2$ (upper panel) and $g = 2$ (lower panel) as a function of energy for different temperatures from $T = 0$ to $T/\mu = 0.5$ with the interval $\Delta T/\mu = 0.05$.

2.4.4 Elementary excitation picture

Now let us return to the Sutherland model and rewrite the thermodynamics in terms of *elementary excitations*. For this purpose, we first consider the ground state. At $T = 0$, the distributions of particles and holes are given by

$$\begin{cases} \rho(p) = 1/\lambda, & \rho^*(p) = 0, & \text{for} \quad |p| < p_F \equiv \mu^{1/2}, \\ \rho(p) = 0, & \rho^*(p) = 1, & \text{for} \quad |p| > p_F. \end{cases} \qquad (2.165)$$

From this result, we obtain $\mu = p_F^2 = (\pi d\lambda)^2$. The ground state is analogous to the Fermi sea. There are two branches of excitations: quasi-particle for $|p| > p_F$ and quasi-hole for $|p| < p_F$.

We introduce $w_{\mathrm{p}} = w$ for $|p| > p_{\mathrm{F}}$ and $w_{\mathrm{h}} = w^{-1}$ for $|p| < p_{\mathrm{F}}$. In terms of w_{p} and w_{h}, (2.147) is rewritten as

$$\epsilon_{\mathrm{p}}(p)/T \equiv \left(p^2 - \mu\right)/T = \log\left(1 + w_{\mathrm{p}}\right) - g_{\mathrm{p}}\log\left(1 + w_{\mathrm{p}}^{-1}\right), \qquad (2.166)$$

$$\epsilon_{\mathrm{h}}(p)/T \equiv \left(\mu - p^2\right)/(\lambda T) = \log\left(1 + w_{\mathrm{h}}\right) - g_{\mathrm{h}}\log\left(1 + w_{\mathrm{h}}^{-1}\right), \qquad (2.167)$$

with $g_{\mathrm{p}} = \lambda$ and $g_{\mathrm{h}} = 1/\lambda$. Using (2.166) and (2.167), the thermodynamic potential is rewritten as

$$\frac{\Omega}{L} = -T\int_{|p|>p_{\mathrm{F}}}\frac{\mathrm{d}p}{2\pi}\ln\left(1 + w_{\mathrm{p}}^{-1}\right) - \frac{T}{\lambda}\int_{|p|<p_{\mathrm{F}}}\frac{\mathrm{d}p}{2\pi}\ln\left(1 + w_{\mathrm{h}}^{-1}\right). \qquad (2.168)$$

In the above equation, the first and second terms represent the contributions of quasi-particles and quasi-holes, respectively.

The characters of elementary excitations are indexed by the charge, statistics, and dispersion relation. The charge can be determined by the coefficient of the chemical potential in the one-particle energy: ϵ_{p} and ϵ_{h}. The charge e_{p} of quasi-particles is unity since the coefficient of μ in ϵ_{p} is -1. Quasi-holes, on the other hand, have the fractional charge $e_{\mathrm{h}} = -1/\lambda$ since the coefficient of μ in ϵ_{h} is $1/\lambda$. If we set $\lambda = 1$, which is the free fermion case, all these relations reduce to the trivial one. In Section 2.6, we shall discuss the proposition that anyonic elementary excitations provide a natural interpretation of dynamical correlation functions.

2.5 Introduction to Jack polynomials

In Section 2.1.3, we have seen that the eigenfunction Φ_κ specified by $\kappa \in \mathcal{L}_N^+$ is given by multiplication of a finite-dimensional matrix. In this subsection, we discuss properties of eigenfunctions in more detail. In the following, we focus on eigenfunctions specified by $\kappa \in \Lambda_N^+$. Otherwise, appropriate Galilean boost realizes a partition μ with $\mu = (\kappa_1 + n, \ldots, \kappa_N + n)$.

In Section 2.1.3, we have used ϕ_κ^{B} defined by (2.27) as the basis function. From now on, we use another basis m_κ, called the monomial symmetric functions. The functions m_κ are the sum of all distinct permutations of the monomial $z_1^{\kappa_1} z_2^{\kappa_2} \cdots$. For example, $m_{21} = \sum_{i \neq j} z_i^2 z_j$. Let us assume a partition κ which has part 1 with multiplicity l_1, part 2 with multiplicity l_2, and so on. Then m_κ is related to ϕ_κ^{B} by

$$m_\kappa = \frac{\phi_\kappa^{\mathrm{B}}}{l_1! l_2! \cdots}. \qquad (2.169)$$

Correspondingly, we consider the eigenfunction J_κ normalized so that

$$J_\kappa = \frac{\Phi_\kappa^{\mathrm{B}}}{l_1! l_2! \cdots}, \qquad (2.170)$$

where Φ_κ^{B} has been defined in (2.41).

Following the procedure in Section 2.1.3, some examples of J_κ in terms of m_κ are listed below:

$$J_{1\cdots 1} = m_{1\cdots 1},$$

$$J_2 = m_2 + \frac{2\lambda}{\lambda+1} m_{11},$$

$$J_3 = m_3 + \frac{3\lambda}{\lambda+2} m_{21} + \frac{6\lambda^2}{(\lambda+1)(\lambda+2)} m_{111},$$

$$J_{21} = m_{21} + \frac{6\lambda}{2\lambda+1} m_{111},$$

$$J_4 = m_4 + \frac{4\lambda}{\lambda+3} m_{31} + \frac{6\lambda(\lambda+1)}{(\lambda+2)(\lambda+3)} m_{22} + \frac{12\lambda^2}{(\lambda+2)(\lambda+3)} m_{211}$$
$$+ \frac{24\lambda^3}{(\lambda+1)(\lambda+2)(\lambda+3)} m_{1111},$$

$$J_{31} = m_{31} + \frac{2\lambda}{\lambda+1} m_{22} + \frac{\lambda(5\lambda+3)}{(\lambda+1)^2} m_{211} + \frac{12\lambda^2}{(\lambda+1)^2} m_{1111},$$

$$J_{22} = m_{22} + \frac{2\lambda}{\lambda+1} m_{211} + \frac{12\lambda^2}{(\lambda+1)(2\lambda+1)} m_{1111},$$

$$J_{211} = m_{211} + \frac{12\lambda}{3\lambda+1} m_{1111}. \qquad (2.171)$$

These polynomials are called Jack polynomials [126, 169], which in general have the form

$$J_\kappa = m_\kappa + \sum_{\mu < \kappa} v_{\mu,\kappa} m_\mu. \qquad (2.172)$$

The Jack polynomials are related to other symmetric polynomials (called also functions) at a specific value of λ. Obviously, J_κ reduces to m_κ when $\lambda = 0$. When $\lambda \to 1$, we obtain from (2.171)

$$J_2 \to m_2 + m_{11},$$
$$J_3 \to m_3 + m_{21} + m_{111},$$
$$J_{21} \to m_{21} + 2m_{111}.$$

When $\lambda = 1$, the Jack polynomials reduce to Schur functions [126], which are defined by

$$s_\kappa = \frac{\det(z_i^{\kappa_j+N-j})}{\det(z_i^{N-j})}$$

$$= \left(\prod_{i<j}(z_i - z_j)\right)^{-1} \sum_P (-1)^P z_{p(1)}^{\kappa_1+N-1} z_{p(2)}^{\kappa_2+N-2} \cdots z_{p(N)}^{\kappa_N}.$$

Here the summation with respect to P runs over the symmetric group S_N of order N.

When $\lambda \to \infty$, J_κ reduces to elementary symmetric functions $e_{\kappa'}$, with the conjugate partition κ' of κ. Here e_κ is defined by [126]

$$e_\kappa = e_{\kappa_1} e_{\kappa_2} \cdots, \quad \text{with} \quad e_k = \sum_{1 \le i_1 < i_2 < \cdots < i_k \le N} z_{i_1} z_{i_2} \cdots z_{i_k}. \tag{2.173}$$

For example, in the limit $\lambda \to \infty$, J_2, J_3 and J_{21} become

$$J_3 \to m_3 + 3m_{21} + 6m_{111} = e_{111},$$
$$J_{21} \to m_{21} + 3m_{111} = e_{21},$$
$$J_{111} = m_{111} = e_3.$$

As a basis set we can use symmetric functions other than m_κ. For example, the power-sum symmetric functions

$$p_\kappa = p_{\kappa_1} p_{\kappa_2} \cdots, \quad \text{with} \quad \kappa \in \Lambda_N^+, \quad p_k = \sum_{i=1}^N z_i^k \tag{2.174}$$

are often used as a basis in the theory of symmetric polynomials; every symmetric function can be written as a linear combination of the power-sum symmetric functions. Furthermore, we can introduce a scalar product on the set of symmetric functions. We seek the scalar product under which both p_κ and J_κ form orthogonal bases.

For a partition κ having part 1 with multiplicity l_1, part 2 with multiplicity l_2 and so on, we define

$$\zeta_\kappa = 1^{l_1} l_1! 2^{l_2} l_2! \cdots. \tag{2.175}$$

Let us define a scalar product $\langle \cdot, \cdot \rangle_c$ by

$$\langle p_\kappa, p_\mu \rangle_c = \delta_{\kappa,\mu} \zeta_\kappa \lambda^{-l(\kappa)}, \tag{2.176}$$

where the suffix c means 'combinatorial', as will be explained in more detail in Section 7.3.6.

The Jack symmetric functions with different partitions turn out to be orthogonal with respect to $\langle \cdot \, , \cdot \rangle_c$. Therefore, the Jack polynomials can also be defined uniquely by the following two conditions [126, 169]:

$$J_\kappa = m_\kappa + \sum_{\mu < \kappa} v_{\mu, \kappa} m_\mu, \tag{2.177}$$

$$\langle J_\kappa, J_\mu \rangle_c = 0 \text{ if } \kappa \neq \mu. \tag{2.178}$$

Actually, we can reproduce (2.171) for J_κ from the definitions (2.177) and (2.178).

Let us give a simple example. For $\kappa = 2, 0, 0, \ldots$, we set

$$J_2 = m_2 + v m_{11}, \tag{2.179}$$

where the coefficient v is to be determined from the orthogonality condition

$$\langle J_2, J_{11} \rangle_c = \langle J_2, m_{11} \rangle_c = 0. \tag{2.180}$$

We note that $m_2 = p_2$ and $m_{11} = (p_{11} - p_2)/2$ and

$$\langle p_{11}, p_{11} \rangle_c = 1^2 2! \lambda^{-2}, \quad \langle p_2, p_2 \rangle_c = 2^1 1! \lambda^{-1}, \quad \langle p_2, p_{11} \rangle_c = 0. \tag{2.181}$$

With these relations and (2.180), the coefficient v in (2.179) is found to be

$$v = \frac{2 \langle p_2, p_2 \rangle_c}{\langle p_2, p_2 \rangle_c + \langle p_{11}, p_{11} \rangle_c} = \frac{2\lambda}{1 + \lambda}, \tag{2.182}$$

which yields the expression for J_2 in (2.171).

The norms of J_{11} and J_2 with respect to (2.176) can be obtained as

$$\langle J_{11}, J_{11} \rangle_c = \frac{1}{4} \left(\langle p_{11}, p_{11} \rangle_c + \langle p_2, p_2 \rangle_c \right) = \frac{(1 + \lambda)}{2\lambda^2} \tag{2.183}$$

and

$$\begin{aligned}
\langle J_2, J_2 \rangle_c &= \langle m_2 + \frac{2\lambda}{1 + \lambda} m_{11}, m_2 + \frac{2\lambda}{1 + \lambda} m_{11} \rangle_c \\
&= \langle \frac{p_2 + \lambda p_{11}}{1 + \lambda}, \frac{p_2 + \lambda p_{11}}{1 + \lambda} \rangle_c \\
&= \frac{\langle p_2, p_2 \rangle_c + \lambda^2 \langle p_{11}, p_{11} \rangle_c}{(1 + \lambda)^2} = \frac{2}{\lambda(1 + \lambda)}.
\end{aligned} \tag{2.184}$$

Generally, the norm of the Jack polynomials with respect to the inner product (2.176) is given by

$$\langle J_\kappa, J_\kappa \rangle_c = \prod_{s \in D(\kappa)} \frac{h_\kappa^*(s)}{h_*^\kappa(s)}, \tag{2.185}$$

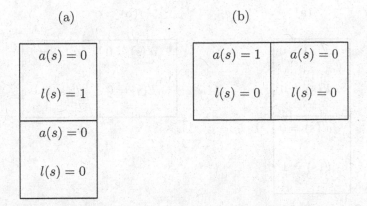

Fig. 2.12. Young diagrams (a) for $\kappa = (1,1)$, (b) for $\kappa = (2,0)$, where the arm length $a(s)$ and leg length $l(s)$ are inscribed.

which will be derived in Section 7.3.6. Here the upper $h_\kappa^*(s)$ and lower $h_*^\kappa(s)$ hook lengths are, respectively, defined as

$$h_\kappa^*(s) = \frac{a(s)+1}{\lambda} + l(s), \quad h_*^\kappa(s) = \frac{a(s)}{\lambda} + l(s) + 1 \qquad (2.186)$$

in terms of arm length $a(s)$ and leg length $l(s)$, which have been introduced in Section 2.3.1. With the help of Fig. 2.12, we obtain

$$\prod_{s \in D(\kappa=(1,1))} \frac{h_\kappa^*(s)}{h_*^\kappa(s)} = \frac{(1/\lambda + 1)(1/\lambda)}{2 \cdot 1} = \frac{1+\lambda}{2\lambda^2},$$

$$\prod_{s \in D(\kappa=(2,0))} \frac{h_\kappa^*(s)}{h_*^\kappa(s)} = \frac{(2/\lambda)(1/\lambda)}{(1/\lambda + 1) \cdot 1} = \frac{2}{\lambda(1+\lambda)},$$

which agree with the earlier results (2.183) and (2.184).

The polynomials J_κ are also orthogonal to each other with respect to another inner product. To define this, we quote the property

$$\int_0^L dx_1 \cdots \int_0^L dx_N J_\kappa^*(z_1, \ldots) J_\mu(z_1, \ldots) |\Psi_{0,N}(x_1, \ldots, x_N)|^2$$

$$= \int_0^L dx_1 \cdots \int_0^L dx_N J_\kappa^*(z_1, \ldots) J_\mu(z_1, \ldots) |\Delta(z_1, \ldots, z_N)|^{2\lambda} \propto \delta_{\kappa,\mu} \qquad (2.187)$$

with the Vandermonde determinant

$$\Delta(z) = \prod_{1 \le i < j \le N} (z_i - z_j). \qquad (2.188)$$

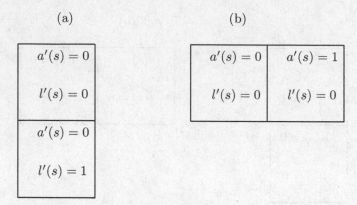

Fig. 2.13. Young diagrams (a) for $\kappa = (1,1)$, (b) for $\kappa = (2,0)$, where the arm colength $a'(s)$ and leg colength $l'(s)$ are inscribed.

Then we define the inner product $\langle f, g \rangle_0$ for functions $f(z)$ and $g(z)$ in complex variables $z = (z_1, \ldots, z_N)$ by

$$\langle f, g \rangle_0 = \prod_{i=1}^{N} \oint_{|z_i|=1} \frac{\mathrm{d}z_i}{2\pi \mathrm{i} z_i} |\Delta(z)|^{2\lambda} f^*(z) g(z), \qquad (2.189)$$

where $f^*(z)$ denotes the complex conjugation of $f(z)$. The LHS of (2.187) is then written as

$$L^N \langle J_\kappa, J_\mu \rangle_0. \qquad (2.190)$$

The orthogonality in this inner product is described as

$$\langle J_\kappa, J_\mu \rangle_0$$
$$= \delta_{\kappa,\mu} c_N(\lambda) \prod_{s \in D(\kappa)} \frac{h_\kappa^*(s) \left(\lambda N + a'(s) - \lambda l'(s)\right)}{h_*^\kappa(s) \left(\lambda N + a'(s) + 1 - \lambda(l'(s) + 1)\right)}, \qquad (2.191)$$

with

$$c_N(\lambda) = \frac{\Gamma(N\lambda + 1)}{\Gamma(\lambda + 1)^N}. \qquad (2.192)$$

The formula (2.191) will be proven in Section 7.3.5. Applying this formula to $\kappa = (1,1)$ and $\kappa = (2,0)$, we obtain with the help of Fig. 2.13

$$\langle J_{11}, J_{11} \rangle_0 = c_N(\lambda) \cdot \frac{1+\lambda}{2\lambda^2} \cdot \frac{N(N-1)\lambda^2}{[(N-1)\lambda + 1][(N-2)\lambda + 1]}, \qquad (2.193)$$

$$\langle J_2, J_2 \rangle_0 = c_N(\lambda) \cdot \frac{2}{\lambda(1+\lambda)} \cdot \frac{N\lambda((N-1)\lambda + 1))}{[(N-1)\lambda + 1][(N-1)\lambda + 2]}. \qquad (2.194)$$

Owing to (2.191), the Jack symmetric polynomials can alternatively be defined by the following two conditions:

$$J_\kappa = m_\kappa + \sum_{\mu < \kappa} v_{\mu,\kappa} m_\mu, \tag{2.195}$$

$$\langle J_\kappa, J_\mu \rangle_0 = 0 \text{ if } \kappa \neq \mu. \tag{2.196}$$

We emphasize that J_κ's form an orthogonal basis set with respect to two different inner products simultaneously. This observation leads to the analytical expression of dynamical correlation functions, as discussed next.

2.6 Dynamics in thermodynamic limit

In Section 2.4, we have seen that thermodynamic quantities are described in terms of elementary excitations such as quasi-particles and quasi-holes, which obey the exclusion statistics. We will see in this section that those anyonic elementary excitations also provide an intuitive interpretation of dynamical properties at zero temperature.

Let us remark on some historical aspects of the dynamical theory. In 1993, Simons *et al.* [164,167] found that the dynamical density–density correlation functions $\langle \hat{\rho}(x,t)\hat{\rho}(0,0) \rangle$ of the Sutherland model at couplings $\lambda = 1/2$ and $\lambda = 2$ are identical to some parametric correlators in random systems. The correlators can be obtained in terms of the so-called supermatrix method. Using a similar method, Haldane and Zirnbauer calculated the hole propagator $\langle \hat{\psi}^\dagger(x,t)\hat{\psi}(0,0) \rangle$ at $\lambda = 2$ [81]. From these results together with knowledge of the non-interacting case ($\lambda = 1$), Haldane conjectured the expression of the dynamical correlation functions at arbitrary rational couplings [83]. His argument is based on consideration of the selection rule for the intermediate states. Subsequently, this conjecture was proved by using the theory of symmetric Jack polynomials [61,62,72,125], where the hole propagator and density correlation functions were obtained. By another approach, which used two-dimensional Yang–Mills theory, the dynamical density correlation functions for rational value of λ were also obtained in [133]. The particle propagator $\langle \hat{\psi}(x,t)\hat{\psi}^\dagger(0,0) \rangle$ at $\lambda = 2$ was calculated in [200] by the supermatrix method. This result was extended to non-negative rational values of λ [160] with the use of symmetric Jack polynomials. The dynamical current–current correlation function was also calculated in [186].

In the present section, we provide exact expressions for propagators and dynamical density correlation functions in the thermodynamic limit. We interpret these results in terms of elementary excitations obeying the exclusion statistics. The derivation makes full use of the theory of Jack polynomials, and is somewhat lengthy. Therefore we relegate the details of derivation to Section 2.7.

2.6.1 Hole propagator $\langle \hat{\psi}^\dagger(x,t)\hat{\psi}(0,0)\rangle$

In this subsection, we consider the hole propagator

$$G^-(x,t) = \frac{\langle g, N|\hat{\psi}^\dagger(x,t)\hat{\psi}(0,0)|g, N\rangle}{\langle g, N|g, N\rangle}, \qquad (2.197)$$

for an integer coupling λ. Here $|g, N\rangle$ represents the ground state for an N-particle system, and $\hat{\psi}^\dagger(x,t)$ denotes the Heisenberg representation of the field operator

$$\hat{\psi}^\dagger(x,t) = e^{i(Ht-\hat{P}x)}\hat{\psi}^\dagger(0,0)e^{-i(Ht-\hat{P}x)}. \qquad (2.198)$$

When the original particles are bosons (fermions), an explicit expression for (2.197) is known for even (odd) integer λ. The expression in the thermodynamic limit is given by [61, 72, 125]

$$G^-(x,t) = \frac{c(\lambda)d}{2} \prod_{k=1}^{\lambda} \int_{-1}^{1} dv_k e^{i(Qx-Et)} \frac{\prod_{i<j}|v_i - v_j|^{2/\lambda}}{\prod_{j=1}^{\lambda}(1-v_j^2)^{1-1/\lambda}}, \qquad (2.199)$$

which will be derived in the next section with use of symmetric Jack polynomials. In (2.199), the constant $c(\lambda)$ is given by

$$c(\lambda) = \prod_{j=1}^{\lambda} \frac{\Gamma(1+1/\lambda)}{\Gamma(j/\lambda)^2}. \qquad (2.200)$$

The excitation energy $\omega = E + \mu$ and momentum are given by

$$\omega = \pi^2 \lambda d^2 \sum_{j=1}^{\lambda}\left(1 - v_j^2\right), \quad Q = \pi d \sum_{j=1}^{\lambda} v_j. \qquad (2.201)$$

By Fourier transform, the spectral function $A^-(Q,\omega)$ is defined as

$$G^-(x,t) = \int_{-\infty}^{\infty} \frac{d\omega}{2\pi} \int_{-\infty}^{\infty} \frac{dQ}{2\pi} A^-(Q,\omega)e^{iQx-i(\omega-\mu)t}. \qquad (2.202)$$

In the case of electron systems in solids, the spectral function can be measured experimentally as the intensity in angle-resolved photo-emission

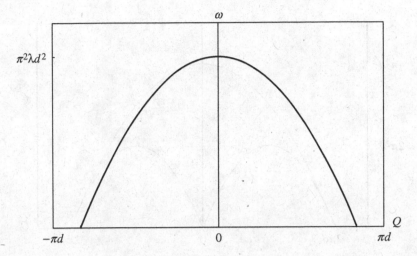

Fig. 2.14. Dispersion of a quasi-hole.

spectroscopy. We obtain from (2.199) the result

$$
A^-(Q,\omega) = 2\pi^2 c(\lambda) d \left(\prod_{i=1}^{\lambda} \int_{-1}^{1} dv_i \right) \frac{\prod_{i<j} |v_i - v_j|^{2/\lambda}}{\prod_{j=1}^{\lambda} (1 - v_j^2)^{1-1/\lambda}}
$$

$$
\times \delta \left(Q - \pi d \sum_{j=1}^{\lambda} v_j \right) \delta \left(\omega - \pi^2 \lambda d^2 \sum_{j=1}^{\lambda} (1 - v_j^2) \right). \qquad (2.203)
$$

In the trivial case of $\lambda = 1$, the system reduces to that of free fermions and the spectral function turns into

$$
A_{\lambda=1}^-(Q,\omega) = 2\pi \delta \left(\omega - (\pi^2 d^2 - Q^2)/2 \right) \theta(\pi d - |Q|).
$$

Here θ is the Heaviside step function:

$$
\theta(x) = \begin{cases} 1, & x \ge 0, \\ 0, & x < 0. \end{cases} \qquad (2.204)
$$

The delta-function peak implies that one-particle removal leaves one hole behind, which has an infinite lifetime. Figure 2.14 shows the dispersion relation of the hole excitation, which in fact applies to a quasi-hole with arbitrary λ according to (2.201).

In the simplest nontrivial case with $\lambda = 2$, the double integrals in (2.203) are performed analytically. The spectral function $A_{\lambda=2}^-(Q,\omega)$ is derived as

$$
A_{\lambda=2}^-(Q,\omega) = \frac{\theta(\omega - \omega_{\mathrm{L}}(Q))\theta(\omega_{\mathrm{U}}(Q) - \omega)}{|\omega + Q^2 - 2\pi d Q|^{1/2}|\omega + Q^2 + 2\pi d Q|^{1/2}}, \qquad (2.205)
$$

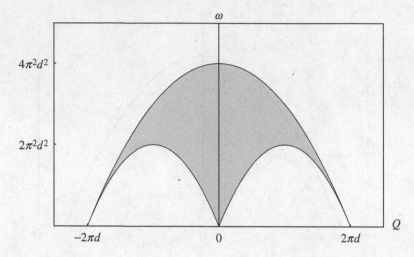

Fig. 2.15. Compact support of the spectral function $A_\lambda^-(Q,\omega)$ for $\lambda = 2$.

with $\omega_U(Q) = 2\pi^2 d^2 - Q^2/2$ and $\omega_L = -Q^2 + 2\pi d|Q|$. We call such a region "support" that has nonzero spectral function $A_\lambda^-(Q,\omega)$. The support of $A_{\lambda=2}^-(Q,\omega)$ is shown in Fig. 2.15. From (2.205) and Fig. 2.15, we observe that the spectral intensity has a continuum instead of the delta function. This feature and (2.201) suggest that one hole breaks into two fractionalized particles, which are nothing but quasi-holes as discussed in Section 2.3.3.

Now we discuss (2.203) for the general integer λ. One hole breaks into λ quasi-holes. The repeated structure of the lower edge of the support can be regarded as evidence for multiple excitations of quasi-holes, an example of which is shown in Fig. 2.16 with $\lambda = 5$. Inspection of (2.205) shows that $A_\lambda^-(Q,\omega)$ diverges at the lower edge as

$$A_\lambda^-(Q,\omega) \sim \begin{cases} \omega^{-\lambda+2/\lambda}, & \text{for } |Q|/(\pi d) = 1, 2, \ldots, \lambda - 2, \\ (\omega - \omega_L)^{\lambda+1/\lambda-3}, & \text{otherwise,} \end{cases}$$

and is discontinuous at the upper edge $\omega_U(Q)$. The singularity of $A_\lambda^-(Q,\omega)$ comes either from the joint density of states of elementary excitations or from the singularity of the integrand of (2.203). When the singularity of spectral functions cannot be attributed to the joint density of states, this singularity is evidence for statistical interaction between anyonic elementary excitations.

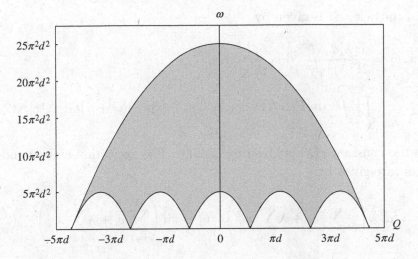

Fig. 2.16. Support of the spectral function $A_\lambda^-(Q,\omega)$ for $\lambda = 5$.

2.6.2 Particle propagator $\langle \hat{\psi}(x,t)\hat{\psi}^\dagger(0,0)\rangle$

We next consider the particle propagator

$$G^+(x,t) = \frac{\langle g, N|\hat{\psi}(x,t)\hat{\psi}^\dagger(0,0)|g, N\rangle}{\langle g, N|g, N\rangle}, \qquad (2.206)$$

which consists of two parts [160]

$$G^+(x,t) = G_1^+(x,t) + G_2^+(x,t).$$

The first part is given by

$$G_1^+(x,t) = \lambda d \int_1^\infty dw \left(\frac{w-1}{w+1}\right)^{\lambda-1} \cos(Qx)e^{-iEt}, \qquad (2.207)$$

where

$$E = \pi^2 d^2 \lambda^2 w^2, \quad Q = \pi d\lambda w. \qquad (2.208)$$

This part is interpreted as the propagation of a single particle over the Fermi sea. For $\lambda = 1$, the result for the free fermions is recovered.

The second part $G_2^+(x,t)$ represents the contribution from one quasi-particle and the minimal number of neutral excitations (λ quasi-holes and a quasi-particle). It is given by

$$G_2^+(x,t) = \frac{c(\lambda)d}{2\Gamma(\lambda)^2} \int_1^\infty dw_1 \int_{-\infty}^{-1} dw_0$$

$$\times \prod_{i=1}^\lambda \int_{-1}^1 dv_i F(w_1, w_0; v_1, \ldots, v_\lambda) \exp[-i(\omega + \mu)t + iQx], \quad (2.209)$$

where the constant $c(\lambda)$ is given by (2.200). The excitation energy and momentum are given by

$$\omega = \pi^2 d^2 \lambda \left(-\sum_{i=1}^\lambda v_i^2 + \lambda \sum_{i=0}^1 w_i^2 \right), \quad Q = \pi d \left(\sum_{i=1}^\lambda v_i + \lambda \sum_{i=0}^1 w_i \right). \quad (2.210)$$

The form factor $F(w_1, w_0; v_1, \ldots, v_\lambda)$ is equal to

$$F(w_1, w_0; v_1, \ldots, v_\lambda) = \frac{\prod_{1 \leq i < j \leq \lambda} |v_i - v_j|^{2/\lambda} |w_1 - w_0|^{2-2\lambda}}{\prod_{i=1}^\lambda (1 - v_i^2)^{1-1/\lambda} \prod_{j=0}^1 (w_j^2 - 1)^{1-\lambda}} \mathcal{K}^2, \quad (2.211)$$

where $\mathcal{K}$ is given by

$$\mathcal{K} = \prod_{k=0}^1 \prod_{l=1}^\lambda (w_k + v_l)^{\lambda-1} \prod_{l<k} (v_l - v_k)^{-1}$$

$$\times \frac{\partial^\lambda}{\partial v_1 \cdots \partial v_\lambda} \left\{ \prod_{l<k} (v_l - v_k) \prod_{k=0}^1 \prod_{l=1}^\lambda (w_k + v_l)^{1-\lambda} \right\}. \quad (2.212)$$

In the case of electrons in solids, the corresponding spectral function $A_\lambda^+(Q, \omega)$ can be measured as the intensity of angle-resolved *inverse* photoemission spectroscopy. The part $A_\lambda^{+,1}(Q, \omega)$ coming from G_1^+ consists of a delta function

$$A_\lambda^{+,1}(Q, \omega) = 2\pi^2 d\lambda \left(\frac{|Q| - \pi\lambda d}{|Q| + \pi\lambda d} \right)^{\lambda-1} \delta\left(\omega - Q^2/2\right) \theta\left(|Q| - \pi d\right). \quad (2.213)$$

Figure 2.17 shows by bold curves the location where $A_\lambda^{+,1}$ becomes nonzero. With $\lambda > 1$, the intensity on those curves vanishes at the two pseudo-Fermi points $|Q| = \pi\lambda d$, as seen from (2.213). Figure 2.17 shows the support of $A_\lambda^{+,2}(Q, \omega)$, which derives from G_2^+, as the shaded region. Contrary to the spectral function of the hole propagator, the support is not compact; the region with nonzero intensity extends infinitely.

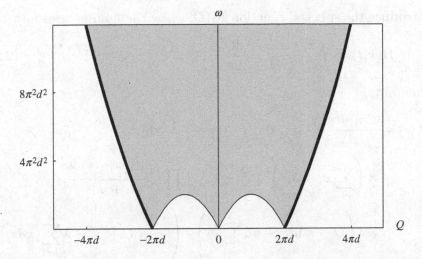

Fig. 2.17. Support of G^+ for $\lambda = 2$. Excitation content consists of four elementary excitations: two quasi-holes and a right-moving quasi-particle and a left-moving quasi-particle. The spectral function has the delta-function peak along the bold curves.

2.6.3 Density correlation function

In the present subsection, we discuss the exact expression for the density correlation function

$$\Pi(x,t) = \frac{\langle g, N | \hat{\rho}(x,t)\hat{\rho}(0,0) | g, N \rangle}{\langle g, N | g, N \rangle} \tag{2.214}$$

of the density fluctuation operator

$$\hat{\rho}(x,0) = \sum_{i=1}^{N} \delta(x - x_i) - d. \tag{2.215}$$

In the thermodynamic limit, $\Pi(x,t)$ is given by [72, 125]

$$\Pi(x,t) = \frac{c(\lambda)d^2}{2\lambda} \int_1^\infty dw \int_{-1}^1 dv_1 \cdots \int_{-1}^1 dv_\lambda \left(\sum_i v_i + \lambda w \right)^2$$

$$\times (w^2 - 1)^{\lambda - 1} \prod_{i=1}^{\lambda} \frac{(1 - v_i^2)^{-1 + 1/\lambda}}{(v_i + w)^2} \prod_{i<j} |v_i - v_j|^{2/\lambda} e^{-i\omega t} \cos(Qx), \tag{2.216}$$

where the constant $c(\lambda)$ is given by (2.200), and

$$\omega = (\pi\lambda dw)^2 - \lambda \sum_{j=1}^{\lambda} (\pi d v_j)^2, \quad Q = \pi\lambda dw + \pi d \sum_{j=1}^{\lambda} v_j. \tag{2.217}$$

We introduce the spectral function $S_\lambda(Q, \omega)$ by the Fourier transform

$$\Pi(x, t) = \int_{-\infty}^{\infty} \frac{d\omega}{2\pi} \int_{-\infty}^{\infty} \frac{dQ}{2\pi} S_\lambda(Q, \omega) e^{-i\omega t} \cos(Qx). \qquad (2.218)$$

We then obtain

$$S_\lambda(Q, \omega) = \frac{(2\pi)^2 d^2 c(\lambda)}{\lambda} \int_1^{\infty} dw \int_{-1}^{1} dv_1 \cdots \int_{-1}^{1} dv_\lambda$$

$$\times \left(\sum_i v_i + \lambda w \right)^2 (w^2 - 1)^{\lambda-1} \prod_{i=1}^{\lambda} \frac{(1 - v_i^2)^{-1+1/\lambda}}{(v_i + w)^2}$$

$$\times \delta \left(Q - \pi\lambda dw + \pi d \sum_{i=1}^{\lambda} v_i \right) \delta \left(\omega - (\pi\lambda dw)^2 - \lambda \sum_{j=1}^{\lambda} (\pi dv_j)^2 \right).$$

$$(2.219)$$

In the trivial case $\lambda=1$, the spectral function reduces to

$$S_{\lambda=1}(Q, \omega)$$

$$= 2\pi^2 d^2 \int_1^{\infty} dw \int_{-1}^{1} dv_1 \delta(Q - \pi dw + \pi dv_1) \delta \left(\omega - \pi^2 d^2 w^2 + \pi^2 d^2 \sum_{j=1}^{\lambda} v_j^2 \right)$$

$$= \frac{2}{|Q|} \theta(\omega_U(Q) - \omega) \theta(\omega - \omega_L(Q)), \qquad (2.220)$$

where the lower and upper edges are given by

$$\omega_L(Q) = |Q(2\pi d - Q)|/2, \quad \omega_U(Q) = |Q(Q + 2\pi d)|/2.$$

Figure 2.18 shows the support for $\lambda = 1$, which reproduces the particle–hole continuum of one-dimensional free fermions.

Now we consider the case for general positive integer λ. The excitation content relevant to $S_\lambda(Q, \omega)$ consists of a quasi-particle plus λ quasi-holes. Figure 2.19 shows the support for the case $\lambda = 3$. By comparing Figs 2.18 and 2.19, one can imagine how the support looks for general integer λ.

The upper edge

$$\omega = \omega_U(Q) = Q(Q + 2\pi\lambda d)/2$$

comes from excited states where λ quasi-holes are located at a pseudo-Fermi point $v_i = 1$, and the momentum w of a quasi-particle takes a value in $[1, \infty)$. Hence $\omega_U(Q)$ represents the momentum-shifted dispersion relation of quasi-particle excitation. On the other hand, one of the lower edges $\omega = \omega_L(Q)$

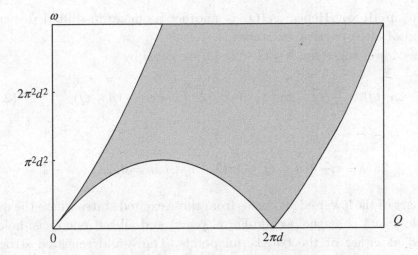

Fig. 2.18. Support of the spectral function $S_{\lambda=1}(Q,\omega)$ of the density correlation function for $\lambda = 1$, that for one-dimensional free fermions.

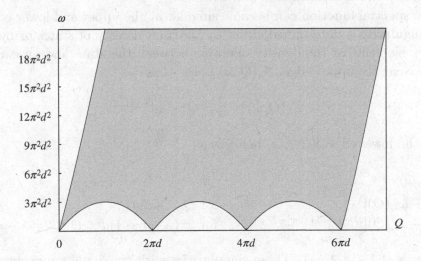

Fig. 2.19. Support of the spectral function $S_{\lambda=3}(Q,\omega)$ of the density correlation function for $\lambda = 3$. Relevant excited states are one quasi-particle/three quasi-hole states.

for $Q > 2\pi\lambda d$ is given by

$$\omega = \omega_{\mathrm{L}}(Q) = Q(Q - 2\pi\lambda d)/2.$$

This energy corresponds to excited states where λ quasi-holes are located at a pseudo-Fermi point $v_i = -1$ and the momentum of a quasi-particle takes a value in $[1, \infty)$. Hence $\omega_{\mathrm{L}}(Q)$ is another momentum-shifted dispersion relation of quasi-particle excitation.

Other lower edges for $0 \leq Q \leq 2\pi\lambda d$ are given by

$$\omega_{\mathrm{L}}(Q) = \frac{\lambda}{2}\left(Q - 2\pi\left(\lambda - r\right)d\right)\left(2\pi\left(\lambda - r - 1\right)d - Q\right), \tag{2.221}$$

for

$$2\pi(\lambda - r - 1)d \leq Q \leq 2\pi(\lambda - r)d, \quad r = 0, 1, \ldots, \lambda - 1.$$

This part of the lower edges comes from those excited states where the quasi-particle $w = 1$ is at the pseudo-Fermi point, and all but one quasi-hole are located at either of the two Fermi points. The λ-fold repeated structure with $0 \leq r \leq \lambda - 1$ is formed by the dispersion relation of a quasi-hole. The repeated structure along the lower edge is evidence for fractionalized particles working as elementary excitations.

The spectral function can become singular at the upper and lower edges; the singularity is determined either by the joint density of states or by the matrix elements of the density operator between the state and an excited state. Near the upper edge, $S_\lambda(Q, \omega)$ behaves as

$$S_\lambda(Q, \omega) \sim |\omega_{\mathrm{U}}(Q) - \omega|^{\lambda - 1}.$$

Near the lower edge $S_\lambda(Q, \omega)$ behaves as

$$S_\lambda(Q, \omega) \sim$$
$$\begin{cases} |\omega - \omega_{\mathrm{L}}(Q)|^{\lambda - 1}, & \text{for } Q > 2\pi\lambda d, \\ |\omega - \omega_{\mathrm{L}}(Q)|^{2\lambda - 4r - 3 + (2r^2 + 2r + 1)/\lambda}, & \text{for } 2\pi(\lambda - r - 1)d \leq Q \leq 2\pi(\lambda - r)d, \end{cases}$$

with $r = 0, 1, \ldots, \lambda - 1$. Those singularities with fractional exponents can also be regarded as evidence of statistical interaction between elementary excitations.

2.7 Derivation of dynamics for finite-sized systems

Throughout this section we take the periodic boundary condition. For simplicity, we focus on bosons with even integer λ as original particles in the Hamiltonian. The final result is also applicable for fermions with odd integer λ.

2.7.1 Hole propagator

In the present subsection, we derive (2.199) in Section 2.7.1. The state $\hat{\psi}(0,0)|g, N\rangle$ is written as

$$\sum_{\kappa \in \mathcal{L}_{N-1}^+} \frac{|\kappa, N-1\rangle \langle \kappa, N-1|\hat{\psi}(0,0)|g, N\rangle}{\langle \kappa, N-1|\kappa, N-1\rangle}, \tag{2.222}$$

where the wave function of the state $|\kappa, N-1\rangle$ is given by

$$J_\kappa(z_1, \ldots, z_{N-1})\Psi_{0,N-1}(z_1, \ldots, z_{N-1}).$$

Then (2.197) has the decomposition

$$G^-(x,t) =$$
$$\sum_{\kappa \in \mathcal{L}_{N-1}^+} \frac{|\langle \kappa, N-1|\hat{\psi}(0,0)|g, N\rangle|^2}{\langle \kappa, N-1|\kappa, N-1\rangle \langle g, N|g, N\rangle} e^{-i(E[\kappa]-E_{0,N})t+iQ_\kappa x}. \tag{2.223}$$

Here the ground state energy $E_{0,N}$ for an N-particle system has been given in (2.12), and

$$E[\kappa] = \left(\frac{2\pi}{L}\right)^2 \sum_{i=1}^{N-1} \left(\kappa_i + \frac{\lambda(N-2i)}{2}\right)^2, \quad Q_\kappa = \frac{2\pi}{L} \sum_{i=1}^{N-1} \kappa_i. \tag{2.224}$$

The wave function of the state $\hat{\psi}(0,0)|g, N\rangle$ is then written as

$$\sqrt{N}\Psi_{0,N}(0, x_1, \ldots, x_{N-1})$$
$$= \sqrt{N}\left(\prod_{i=1}^{N-1} (z_i-1)^\lambda z_i^{-\lambda/2}\right) \Psi_{0,N-1}(x_1, \ldots, x_{N-1}). \tag{2.225}$$

Here the factor $\prod_{i=1}^{N-1}(z_i-1)^\lambda$ is decomposed with J_κ as [94]

$$\prod_{i=1}^{N-1} (z_i-1)^\lambda = \sum_{\mu \in \Lambda_{N-1}^+} (-1)^{\lambda(N-1)+|\mu|} b_\mu J_\mu(z_1, \ldots, z_{N-1}), \tag{2.226}$$

where

$$b_\mu = \prod_{s \in D(\mu)} \frac{-a'(s)/\lambda + l'(s) + 1}{(a(s)+1)/\lambda + l(s)}. \tag{2.227}$$

The relation (2.226) with (2.227) is called the binomial formula and will be derived in Section 7.3.7.

From (2.225) and (2.226), we obtain

$$\hat{\psi}(0,0)|g,N\rangle = \sqrt{N} \sum_{\mu \in \Lambda_{N-1}^+} (-1)^{\lambda(N-1)+|\mu|} b_\mu |\mu - \lambda/2, N-1\rangle, \qquad (2.228)$$

where the Galilean-shifted momentum is given by

$$\mu - \lambda/2 = (\mu_1 - \lambda/2, \ldots, \mu_{N-1} - \lambda/2) \in \mathcal{L}_{N-1}^+.$$

The Galilean shift is necessary since the physical momentum can become negative, while μ is a partition with $\mu_{N-1} \geq 0$. The propagator is given in terms of b_μ by

$$G^-(x,t) = N \sum_{\mu \in \Lambda_{N-1}^+} \frac{\langle \mu, N-1 | \mu, N-1 \rangle}{\langle g, N | g, N \rangle} |b_\mu|^2$$

$$\times \sum_{\kappa \in \mathcal{L}_{N-1}^+} \delta_{\mu,\kappa+\lambda/2} \exp\left[-\mathrm{i}(E_\kappa - E_{0,N}) + \mathrm{i}Q_\kappa x\right]. \quad (2.229)$$

Let us analyze the matrix elements b_μ in detail. The numerator of (2.227) is rewritten as

$$\prod_{s \in \mathcal{D}(\mu)} \left(-\frac{a'(s)}{\lambda} + l'(s) + 1\right) = \prod_{(i,j) \in D(\mu)} \left(i + \frac{1-j}{\lambda}\right), \qquad (2.230)$$

which is nonzero only if

$$\mu_1 \leq \lambda. \qquad (2.231)$$

Therefore, only such excited states μ that satisfy (2.231) contribute to the summation in (2.229). A typical diagram of μ is shown in Fig. 2.7. These quasi-hole states are described in terms of rapidities $\dot{\mu}_1, \ldots, \dot{\mu}_\lambda$ defined by (2.110). When we write (2.230) in terms of $\dot{\mu}_j$, it is convenient to decompose $D(\mu)$ into λ columns as shown in Fig. 2.20(b). The decomposition is in general useful when we calculate the product of expressions such as

$$\prod_{s \in D(\mu)} \left(\alpha a'(s) + \beta l'(s) + \gamma\right) \qquad (2.232)$$

over $D(\mu)$ of the Young diagram of quasi-hole states.

We first calculate the relevant product within each column, and then take the product of contributions from each column. Correspondingly, we rewrite

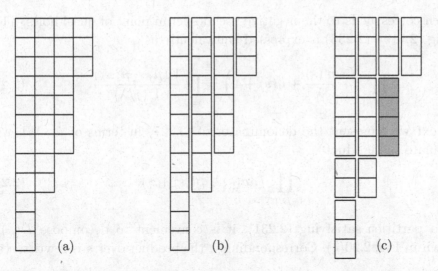

Fig. 2.20. A quasi-hole state (a) is decomposed as (b) when a product containing $a'(s)$ and $l'(s)$ is calculated, or as (c) when a product containing $a(s)$ and $l(s)$ is calculated. Shaded squares represent D_{jk} for $(j,k) = (2,3)$.

(2.230) as

$$\prod_{s \in \mathcal{D}(\mu)} \left(-\frac{a'(s)}{\lambda} + l'(s) + 1 \right) = \prod_{j=1}^{\lambda} \prod_{i=1}^{\mu'_j} \left(-\frac{j-1}{\lambda} + i \right)$$

$$= \prod_{j=1}^{\lambda} \frac{\Gamma\left(\mu'_j + 1 + (1-j)/\lambda \right)}{\Gamma(j/\lambda)}. \tag{2.233}$$

In the last equality, we have used the following relations:

$$\prod_{i=n_1}^{n_2} (i + a) = \frac{\Gamma(n_2 + a + 1)}{\Gamma(n_1 + a)}, \tag{2.234}$$

for $n_1 \leq n_2 \in \mathbf{Z}$, and

$$\prod_{j=1}^{\lambda} \Gamma\left(1 + (1-j)/\lambda \right) = \prod_{j=1}^{\lambda} \Gamma\left(j/\lambda \right).$$

We introduce an auxiliary rapidity:

$$\dot{\mu}_{1+\lambda} = \underbrace{\mu'_{1+\lambda}}_{0} - \frac{N-1}{2} + \frac{\lambda + 1 - 2j}{2\lambda} \bigg|_{j=\lambda+1} = -\frac{N}{2} - \frac{1}{2\lambda}, \tag{2.235}$$

which corresponds to the negative pseudo-Fermi point of quasi-holes. Then, using (2.110), (2.233) is expressed equivalently as

$$\prod_{s \in \mathcal{D}(\mu)} \left(-\frac{a'(s)}{\lambda} + l'(s) + 1 \right) = \prod_{j=1}^{\lambda} \frac{\Gamma\left(\dot\mu_j - \dot\mu_{1+\lambda}\right)}{\Gamma\left(j/\lambda\right)}. \tag{2.236}$$

Next we represent the denominator of (2.227) in terms of $\dot\mu_i$. When we calculate the product

$$\prod_{s \in D(\mu)} \left(\alpha a(s) + \beta l(s) + \gamma \right), \tag{2.237}$$

for a partition satisfying (2.231), it is convenient to decompose $D(\mu)$ as shown in Fig. 2.20(c). Correspondingly, the product over s is rewritten as

$$\prod_{s \in D(\mu)} \left(\alpha a(s) + \beta l(s) + \gamma \right) = \prod_{j=1}^{\lambda} \prod_{k=j}^{\lambda} \prod_{s \in D_{jk}} \left(\alpha a(s) + \beta l(s) + \gamma \right) \tag{2.238}$$

with

$$D_{jk} = \left\{ s = (i,j) | i \in [1 + \mu'_{k+1}, \mu'_k] \right\}. \tag{2.239}$$

An example of D_{jk} is shown in Fig. 2.20(c) by shaded squares. Since $a(s)$ is constant $(= k - j)$ within D_{jk}, it is possible to calculate the product with respect to $s \in D_{jk}$. When $\alpha = \gamma = 1/\lambda, \beta = 1$, for example, we obtain

$$\prod_{s \in D_{jk}} \left(\frac{a(s) + 1}{\lambda} + l(s) \right) = \prod_{i=1+\mu'_{1+k}}^{\mu'_k} \left(\frac{\mu_i - j + 1}{\lambda} + \mu'_j - i \right)$$

$$= \frac{\Gamma\left((k - j + 1)/\lambda + \mu'_j - \mu'_{k+1} \right)}{\Gamma\left((k - j + 1)/\lambda + \mu'_j - \mu'_k \right)}$$

$$= \frac{\Gamma\left(\dot\mu_j - \dot\mu_{k+1} \right)}{\Gamma\left(\dot\mu_j - \dot\mu_k + 1/\lambda \right)}. \tag{2.240}$$

We have used (2.234) in the second equality, while we used (2.110) in the last equality. From (2.238) and (2.240), the denominator of (2.227) is expressed as

$$\prod_{s \in D_{jk}} \left(\frac{a(s) + 1}{\lambda} + l(s) \right) = \frac{\prod_{j=1}^{\lambda} \Gamma\left(\dot\mu_j - \dot\mu_{1+\lambda} \right)}{\Gamma\left(1/\lambda \right)^{\lambda}} \prod_{j<k}^{\lambda} \frac{\Gamma\left(\dot\mu_j - \dot\mu_k \right)}{\Gamma\left(\dot\mu_j - \dot\mu_k + 1/\lambda \right)}. \tag{2.241}$$

Combining (2.236) and (2.241), we finally obtain

$$b_\mu = \frac{\Gamma(1/\lambda)^\lambda}{\prod_{j=1}^\lambda \Gamma(j/\lambda)} \prod_{j<k} \frac{\Gamma(\dot{\mu}_j - \dot{\mu}_k + 1/\lambda)}{\Gamma(\dot{\mu}_j - \dot{\mu}_k)}. \tag{2.242}$$

We now consider the norm written as

$$\langle \mu, N-1 | \mu, N-1 \rangle = L^{N-1} \langle J_\mu, J_\mu \rangle_{0,N-1}, \tag{2.243}$$

where the inner product $\langle \cdot, \cdot \rangle_{0,N-1}$ has been defined in (2.189). Here we have included the number $N-1$ of particles explicitly. Similarly we obtain

$$\langle g, N | g, N \rangle = L^N \langle 1, 1 \rangle_0 = c_N(\lambda) = \left(\frac{L}{\lambda!} \right)^N \Gamma(N\lambda + 1), \tag{2.244}$$

where $c_N(\lambda)$ has been introduced by (2.192). Combining (2.243) and (2.244), we obtain

$$\frac{\langle \mu, N-1 | \mu, N-1 \rangle}{\langle g, N | g, N \rangle}$$
$$= \frac{c_{N-1}(\lambda)}{L c_N(\lambda)} \prod_{s \in D(\mu)} \frac{[(a(s)+1)/\lambda + l(s)] [a'(s)/\lambda - l'(s) + N - 1]}{[a(s)/\lambda + l(s) + 1] [(a'(s)+1)/\lambda - l'(s) + N - 2]}, \tag{2.245}$$

where (2.191) has been used for norms of symmetric Jack polynomials.

By using the decompositions in Fig. 2.20(b) and (c), we further proceed to express (2.245) in terms of the rapidity of quasi-holes as

$$\frac{\langle \mu, N-1 | \mu, N-1 \rangle}{\langle g, N | g, N \rangle} = \frac{\Gamma(1+\lambda)}{NL\lambda^\lambda (\Gamma(1/\lambda))^\lambda}$$
$$\times \prod_{1 \leq j < k \leq \lambda} \frac{\Gamma(\dot{\mu}_j - \dot{\mu}_k) \Gamma(\dot{\mu}_j - \dot{\mu}_k + 1)}{\Gamma(\dot{\mu}_j - \dot{\mu}_k + 1/\lambda) \Gamma(\dot{\mu}_j - \dot{\mu}_k + 1 - 1/\lambda)}$$
$$\times \prod_{j=1}^\lambda \frac{\Gamma(\dot{\mu}_0 - \dot{\mu}_j) \Gamma(\dot{\mu}_j - \dot{\mu}_{1+\lambda})}{\Gamma(\dot{\mu}_0 - \dot{\mu}_j + 1 - 1/\lambda) \Gamma(\dot{\mu}_j - \dot{\mu}_{1+\lambda} + 1 - 1/\lambda)}. \tag{2.246}$$

Here we have introduced another auxiliary rapidity

$$\dot{\mu}_0 = \frac{N}{2} + \frac{1}{2\lambda}, \tag{2.247}$$

which corresponds to a positive pseudo-Fermi point of quasi-holes. Combining (2.242), (2.246), (2.109), and (2.112), we obtain

$$G^-(x,t) = \frac{\lambda! c(\lambda)}{L} \sum_{0 \leq \mu'_\lambda \leq \cdots \leq \mu'_1 \leq N-1} \exp[-\mathrm{i}(E_\kappa - E_{0,N})t + \mathrm{i}Q_\kappa x]$$

$$\times F_{0,\lambda}(\dot{\mu}_1, \ldots, \dot{\mu}_\lambda), \tag{2.248}$$

where $F_{0,\lambda}(\dot{\mu}_1, \ldots, \dot{\mu}_\lambda)$ is called the form factor, and is given by

$$F_{0,\lambda}(\dot{\mu}_1, \ldots, \dot{\mu}_\lambda) = \prod_{1 \leq j < k \leq \lambda} \frac{\Gamma(\dot{\mu}_j - \dot{\mu}_k + 1/\lambda)\,\Gamma(\dot{\mu}_j - \dot{\mu}_k + 1)}{\Gamma(\dot{\mu}_j - \dot{\mu}_k)\,\Gamma(\dot{\mu}_j - \dot{\mu}_k + 1 - 1/\lambda)}$$

$$\times \prod_{j=1}^{\lambda} \frac{\Gamma(\dot{\mu}_j - \dot{\mu}_{1+\lambda})\,\Gamma(\dot{\mu}_0 - \dot{\mu}_j)}{\Gamma(\dot{\mu}_j - \dot{\mu}_{1+\lambda} + 1 - 1/\lambda)\,\Gamma(\dot{\mu}_0 - \dot{\mu}_j + 1 - 1/\lambda)}. \tag{2.249}$$

The constant $c(\lambda)$ in (2.248) has been defined in (2.200).

We now consider the thermodynamic limit (t.d.l.) by taking

$$\text{t.d.l.:} \quad N \to \infty, \quad L \to \infty, \quad N/L = d \text{ (fixed)}.$$

Let us introduce the normalized rapidity $v_j = 2\dot{\mu}_j/N$ for quasi-holes $j \in [1, \lambda]$. Then the main contribution to the summation in (2.248) comes only from configurations satisfying

$$v_j - v_{j+1} = \mathcal{O}(1), \quad j \in [0, \lambda]. \tag{2.250}$$

For such configurations, we obtain

$$\lim_{\text{t.d.l.}} (E[\kappa] - E_{0,N}) = -\pi^2 \lambda d^2 \sum_{j=1}^{\lambda} v_j^2, \quad \lim_{\text{t.d.l.}} Q_\kappa = \pi d \sum_{j=1}^{\lambda} v_j \tag{2.251}$$

and the form factor is drastically simplified as

$$\lim_{\text{t.d.l.}} \left(\frac{N}{2}\right)^{\lambda-1} F_{0,\lambda}(\dot{\mu}_1, \ldots, \ldots, \dot{\mu}_\lambda) = \prod_{i=1}^{\lambda} (1 - v_i^2)^{-1+1/\lambda} \prod_{i<j} |v_i - v_j|^{2/\lambda}. \tag{2.252}$$

The result (2.252) is derived with use of the property

$$\frac{\Gamma(n+a)}{\Gamma(n+b)} \sim n^{a-b}, \quad \text{for } n \gg 1, \tag{2.253}$$

which follows from the Stirling formula

$$\Gamma(n+1)|_{n\to\infty} \simeq \sqrt{2\pi}\, n^{n+1/2}\, \mathrm{e}^{-n}. \tag{2.254}$$

In the thermodynamic limit, the summations over $(\mu_1', \ldots, \mu_\lambda')$ in (2.248) reduce to the following integrals:

$$\left(\frac{2}{N}\right)^\lambda \sum_{\mu_1'=0}^{N-1} \sum_{\mu_2'=0}^{\mu_1'} \cdots \sum_{\mu_\lambda'=0}^{\mu_{\lambda-1}'} \to \int_{-1}^{1} dv_1 \int_{-1}^{v_1} dv_2 \cdots \int_{-1}^{v_{\lambda-1}} dv_\lambda. \qquad (2.255)$$

From (2.251), (2.252), and (2.255), we obtain

$$\lim_{\text{t.d.l.}} G^-(x,t) = \frac{c(\lambda)d\lambda!}{2} \int_{-1}^{1} dv_1 \int_{-1}^{v_1} dv_2 \cdots \int_{-1}^{v_{\lambda-1}} dv_\lambda$$

$$\times \exp\left[i \sum_j (\pi d v_j x - \pi^2 \lambda d^2 v_j^2 t)\right] \prod_{i=1}^{\lambda} (1 - v_i^2)^{-1+1/\lambda} \prod_{i<j} (v_i - v_j)^{2/\lambda}.$$
$$(2.256)$$

We finally arrive at (2.199) by symmetrizing the integrand with respect to v_j, so that all integrals are to be performed for the range $v_j \in [-1, 1]$.

2.7.2 *Particle propagator

In the present subsection, we derive the particle propagator for finite-sized systems. The action of the annihilation operator $\hat{\psi}(0,0)$ on wave functions is easier to handle than that of the creation operator $\hat{\psi}^\dagger(0,0)$. Then in (2.206) we deal with the matrix element

$$\langle g, N | \hat{\psi}(0,0) | \kappa, N+1 \rangle, \qquad (2.257)$$

where $|\kappa, N+1\rangle$ is the $(N+1)$-particle state whose wave function is given by $J_\kappa \Psi_{0,N+1}$. Hence we obtain the wave function for $\hat{\psi}(0,0)|\kappa, N+1\rangle$:

$$\sqrt{N+1} J_\kappa(z_1, \ldots, z_N, 1) \Psi_{0,N+1}(1, z_1, \ldots, z_N)$$

$$= \sqrt{N+1} \prod_{i=1}^{N} (z_i - 1)^\lambda z_i^{-\lambda/2} J_\kappa(z_1, \ldots, z_N, 1) \Psi_{0,N}(z_1, \ldots, z_N). \qquad (2.258)$$

The polynomial $J_\kappa(z_1, \ldots, z_N, 1)$ is expanded with $J_\nu(z_1, \ldots, z_N)$ as

$$J_\kappa(z_1, \ldots, z_N, 1) = \sum_{\nu \in \mathcal{L}_N^+} f_{\kappa,\nu} J_\nu(z_1, \ldots, z_N), \qquad (2.259)$$

where the coefficient $f_{\kappa,\nu}$ is given by

$$f_{\kappa,\nu} = \prod_{s \in C_{\kappa/\nu} \backslash R_{\kappa/\nu}} \left(\frac{h_\kappa^*}{h_*^\kappa}\right) \left(\frac{h_*^\nu}{h_\nu^*}\right), \qquad (2.260)$$

when both κ and ν are partitions.

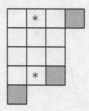

Fig. 2.21. An example of $\mu = (3,3,3,2,0)$ contributing to the summation (2.259) where $\kappa = (4,3,3,2,1)$. $C_{\kappa/\mu}$ is $\{2\}$ and $R_{\kappa/\mu}$ is $\{2,3\}$. The squares belonging to $C_{\kappa/\mu} \setminus R_{\kappa/\mu}$ are marked by $*$.

Here upper and lower hook lengths are defined by (2.186), and the notation $A \setminus B$ means the complementary set which is sometimes written as $A - B$. Namely, $C_{\kappa/\nu}$ denotes the columns $\{j\}$ where $\kappa'_j = \nu'_j$ and $R_{\kappa/\nu}$ denotes the rows $\{i\}$ where $\kappa_i = \nu_i$. Actually $f_{\kappa,\nu}$ is nonzero only when κ/ν is a horizontal strip, where $\kappa'_j - \nu'_j$ is 0 or 1 for each column of the Young diagram. Figure 2.21 shows an example where κ/ν is a horizontal strip. We shall derive (2.260) in Section 7.3.10.

When $\kappa_{N+1} < 0$, we use the relation

$$f_{\kappa,\nu} = f_{\kappa+n,\nu+n}, \tag{2.261}$$

where n is an integer larger than $-\kappa_{N+1}$ and

$$\kappa + n = (\kappa_1 + n, \ldots, \kappa_{N+1} + n) \in \Lambda^+_{N+1},$$
$$\nu + n = (\nu_1 + n, \ldots, \nu_N + n) \in \Lambda^+_N.$$

The right-hand side of (2.261) is given by (2.260).

We rewrite the binomial expansion in (2.258) as

$$\prod_{i=1}^{N} (z_i - 1)^\lambda z_i^{-\lambda/2} = \prod_{i=1}^{N} \left(1 - z_i^{-1}\right)^\lambda z_i^{\lambda/2}$$

$$= \left(\prod_{i=1}^{N} z_i^{\lambda/2}\right) \sum_{\nu \in \Lambda^+_N} (-1)^{|\nu|} b_\nu J_\nu(z_1^{-1}, \ldots, z_N^{-1}), \tag{2.262}$$

using (2.226). Combining (2.258), (2.260), and (2.262), we obtain

$$\langle g, N | \psi(0,0) | \kappa, N+1 \rangle = \sqrt{N+1} \sum_{\nu \in \Lambda^+_N} \langle J_\nu, J_\nu \rangle_0 (-1)^{|\nu|} b_\nu f_{\kappa,\nu-\lambda/2}, \tag{2.263}$$

where $\nu - \lambda/2$ denotes $(\nu_1 - \lambda/2, \ldots, \nu_N - \lambda/2) \in \mathcal{L}^+_N$. Inserting (2.263) into (2.257), we obtain the matrix element for $G^+(x,t)$ [160].

As we have seen in Section 2.7.1, the factor b_ν is nonzero only if

$$\lambda \geq \nu_2 \geq \cdots \geq \nu_N \geq 0. \tag{2.264}$$

There is no upper limit on the magnitude of ν_1. Hence in the matrix element in (2.257), the nonzero contribution comes only from $\kappa \in \mathcal{L}_{N+1}^+$ with

$$\frac{\lambda}{2} \geq \kappa_2 \geq \cdots \geq \kappa_N \geq -\frac{\lambda}{2}, \tag{2.265}$$

but without upper limit for κ_1. From (2.264), we obtain the condition $\kappa_{N+1} \leq \lambda/2$, but without lower limit.

For the case $-\lambda/2 \leq \kappa_{N+1} \leq \lambda/2$, we consider the skew Young diagram κ_+/ν with $\kappa_+ = \kappa + \lambda/2$. Here a skew Young diagram κ_+/ν for two partitions $\kappa_+ \supset \nu$ consists of cells $s \in \kappa_+$ but $s \notin \nu$. Figure 2.22 shows an example for a case where κ_+/ν is a horizontal strip which does not have a column of κ_+/ν with more than one square.

For another case $\kappa_{N+1} \leq -\lambda/2$, we perform a Galilean shift

$$\kappa_{++} = \kappa - \kappa_{N+1}, \quad \nu_+ = \nu - \kappa_{N+1} - \lambda/2,$$

in order to make both κ_{++} and ν_+ partitions. The coefficient

$$f_{\kappa,\nu-\lambda/2} = f_{\kappa_{++},\nu_+} \tag{2.266}$$

is zero unless ν_+ is a partition, since κ_{++} is a partition.

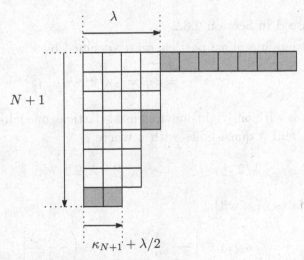

Fig. 2.22. Young diagram of κ_+ for the case $-\lambda/2 \leq \kappa_{N+1} \leq \lambda/2$. Unshaded squares represent Young diagram of ν.

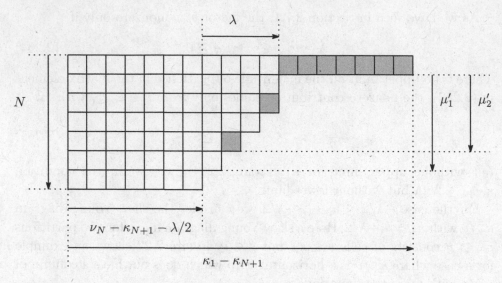

Fig. 2.23. The Young diagram of $(\kappa - \kappa_{N+1})$ when $\kappa_{N+1} < \lambda/2$ is satisfied. Unshaded squares represent the Young diagram of $\nu - \kappa_{N+1} - \lambda/2$.

In this case we consider the skew Young diagram where κ_{++}/ν_+ is the horizontal strip. Figure 2.23 shows an example.

In this way the states κ with (2.265) are classified into:

(i) One right-moving quasi-particle state specified by κ with

$$\kappa_1 > \kappa_2 = \cdots = \kappa_{N+1} = \lambda/2, \qquad (2.267)$$

as discussed in Section 2.3.2.

(ii) One left-moving quasi-particle state specified by κ with

$$\kappa_1 = \cdots = \kappa_N = -\lambda/2 > \kappa_{N+1}. \qquad (2.268)$$

(iii) The states with one right-moving quasi-particle, one left-moving quasi-particle, and λ quasi-holes with κ where

$$\kappa_1 > \lambda/2 \geq \kappa_2 \geq \cdots \geq \kappa_N \geq -\lambda/2 > \kappa_{N+1}. \qquad (2.269)$$

(iv) Other states of κ with

$$\kappa_{N+1} \in \left[-\frac{\lambda}{2}, \frac{\lambda}{2}\right), \text{ or } \kappa_1 \in \left(-\frac{\lambda}{2}, \frac{\lambda}{2}\right]. \qquad (2.270)$$

States (i) and (ii) yield $G_1^+(x,t)$ and states (iii) give $G_2^+(x,t)$, as we show below. Other states (iv) do not contribute in the thermodynamic limit; the contribution to $G^+(x,t)$ from the states with κ satisfying

$$\kappa_{N+1} = \frac{\lambda}{2} - p, \quad p = \{1, 2, \ldots, \lambda\}$$

can be shown to be zero for $\lambda = 1$ or $\mathcal{O}(N^{-p^2/\lambda})$ for $\lambda \neq 1$, in a way similar to the derivation of $G_2^+(x,t)$.

For one right-moving quasi-particle state (i), only $\nu = \lambda^N \equiv (\lambda, \ldots, \lambda)$ contributes to (2.263). Then both b_ν and $f_{\kappa+\lambda/2,\nu}$ become unity. The excitation spectrum has already been given in (2.94). The norm $\langle J_\kappa, J_\kappa \rangle_0$ can be derived from (2.191). The term on the RHS of (2.263) coming from the one right-moving quasi-particle state is given by

$$G_{1R}^+(x,t) = \frac{1}{L} \sum_{\kappa_1 > \lambda/2}^{\infty} \frac{\Gamma(\tilde{\kappa}_1 - \tilde{\kappa}_{1,0})\,\Gamma(\tilde{\kappa}_1 - \tilde{\kappa}_{N,0} + 1)}{\Gamma(\tilde{\kappa}_1 - \tilde{\kappa}_{1,0} + 1 - \lambda)\,\Gamma(\tilde{\kappa}_1 - \tilde{\kappa}_{N,0} + \lambda)}$$

$$\times \exp\left[-i\left(\frac{2\pi\tilde{\kappa}_1}{L}\right)^2 t + i\left(\frac{2\pi\tilde{\kappa}_1}{L}\right)x\right], \qquad (2.271)$$

where

$$\tilde{\kappa}_1 = \kappa_1 + \lambda N/2, \quad \tilde{\kappa}_{1,0} = -\tilde{\kappa}_{N,0} = \lambda(N-1)/2. \qquad (2.272)$$

Similarly, the contribution $G_{1L}^+(x,t)$ to (2.263) from one left-moving quasi-particle state (ii) comes only from $\nu = 0^N \equiv (0, \ldots, 0)$, and we obtain

$$G_{1L}^+(x,t) = \frac{1}{L} \sum_{\kappa_{N+1}=-\infty}^{-\lambda/2} \frac{\Gamma(\tilde{\kappa}_{N,0} - \tilde{\kappa}_{N+1})\,\Gamma(\tilde{\kappa}_{N,0} - \tilde{\kappa}_{N+1} + 1)}{\Gamma(\tilde{\kappa}_{N,0} - \tilde{\kappa}_{N+1} + 1 - \lambda)\,\Gamma(\tilde{\kappa}_{N,0} - \tilde{\kappa}_{N+1} + \lambda)}$$

$$\times \exp\left[-i\left(\frac{2\pi\tilde{\kappa}_{N+1}}{L}\right)^2 t + i\left(\frac{2\pi\tilde{\kappa}_{N+1}}{L}\right)x\right], \qquad (2.273)$$

with $\tilde{\kappa}_{N+1} = \kappa_{N+1} - \lambda N/2$.

In (2.271) and (2.273), we put

$$\frac{2\tilde{\kappa}_1}{\lambda N} = -\frac{2\tilde{\kappa}_{N+1}}{\lambda N} = w$$

and take the thermodynamic limit. We then obtain (2.207) for $G_1^+(x,t)$ in Section 2.6.2.

For a state belonging to (iii), the energy $E[\kappa]$ is given under the condition (2.269) by

$$\left(\frac{L}{2\pi}\right)^2 E[\kappa] = \sum_{i=1}^{N+1}\left(\kappa_i + \frac{\lambda}{2}(N+2-2i)\right)^2$$

$$= \tilde{\kappa}_1^2 + \tilde{\kappa}_{N+1}^2 + \sum_{i=2}^{N}\left(\kappa_i + \frac{\lambda}{2}(N+2-2i)\right)^2. \qquad (2.274)$$

We introduce $\mu \in \Lambda_{N-1}^+$ as

$$\mu = (\kappa_2 + \lambda/2, \ldots, \kappa_N + \lambda/2) \qquad (2.275)$$

and the quasi-hole rapidity as

$$\dot{\mu}_j = \mu_j' - \frac{N-1}{2} + \frac{\lambda + 1 - 2j}{2\lambda}, \qquad (2.276)$$

for $j \in [1, \lambda]$. Then the excitation energy $\mathcal{E}$ is rewritten from (2.274) as

$$\mathcal{E} = \tilde{\kappa}_1^2 + \tilde{\kappa}_{N+1}^2 - \lambda \sum_{j=1}^{\lambda} \dot{\mu}_j^2 + \frac{\lambda}{12}(\lambda^2 - 1). \qquad (2.277)$$

Next we consider the matrix element. Let I be the subset of $[1, 2, \ldots, \lambda]$ such that

$$\nu_j' = \begin{cases} \mu_j' + 1, & j \in I, \\ \mu_j', & j \in J = [1, 2, \ldots, \lambda] \setminus I, \end{cases} \qquad (2.278)$$

where $A \setminus B$ is the complementary set of B in A. For example, Fig. 2.23 shows the case where $\lambda = 4$ and $I = \{1, 3\}$ (and hence $J = \{2, 4\}$). In the expression

$$\sum_{\nu \in \Lambda_N^+} \langle J_\nu, J_\nu \rangle_0 (-1)^{|\nu|} b_\nu f_{\kappa+\lambda/2, \nu}, \qquad (2.279)$$

the summation over ν can be expressed as that over all subsets I of $[1, 2, \ldots, \lambda]$. The quantity $f_{\kappa+\lambda/2, \nu}$ in (2.279) is obtained from (2.260). Namely, we make the product of $h_{\kappa-\kappa_{N+1}}^* / h_*^{\kappa-\kappa_{N+1}}$ and $h_*^{\nu-\kappa_{N+1}-\lambda/2} / h_{\nu-\kappa_{N+1}-\lambda/2}^*$ on

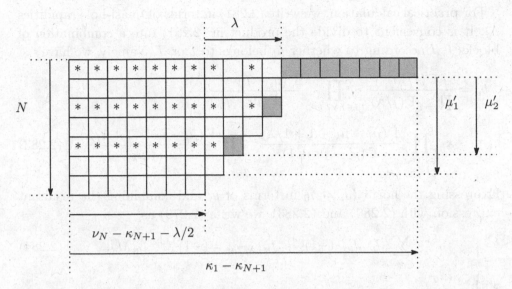

Fig. 2.24. The same Young diagram as that shown in Fig. 2.23. The squares marked by * represent those squares that should be evaluated in the calculation of $f_{\kappa+\lambda/2,\nu}$.

the squares marked by * in Fig. 2.24. The result is given by

$$
\begin{aligned}
&f_{\kappa+\lambda/2,\nu} \\
&= \frac{\Gamma(N\lambda+1)\Gamma(\tilde{\kappa}_1-\tilde{\kappa}_{N+1}+1-\lambda)\Gamma(\tilde{\kappa}_1-\tilde{\kappa}_{N,0}+\lambda)\Gamma(\tilde{\kappa}_{1,0}-\tilde{\kappa}_{N+1}+\lambda)}{\Gamma((N+1)\lambda)\Gamma(\tilde{\kappa}_1-\tilde{\kappa}_{N+1})\Gamma(\tilde{\kappa}_1-\tilde{\kappa}_{N,0}+1)\Gamma(\tilde{\kappa}_{1,0}-\tilde{\kappa}_{N+1}+1)} \\
&\quad \times \prod_{l\in I}\frac{(\tilde{\kappa}_1+\lambda\dot{\mu}_l+(1-\lambda)/2)(\dot{\mu}_l-\dot{\mu}_{1+\lambda}+1-1/\lambda)}{(\tilde{\kappa}_1+\lambda\dot{\mu}_l+(\lambda-1)/2)(\dot{\mu}_l-\dot{\mu}_{1+\lambda})} \\
&\quad \times \prod_{l\in J}\frac{(\tilde{\kappa}_{N+1}+\lambda\dot{\mu}_l+(\lambda-1)/2)(\dot{\mu}_0-\dot{\mu}_l+1-1/\lambda)}{(\tilde{\kappa}_{N+1}+\lambda\dot{\mu}_l+(1-\lambda)/2)(\dot{\mu}_0-\dot{\mu}_l)} \\
&\quad \times \prod_{\substack{l\in I,k\in J;\\ \text{s.t.}\,l<k}}\frac{(\dot{\mu}_l-\dot{\mu}_k+1/\lambda)(\dot{\mu}_l-\dot{\mu}_k+1-1/\lambda)}{(\dot{\mu}_l-\dot{\mu}_k)(\dot{\mu}_l-\dot{\mu}_k+1)}.
\end{aligned} \tag{2.280}
$$

The binomial coefficient b_ν is given by

$$
b_\nu = \frac{\Gamma(1/\lambda)^\lambda}{\prod_{j=1}^{\lambda}\Gamma(j/\lambda)}\prod_{l<k}\frac{\Gamma(\dot{\nu}_l-\dot{\nu}_k+1/\lambda)}{\Gamma(\dot{\nu}_l-\dot{\nu}_k)}, \tag{2.281}
$$

where $\dot{\nu}_k$ and $\dot{\nu}_l$ are obtained from the conjugate partition ν' by

$$
\dot{\nu}_j = \nu'_j - \frac{N-1}{2} + \frac{\lambda+1-2j}{2\lambda}, \quad \text{for } j\in[1,\lambda]. \tag{2.282}
$$

For practical calculation, we write (2.281) in terms of quasi-hole rapidities $\dot\mu_j$. It is convenient to divide the product in (2.281) into a combination of blocks I, J according to whether $\dot\mu_l$ belongs to I or J. Namely, we have

$$
b_\nu = \frac{\Gamma(1/\lambda)^\lambda}{\prod_{j=1}^\lambda \Gamma(j/\lambda)} \prod_{\substack{l,k\in I \text{ or } l,k\in J \\ l<k}} \frac{\Gamma(\dot\mu_l - \dot\mu_k + 1/\lambda)}{\Gamma(\dot\mu_l - \dot\mu_k)}
$$

$$
\times \prod_{\substack{l\in I, k\in J \\ l<k}} \frac{\Gamma(\dot\mu_l - \dot\mu_k + 1 + 1/\lambda)}{\Gamma(\dot\mu_l - \dot\mu_k + 1)} \prod_{\substack{l\in J, k\in I \\ l<k}} \frac{\Gamma(\dot\mu_l - \dot\mu_k - 1 + 1/\lambda)}{\Gamma(\dot\mu_l - \dot\mu_k - 1)}. \quad (2.283)
$$

Expressing the norm $\langle J_\nu, J_\nu\rangle_0$ in terms of $\dot\mu_l$ and combining the resultant expression with (2.280) and (2.283), we write (2.279) as

$$
\sum_\nu \langle J_\nu, J_\nu\rangle_0 (-1)^{|\nu|} b_\nu f_{\kappa+\lambda/2,\nu} = (-1)^{\lambda+|\mu|} L_\kappa M_\kappa, \quad (2.284)
$$

with

$$
L_\kappa = \frac{c_N(\lambda)}{\lambda^{\lambda-1}\prod_{j=1}^\lambda \Gamma(j/\lambda)} \frac{\Gamma(\tilde\kappa_1 - \tilde\kappa_{N,0} + \lambda)\Gamma(\tilde\kappa_{1,0} - \tilde\kappa_{N+1} + \lambda)}{\Gamma(\tilde\kappa_1 - \tilde\kappa_{N,0} + 1)\Gamma(\tilde\kappa_{1,0} - \tilde\kappa_{N+1} + 1)}
$$

$$
\times \prod_{l=1}^\lambda \frac{\Gamma(\dot\mu_l - \dot\mu_{1+\lambda})\Gamma(\dot\mu_0 - \dot\mu_l)}{\Gamma(\dot\mu_l - \dot\mu_{1+\lambda} + 1 - 1/\lambda)\Gamma(\dot\mu_0 - \dot\mu_l + 1 - 1/\lambda)}
$$

$$
\times \prod_{l=1}^\lambda \frac{(\tilde\kappa_1 + \lambda\dot\mu_l)(\tilde\kappa_{N+1} + \lambda\dot\mu_l)}{(\tilde\kappa_{N+1} + \lambda\dot\mu_l)(\tilde\kappa_{N+1} + \lambda\dot\mu_l + (1-\lambda)/2)} \prod_{l<k} \frac{\Gamma(\dot\mu_l - \dot\mu_k + 1)}{\Gamma(\dot\mu_l - \dot\mu_k + 1 - 1/\lambda)}
$$

$$
\tag{2.285}
$$

and

$$
M_\kappa = \sum_{I\subset[1,2,\ldots,\lambda]} (-1)^{\lambda-|I|} \prod_{l\in I} \left(1 + \frac{1-\lambda}{2(\tilde\kappa_1 + \lambda\dot\mu_l)}\right)\left(1 + \frac{1-\lambda}{2(\tilde\kappa_{N+1} + \lambda\dot\mu_l)}\right)
$$

$$
\times \prod_{l\in J} \left(1 + \frac{\lambda-1}{2(\tilde\kappa_1 + \lambda\dot\mu_l)}\right)\left(1 + \frac{\lambda-1}{2(\tilde\kappa_{N+1} + \lambda\dot\mu_l)}\right)
$$

$$
\times \prod_{\substack{l\in I, k\in J; \\ \text{s.t.} l<k}} \left(1 + \frac{1}{\lambda(\dot\mu_l - \dot\mu_k)}\right) \prod_{\substack{l\in J, k\in I; \\ \text{s.t.} l<k}} \left(1 - \frac{1}{\lambda(\dot\mu_l - \dot\mu_k)}\right). \quad (2.286)
$$

Here $|I|$ is the number of elements in I.

Let us introduce notations w_1, w_0, v_1, ..., v_λ through the following relations:

$$
w_1 = \frac{2\tilde\kappa_1}{N\lambda}, \quad w_0 = \frac{2\tilde\kappa_{N+1}}{N\lambda}, \quad v_l = \frac{2\dot\mu_l}{N}, \quad (2.287)
$$

for $l = 1, \ldots, \lambda$, and consider the thermodynamic limit. The factor $\langle J_\kappa, J_\kappa \rangle_0$ can be expressed in terms of $\tilde{\kappa}_1, \tilde{\kappa}_{N+1}$, and $\dot{\mu}_j$ in a way similar to Section 2.7.1. The expression

$$\frac{L_\kappa^2}{\langle J_\kappa, J_\kappa \rangle_0 \langle 1, 1 \rangle_0}$$

has the asymptotic behavior

$$\frac{\lambda^{2\lambda} c(\lambda)}{(N+1)\Gamma(\lambda+1)} \left(\frac{N}{2}\right)^{\lambda-1} \left(w_1^2 - 1\right)^{\lambda-1} \left(w_0^2 - 1\right)^{\lambda-1} (w_1 - w_0)^{2(1-\lambda)}$$

$$\times \prod_{l=1}^{\lambda} (1 - v_l^2)^{1/\lambda - 1} \prod_{l<k} (v_k - v_l)^{2/\lambda} + O\left(\frac{1}{N}\right), \tag{2.288}$$

where the notation $c(\lambda)$ has been given by (2.200). The asymptotic behavior of M_κ is given by

$$M_\kappa \to \left(\frac{2}{N\lambda}\right)^\lambda \frac{\mathcal{D}_\lambda \left\{ \prod_{k<l} (v_l - v_k) \prod_{k=0}^{1} \prod_{l=1}^{\lambda} (w_k + v_l)^{1-\lambda} \right\}}{\left\{ \prod_{k<l} (v_l - v_k) \prod_{k=0}^{1} \prod_{l=1}^{\lambda} (w_k + v_l)^{1-\lambda} \right\}}, \tag{2.289}$$

where $\mathcal{D}_\lambda \equiv \partial^\lambda / \partial v_1 \cdots \partial v_\lambda$. In the following, we derive (2.289).

The expression (2.286) for M_κ is rewritten as

$$M_\kappa = \sum_{\alpha_1 = 1, -1} \cdots \sum_{\alpha_\lambda = 1, -1} \left(\prod_{l=1}^{\lambda} \alpha_l\right) \mathcal{M}_\kappa \left(\frac{\alpha_1}{2\lambda}, \ldots, \frac{\alpha_\lambda}{2\lambda}\right), \tag{2.290}$$

with

$$\mathcal{M}_\kappa(\alpha_1, \ldots, \alpha_\lambda)$$

$$= \prod_{l<k} \left(1 + \frac{\alpha_l - \alpha_k}{(\dot{\mu}_l - \dot{\mu}_k)}\right) \prod_{l=1}^{\lambda} \left(1 + \frac{(1-\lambda)\alpha_l}{(\tilde{\kappa}_1/\lambda + \dot{\mu}_l)}\right) \left(1 + \frac{(1-\lambda)\alpha_l}{(\tilde{\kappa}_{N+1}/\lambda + \dot{\mu}_l)}\right). \tag{2.291}$$

By setting $\alpha_\ell = \pm 1$ depending on $l \in I$ or $l \in J$, we can confirm the equivalence between (2.290) and (2.286). $\mathcal{M}_\kappa$ is a polynomial of $(\alpha_1, \ldots, \alpha_\lambda)$ and hence it can be expanded as

$$\mathcal{M}_\kappa(\alpha_1, \ldots, \alpha_\lambda) = \sum_n c_n(\tilde{\kappa}_1, \tilde{\kappa}_{N+1}; \dot{\mu}_1, \ldots, \dot{\mu}_\lambda) \alpha_1^{n_1} \cdots \alpha_\lambda^{n_\lambda}, \tag{2.292}$$

where $n = (n_1, \ldots, n_\lambda)$ is the set of integers. We then express M_κ as

$$M_\kappa = 2^\lambda \sum_{n_1 = 1, 3, \ldots} \cdots \sum_{n_\lambda = 1, 3, \ldots} (2\lambda)^{-|n|} c_n(\tilde{\kappa}_1, \tilde{\kappa}_{N+1}; \dot{\mu}_1, \ldots, \dot{\mu}_\lambda), \tag{2.293}$$

where n_l are positive odd integers. The symbol $|n|$ denotes $\sum_{l=1}^{\lambda} n_l$.

For the rapidities with

$$\dot{\mu}_j - \dot{\mu}_{j+1} \gg 1, \quad \dot{\mu}_j + \tilde{\kappa}_1/\lambda \gg 1, \quad |\dot{\mu}_j + \tilde{\kappa}_{N+1}/\lambda| \gg 1, \qquad (2.294)$$

the leading contribution in (2.293) comes from the term with $n = (1, \ldots, 1)$:

$$M_\kappa \sim \lambda^{-\lambda} c_{(1,\ldots,1)}(\tilde{\kappa}_1, \tilde{\kappa}_{N+1}; \dot{\mu}_1, \ldots, \dot{\mu}_\lambda). \qquad (2.295)$$

In order to obtain $c_{(1,\ldots,1)}(\tilde{\kappa}_1, \tilde{\kappa}_{N+1}; \dot{\mu}_1, \ldots, \dot{\mu}_\lambda)$, we introduce

$$W(\dot{\mu}_1, \ldots, \dot{\mu}_\lambda) = \prod_{l<k}(\dot{\mu}_l - \dot{\mu}_k), \qquad (2.296)$$

$$W'(\tilde{\kappa}_1, \tilde{\kappa}_{N+1}; \dot{\mu}_1, \ldots, \dot{\mu}_\lambda) = \prod_{l=1}^{\lambda}(\dot{\mu}_l + \tilde{\kappa}_1/\lambda)^{1-\lambda}(\dot{\mu}_l + \tilde{\kappa}_{N+1}/\lambda)^{1-\lambda}. \quad (2.297)$$

Then each factor in $\mathcal{M}_\kappa$ is written as

$$\prod_{l<k}\left(1 + \frac{\alpha_l - \alpha_k}{\dot{\mu}_l - \dot{\mu}_k}\right) = \frac{1}{W}\left(1 + \sum_{l=1}^{\lambda}\alpha_l\frac{\partial}{\partial\dot{\mu}_l} + \sum_{l<k}\alpha_l\alpha_k\frac{\partial^2}{\partial\dot{\mu}_l\partial\dot{\mu}_k} + \cdots\right)W,$$
$$(2.298)$$

$$\prod_{l=1}^{\lambda}\left(1 + \frac{(1-\lambda)\alpha_l}{(\tilde{\kappa}_1/\lambda + \dot{\mu}_l)}\right)\left(1 + \frac{(1-\lambda)\alpha_l}{(\tilde{\kappa}_{N+1}/\lambda + \dot{\mu}_l)}\right)$$

$$= \frac{1}{W'}\left(1 + \sum_{l=1}^{\lambda}\alpha_l\frac{\partial}{\partial\dot{\mu}_l} + \sum_{l<k}\alpha_l\alpha_k\frac{\partial^2}{\partial\dot{\mu}_l\partial\dot{\mu}_k} + \cdots\right)W'. \qquad (2.299)$$

From (2.298) and (2.299), it follows that

$$c_{(1,\ldots,1)}(\tilde{\kappa}_1, \tilde{\kappa}_{N+1}; \dot{\mu}_1, \ldots, \dot{\mu}_\lambda) = \frac{1}{WW'}\frac{\partial^\lambda WW'}{\partial\dot{\mu}_1\cdots\partial\dot{\mu}_\lambda}. \qquad (2.300)$$

Combining (2.295) and (2.300) and using the notations (2.287), we obtain the relation (2.289).

In the thermodynamic limit, we use (2.288), (2.289), and

$$\frac{2}{\lambda N}\sum_{\kappa_1} \to \int_1^\infty \mathrm{d}w_1, \quad \frac{2}{\lambda N}\sum_{\kappa_{N+1}} \to \int_{-\infty}^{-1} \mathrm{d}w_0, \quad \frac{2}{N}\sum_{\mu'_l} \to \int_{-1}^1 \mathrm{d}v_l, \quad (2.301)$$

for $l = 1, \ldots, \lambda$. Then we finally arrive at (2.209)–(2.212) for $G_2^+(x,t)$

2.7.3 Density correlation function

In this subsection, we derive the dynamical density correlation function for a finite-size system; the resultant expression reduces to (2.216) in the thermodynamic limit. In contrast with the propagator, the results for the density correlation function apply to both bosons and fermions with non-negative integer λ.

The expression (2.214) is written as

$$\Pi(x,t) = \sum_{\kappa \in \mathcal{L}_N^+} \frac{|\langle \kappa, N | \hat{\rho}(0,0) | g, N \rangle|^2}{\langle \kappa, N | \kappa, N \rangle \langle g, N | g, N \rangle} e^{-i(E[\kappa] - E_{0,N})t + iQ_\kappa x} \qquad (2.302)$$

by inserting the complete set of N-particle states

$$1 = \sum_{\kappa \in \mathcal{L}_N^+} \frac{|\kappa, N\rangle \langle \kappa, N|}{\langle \kappa, N | \kappa, N \rangle}. \qquad (2.303)$$

Here $|\kappa, N\rangle$ represents the N-particle state whose wave function is given by

$$J_\kappa(z_1, \ldots, z_N) \Psi_{0,N}(z_1, \ldots, z_N).$$

Owing to the symmetry with respect to spatial inversion I, the energy, momentum, and matrix element of the state $|I\kappa, N\rangle$ with

$$I\kappa = (-\kappa_N, \ldots, -\kappa_1)$$

are given by

$$E_{I\kappa} = E[\kappa], \quad Q_{I\kappa} = -Q_\kappa, \quad \langle \kappa, N | \hat{\rho}(0,0) | g, N \rangle = \langle I\kappa, N | \hat{\rho}(0,0) | g, N \rangle. \qquad (2.304)$$

Using (2.304), the density correlation function (2.302) is given only by the non-negative momentum Q_κ. Namely, writing such a summation as $\sum_\kappa'$, we obtain

$$\Pi(x,t) = 2 \sum_\kappa' \frac{|\langle \kappa, N | \hat{\rho}(0,0) | g, N \rangle|^2}{\langle \kappa, N | \kappa, N \rangle \langle g, N | g, N \rangle} e^{-i(E[\kappa] - E_{0,N})t} \cos Q_\kappa x. \qquad (2.305)$$

Under the periodic boundary condition, the density fluctuation operator (2.215) is expanded in a Fourier series:

$$\hat{\rho}(x,0) = \frac{1}{L} \sum_{n \neq 0} e^{-2\pi i n x/L} \sum_{i=1}^{N} z_i^n, \qquad (2.306)$$

with $z_i = \exp(2\pi i x_i / L)$. Then we obtain

$$\hat{\rho}(0,0) = \frac{1}{L} \sum_{n>0} p_n(\{z_i\}) + \text{complex conjugate}. \qquad (2.307)$$

Here $p_n(\{z_i\})$ is the power-sum symmetric function defined in Section 2.5. The function $p_n(\{z_i\})$ is expanded in terms of the symmetric Jack polynomials as [84]

$$p_n(\{z_i\}) = \frac{n}{\lambda} \sum_{\substack{\kappa \in \Lambda_N^+; \\ \text{s.t.} |\kappa|=n}} \vartheta_\kappa (J_\kappa(\{z_i\}), \qquad (2.308)$$

with

$$\vartheta_\kappa = \frac{\prod_{\substack{s \in \mathcal{D}(\kappa) \\ (\neq (1,1))}} (a'(s)/\lambda - l'(s))}{\prod_{s \in \mathcal{D}(\kappa)} h_\kappa^*(s)}. \qquad (2.309)$$

Here the upper hook length $h_\kappa^*(s)$ has been defined in (2.186). The property (2.308) will be derived in Section 7.3.8.

Owing to (2.307) and (2.308), only $\kappa \in \Lambda_N^+$ contributes to the summation in (2.305) and hence (2.305) becomes

$$\Pi(x,t) = \frac{2}{\lambda^2 L^2} \sum_{\kappa \in \Lambda_N^+} \frac{\langle \kappa, N | \kappa, N \rangle}{\langle g, N | g, N \rangle} |\kappa|^2 \vartheta_\kappa^2 e^{-i(E[\kappa] - E_{0,N})t} \cos Q_\kappa x, \qquad (2.310)$$

where the norm part is given by

$$\frac{\langle \kappa, N | \kappa, N \rangle}{\langle g, N | g, N \rangle} = \prod_{s \in D(\kappa)} \frac{h_\kappa^*(s)\, (a'(s)/\lambda - l'(s) + N)}{h_*^\kappa(s)\, ((a'(s)+1)/\lambda - l'(s) + N - 1)} \qquad (2.311)$$

from (2.191).

In (2.310), ϑ_κ vanishes if κ contains $(i,j) = (2, \lambda+1)$, as seen from (2.309). A typical diagram for partition κ contributing to the summation in (2.310) is shown in Fig. 2.8. Other factors in (2.310) have the forms

$$\prod_{s \in D(\kappa)} (\alpha a'(s) + \beta l'(s) + \gamma), \qquad (2.312)$$

$$\prod_{s \in D(\kappa)} (\alpha a(s) + \beta l(s) + \gamma), \qquad (2.313)$$

which are given in terms of rapidities $(\tilde{\kappa}_1, \dot{\mu}_1, \ldots, \dot{\mu}_\lambda)$. They have been defined in (2.64) and (2.110) as

$$\tilde{\kappa}_1 = \kappa_1 + \frac{\lambda(N-1)}{2}, \quad \dot{\mu}_j = \mu_j' - \frac{N-1}{2} + \frac{\lambda + 1 - 2j}{2\lambda}. \qquad (2.314)$$

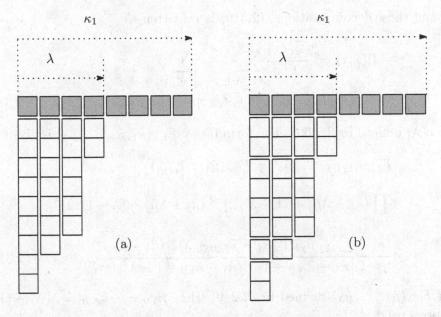

Fig. 2.25. Decomposition of Young diagram of partition contributing to the density correlation function.

In evaluating (2.312), we decompose $D(\kappa)$ into blocks as shown in Fig. 2.25(a) and rewrite (2.312) as

$$\prod_{s\in D(\kappa)} \left(\alpha a'(s) + \beta l'(s) + \gamma\right)$$

$$= \prod_{j=1}^{\kappa_1} \left(\alpha a'(1,j) + \beta l'(1,j) + \gamma\right) \prod_{j=1}^{\lambda}\prod_{i=1}^{\kappa'_j} \left(\alpha a'(s) + \beta l'(s) + \gamma\right).$$

In evaluating (2.313), on the other hand, we decompose $D(\kappa)$ into blocks as shown in Fig. 2.25(b) and rewrite (2.313) as

$$\prod_{s\in D(\kappa)} \left(\alpha a(s) + \beta l(s) + \gamma\right)$$

$$= \prod_{j=1}^{\kappa_1} \left(\alpha a(1,j) + \beta l(1,j) + \gamma\right) \prod_{s\in D(\mu)} \left(\alpha a(s) + \beta l(s) + \gamma\right), \qquad (2.315)$$

with $\mu = (\mu_1 \cdots \mu_{N-1}) = (\kappa_2 \cdots \kappa_N)$. Then the second product in (2.315) is evaluated as in (2.238).

Using these decompositions, (2.310) is rewritten as

$$\Pi(x,t) = \frac{2\lambda! c(\lambda)}{L^2} \sum_{\kappa_1=\lambda}^{\infty} \sum_{0 \leq \mu'_\lambda \leq \cdots \leq \mu'_1 \leq N-1}$$
$$\exp[-\mathrm{i}(E_\kappa - E_{0,N})t + \mathrm{i}Q_\kappa x] F_{1,\lambda}(\tilde{\kappa}_1; \dot{\mu}_1, \ldots, \dot{\mu}_\lambda), \tag{2.316}$$

with $c(\lambda)$ defined in (2.200). The form factor $F_{1,\lambda}(\tilde{\kappa}_1; \dot{\mu}_1, \ldots, \dot{\mu}_\lambda)$ is given by

$$F_{1,\lambda}(\tilde{\kappa}_1; \dot{\mu}_1, \ldots, \dot{\mu}_\lambda) = F_{0,\lambda}(\dot{\mu}_1, \ldots, \dot{\mu}_\lambda)$$
$$\times \prod_{j=1}^{\lambda} (\tilde{\kappa}_1 + \lambda\dot{\mu}_j + (1-\lambda)/2)^{-1} (\tilde{\kappa}_1 + \lambda\dot{\mu}_j + (\lambda-1)/2)^{-1}$$
$$\times \frac{(\tilde{\kappa}_1 + \sum_{j=1}^{\lambda} \dot{\mu}_j)^2 \Gamma(\tilde{\kappa}_1 - \tilde{\kappa}_{N,0} + \lambda) \Gamma(\tilde{\kappa}_1 - \tilde{\kappa}_{1,0})}{\Gamma(\tilde{\kappa}_1 - \tilde{\kappa}_{N,0} + 1) \Gamma(\tilde{\kappa}_1 - \tilde{\kappa}_{1,0} + 1 - \lambda)}, \tag{2.317}$$

with $F_{0,\lambda}(\dot{\mu}_1, \ldots, \dot{\mu}_\lambda)$ defined in (2.249). Here $\tilde{\kappa}_{1,0} = -\tilde{\kappa}_{N,0} = \lambda(N-1)/2$ has been used.

We take the thermodynamic limit in a way similar to that used in Section 2.7.1 by setting

$$w = \frac{2\tilde{\kappa}_1}{\lambda N}, \quad v_j = \frac{2\dot{\mu}_j}{N}, \quad j = [1, \lambda]. \tag{2.318}$$

Then we finally arrive at (2.216).

2.8 *Reduction to Tomonaga–Luttinger liquid

In this section, low-energy properties of the Sutherland model are described as a Tomonaga–Luttinger liquid (TLL). The TLL is a typical one-dimensional quantum system where the excitation spectrum is linear with respect to momentum. Many systems of interacting bosons or fermions in one dimension have the excitation spectrum linear with respect to momentum at low energy. The TLL is the effective theory of those various models at low energy. The single-component Sutherland model is the simplest model of the TLL in the sense that various physical quantities are derived explicitly.

In Section 2.8.1, we derive the long-distance or long-time asymptotic form of dynamical correlation functions from the exact results in Section 2.6. In Section 2.8.2, we consider the finite-size spectrum in the TLL and the Sutherland model. It is shown that the parameters in the TLL are determined by the asymptotic behavior derived in Section 2.8.1.

2.8.1 Asymptotic behavior of correlation functions

We shall show that the long-distance and long-time asymptotics of the hole propagator (2.199) are given by

$$G^-(x,t) \sim \exp\left[-i\pi\lambda dx + i\mu t\right] \sum_{r=0}^{\lambda} \frac{A_r \exp\left[i2\pi drx\right]}{X_{\mathrm{R}}^{(\lambda-r)^2/\lambda} X_{\mathrm{L}}^{r^2/\lambda}}. \qquad (2.319)$$

Here A_r is a constant and $\mu = \pi^2\lambda^2 d^2$ is the chemical potential, and we have introduced

$$X_{\mathrm{L}} = x + 2\pi d\lambda t, \quad X_{\mathrm{R}} = x - 2\pi d\lambda t.$$

The result (2.319) has a typical form of correlation functions in the TLL.

In the asymptotic behavior of correlation functions, the main contributions come from the low-energy excited states where all quasi-holes have the rapidity $v_j \sim \pm 1$. For the hole propagator (2.199), low-energy states are classified into $\lambda + 1$ sectors. Each sector is specified by an integer $r \in [0, \lambda]$, and the rapidities $\{v_j\}$ in the rth sector are distributed as

$$v_j \sim \begin{cases} 1, & j \in [1, r], \\ -1, & j \in [r+1, \lambda]. \end{cases} \qquad (2.320)$$

Figure 2.26 illustrates the distribution. We calculate the contribution from the rth sector (2.320) to the asymptotics (2.319). Introducing variables

$$\delta v_j \sim \begin{cases} v_j - 1, & j \in [1, r] \\ v_j + 1, & j \in [r+1, \lambda] \end{cases}, \qquad (2.321)$$

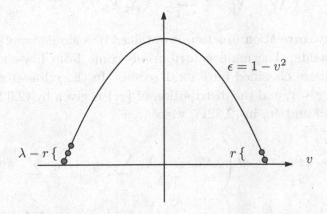

Fig. 2.26. Graphical description of the distribution of the $\{v_j\}$ in rth sector of low-energy states for $\lambda = 5$ and $r = 2$. Solid circles represent quasi-holes.

we rewrite as follows:

$$Qx - Et = \sum_{j=1}^{\lambda} (\pi dv_j x - \pi^2 \lambda d^2 v_j^2 t)$$

$$= (2r - \lambda)\pi dx + \mu t + \pi X_{\mathrm{L}} d \sum_{j=1}^{r} \delta v_j + \pi X_{\mathrm{R}} d \sum_{j=r+1}^{\lambda} \delta v_j, \quad (2.322)$$

with the correction of $\mathcal{O}((\delta v_j)^2)$. Other factors in (2.199) are replaced as

$$\prod_{1 \leq i < j \leq \lambda} |v_i - v_j|^{2/\lambda} \sim \prod_{1 \leq i < j \leq r} |v_i - v_j|^{2/\lambda} \prod_{r+1 \leq i < j \leq \lambda} |v_i - v_j|^{2/\lambda},$$

$$\prod_{j=1}^{\lambda} (1 - v_j^2)^{-1+1/\lambda} \sim \prod_{j=1}^{r} (-\delta v_j)^{-1+1/\lambda} \prod_{j=r+1}^{\lambda} (\delta v_j)^{-1+1/\lambda},$$

$$\int_{-1}^{1} \mathrm{d}v_1 \cdots \int_{-1}^{1} \mathrm{d}v_\lambda \sim \prod_{j=1}^{r} \int_{-\infty}^{0} \mathrm{d}\delta v_j \prod_{j=r+1}^{\lambda} \int_{0}^{\infty} \mathrm{d}\delta v_j. \quad (2.323)$$

We change the variables as

$$\delta v_j \to \begin{cases} X_{\mathrm{L}}^{-1} \delta v_j, & j \in [1, r], \\ X_{\mathrm{R}}^{-1} \delta v_j, & j \in [r+1, \lambda]. \end{cases} \quad (2.324)$$

Then we obtain the contribution to (2.319) from the rth sector.

The asymptotic form

$$\Pi(x, t) = A \left(\frac{1}{X_{\mathrm{L}}^2} + \frac{1}{X_{\mathrm{R}}^2} \right) + \sum_{r=1}^{\lambda} A_r \left(\frac{1}{X_{\mathrm{L}} X_{\mathrm{R}}} \right)^{r^2/\lambda} \cos[2\pi drx] \quad (2.325)$$

of the density correlation function is obtained in a similar way. Here A and A_r are constants. The main contributions come from low-energy excited states, which are classified into $\lambda + 1$ sectors. In the rth sector, the quasi-particle has $w \sim 1$, and the distribution of $\{v_j\}$ is given by (2.320). In terms of $\delta w = w - 1$ and δv_j in (2.321), we obtain

$$-\omega t + Qx \sim 2\pi d\lambda x + \pi d \left(\lambda X_{\mathrm{R}} \delta w + X_{\mathrm{L}} \sum_{j=1}^{r} \delta v_j - X_{\mathrm{R}} \sum_{j=r+1}^{\lambda} \delta v_j \right)$$

$$\omega t + Qx \sim 2\pi d\lambda x + \pi d \left(\lambda X_{\mathrm{L}} \delta w + X_{\mathrm{R}} \sum_{j=1}^{r} \delta v_j - X_{\mathrm{L}} \sum_{j=r+1}^{\lambda} \delta v_j \right). \quad (2.326)$$

In the $r(\neq 0)$th sector, we rewrite each factor in (2.216) as

$$\int_1^\infty dw \sim \int_0^\infty d\delta w, \quad (w^2 - 1)^{\lambda-1} \sim (\delta w)^{\lambda-1}, \tag{2.327}$$

$$\prod_{j=1}^\lambda (v_j + w)^{-2} \sim \prod_{j=r+1}^\lambda (\delta v_j + \delta w)^{-2}, \quad \left(\sum_{j=1}^\lambda v_j + \lambda w\right) \sim 1. \tag{2.328}$$

Other factors are rewritten as (2.323). We rewrite

$$\exp(-i\omega t)\cos(Qx) = \left[\exp(-i\omega t + iQx) + \exp(-i\omega t - iQx)\right]/2,$$

and change the variables $\delta w \to X_R^{-1}\delta w$ as in (2.324). Then the contribution from the part containing $\exp(-i\omega t + iQx)$ in the rth sector is given in the form

$$\text{constant} \times \frac{\exp(i2\pi d\lambda x)}{(X_R X_L)^{r^2/\lambda}}. \tag{2.329}$$

On the other hand, the contribution from the part containing $\exp(-i\omega t - iQx)$ in the rth sector has the form

$$\text{constant} \times \frac{\exp(-i2\pi d\lambda x)}{(X_R X_L)^{r^2/\lambda}}. \tag{2.330}$$

Here we have changed the variables as $\delta w \to X_L^{-1}\delta w$ and

$$\delta v_j \to \begin{cases} X_R^{-1}\delta v_j, & j \in [1, r], \\ X_L^{-1}\delta v_j, & j \in [r+1, \lambda]. \end{cases} \tag{2.331}$$

The constant factors in (2.329) and (2.330) are common. Thus these two terms give the summand for $r \neq 0$ in (2.325). The non-oscillating term on the RHS of (2.325) can be obtained similarly. The only exception is that

$$\left(\sum_{j=1}^\lambda v_j + \lambda w\right)^2 \sim \left(\sum_{j=1}^\lambda \delta v_j + \lambda \delta w\right)^2 \tag{2.332}$$

causes a singular contribution to either $1/X_R^2$ or $1/X_L^2$. Hence we arrive at (2.325).

2.8.2 *Finite-size corrections*

In the TLL, each eigenstate is specified by a set of quantum numbers $(\Delta N, \Delta D, N^+, N^-)$. The ground state is specified by $(0, 0, 0, 0)$. ΔN is the difference in particle number between the ground state and excited states

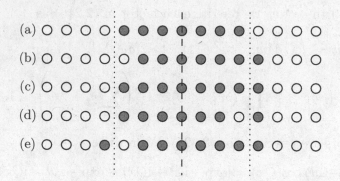

Fig. 2.27. Energy spectrum for fermions with $N = 7$. Dashed line and dotted lines represent zero momentum and Fermi points, respectively. Solid circles represent occupied one-particle states and open circles vacant one-particle states. (a) The ground state ($\Delta N = 0, \Delta D = 0, N^+ = 0, N^- = 0$), (b) a current excited state ($\Delta N = 0, \Delta D = 1, N^+ = 0, N^- = 0$), and (c) a charge excitation ($\Delta N = 1, \Delta D = 1/2, N^+ = 0, N^- = 0$). These three are primary states. (d) ($\Delta N = 0, \Delta D = 0, N^+ = 1, N^- = 0$) and (e) ($\Delta N = 0, \Delta D = 0, N^+ = 0, N^- = 1$) describe particle–hole excitations or descendant states above the ground state.

and ΔD is the number of differences between right-moving particles and left-moving particles For bosons, ΔD and ΔN satisfy

$$\Delta D = \text{integer} \quad \Delta N = \text{integer}, \tag{2.333}$$

while for fermions, we obtain

$$\Delta D = \Delta N/2 + \text{integer}. \tag{2.334}$$

A non-negative integer N^+ (N^-) is the quantum number of right (left)-moving particle–hole excitations which carry small energy and small momentum. Some examples of excited states for fermions are given in Fig. 2.27.

The excitation energy and momentum of excited states are given by

$$\Delta E = \mu \Delta N + \frac{2\pi v}{L} \left[\frac{(\Delta N)^2}{4\xi_c^2} + \xi_c^2 (\Delta D)^2 + N^+ + N^- \right] \tag{2.335}$$

and

$$Q = 2\pi d \Delta D + \frac{2\pi}{L} \left(\Delta N \Delta D + N^+ - N^- \right). \tag{2.336}$$

Here $d = N/L$ and the parameter ξ_c is called dressed charge, which describes interaction effects between original particles. The spectrum (2.335) and (2.336) of the TLL is shown in Fig. 2.28.

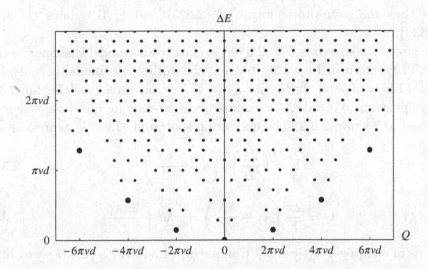

Fig. 2.28. Energy spectrum of Tomonaga–Luttinger liquid described by (2.336) and (2.335) for $\xi_c = 1/\sqrt{2}$, $N = 7$, and $\Delta N = 0$. The unit of the vertical scale is $2\pi v N/L$, and that of the horizontal scale is $2\pi N/L$. Solid circles correspond to primary states $N^+ = N^- = 0$ in conformal field theory. Dots represent descendant states $N^+ \neq 0$ or $N^- \neq 0$.

Using an analogy to free fermions, the excitations with $\Delta D = 1$ have the momentum $2\pi d = 2k_{\mathrm{F}}$ and describe the transition from left pseudo-Fermi point to the other pseudo-Fermi point. The spectrum (2.335) and (2.336) is expressed as

$$\Delta E = \frac{2\pi v}{L}\left(\Delta^+ + \Delta^-\right) + \mu\Delta N, \quad Q = \frac{2\pi}{L}\left(\Delta^+ - \Delta^-\right) + 2\pi d\Delta D \quad (2.337)$$

in terms of

$$\Delta^\pm = \Delta^\pm(\Delta N, \Delta D, N^+, N^-) = \frac{1}{2}\left(\frac{\Delta N}{2\xi_c} \pm \xi_c\Delta D\right)^2 + N^\pm. \quad (2.338)$$

The quantities (2.338) are called conformal weights of the compactified Gaussian model in CFT. The state $(\Delta N, \Delta D, 0, 0)$ corresponds to the primary state and the states $(\Delta N, \Delta D, N^+, N^-)$ with $N^+ \neq 0$ or $N^- \neq 0$ correspond to the descendant states. We see in Fig. 2.28 that excited states with common values of $\Delta N, \Delta D$ form a tower structure, which is called a conformal tower in CFT.

The dynamical correlation function of a local operator $\hat{\mathcal{O}}$ in the TLL is expressed in the following form:

$$\langle\hat{\mathcal{O}}(x,t)\hat{\mathcal{O}}(0,0)\rangle \sim \sum_{\Delta N,\Delta D,N^+,N^-} \frac{A(\Delta N, \Delta D, N^+, N^-)\exp\left[-\mathrm{i}\mu\Delta Nt + \mathrm{i}2\pi d\Delta Dx\right]}{(x - vt)^{2\Delta^+}(x + vt)^{2\Delta^-}}.$$

$$(2.339)$$

Note that the asymptotic forms of (2.325) and (2.319) have the form of (2.339).

The spectrum shown in Fig. 2.3 of the Sutherland model is similar to that in the TLL shown in Fig. 2.28. Indeed, the Sutherland model is equivalent to the TLL with $v = 2\pi d\lambda$ and $\xi_c = 1/\lambda^{1/2}$, as seen from its finite-size energy spectrum. We take, for example, a right-moving quasi-particle state discussed in Section 2.3.2. The excitation spectrum (2.94) is expressed as

$$\Delta E = \left(\frac{2\pi}{L}\right)^2 \left(\kappa_1 + \frac{\lambda N}{2}\right)^2, \tag{2.340}$$

$$Q = \frac{2\pi}{L}\left(\kappa_1 + \frac{\lambda N}{2}\right) = \pi\lambda d + \frac{2\pi\kappa_1}{L} \tag{2.341}$$

in terms of an integer $\kappa_1(\geq \lambda/2)$. Comparing (2.341) with (2.337), we obtain

$$\Delta N = 1, \quad \Delta D = \frac{\lambda}{2}, \quad \kappa_1 = \frac{\lambda}{2} + N^+ - N^-. \tag{2.342}$$

The excitation energy (2.340) for $\kappa_1 = \mathcal{O}(1)$ becomes

$$\Delta E = \pi^2\lambda^2 d^2 + \frac{2\pi v}{L}\left(\frac{\lambda}{2} + N^+ - N^-\right) + \mathcal{O}(L^{-2}), \tag{2.343}$$

with $v = 2\pi\lambda d$. In order for (2.343) to coincide with (2.337), we need $\xi_c = \lambda^{-1/2}$.

In the case where $\hat{\mathcal{O}}(x,0) = \hat{\rho}(x,0)$, excited states contribute to $\Pi(x,t)$ only if $\Delta N = 0$. Then ΔD is an integer regardless of bosons or fermions from the selection rules (2.333) and (2.334). Owing to the inversion symmetry $\Pi(x,t) = \Pi(-x,t)$, we have the relation

$$A(0, \Delta D, N^+, N^-) = A(0, -\Delta D, N^-, N^+).$$

Thus we obtain [74]

$$\begin{aligned}
\Pi(x,t) &\sim \sum_{\Delta D, N^+, N^-} \frac{A(0, \Delta D, N^+, N^-)\cos\left[2\pi d\Delta D x\right]}{(x - vt)^{(\Delta D)^2/\lambda + 2N^+}(x + vt)^{(\Delta D)^2/\lambda + 2N^-}} \\
&= \frac{A(0,0,1,0)}{(x-vt)^2} + \frac{A(0,0,0,1)}{(x+vt)^2} + \frac{A(0,1,0,0)\cos\left[2\pi dx\right]}{(x-vt)^{1/\lambda}(x+vt)^{1/\lambda}} \cdots,
\end{aligned} \tag{2.344}$$

where the sums are taken over integers. Comparing (2.325) with (2.344), we see that ΔD corresponds to the number r of quasi-holes with rapidity v_j close to unity. Further we see that the oscillating terms in (2.325) come from the primary states. The non-oscillating terms come from the descendant states whose primary state is the ground state. Although the TLL has an infinite number of primary states, only the states with $\Delta D \in [0, \lambda]$ contribute to (2.325). This selection rule is specific to the Sutherland model with integer λ.

Next we consider the asymptotic form of the hole propagator. The field operator $\hat{\psi}(0,0)$ annihilates a particle, and hence $\Delta N = -1$ for states relevant to the hole propagator. We obtain

$$G^-(x,t) \sim \sum_{\Delta D, N^+, N^-} \frac{A(-1, \Delta D, N^+, N^-) \exp\left[i2\pi d\Delta Dx\right]}{(x - vt)^{\lambda(\frac{1}{2} - \Delta D/\lambda)^2 + 2N^+} (x + vt)^{\lambda(\frac{1}{2} + \Delta D/\lambda)^2 + 2N^-}}.$$

(2.345)

For bosons, ΔD is an integer from the selection rule (2.333). We then obtain

$$G^-(x,t) = G_B^-(x,t)$$

$$\sim \frac{A(-1,0,0,0)}{(x - vt)^{\lambda/4} (x + vt)^{\lambda/4}} + \frac{A(-1,0,1,0)}{(x - vt)^{\lambda/4+2} (x + vt)^{\lambda/4}}$$

$$+ \frac{A(-1,0,0,1)}{(x - vt)^{\lambda/4} (x + vt)^{\lambda/4+2}} + \frac{A(-1,1,0,0) \exp\left[i2\pi dx\right]}{(x - vt)^{\lambda(\frac{1}{2} - 1/\lambda)^2} (x + vt)^{\lambda(\frac{1}{2} + 1/\lambda)^2}}$$

$$+ \frac{A(-1,-1,0,0) \exp\left[-i2\pi dx\right]}{(x - vt)^{\lambda(\frac{1}{2} + 1/\lambda)^2} (x + vt)^{\lambda(\frac{1}{2} - 1/\lambda)^2}} + \cdots,$$

(2.346)

which is consistent with (2.319) for even λ since $\Delta D = r - \lambda/2$ is an integer.

When particles are fermions, the selection rule (2.334) requires that ΔD is a half odd integer. The hole propagator $G^-(x,t) = G_F^-(x,t)$ then has the form

$$G^-(x,t) = G_F^-(x,t) \sim \frac{A(-1,1/2,0,0) \exp\left[i\pi dx\right]}{(x - vt)^{\lambda(1 - 1/\lambda)^2/4} (x + vt)^{\lambda(1 + 1/\lambda)^2/4}}$$

$$+ \frac{A(-1,3/2,0,0) \exp\left[i3\pi dx\right]}{(x - vt)^{\lambda(1 - 3/\lambda)^2/4} (x + vt)^{\lambda(1 + 3/\lambda)^2/4}} + \cdots.$$

(2.347)

Note that there are no non-oscillating terms. This is consistent with (2.319) for odd λ. In the fermion models, ΔD corresponds to $r - \lambda/2$ which is a half odd integer. We thus confirm the consistency of the dynamical correlation functions with the TLL.

3

Multi-component Sutherland model

The Sutherland model has a number of variants. One of them is the U(K) Sutherland model [71, 85, 86, 132]. This model describes N particles moving along a circle of perimeter L, and each particle possesses an internal degree of freedom with K possible values. This corresponds to spin with $K = 2$, and more generally a color. In the U(K) Sutherland model, all particles obey common statistics: bosonic or fermionic. We can generalize the model further. The U(K_B, K_F) Sutherland model [177] consists of bosons having K_B possible colors and fermions having K_F colors.

The multi-component Sutherland model has a degeneracy in energy levels which is described in terms of a Yangian. The Yangian is an algebra related to quantum groups [43, 44]. The Yangian is nicely realized by variants of Jack polynomials which are modified so as to conform to the internal symmetry. Elementary excitations in the multi-component Sutherland model are described in a few alternative ways: interacting bosonic or fermionic particles; or non-interacting particles obeying generalized exclusion statistics. Furthermore, the lattice models such as the Haldane–Shastry models [77, 161] and $1/r^2$ supersymmetric t–J model [119] are obtained in the strong coupling limit of U(2) and U(2,1) Sutherland models, respectively. The Sutherland models in the continuum space are much more tractable mathematically than the corresponding lattice models. Hence, the mapping to lattice models turns out to be useful to derive the explicit results on thermodynamics and dynamics in lattice models.

In the present chapter, we extend our treatment for the single-component Sutherland model in order to include the internal degrees of freedom. We shall discuss the energy spectrum, thermodynamics, and dynamical correlation functions. In the course of the discussion, we introduce non-symmetric Jack polynomials, which turn out to be the most fundamental quantity in the family of Jack polynomials.

3.1 Triangular form of Hamiltonian

We consider N quantum particles with internal degrees of freedom in the continuum one-dimensional space. The particles can either be fermions or bosons, or even their mixtures. The Fock condition for the wave function is given by

$$\Psi(\ldots, x_i, \sigma_i, \ldots, x_j, \sigma_j, \ldots)$$
$$= \begin{cases} -\Psi(\ldots, x_j, \sigma_j, \ldots, x_i, \sigma_i, \ldots), & \text{(exchange of two fermions)}, \\ \Psi(\ldots, x_j, \sigma_j, \ldots, x_i, \sigma_i, \ldots), & \text{(otherwise)}. \end{cases} \tag{3.1}$$

Here x_i and σ_i are, respectively, spatial and spin coordinates of the ith particle. When the ith particle has $\mathrm{SU}(K)$ internal degrees of freedom, σ_i takes a value in $[1, K]$. If the ith particle is a fermion and the jth one is a boson, the wave function is symmetric under their exchange according to (3.1). Note that we can fix the mutual statistics between bosons and fermions without loss of generality. In order to deal with the mixtures of fermions and bosons, we use the graded permutation operator $\tilde{P}_{ij}$. The grading means the following:

$$\tilde{P}_{ij} = \begin{cases} -P_{ij}, & \text{(spin exchange of two fermions)}, \\ P_{ij}, & \text{(otherwise)}, \end{cases} \tag{3.2}$$

where P_{ij} is the spin exchange operator:

$$P_{ij}\Psi(\ldots, x_i, \sigma_i, \ldots, x_j, \sigma_j, \ldots) = \Psi(\ldots, x_i, \sigma_j, \ldots, x_j, \sigma_i, \ldots).$$

We now introduce the coordinate exchange operator K_{ij}:

$$K_{ij}\Psi(\ldots, x_i, \sigma_i, \ldots, x_j, \sigma_j, \ldots) = \Psi(\ldots, x_j, \sigma_i, \ldots, x_i, \sigma_j, \ldots).$$

The Fock condition is then represented as

$$\tilde{P}_{ij}K_{ij} = 1 \tag{3.3}$$

for any combination of fermions and bosons. Throughout this chapter, we take the periodic boundary condition

$$\Psi(\ldots, x_i + L, \sigma_i, \ldots) = \Psi(\ldots, x_i, \sigma_i, \ldots) \tag{3.4}$$

regardless of ζ, λ, N, and the statistics of the original particles. The Sutherland model for the multi-component system is written in two different but equivalent forms in this Fock space. The first form is given by

$$H_P = -\sum_{i=1}^{N} \frac{\partial^2}{\partial x_i^2} + 2\left(\frac{\pi}{L}\right)^2 \sum_{1 \le i < j \le N} \frac{\lambda(\lambda - \zeta\tilde{P}_{ij})}{\sin^2(\pi(x_i - x_j)/L)}, \tag{3.5}$$

where $\zeta = \pm 1$ is chosen according to the statistics of the original particles. Under condition (3.1), the Hamiltonian (3.5) is equivalent to

$$H_K = -\sum_{i=1}^{N} \frac{\partial^2}{\partial x_i^2} + 2\left(\frac{\pi}{L}\right)^2 \sum_{1 \le i < j \le N} \frac{\lambda(\lambda - \zeta K_{ij})}{\sin^2\left[\pi(x_i - x_j)/L\right]}. \tag{3.6}$$

With $\zeta = 1$ in (3.6), the bosonic single-component model (2.1) is reproduced since $K_{ij} = 1$ in this case. With $\zeta = -1$, on the other hand, the fermionic single-component model (2.1) with the same parameter λ is reproduced. Therefore, Jastrow-type wave functions Ψ_0^B and Ψ_0^F, which have been defined by (2.5) and (2.15), respectively, are the eigenfunction of (3.6) with eigenvalue $E_{0,N}$ (2.12).

Following Chapter 2, we seek eigenfunctions of (3.6) of the form

$$\Psi(\{x_i, \sigma_i\}) = \begin{cases} \Phi(\{x_i, \sigma_i\})\Psi_0^B, & (\zeta = 1), \\ \Phi(\{x_i, \sigma_i\})\Psi_0^F, & (\zeta = -1). \end{cases} \tag{3.7}$$

We call Ψ_0^B and Ψ_0^F the absolute ground states, since they have the lowest energy of (3.6) if no conditions are imposed for the symmetry of wave functions and spins. The multiplying function Φ describes not only the excitations in the system, but is necessary to fulfill the Fock condition (3.1). In some cases, the periodic boundary condition also needs Φ. For example, Φ in the U(K) fermionic Sutherland model should obey the conditions

$$\Phi(\dots, x_i, \sigma_i, \dots, x_j, \sigma_j, \dots) = -\zeta\Phi(\dots, x_j, \sigma_j, \dots, x_i, \sigma_i, \dots) \tag{3.8}$$

and

$$\Phi(\dots, x_i + L, \sigma_i, \dots) = \zeta^{N-1}\Phi(\dots, x_i, \sigma_i \dots) \tag{3.9}$$

in order that Ψ satisfies the Fock condition (3.1) and the periodic boundary condition (3.4). When we have $\zeta = -1$ and N even, Φ should satisfy the anti-periodic boundary condition since the absolute ground state Ψ_0^F in this case obeys the antiperiodic boundary condition.

Let us first assume $\zeta = 1$. Following an argument similar to that in Section 2.1.1, the eigenvalue problem for Φ is derived. We note that the Hamiltonian (3.6) is written as

$$H_K = H + 2\left(\frac{\pi}{L}\right)^2 \sum_{1 \le i < j \le N} \frac{\lambda(1 - K_{ij})}{\sin^2(\pi(x_i - x_j)/L)}. \tag{3.10}$$

The first term H on the RHS is the Hamiltonian for the single-component model defined by (2.1). The second term on the RHS is rewritten as

$(2\pi/L)^2\lambda\mathcal{H}^{(3)}$ with

$$\mathcal{H}^{(3)} = -2 \sum_{1 \leq i < j \leq N} \frac{z_i z_j}{(z_i - z_j)^2}(1 - K_{ij}), \qquad (3.11)$$

in terms of $z_i = \exp(2\pi i x_i/L)$. Then we obtain

$$H_K \Phi \Psi_0^B = H\Phi\Psi_0^B + (2\pi/L)^2 \lambda \Psi_0^B \mathcal{H}^{(3)}\Phi. \qquad (3.12)$$

The first term on the RHS has been written as

$$H\Phi\Psi_0^B = \Psi_0^B \left[E_{0,N}\Phi + (2\pi/L)^2 \left(\mathcal{H}^{(1)} + \lambda\mathcal{H}^{(2)} \right)\Phi \right] \qquad (3.13)$$

in (2.1). Here $\mathcal{H}^{(1)}$ and $\mathcal{H}^{(2)}$ have been defined in (2.26). From (3.12) and (3.13), the eigenvalue problem for Φ is written as

$$\mathcal{H}\Phi \equiv \left[\mathcal{H}^{(1)} + \lambda(\mathcal{H}^{(2)} + \mathcal{H}^{(3)}) \right]\Phi = \mathcal{E}\Phi, \qquad (3.14)$$

with

$$\mathcal{E} = (L/(2\pi))^2(E - E_{0,N}). \qquad (3.15)$$

For later convenience, we rewrite $\mathcal{H}$ as [22, 99]

$$\mathcal{H} = \sum_j \left(z_j \frac{\partial}{\partial z_j} \right)^2 + \lambda(N-1) \sum_j z_j \frac{\partial}{\partial z_j} + 2\lambda \sum_{j<k} \mathcal{H}'_{jk}, \qquad (3.16)$$

with

$$\mathcal{H}'_{jk} = \frac{z_j z_k}{z_j - z_k} \left[\frac{\partial}{\partial z_j} - \frac{\partial}{\partial z_k} - \frac{1}{z_j - z_k}(1 - K_{jk}) \right]. \qquad (3.17)$$

Here we have used the relation

$$\mathcal{H}^{(2)} = (N-1) \sum_{i=1}^{N} z_i \frac{\partial}{\partial z_i} + 2\sum_{i<j} \frac{z_i z_j}{z_i - z_j} \left(\frac{\partial}{\partial z_i} - \frac{\partial}{\partial z_j} \right). \qquad (3.18)$$

Since the Hamiltonian (3.16) does not contain spin variables, we neglect the Fock condition for the moment, and look for a wave function of the form

$$\Phi(\{x_i, \sigma_i\}) = \Phi(\{x_i\})\varphi(\{\sigma_i\}). \qquad (3.19)$$

Eigenfunctions of (3.14) satisfying the Fock condition (3.1) can be recovered by symmetrization or antisymmetrization of $\Phi\varphi$ in (3.19) with respect to particle exchanges.

Since Φ obeys the periodic boundary condition, we take plane waves as a basis,

$$\phi_\eta = z_1^{\eta_1} z_2^{\eta_2} \ldots z_N^{\eta_N}, \qquad (3.20)$$

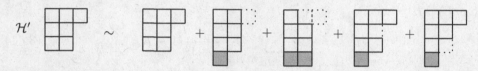

Fig. 3.1. Graphical representation of (3.23). The action of $\mathcal{H}'$ on ϕ_η generates $\phi_{\eta'}$ only if η' is obtained by moving a square (squares) from a row j to another row k. The dotted squares and shaded ones represent removed and added squares, respectively.

where $\eta = (\eta_1, \eta_2, \ldots, \eta_N)$ is a set of integers. Note that we are dealing with non-symmetric functions. If every η_i is non-negative, the set η is called a *composition*. A *partition*, which we have used extensively in Chapter 2, is a special case of a composition with the condition $\eta_1 \geq \eta_2 \geq \cdots \geq \eta_N$.

The action of $\mathcal{H}$ on ϕ_η is given by

$$\mathcal{H}\phi_\eta = \sum_{i=1}^{N} \left[\eta_i^2 + \lambda(N-1)\eta_i \right] \phi_\eta + 2\lambda \sum_{j<k} \mathcal{H}'_{jk}\phi_\eta. \qquad (3.21)$$

After some calculation, we obtain

$$\mathcal{H}'_{jk} z_j^{\eta_j} z_k^{\eta_k}$$

$$= \begin{cases} -\eta_k z_j^{\eta_j} z_k^{\eta_k} + \displaystyle\sum_{l=1}^{\eta_j - \eta_k - 1} (\eta_j - \eta_k - l)\, z_j^{\eta_j - l} z_k^{\eta_k + l}, & \eta_k + 2 \leq \eta_j, \\[2ex] -\eta_j z_j^{\eta_j} z_k^{\eta_k} + \displaystyle\sum_{l=1}^{\eta_k - \eta_l - 1} (\eta_k - \eta_l - l)\, z_j^{\eta_j + l} z_k^{\eta_k - l}, & \eta_j + 2 \leq \eta_k, \\[2ex] -\eta_k z_j^{\eta_j} z_k^{\eta_k}, & \eta_j = \eta_k, \eta_k + 1, \\[1ex] -\eta_j z_j^{\eta_j} z_k^{\eta_k}, & \eta_j = \eta_k - 1. \end{cases}$$

$$(3.22)$$

Let us consider the case with $N = 4$ and $\eta = (3, 2, 2, 0)$, for example. Writing $\mathcal{H}' = \sum_{j<k} \mathcal{H}'_{jk}$, we obtain

$$\mathcal{H}'\phi_{3,2,2,0} = -6\phi_{3,2,2,0} + 2\phi_{2,2,2,1} + \phi_{1,2,2,2} + \phi_{3,1,2,1} + \phi_{3,2,1,1}, \qquad (3.23)$$

which is described graphically in Fig. 3.1.

The action of $\mathcal{H}'$ on ϕ_η generates $\phi_{\eta'}$ only if η' is obtained from η by squeezing, as in the case (2.39) with $K_{ij} = 1$. Then the operation of $\mathcal{H}$ (3.16) on the basis (3.20) can be arranged in the form of a triangular matrix, which displays eigenvalues as its diagonal elements. An example of the action of $\mathcal{H}$

for the three-particle system is given by

$$
\mathcal{H}\begin{pmatrix} \phi_{2,0,0} \\ \phi_{1,1,0} \\ \phi_{1,0,1} \end{pmatrix} = \begin{pmatrix} 4+4\lambda & 2\lambda & 2\lambda \\ 0 & 2+4\lambda & 0 \\ 0 & 0 & 2+4\lambda \end{pmatrix} \begin{pmatrix} \phi_{2,0,0} \\ \phi_{1,1,0} \\ \phi_{1,0,1} \end{pmatrix} \tag{3.24}
$$

using (3.21) and (3.22). Following the method in Section 2.5, we obtain

$$
\Phi_{2,0,0} = \phi_{2,0,0} + \lambda\left(\phi_{1,1,0} + \phi_{1,0,1}\right),
$$
$$
\Phi_{1,1,0} = \phi_{1,1,0},
$$
$$
\Phi_{1,0,1} = \phi_{1,0,1}
$$

as eigenfunctions of $\mathcal{H}$. The eigenvalues $4+4\lambda, 2+4\lambda, 2+4\lambda$ are obtained from (3.24). Generally, we define Φ_η as the eigenfunction of $\mathcal{H}$ of the form

$$
\Phi_\eta = \phi_\eta + \sum_{\eta'} a_{\eta'}\phi_{\eta'}, \tag{3.25}
$$

where the summation is taken over η' generated by successive squeezing of η. The function (3.25) yields an eigenfunction of (3.6) through (3.19).

In order to describe the diagonal elements of the action of $\mathcal{H}'_{jk}$ in (3.22), we introduce the following notation:

$$
k'_i = \sharp\{j \in \{1,\ldots,i-1\}|\eta_j \geq \eta_i\}, \quad k''_i = \sharp\{j \in \{i+1,\ldots,N\}|\eta_j > \eta_i\}, \tag{3.26}
$$

where $\sharp\{\cdot\}$ represents a number of elements in the set $\{\cdot\}$. For example, k'_i and k''_i are given by

$$
(k'_1, k'_2, k'_3) = (0,1,2), \quad (k''_1, k''_2, k''_3) = (0,0,0), \quad \text{for} \quad \eta = (2,1,0)
$$

and

$$
(k'_1, k'_2, k'_3) = (0,0,2), \quad (k''_1, k''_2, k''_3) = (1,0,0), \quad \text{for} \quad \eta = (1,2,0).
$$

The sum $k'_i + k''_i$ represents the ranking of η_i when counted from the longest row.

Let us now express the eigenvalue of (3.14) as

$$
\mathcal{E} = \sum_{i=1}^{N} \eta_i^2 + \lambda \sum_{i=1}^{N} \left(N - 1 - 2k'_i - 2k''_i\right)\eta_i. \tag{3.27}
$$

The energy (3.27) for another composition $\tilde{\eta}$ is the same as that for η, provided $\tilde{\eta}$ can be obtained by sorting η. For example, we can easily confirm that the sum

$$
\sum_{i=1}^{3}(k'_i + k''_i)\eta_i
$$

takes the same value ($= 1$) for both $\eta = (2,1,0)$ and $(1,2,0)$. Let

$$\eta^+ = (\eta_1^+, \eta_2^+, \ldots, \eta_N^+) \in \mathcal{L}_N^+$$

be a Galilean-shifted partition $\eta_1^+ \geq \eta_2^+ \geq \cdots \geq \eta_N^+$, which is the rearrangement of η. Then we obtain

$$\mathcal{E} = \sum_{i=1}^{N} \left(\eta_i^+\right)^2 + \lambda \sum_{i=1}^{N} (N+1-2i)\,\eta_i^+. \tag{3.28}$$

This expression is the same as that for the single-component Sutherland model (2.62). The argument from (3.16) to (3.28) is applicable when $\zeta = 1$. When $\zeta = -1$, the argument is still applicable with the following modifications:

- We replace the absolute ground state Ψ_0^{B} by Ψ_0^{F} in (3.16). Then we obtain the same $\mathcal{H}$ given by (3.16).
- When N is even, η_i should be taken to be a half odd integer.

From the result in the present subsection, we observe the following:

(i) There is a one-to-one correspondence between Φ_η and ϕ_η.
(ii) Both Φ_η and ϕ_η are invariant under the action of $K_{i,i+1}$ when $\eta_i = \eta_{i+1}$.

Owing to these properties, we can construct the eigenstates of the multi-component Sutherland model from Φ_η in a way similar to the construction of eigenstates of multi-component free particles from ϕ_η. The procedure will be shown in the next section.

3.2 Energy spectrum of multi-component fermionic model

3.2.1 *Eigenstates of identical particles*

Now we consider the solution of (3.14) satisfying the Fock condition (3.1). First we consider the case $\zeta = 1$, where Φ satisfies the fermionic Fock condition. We introduce a basis of the one-particle spin wave function

$$v_\alpha(\sigma) = \delta_{\alpha\sigma}, \quad \alpha \in [1, K]. \tag{3.29}$$

A basis of the N-particle spin wave function is introduced as

$$\begin{aligned} v_\alpha(\{\sigma_i\}) &= v_\alpha(\sigma_1, \ldots, \sigma_N) \\ &= v_{\alpha_1}(\sigma_1) v_{\alpha_2}(\sigma_2) \cdots v_{\alpha_N}(\sigma_N), \end{aligned} \tag{3.30}$$

with $\alpha \in [1, K]^N$.

Let us start from the case with $\lambda = 0$, i.e., U(K) free fermions. The energy eigenfunction is given by

$$\text{Asym } \phi_\kappa v_\alpha, \tag{3.31}$$

where "Asym" represents the antisymmetrization with respect to particle exchange. Namely, we define

$$\text{Asym } \Psi(z_1, \sigma_1, \ldots, z_N, \sigma_N)$$
$$= \sum_{p \in S_N} \text{sgn}(p)\Psi(z_{p(1)}, \sigma_{p(1)}, \ldots, z_{p(N)}, \sigma_{p(N)}), \tag{3.32}$$

where $\text{sgn}(p)$ denotes the sign of the permutation P in the symmetry group S_N. In the U(K) free fermions, a momentum state accommodates at most K particles. Hence κ in (3.31) is restricted to the elements of $\mathcal{L}_{N,K}^+$, which is defined as

$$\mathcal{L}_{N,K}^+ = \left\{ \kappa = (\kappa_1, \kappa_2, \ldots, \kappa_N) \in \mathcal{L}_N^+ | \sharp \{\kappa_i \mid \kappa_i = \forall s\} \leq K \right\}. \tag{3.33}$$

In $\mathcal{L}_{N,K}^+$, at most K elements can take a common value. For a given set of momentum $\kappa \in \mathcal{L}_{N,K}^+$, each spin configuration is specified by the element of W_κ defined as

$$W_\kappa = \left\{ \alpha = (\alpha_1, \ldots, \alpha_N) \in [1, K]^N | \alpha_i < \alpha_{i+1} \text{ if } \kappa_i = \kappa_{i+1} \right\}. \tag{3.34}$$

For $N = 3$ and $\kappa = (\kappa_1, \kappa_2, \kappa_3)$ satisfying $\kappa_1 = \kappa_2 > \kappa_3$, W_κ is given by

$$(1, 2, 1), \ (1, 2, 2) \tag{3.35}$$

for $K = 2$ and

$$\begin{array}{lll}
(1, 2, 1), & (1, 2, 2), & (1, 2, 3), \\
(1, 3, 1), & (1, 3, 2), & (1, 3, 3), \\
(2, 3, 1), & (2, 3, 2), & (2, 3, 3)
\end{array} \tag{3.36}$$

for $K = 3$, respectively.

For non-negative λ (and $\zeta = 1$), the eigenfunction Φ of (3.14) satisfying (3.8) is then given by

$$\text{Asym } \Phi_\kappa v_\alpha, \quad \kappa \in \mathcal{L}_{N,K}^+ \text{ and } \alpha \in W_\kappa. \tag{3.37}$$

The corresponding eigenvalue $\mathcal{E}$ is given by

$$\mathcal{E} = \sum_{i=1}^{N} (\kappa_i)^2 + \lambda \sum_{i=1}^{N} (N + 1 - 2i)\kappa_i, \tag{3.38}$$

with use of (3.28). Note that (3.38) has the same form as (2.63) for the single-component model, and is rewritten as

$$\mathcal{E} = \sum_{i=1}^{N} (\kappa_i)^2 + \frac{\lambda}{2} \sum_{i=1}^{N} \sum_{j=1}^{N} |\kappa_i - \kappa_j|, \tag{3.39}$$

using (2.61). The energy spectrum of the U(K) model is, however, different from that of the single-component model in the following sense. First, the set of momenta $\kappa = (\kappa_1, \kappa_2, \ldots, \kappa_N)$ in (3.38) and (3.39) is restricted to $\mathcal{L}_{N,K}^+$ while each eigenstate of the single-component model is specified uniquely by κ in $\mathcal{L}_N^+$. Second, for a given $\kappa \in \mathcal{L}_{N,K}^+$, various spin configurations are allowed. The corresponding degeneracy is given by $\sharp W_\kappa$.

Each state specified by $(\kappa, \alpha) \in (\mathcal{L}_{N,K}^+, W_\kappa)$ is uniquely specified also by the momentum distribution function of particles with spin $\sigma (= 1, \ldots, K)$:

$$\nu_\sigma(\kappa) = \sum_{i=1}^{N} \delta(\sigma, \sigma_i) \delta(\kappa, \eta_i), \tag{3.40}$$

which is either 1 or 0. We can rewrite (3.28) as

$$\mathcal{E} = \sum_{\kappa=-\infty}^{\infty} \kappa^2 \nu(\kappa) + \frac{\lambda}{2} \sum_{\kappa=-\infty}^{\infty} \sum_{\kappa'=-\infty}^{\infty} |\kappa - \kappa'| \nu(\kappa) \nu(\kappa'), \tag{3.41}$$

in terms of the component sum

$$\nu(\kappa) = \sum_{\sigma=1}^{K} \nu_\sigma(\kappa). \tag{3.42}$$

Namely, the eigenenergy depends on $\nu_\sigma(\kappa)$ only through $\nu(\kappa)$. Thus the eigenenergy (3.41) of the U(K) fermionic Sutherland model has the same form as (2.125) for the single-component bosonic Sutherland model. In the present fermionic model, however, $\nu(\kappa)$ takes an integer value among $0, 1, 2, \ldots, K$, while in the single-component bosonic Sutherland model, $\nu(\kappa)$ takes arbitrary non-negative integer values.

It is often useful to specify each eigenfunction of the U(K) fermionic Sutherland model with the number N_σ of particles with spin σ fixed. We define a subset $\mathcal{L}_N^{+>}$ of $\mathcal{L}_N^+$ as

$$\mathcal{L}_N^{+>} \equiv \{\kappa = (\kappa_1, \ldots, \kappa_N) \in \mathcal{L}_N^+ | \kappa_1 > \kappa_2 > \cdots > \kappa_N\}. \tag{3.43}$$

Each eigenfunction of the U(K) fermionic Sutherland model is specified by the set of momenta of particles with spin (color) σ

$$\kappa^{(\sigma)} = (\kappa_1^{(\sigma)}, \ldots, \kappa_{N_\sigma}^{(\sigma)}) \in \mathcal{L}_{N_\sigma}^{+>}. \tag{3.44}$$

The wave function of the state is given by

$$\text{Asym } \Phi_\kappa v_\alpha, \tag{3.45}$$

with

$$\kappa = (\kappa^{(1)}, \ldots, \kappa^{(K)})$$
$$= (\kappa_1^{(1)}, \ldots, \kappa_{N_1}^{(1)}, \kappa_1^{(2)}, \ldots, \kappa_{N_2}^{(2)}, \ldots, \kappa_1^{(K)}, \ldots, \kappa_{N_K}^{(K)})$$
$$\in \mathcal{L}_{N_1}^{+>} \otimes \cdots \otimes \mathcal{L}_{N_K}^{+>}, \tag{3.46}$$

$$\alpha = (\overbrace{1, \ldots, 1}^{N_1}, \overbrace{2, \ldots, 2}^{N_2}, \ldots, \overbrace{K, \ldots, K}^{N_K}). \tag{3.47}$$

In summary, the eigenstates of the $U(K)$ fermionic Sutherland model with $\zeta = 1$ are specified in the following three equivalent ways:

(i) (κ, α), $\quad \kappa \in \mathcal{L}_{N,K}^+$, $\quad \alpha \in W_\kappa$;
(ii) $\{\nu_\sigma(\kappa)\}$, $\quad \kappa \in \mathbf{Z}$, $\quad \sigma \in [1, K]$, $\quad \nu_\sigma(\kappa) = 0, 1$;
(iii) $(N_1, N_2, \ldots, N_K, \kappa)$, $\quad \kappa \in \mathcal{L}_{N_1}^{+>} \otimes \cdots \otimes \mathcal{L}_{N_K}^{+>}$.

Each description of an eigenstate has advantages: (i) it is useful for calculation of eigenenergy, and hence for discussion of the spectrum of elementary excitations in Section 3.4. We use (i) to construct an orthogonal set of eigenstates (Yangian Gelfand–Zetlin basis) in Chapter 8. On the other hand, (ii) it is useful to discuss thermodynamic properties at finite temperatures as shown in Section 3.5. Finally, (iii) it is useful to derive the wave function of the ground state, and is related to the U(2) Jack polynomial, which will appear in Section 3.6.2. We use (iii) to calculate the dynamical correlation functions presented in Section 3.7.

3.2.2 Wave function of ground state

Let us construct the wave function of the ground state. Under the condition

$$\sum_\kappa \nu_\sigma(\kappa) = N_\sigma \tag{3.48}$$

with odd integers N_σ, we take the eigenstate corresponding to

$$\nu_\sigma(\kappa) = \theta(N_\sigma/2 - |\kappa|), \tag{3.49}$$

where $\theta(\kappa)$ is the step function

$$\theta(\kappa) = \begin{cases} 1, & \kappa \geq 0, \\ 0, & \kappa < 0. \end{cases}$$

It is obvious that (3.49) yields the minimum for both the first and second terms in (3.41). Thus, (3.49) describes the ground state of the U(K) Sutherland model as well as the U(K) free fermion under a given set of $\{N_\sigma\}$. The wave function of the state (3.49) is given by (3.45), with

$$\kappa^{(\sigma)} = \left(\frac{N_\sigma - 1}{2}, \frac{N_\sigma - 3}{2}, \cdots, \frac{-N_\sigma + 3}{2}, \frac{-N_\sigma + 1}{2} \right) \in \mathcal{L}_{N_\sigma}^{+>} \tag{3.50}$$

and α given by (3.47). We consider the wave function of the ground state. As an example, we take a simple case $K = 2$, $N = 6$, $N_1 = N_2 = 3$. By squeezing $\kappa = (\kappa^{(1)}, \kappa^{(2)}) = (1, 0, -1, 1, 0, -1)$, the following are generated:

$$\eta' = (0, 0, 0, 1, 0, -1), \quad (1, 0, 0, 0, 0, -1),$$
$$(0, 0, -1, 1, 0, 0), \quad (1, 0, -1, 0, 0, 0).$$

The second squeezing generates $\eta'' = (0, 0, 0, 0, 0)$. Therefore, the function Φ_κ is in the form

$$\Phi_\kappa = \phi_\kappa + \left(\sum_{\eta'} a_{\eta'} \phi_{\eta'} \right) + a_{\eta''} \phi_{\eta''}. \tag{3.51}$$

We then multiply Φ_κ by the spin function

$$v_\alpha(\sigma_1, \sigma_2, \sigma_3, \sigma_4, \sigma_5, \sigma_6), \quad \alpha = (1, 1, 1, 2, 2, 2)$$

and antisymmetrize the resultant wave function. We then see

$$\text{Asym } \phi_{\eta'} v_\alpha = 0, \quad \text{Asym } \phi_{\eta''} v_\alpha = 0, \tag{3.52}$$

because η' and η'' involve the same momentum in the same spin. Hence we obtain

$$\text{Asym } \Phi_\kappa v_\alpha = \text{Asym } \phi_\kappa v_\alpha. \tag{3.53}$$

Namely, in the antisymmetrization of $\Phi_\kappa v_\alpha$, all terms but ϕ_κ vanish.

The property illustrated above holds for a general case of $N_\sigma \equiv N(\sigma)$. We define the sets

$$I_1 = [1, 2, \ldots, N(\sigma_1)],$$
$$I_2 = [N(\sigma_1) + 1, N(\sigma_1) + 2, \ldots, N(\sigma_1) + N(\sigma_2)], \ldots,$$
$$I_K = \left[\sum_i^{K-1} N(\sigma_i) + 1, \ldots, N \right]. \tag{3.54}$$

We write Φ_κ as the sum of the ϕ_κ and the squeezed terms $\phi_{\eta'}$. The latter always contains a factor $z_i^{\eta'_i} z_j^{\eta'_j}$ with $\eta'_i = \eta'_j$ for $i, j \in I_\sigma$. Then all $\phi_{\eta'}$ vanish by antisymmetrization of spatial coordinates within each spin species σ. After antisymmetrization within each species, the wave function $\phi_\kappa v_\alpha$ becomes

$$\prod_{\sigma=1}^{K} \prod_{i \in I_\sigma} z_i^{-(N_\sigma-1)/2} \prod_{(i<j) \in I_\sigma} (z_i - z_j), \tag{3.55}$$

which is equivalently written as

$$\Phi_{0,\mathrm{F}} = \prod_{1 \le i \le N} z_i^{-[N(\sigma_i)-1]/2} \prod_{1 \le i < j \le N} (z_i - z_j)^{\delta(\sigma_i, \sigma_j)}. \tag{3.56}$$

This wave function is antisymmetric against particle exchange with the same spin, and symmetric against exchange with different spins. Thus, we multiply (3.56) by a spin function

$$\chi_0 = P(\{N_\sigma\}, \{\sigma_i\}) \prod_{1 \le i \le N} \exp\left[\frac{\mathrm{i}\pi}{2} \mathrm{sgn}(\sigma_i - \sigma_j)\right], \tag{3.57}$$

where the first factor is the projection to a given spin polarization N_σ given by

$$P(\{N_\sigma\}, \{\sigma_i\}) = \prod_{\sigma=1}^{K} \delta\left(N_\sigma, \sum_{i=1}^{N} \delta(\sigma, \sigma_i)\right). \tag{3.58}$$

Note that χ_0 is antisymmetric for $\sigma_i \ne \sigma_j$, and symmetric for $\sigma_i = \sigma_j$. As a result, we obtain the fermionic ground state [71, 99]

$$\Psi_{\mathrm{g,F}} = \Psi_0^{\mathrm{B}} \Phi_{0,\mathrm{F}} \chi_0. \tag{3.59}$$

3.2.3 Eigenstates with bosonic Fock condition

When $\zeta = -1$, the Fock condition (3.8) for Φ is bosonic. In the present case, the eigenstates are specified in a way similar to those of the U(K) free bosons. Then Φ obeys the periodic (antiperiodic) boundary condition when N is odd (even).

When N is odd, each eigenstate of the U(K) fermionic Sutherland model with $\zeta = -1$ can be specified by (κ, α) with $\kappa \in \mathcal{L}_N^+$ and $\alpha \in W_\kappa^{\mathrm{B}}$, where

$$W_\kappa^{\mathrm{B}} = \left\{\alpha = (\alpha_1, \ldots, \alpha_N) \in [1, K]^N | \alpha_i \le \alpha_{i+1} \text{ if } \kappa_i = \kappa_{i+1}\right\}. \tag{3.60}$$

For $N = 3$ and $\kappa = (\kappa_1, \kappa_2, \kappa_3)$ with $\kappa_1 = \kappa_2 > \kappa_3$, W_κ^{B} is given by

$$(1,1,1), (1,1,2), (1,2,1), (1,2,2), (2,2,1), (2,2,2) \tag{3.61}$$

for $K = 2$ and

$$
\begin{array}{lll}
(1,1,1), & (1,1,2), & (1,1,3), \\
(1,2,1), & (1,2,2), & (1,2,3), \\
(1,3,1), & (1,3,2), & (1,3,3), \\
(2,2,1), & (2,2,2), & (2,2,3), \\
(2,3,1), & (2,3,2), & (2,3,3), \\
(3,3,1), & (3,3,2), & (3,3,3)
\end{array}
\tag{3.62}
$$

for $K = 3$, respectively.

The wave function of a state specified by $(\kappa, \alpha) \in \mathcal{L}_N^+ \otimes W_\kappa^B$ is given by

$$
\Phi = \mathrm{Sym}\ (\Phi_\kappa(\{z_i\}) v_\alpha(\{\sigma_i\})),
\tag{3.63}
$$

where "Sym" is the symmetrization operator defined by

$$
\mathrm{Sym}\ \Phi(z_1, \sigma_1, \ldots, z_N, \sigma_N) = \sum_{P \in S_N} \Phi(z_{p(1)}, \sigma_{p(1)}, \ldots, z_{p(N)}, \sigma_{p(N)}).
\tag{3.64}
$$

The eigenenergy is given by (3.38) or (3.39). The degeneracy for a given κ is the number of elements in W_κ^B.

The eigenstates of the U(K) fermionic model with $\zeta = -1$ are specified in another way. There is a one-to-one correspondence between $(\kappa, \alpha) \in \mathcal{L}_N^+ \otimes W_\kappa^B$ and the momentum distribution $\{\nu_\sigma(\kappa)\}$ for $\kappa \in \mathbf{Z}$, $\sigma \in [1, K]$, and $\nu_\sigma(\kappa) = 0, 1, 2 \ldots$ In terms of $\{\nu_\sigma(\kappa)\}$, the eigenenergy of (3.63) for (3.6) is given by the same expression as (3.41), but the momentum distribution function $\nu_\sigma(\kappa)$ with species σ can take arbitrary integer values.

With the number of particles N_σ fixed for $\sigma \in [1, K]$, the eigenstates are uniquely specified by

$$
\kappa = (\kappa^{(1)}, \ldots, \kappa^{(K)}), \quad \kappa^{(\sigma)} \in \mathcal{L}_{N_\sigma}^+.
\tag{3.65}
$$

The wave function of the eigenstate is given by

$$
\Phi = \mathrm{Sym}\ (\Phi_\kappa(\{z_i\}) v_\alpha(\{\sigma_i\})),
\tag{3.66}
$$

with $\alpha = (\overbrace{1, \ldots, 1}^{N_1}, \overbrace{2, \ldots, 2}^{N_2}, \ldots, \overbrace{K, \ldots, K}^{N_K})$. The ground state is specified by $\kappa = (0, \ldots, 0) \in \mathbf{Z}^N$ or

$$
\nu_\sigma(\kappa) = N_\sigma \delta(\kappa, 0), \quad \sigma \in [1, K].
\tag{3.67}
$$

So far we have considered the case with N being odd. As a result of the antiperiodic boundary condition on Φ, the eigenstates for even N are described in the following three ways:

(i) The wave function is given by (3.63) and the eigenvalue $\mathcal{E}$ for (3.6) is given by (3.38). Here $\kappa + 1/2 = (\kappa_1 + 1/2, \ldots, \kappa_N + 1/2)$ belongs to $\mathcal{L}_N^+$ and $\alpha \in W_\kappa$.

(ii) Each eigenstate is specified by $\{\nu_\sigma(\kappa)\}$ with $\kappa \in 1/2 + \mathbf{Z}$, $\sigma \in [1, K]$ and $\nu_\sigma(\kappa) = 0, 1, 2, \ldots$ The eigenenergy for $\mathcal{H}$ (3.6) is given by

$$\mathcal{E} = \sum_{\kappa=-\infty}^{\infty} \kappa^2 \nu(\kappa) + \frac{\lambda}{2} \sum_{\kappa=-\infty}^{\infty} \sum_{\kappa'=-\infty}^{\infty} |\kappa - \kappa'| \nu(\kappa) \nu(\kappa').$$

Here κ is a half-odd integer.

(iii) The wave function is given by (3.66) with $\kappa + 1/2 \in \mathcal{L}_{N_1}^+ \otimes \cdots \otimes \mathcal{L}_{N_K}^+$.

In the ground state, all particles have momentum either $1/2$ or $-1/2$.

3.3 Energy spectrum with most general internal symmetry

In the case where particles are $U(K)$ bosonic models, the eigenstates can be discussed in a way similar to that in Section 3.2. In the present section, therefore, we consider the most general $U(K_B, K_F)$ model.

First we consider the model with $\zeta = 1$, where the absolute ground state is Ψ_0^B and Φ obeys the Fock condition

$$\Phi(\ldots, x_i, \sigma_i, \ldots, x_j, \sigma_j, \ldots)$$
$$= \begin{cases} -\Phi(\ldots, x_j, \sigma_j, \ldots, x_i, \sigma_i, \ldots), & \text{(exchange of two fermions)}, \\ \Phi(\ldots, x_j, \sigma_j, \ldots, x_i, \sigma_i, \ldots), & \text{(otherwise)} \end{cases} \quad (3.68)$$

and the periodic boundary condition. The energy eigenstate of $U(K_B, K_F)$ is specified uniquely by

$$\{\nu_\sigma^B(\kappa)\}, \text{ for } \kappa \in \mathbf{Z}, \quad \sigma \in [1, K_B], \quad \nu_\sigma^B(\kappa) = 0, 1, 2, \ldots \quad (3.69)$$

and

$$\{\nu_\sigma^F(\kappa)\}, \text{ for } \kappa \in \mathbf{Z}, \quad \sigma \in [1, K_F], \quad \nu_\sigma^F(\kappa) = 0, 1. \quad (3.70)$$

The eigenvalue for (3.6) is given by (3.41) but $\nu(\kappa)$ should read

$$\nu(\kappa) = \sum_{\sigma=1}^{K_B} \nu_\sigma^B(\kappa) + \sum_{\sigma=1}^{K_F} \nu_\sigma^F(\kappa). \quad (3.71)$$

Alternatively, we can specify the energy eigenstate of the $U(K_B, K_F)$ model by the set of momenta for each species

$$\kappa = (\kappa^{(B,1)}, \ldots, \kappa^{(B,K_B)}, \kappa^{(F,1)}, \ldots, \kappa^{(F,K_F)})$$

with

$$\kappa^{(B,\sigma)} \in \mathcal{L}^+_{N(B,\sigma)}, \quad \kappa^{(F,\sigma)} \in \mathcal{L}^{+>}_{N(F,\sigma)}. \tag{3.72}$$

Here $N(X,\sigma) = N^X_\sigma$ for $X = B, F$.

In order to construct the wave function that is symmetric with respect to exchange between bosons and fermions, we introduce a one-particle wave function for an internal symmetry as

$$w_B(\tau) = \begin{cases} 1, & (\tau = 0), \\ 0, & (\tau = 1), \end{cases} \quad w_F(\tau) = \begin{cases} 0, & (\tau = 0), \\ 1, & (\tau = 1), \end{cases} \tag{3.73}$$

where $\tau = 0$ corresponds to a boson and $\tau = 1$ to a fermion. A basis of N-particle wave functions with internal symmetry is introduced as

$$
\begin{aligned}
w_\beta(\{\tau_i\}) &= w_\beta(\tau_1, \ldots, \tau_N) \\
&= w_{\beta_1}(\tau_1) w_{\beta_2}(\tau_2) \cdots w_{\beta_N}(\tau_N)
\end{aligned}
\tag{3.74}
$$

with $\beta \in \{B, F\}^{\otimes N}$.

The wave function satisfying the condition (3.68) is derived in two steps. First we construct the wave function $\tilde{\Phi}$ symmetric with respect to the exchange of two bosons and antisymmetric with respect to two fermions:

$$\tilde{\Phi}(\{\gamma_i\}) = \mathrm{Sym}_B \mathrm{Asym}_F \, \Phi_\kappa(\{z_i\}) v_\alpha(\{\sigma_i\}) \tag{3.75}$$

with

$$\alpha = (1^{N(B,1)}, 2^{N(B,2)}, \ldots, K_B^{N(B,K_B)}, 1^{N(F,1)}, \ldots, K_F^{N(F,K_F)}). \tag{3.76}$$

Here we have used the notation $\gamma_i = (z_i, \sigma_i)$, and $1^{N(B,1)}$ stands for $\overbrace{1, \ldots, 1}^{N(B,1)}$, for example. In (3.75), Sym_B denotes the symmetrization operator with respect to the variables γ_i for $i \in [1, N_B]$, and Asym_F donotes the antisymmetrization operator with respect to the variables γ_i for $i \in [N_B + 1, N]$.

Let I be the subset of $[1, N]$ defined as

$$I = \{(i(1), i(2), \ldots, i(N_B)) \in \mathbf{Z}^{N_B} | 1 \le i(1) < i(2) \cdots < i(N_B) \le N\} \tag{3.77}$$

and J be the complementary set of I in $[1, N]$ (i.e., $J = [1, N] \setminus I$):

$$J = \{(j(1), j(2), \ldots, j(N_F)) \in \mathbf{Z}^{N_F} | 1 \le j(1) < j(2) \cdots < j(N_F) \le N\}. \tag{3.78}$$

We define the wave function $\Phi(\{\gamma_i, \tau_i\})$ as

$$
\begin{aligned}
\Phi(\{\gamma_i, \tau_i\}) = \sum_I \tilde{\Phi} \left(\gamma_{i(1)}, \ldots, \gamma_{i(N_B)}, \gamma_{j(1)}, \ldots, \gamma_{j(N_F)} \right) \\
\times w_\beta \left(\tau_{i(1)}, \ldots, \tau_{i(N_B)}, \tau_{j(1)}, \ldots, \tau_{i(N_F)} \right)
\end{aligned}
\tag{3.79}
$$

with

$$\beta = (\overbrace{B\cdots B}^{N_B}, \overbrace{F\cdots F}^{N_F}). \tag{3.80}$$

This function has the required property

$$\Phi(\ldots, \gamma_i, 0, \ldots, \gamma_j, 1, \ldots) = \Phi(\ldots, \gamma_j, 1, \ldots, \gamma_i, 0, \ldots) \tag{3.81}$$

with respect to exchange $(\gamma_i, \tau_i = 0) \leftrightarrow (\gamma_j, \tau_j = 1)$. When N_σ^F is odd for $\sigma \in [1, K_F]$, the ground state under a given set of $(\{N_\sigma^B\}, \{N_\sigma^F\})$ is non-degenerate, and is given by

$$\nu_\sigma^B(\kappa) = N_\sigma^B \delta(\kappa, 0), \quad \nu_\sigma^F(\kappa) = \theta(N_\sigma^F/2 - |\kappa|). \tag{3.82}$$

The ground state is also specified by

$$\kappa^{(B,\sigma)} = (\overbrace{0, \ldots, 0}^{N(B,\sigma)}), \quad \text{for } \sigma \in [1, K_B] \tag{3.83}$$

and

$$\kappa^{(F,\sigma)} = \left(\frac{N_\sigma^F - 1}{2}, \frac{N_\sigma^F - 3}{2}, \cdots, \frac{-N_\sigma^F + 3}{2}, \frac{-N_\sigma^F + 1}{2}\right) \tag{3.84}$$

for $\sigma \in [1, K_F]$. It is convenient to use $\tilde{\Phi}$ given by (3.75) instead of (3.79) to express the wave function of the ground state. We change the complex spatial coordinates as

$$\xi_i = z_i, \quad i \in [1, N_B], \quad \omega_i = z_{i+N_B}, \quad i \in [1, N_F], \tag{3.85}$$

and write the absolute ground state as

$$\Psi_0^B = \prod_{1 \le j < k \le N_B} |\xi_j - \xi_k|^\lambda \prod_{1 \le j < k \le N_F} |\omega_j - \omega_k|^\lambda \prod_{j=1}^{N_B} \prod_{k=1}^{N_F} |\xi_j - \omega_k|^\lambda. \tag{3.86}$$

The corresponding ground state is obtained as

$$\Psi_{g,B+F} = \Psi_0^B \tilde{\Phi} = \Psi_0^B(\xi, \omega) \Phi_{0,F}(\omega) \chi_0(\sigma^F) P(\{N_\sigma^B\}, \{\sigma_i^B\}). \tag{3.87}$$

The eigenstate and eigenenergy of the U(K_B, K_F) Sutherland model with $\zeta = -1$ can be derived similarly. In the model with $\zeta = -1$, the absolute ground state is antisymmetric with respect to particle exchange and, as a result, the spectrum of the U(K_B, K_F) model with $\zeta = -1$ is similar to that of the U(K_F, K_B) model with $\zeta = 1$.

3.4 Elementary excitations

3.4.1 Quasi-particles

We consider the U(2) Sutherland model with fermions for even λ and $\zeta = 1$, or bosons for odd λ and $\zeta = -1$. The boundary condition on Ψ is taken to be periodic. We take $N/2 = M$ to be odd (even) for even (odd) λ, so that the ground state is non-degenerate. The set of momenta in the N-particle ground state is described by

$$\kappa_{2i-1,0} = \kappa_{2i,0} = \frac{M+1}{2} - i, \quad i \in [1, M]. \tag{3.88}$$

The distribution of κ_i is the same as that in the ground state of the free fermions with spin $1/2$.

A quasi-particle state is constructed by adding a particle outside the Fermi sea with the Fermi momenta fixed. The right-moving quasi-particle state has rapidities $(\tilde{\kappa}_1, \ldots, \tilde{\kappa}_{N+1})$ satisfying

$$\tilde{\kappa}_{i+1} = \tilde{\kappa}_{0,i}, \quad i \in [1, N], \tag{3.89}$$

where the rapidities are related to momenta through

$$\tilde{\kappa}_{i,0} = \kappa_{i,0} + \frac{\lambda}{2}(N+1-2i), \quad i \in [1, N], \tag{3.90}$$

$$\tilde{\kappa}_i = \kappa_i + \frac{\lambda}{2}(N+2-2i), \quad i \in [1, N+1]. \tag{3.91}$$

The excitation energy $\Delta E = E[\kappa] - E_{g,N} \equiv (L/2\pi)^2 \mathcal{E}$ and the momentum $Q \equiv (L/2\pi)\mathcal{Q}$ of the quasi-particle state are given by

$$\mathcal{E} = \sum_{i=1}^{N+1} \tilde{\kappa}_i^2 - \sum_{i=1}^{N} (\tilde{\kappa}_{i,0})^2 = \tilde{\kappa}_1^2, \quad \mathcal{Q} = \tilde{\kappa}_1 \tag{3.92}$$

with

$$\tilde{\kappa}_1 \geq \frac{1}{2}(M - 1 + \lambda) + \frac{\lambda}{2}N.$$

We demonstrate that the statistical interaction between quasi-particles depends on spin. Let us consider $(N + 2)$-particle states with momenta satisfying

$$\tilde{\kappa}_{i+2} = \tilde{\kappa}_{0,i}, \quad i \in [1, N],$$

which are identified with two (right-moving) quasi-particle states. When the particle with κ_i ($i = 1$ or 2) has the spin σ_i, we obtain

$$\kappa_1 \geq \kappa_2 + \delta(\sigma_1, \sigma_2).$$

From this inequality, the relation

$$\tilde{\kappa}_1 \geq \tilde{\kappa}_2 + \lambda + \delta(\sigma_1, \sigma_2) \tag{3.93}$$

follows. The statistical interaction $g^{\mathrm{p}}_{\sigma_1,\sigma_2}$ between quasi-particles is then given by

$$g^{\mathrm{p}}_{\sigma_1,\sigma_2} = \lambda + \delta(\sigma_1, \sigma_2). \tag{3.94}$$

We interpret the second term as the extra exclusion originating from the fermionic Pauli principle.

3.4.2 Quasi-holes

A quasi-hole state is obtained by removing one particle while keeping the Fermi surface fixed. We shall show in this subsection that the statistical interaction between quasi-holes is given by

$$g^{\mathrm{h}}_{\sigma,\sigma'} = \delta(\sigma, \sigma') - \frac{\lambda}{1 + 2\lambda}. \tag{3.95}$$

The result can be written in the matrix form

$$\mathbf{g}_{\mathrm{h}} \equiv \begin{pmatrix} g_{\mathrm{h}}^{\uparrow\uparrow} & g_{\mathrm{h}}^{\uparrow\downarrow} \\ g_{\mathrm{h}}^{\downarrow\uparrow} & g_{\mathrm{h}}^{\downarrow\downarrow} \end{pmatrix} = \frac{1}{2\lambda + 1} \begin{pmatrix} \lambda + 1 & -\lambda \\ -\lambda & \lambda + 1 \end{pmatrix} = \mathbf{g}_{\mathrm{p}}^{-1}, \tag{3.96}$$

where $\mathbf{g}_{\mathrm{p}}$ is the matrix form of (3.94). Hence, the duality relation $g_{\mathrm{h}} = g_{\mathrm{p}}^{-1}$ is naturally generalized to the U(2) case. Note that the single-component case has been discussed in Section 2.4.3. In the following we shall derive (3.95).

Let us consider redistribution of rapidities after the removal of a particle at $\tilde{\kappa}_{0,i}$ with $i \sim \mathcal{O}(N/2)$ from the N-particle ground state. The rapidity $\tilde{\kappa}_j$ for $j \sim 1$ or $j \sim \mathcal{O}(N)$ has the following asymptotics:

$$\tilde{\kappa}_1 = \tilde{\kappa}_{1,0}, \quad \tilde{\kappa}_2 = \tilde{\kappa}_{2,0}, \dots, \tilde{\kappa}_{N-2} = \tilde{\kappa}_{N-1,0}, \quad \tilde{\kappa}_{N-1} = \tilde{\kappa}_{N,0}. \tag{3.97}$$

Using the relation $\tilde{\kappa}_i = \kappa_i + \lambda(N/2 - i)$ and (3.88), we rewrite (3.97) as

$$\kappa_1 = \kappa_{1,0} + \frac{\lambda}{2}, \quad \kappa_2 = \kappa_{2,0} + \frac{\lambda}{2}, \dots,$$

$$\kappa_{N-2} = \kappa_{N-1,0} - \frac{\lambda}{2}, \quad \kappa_{N-1} = \kappa_{N,0} - \frac{\lambda}{2}. \tag{3.98}$$

Namely, the momentum κ_i is shifted by $\lambda/2$ if κ_i is much larger than the removed momentum, while it is shifted by $-\lambda/2$ if much smaller.

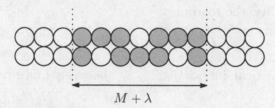

$$M + \lambda$$

Fig. 3.2. An example of the set of momenta satisfying (3.98) for $M = N/2 = 6$, $\lambda = 1$, and $\zeta = -1$ under the periodic boundary condition. The number of vacancies is $2\lambda + 1 (=3$ in the present example).

We shall show that $(N-1)$-particle states satisfying (3.98) are characterized by $2\lambda + 1$ quasi-holes. An example of the states (3.98) for $N = 12$, $\lambda = 1$, and $\zeta = -1$ is shown in Fig. 3.2. The solid circles in the upper (lower) row represent the momenta occupied by particles with spin up (down). The open circles represent unoccupied momenta. The state given by (3.98) is obtained by making $2\lambda + 1$ vacancies in the one-particle states with momentum in the range $[-(M-1+\lambda)/2, (M-1+\lambda)/2]$ with $M = N/2$.

It is convenient to study the quasi-hole states in terms of Young diagrams. Let us introduce the partition $\mu \in \Lambda_{N-1}^+$ defined by the Galilean shift

$$\mu_i = \kappa_i + \frac{1}{2}(M + \lambda + 1). \tag{3.99}$$

Then the Young diagram $D(\mu)$ for μ, which is related to (3.98) through (3.99), is constructed in the following way.

(i) Prepare a Young diagram $D(\gamma)$ of $\gamma \in \Lambda_{2(M+\lambda)}^+$ that corresponds to the singlet ground state of $N + 2\lambda$ fermions. Namely, we have here

$$\gamma_{2i-1} = \gamma_{2i} = M + \lambda + 1 - i, \quad i \in [1, M + \lambda]. \tag{3.100}$$

(ii) Select a set of rows $r(1), r(2), \ldots, r(2\lambda + 1)$ such that

$$1 \leq r(1) < r(2) < \cdots < r(2\lambda + 1) \leq 2(M + \lambda). \tag{3.101}$$

(iii) Assign the momentum ζ_j and spin σ_j to the jth quasi-hole so that

$$\zeta_j = \gamma_{r(j)} \text{ for } j \in [1, 1 + 2\lambda], \quad \sigma_j = \downarrow (\uparrow)$$

when $r(j)$ is odd (even).

(iv) Remove the $2\lambda + 1$ rows corresponding to ζ_j from $D(\gamma)$.

(v) Push the remaining rows upward so that all vacant rows are eliminated except for the lowest one. As a result, a new Young diagram $D(\mu)$ with $N - 1$ rows appears.

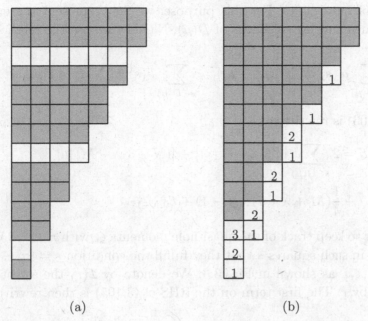

(a)   (b)

Fig. 3.3. Illustration of how to construct the Young diagram $D(\mu)$ of shifted momenta μ from $D(\gamma)$ for $M = 6$ and $\lambda = 1$. (a) Young diagram $D(\gamma)$ for γ in (3.100). Unshaded squares represent the rows to be removed. Here $(\zeta_1, \zeta_2, \zeta_3) = (6, 4, 1)$ and $(\sigma_1, \sigma_2, \sigma_3) = (\downarrow, \uparrow, \downarrow)$. (b) Young diagram $D(\mu)$ as shown by shaded squares. Inscribed squares belonging to $D(\gamma) \setminus D(\mu)$ are also shown.

Figure 3.3 shows an example of $D(\mu)$ and $D(\gamma)$. Then the quasi-hole state, or equivalently μ, is parameterized by

$$(\zeta_1, \zeta_2, \ldots, \zeta_{2\lambda+1}) \in \mathbf{Z}_{>0}^{2\lambda+1}, \quad (\sigma_1, \sigma_2, \ldots, \sigma_{2\lambda+1}),$$

where $\mathbf{Z}_{>0}$ is the set of positive integers. The relation

$$\zeta_j \geq \zeta_{j+1} + \delta(\sigma_j, \sigma_{j+1}) \tag{3.102}$$

is related to the exclusion between quasi-holes, as shown below.

The energy $E[\kappa] = (2\pi/L)^2 \mathcal{E}[\kappa]$ of the $(N - 1)$-particle state is given by

$$\mathcal{E}[\kappa] = \sum_{i=1}^{N-1} \left[\kappa_i^2 + \lambda\,(N - 2i)\,\kappa_i \right] + \mathcal{E}_{0,N-1}, \tag{3.103}$$

where $\mathcal{E}_{0,N-1} = \lambda^2 N(N-1)(N-2)/12$ is given from $\mathcal{E}_{0,N} \equiv (L/(2\pi))^2 E_{0,N}$ defined by (2.12) with replacement of N by $N - 1$. We want to rewrite $\mathcal{E}[\kappa]$

in terms of quasi-holes. For this purpose, we first represent the energy in terms of arm and leg colengths of $D(\mu)$. Namely, we use the relation

$$2 \sum_{s \in D(\mu)} a'(s) = \sum_{i=1}^{N-1} (\mu_i^2 - \mu_i), \qquad \sum_{s \in D(\mu)} l'(s) = \sum_{i=1}^{N-1} (i-1)\mu_i. \qquad (3.104)$$

Then (3.103) is rewritten as

$$\mathcal{E} = 2 \sum_{s \in D(\mu)} \left[a'(s) - \lambda l'(s) \right] + \left[(N-3)\lambda - M \right] |\mu| \qquad (3.105)$$

$$+ \frac{1}{4}(M + \lambda + 1)^2 (N-1) + \mathcal{E}_{0,N-1}. \qquad (3.106)$$

In order to keep track of the quasi-hole momenta ζ_r with $r \in [1, 2\lambda+1]$, we inscribe r in such squares s that they fulfill the condition $s = (\gamma'_j - r + 1, j)$ for $j \in [1, \zeta_r]$, as shown in Fig. 3.3. We denote by $\mathcal{I}(r)$ the set of squares inscribed by r. The first term on the RHS of (3.105) is then rewritten as

$$\sum_{s \in D(\mu)} \left[a'(s) - \lambda l'(s) \right] = \sum_{s \in D(\gamma)} \left[a'(s) - \lambda l'(s) \right] - \sum_{r=1}^{2\lambda+1} \sum_{s \in \mathcal{I}(r)} \left[a'(s) - \lambda l'(s) \right].$$
$$(3.107)$$

The first term on the RHS is calculated as

$$\sum_{s \in D(\gamma)} \left[a'(s) - \lambda l'(s) \right]$$

$$= \sum_{j=1}^{M+\lambda} (j-1)\gamma'_j - \lambda \sum_{i=1}^{2M+2\lambda} (i-1)\gamma_i$$

$$= \sum_{j=1}^{M+\lambda} (j-1)\gamma'_j - \lambda \left[\sum_{i=1}^{M+\lambda} (2i-1)\gamma_{2i} + \sum_{i=1}^{M+\lambda} (2i-2)\gamma_{2i-1} \right]$$

$$= \sum_{j=1}^{M+\lambda} (2j - 2 - 4\lambda j + 3\lambda)(M + \lambda + 1 - j)$$

$$= -\frac{1}{6}(M + \lambda)(M + \lambda + 1)(2 - 3\lambda + 4\lambda^2 - 2M + 4M\lambda), \qquad (3.108)$$

where

$$\gamma'_j = 2(M + \lambda + 1 - j) \qquad (3.109)$$

denotes the length of the jth column in $D(\gamma)$ with $j \in [1, M + \lambda]$. Each summand with respect to r in the second term on the RHS of (3.107) is rewritten as

$$\sum_{j=1}^{\zeta_r} \left[a'(s) - \lambda l'(s) \right] \Big|_{s=(\gamma'_j - r + 1, j)} = \sum_{j=1}^{\zeta_r} \left[(j-1) - \lambda(\gamma'_j - r) \right]. \qquad (3.110)$$

Substituting (3.109) into (3.110), it follows that

$$2 \sum_{s \in \mathcal{I}(r)} \left[a'(s) - \lambda l'(s) \right] = \sum_{j=1}^{\zeta_r} \left[(1 + 2\lambda)j + \lambda(r - 2M - 2\lambda - 2) - 1 \right]$$

$$= (1 + 2\lambda)\zeta_r^2 - \left[1 + 2\lambda(2M + 2\lambda + 1 - r) \right] \zeta_r. \qquad (3.111)$$

The term $((N-3)\lambda - M)|\mu|$ in (3.105) is also written as the sum of ζ_r-dependent terms and a constant:

$$[(N-3)\lambda - M] |\mu| = ((N-3)\lambda - M) \left(|\gamma| - \sum_{r=1}^{2\lambda+1} \zeta_r \right)$$

$$= [(N-3)\lambda - M] \left[(M+\lambda)(M+\lambda+1) - \sum_{r=1}^{2\lambda+1} \zeta_r \right]. \qquad (3.112)$$

We then obtain $\mathcal{E}[\kappa]$ in the form

$$\mathcal{E}[\kappa] = -(1 + 2\lambda) \sum_{r=1}^{2\lambda+1} \left[\zeta_r^2 - \left((M+\lambda+1) + \frac{2\lambda(\lambda+1-r)}{1+2\lambda} \right) \zeta_r \right] + \mathcal{E}_{\text{const}}. \qquad (3.113)$$

The constant term on the RHS is the sum of constants obtained so far, which is given by

$$\mathcal{E}_{\text{const}} = 2 \sum_{s \in D(\gamma)} \left[a'(s) - \lambda l'(s) \right] + \mathcal{E}_{0,N-1} + (M+\lambda+1)^2(N-1)/4$$

$$+ \left[(N-3)\lambda - M \right] (M+\lambda)(M+\lambda+1). \qquad (3.114)$$

The excitation energy $\Delta\mathcal{E}$ is the difference between $\mathcal{E}[\kappa]$ and the ground-state energy $\mathcal{E}_{\text{g},N} \equiv (L/(2\pi))^2 E_{\text{g},N}$. The latter in the U(2) system is given by

$$\mathcal{E}_{\text{g},N} = \sum_{i=1}^{N} \left(\kappa_{i,0} + \frac{\lambda(N+1-2i)}{2} \right)^2$$

$$= \frac{1}{6}(1+2\lambda)^2(M^2-1)M + \frac{1}{2}\lambda^2 M, \qquad (3.115)$$

which can be obtained from (3.88). By a straightforward calculation, the excitation energy of the quasi-hole state is calculated as

$$\Delta \mathcal{E} = -(1 + 2\lambda) \sum_{r=1}^{2\lambda+1} \tilde{\zeta}_r^2 + \frac{1}{3}\lambda(\lambda + 1), \tag{3.116}$$

where we have introduced the notation

$$\tilde{\zeta}_r = \zeta_r - \frac{1}{2}(M + \lambda + 1) - \frac{\lambda}{1 + 2\lambda}(\lambda + 1 - r). \tag{3.117}$$

From this expression, we interpret $\tilde{\zeta}_r$ as the rapidity of a quasi-hole. The negative curvature of the spectrum in (3.116) is a hallmark of hole-like excitations. By comparing with (2.90), the prefactor $-(1 + 2\lambda)$ in (3.116) shows that a quasi-hole has the charge $-1/(1 + 2\lambda)$ compared with the unit charge of the original particles. From (3.102) and (3.117), the inequality

$$\tilde{\zeta}_r \geq \tilde{\zeta}_{r+1} + \delta(\sigma_r, \sigma_{r+1}) - \frac{\lambda}{1 + 2\lambda} \tag{3.118}$$

follows. Hence we obtain (3.95).

3.5 Thermodynamics

The thermodynamics of multi-component Sutherland models can be formulated in a way parallel to that of the single-component model. The multi-component Sutherland model describes free particles obeying multi-component exclusion statistics. In this section we assume $\zeta = 1$, but the model with $\zeta = -1$ can be discussed in a similar way.

3.5.1 *Multi-component bosons and fermions*

Here we derive thermodynamics for the most general case: the $U(K_B, K_F)$ model [163]. First we start with the energy functional (3.41). For the $U(K_B, K_F)$ model, the distribution function should read

$$\nu(\kappa) = \sum_{\sigma=1}^{K_B} \nu_\sigma^B(\kappa) + \sum_{\sigma=1}^{K_F} \nu_\sigma^F(\kappa),$$

where $\nu_\sigma^B(\kappa) = 0, 1, 2, \ldots$ and $\nu_\sigma^F(\kappa) = 0, 1$. Then the thermodynamic potential is given by

$$\Omega = E - TS - L\sum_{\sigma=1}^{K_B} \mu_\sigma^B \nu_\sigma^B(\kappa) - L\sum_{\sigma=1}^{K_F} \mu_\sigma^F \nu_\sigma^F(\kappa). \tag{3.119}$$

Here μ_σ^B and μ_σ^F are chemical potentials of bosons or fermions with species σ. In the thermodynamic limit, it is convenient to use the momentum $k = 2\pi\kappa/L$ and then we write $\nu_\sigma^{B(F)}(\kappa) \to \nu_\sigma^{B(F)}(k)$. We obtain from (3.119) the following:

$$
\Omega[\nu(k)]/L
$$

$$
= \frac{1}{2\pi} \int_{-\infty}^{\infty} dk k^2 \nu(k) - \frac{1}{2\pi} \int_{-\infty}^{\infty} dk \left(\sum_{\sigma=1}^{K_B} \mu_\sigma^B \nu_\sigma^B(k) + \sum_{\sigma=1}^{K_F} \mu_\sigma^F \nu_\sigma^F(k) \right)
$$

$$
+ \frac{\lambda}{4\pi} \int_{-\infty}^{\infty} dk \int_{-\infty}^{\infty} dk' |k - k'| \nu(k)\nu(k') + \frac{\pi^2 \lambda^2 d^3}{3}
$$

$$
- \frac{T}{2\pi} \sum_{\sigma=1}^{K_B} \int_{-\infty}^{\infty} dk \left[\left(\nu_\sigma^B(k) + 1 \right) \ln \left(\nu_\sigma^B(k) + 1 \right) - \nu_\sigma^B(k) \ln \nu_\sigma^B(k) \right]
$$

$$
+ \frac{T}{2\pi} \sum_{\sigma=1}^{K_F} \int_{-\infty}^{\infty} dk \left[\left(1 - \nu_\sigma^F(k) \right) \ln \left(1 - \nu_\sigma^F(k) \right) + \nu_\sigma^F(k) \ln \nu_\sigma^F(k) \right], \quad (3.120)
$$

which should be minimized with respect to variations of $\nu_\sigma^B(k)$ and $\nu_\sigma^F(k)$. The solution of the stationary conditions

$$
\frac{\delta\Omega}{\delta\nu_\sigma^B(k)} = 0, \quad \frac{\delta\Omega}{\delta\nu_\sigma^F(k)} = 0
$$

is given by

$$
\nu_\sigma^B(k) = \left\{ \exp\left[\left(\epsilon(k) - \mu_\sigma^B \right) / T \right] - 1 \right\}^{-1}, \quad (3.121)
$$

$$
\nu_\sigma^F(k) = \left\{ \exp\left[\left(\epsilon(k) - \mu_\sigma^F \right) / T \right] + 1 \right\}^{-1}, \quad (3.122)
$$

where $\epsilon(k)$ is self-consistently determined by

$$
\epsilon(k) = k^2 + \lambda \int_{-\infty}^{\infty} dk' |k - k'| \nu(k') + \pi^2 \lambda^2 d^2. \quad (3.123)
$$

For a given set of $k, T, \{\mu_\sigma^B\}, \{\mu_\sigma^F\}$, we determine $\nu_\sigma^F(k)$ and $\nu_\sigma^B(k)$ by inserting $\epsilon(k)$ into (3.121) and (3.122). In Chapter 2, we have already discussed $\epsilon(k)$ in (2.135).

Let us introduce the rapidity

$$
p = \frac{1}{2} \frac{\partial\epsilon(k)}{\partial k} = k + \frac{\lambda}{2} \int_{-\infty}^{\infty} dk' \mathrm{sgn}\left(k - k' \right) \nu(k'). \quad (3.124)
$$

Further differentiation of (3.124) with k gives

$$
\frac{\partial p(k)}{\partial k} = 1 + \lambda\nu(k). \quad (3.125)
$$

By multiplying both sides of (3.125) by $v(k)$ and integrating them over k, we obtain

$$p^2 = \epsilon - T\lambda \ln \sum_{\sigma=1}^{K_B} \left(1 + v_\sigma^B\right) + T\lambda \sum_{\sigma=1}^{K_F} \ln \left(1 - v_\sigma^F\right) + c_0, \qquad (3.126)$$

where c_0 is a constant which should be determined.

In order to determine c_0, we consider the limit $k \to \infty$ of (3.123) and (3.126). Using the fact that $\nu(k)$ is an even function of k, we rewrite (3.123) for $k > 0$ as

$$\epsilon(k) = k^2 + \lambda k \int_{-k}^{k} dk' \nu(k') + 2\lambda \int_{k}^{\infty} dk' k' \nu(k') + \pi^2 \lambda^2 d^2. \qquad (3.127)$$

In the limit $k \to \infty$, the third term on the RHS of (3.127) vanishes and we obtain the asymptotic form

$$\epsilon(k) \sim k^2 + \lambda k \underbrace{\int_{-\infty}^{\infty} dk \nu(k)}_{2\pi d} + \pi^2 \lambda^2 d^2 = (k + \pi \lambda d)^2. \qquad (3.128)$$

In the limit $k \to \infty$, (3.126) becomes

$$p = k + \frac{\lambda}{2} \underbrace{\int_{-\infty}^{k} dk' \nu(k')}_{\to 2\pi d} - \frac{\lambda}{2} \underbrace{\int_{k}^{\infty} dk' \nu(k')}_{\to 0} \sim k + \pi \lambda d. \qquad (3.129)$$

From these results and $\nu(\infty) = 0$, we obtain $c_0 = 0$.

We introduce the distribution function $\rho_\sigma^B(p)$, $\rho_\sigma^F(p)$ as

$$\rho_\sigma^B(p) = \frac{v_\sigma^B(k)}{1 + \lambda v(k)}, \qquad \rho_\sigma^F(p) = \frac{v_\sigma^F(k)}{1 + \lambda v(k)}, \qquad (3.130)$$

so that the density of each species $X = B, F$ is given by

$$d_\sigma^X = \int_{-\infty}^{\infty} \frac{dk}{2\pi} v_\sigma^X(k) = \int_{-\infty}^{\infty} \frac{dp}{2\pi} \rho_\sigma^X(p), \quad v_\sigma^X(k) = \int_{-\infty}^{\infty} \frac{dp}{2\pi} \rho_\sigma^X(p). \qquad (3.131)$$

The internal energy is rewritten in terms of p as

$$E/L = \int_{-\infty}^{\infty} \frac{dp}{2\pi} p^2 \left(\sum_{\sigma=1}^{K_B} \rho_\sigma^B(p) + \sum_{\sigma=1}^{K_F} \rho_\sigma^F(p) \right), \qquad (3.132)$$

as a generalization of the single-component result given by (2.142).

Each contribution to the entropy is given by

$$S_{\rm B} = \int_{-\infty}^{\infty} \frac{dk}{2\pi} \left[\left(1 + \nu_\sigma^{\rm B}\right) \ln \left(1 + \nu_\sigma^{\rm B}\right) - \nu_\sigma^{\rm B} \ln \nu_\sigma^{\rm B} \right] \tag{3.133}$$

$$= \int_{-\infty}^{\infty} \frac{dp}{2\pi} \rho_\sigma^{\rm B} \left[\left(1 + w_\sigma^{\rm B}\right) \ln \left(1 + w_\sigma^{\rm B}\right) - w_\sigma^{\rm B} \ln w_\sigma^{\rm B} \right] \tag{3.134}$$

for bosons, and

$$S_{\rm F} = -\int_{-\infty}^{\infty} \frac{dk}{2\pi} \left[\left(1 - \nu_\sigma^{\rm F}\right) \ln \left(1 - \nu_\sigma^{\rm F}\right) + \nu_\sigma^{\rm F} \ln \nu_\sigma^{\rm F} \right] \tag{3.135}$$

$$= \int_{-\infty}^{\infty} \frac{dp}{2\pi} \rho_\sigma^{\rm F} \left[\left(1 + w_\sigma^{\rm F}\right) \ln \left(1 + w_\sigma^{\rm F}\right) - w_\sigma^{\rm B} \ln w_\sigma^{\rm F} \right] \tag{3.136}$$

for fermions. Here we have defined

$$w_\sigma^{\rm B}(p) = 1/\nu_\sigma^{\rm B}(k), \quad w_\sigma^{\rm F}(p) = 1/\nu_\sigma^{\rm F}(k) - 1. \tag{3.137}$$

If we set $K_{\rm B} = 1$ and $K_{\rm F} = 0$, the above results reduce to the thermodynamics of free particles with statistical parameter $g = \lambda$. If we set $K_{\rm B} = 0$ and $K_{\rm F} = 1$, the above results reduce to the thermodynamics of free particles with statistical parameter $g = \lambda + 1$.

3.5.2 Explicit results for U(2) anyons

So far we have regarded the system as multi-component interacting bosons and fermions. Instead, we can formulate the thermodynamics of those systems in terms of multi-component free particles obeying exclusion statistics. We use the notation α as the index of species $(\rm B, \sigma)$ or $(\rm F, \sigma)$. From (3.126) and (3.137), w_α is given as the positive real solution of the following equation:

$$\exp\left[\left(p^2 - \mu_\alpha\right)/T\right] = (1 + w_\alpha) \prod_{\alpha'} \left(\frac{w_{\alpha'}}{w_{\alpha'} + 1}\right)^{g_{\alpha'\alpha}}, \tag{3.138}$$

where the statistical parameter $g_{\alpha'\alpha}$ is given by

$$g_{\alpha'\alpha} = \begin{cases} \delta_{\alpha\alpha'} + \lambda, & \text{for } \alpha \text{ or } \alpha' \in {\rm F}, \\ \lambda, & \text{otherwise.} \end{cases} \tag{3.139}$$

The thermodynamic potential is expressed as

$$\Omega/L = -\frac{T}{2\pi} \int_{-\infty}^{\infty} dp \sum_\alpha \ln\left(1 + w_\alpha^{-1}\right). \tag{3.140}$$

Thus the thermodynamics of the multi-component Sutherland model are equivalent to those of free particles with the statistical parameter $g_{\alpha\alpha'}$. From

(3.138), (3.139), and (3.140), we can calculate all the thermodynamic quantities such as heat capacity and particle density for each species as the derivatives of the thermodynamic potential.

We now consider the system of particles with SU(2) spin. The $U(K_F = 2)$ model with coupling parameter λ, and the $U(K_B = 2)$ model with $\lambda + 1$ are equivalent in thermodynamics. We can therefore take the original particles as fermions without loss of generality. The chemical potential for particles with spin $\sigma = \pm 1$ is rewritten as

$$\mu_\uparrow = \mu - h, \quad \mu_\downarrow = \mu + h$$

in terms of magnetic field h and chemical potential μ. The thermodynamic potential is given by

$$\Omega = -T \int_{-\infty}^{\infty} \frac{dp}{2\pi} \sum_{\sigma=\uparrow,\downarrow} \ln\left(1 + w_\sigma^{-1}\right), \tag{3.141}$$

where w_σ is the real solution of the following equations:

$$\epsilon_{p\sigma}(p)/T \equiv \left(p^2 - \mu - \sigma h\right)/T$$
$$= \ln\left(1 + w_\sigma\right) - \sum_{\sigma'} g_p^{\sigma,\sigma'} \ln\left(1 + w_{\sigma'}^{-1}\right), \tag{3.142}$$

where $g_p^{\sigma,\sigma'}$ is the statistical interaction given by (3.94). Let us see the effect of internal symmetry on thermodynamics. When $h = 0$, (3.142) reduces to a single equation

$$\frac{p^2 - \mu}{\lambda T} = \frac{1 + 2\lambda}{\lambda} \ln\left(1 + w\right) - \frac{1 + \lambda}{\lambda} \ln\left(1 + w^{-1}\right), \tag{3.143}$$

for $w = w_\uparrow = w_\downarrow$. This equation describes free particles with statistical parameter $2 + 1/\lambda$ and energy $(p^2 - \mu)/\lambda$. Thus, provided that the magnetic field is much weaker than the temperature T, the thermodynamic properties are described in terms of an approximate single-component picture. Figure 3.4 shows the rapidity distribution of $\rho_\uparrow(p) = \rho_\downarrow(p)$ of the U(2) fermionic model with $\lambda = 1$ without a magnetic field ($h = 0$) for various T. We see that $\rho_\uparrow, \rho_\downarrow$ are bounded by $1/(1 + 2\lambda) = 1/3$.

Next we consider the case $h \gg T$. For that case, Fig. 3.5 shows the profile of $\rho_\uparrow(p)$ and $\rho_\downarrow(p)$. We see that there are two pseudo-Fermi momenta $p_\uparrow = \sqrt{\mu_\uparrow}$ and $p_\downarrow = \sqrt{\mu_\downarrow}$. The presence of one component renormalizes the statistics of the other component of particles. The distribution function $\rho_\downarrow(p)$ for the minor component looks similar to Fig. 3.4. This means that the distribution can be approximated by a single-component particle obeying exclusion statistics. The distribution $\rho_\uparrow(p)$ of the major component behaves

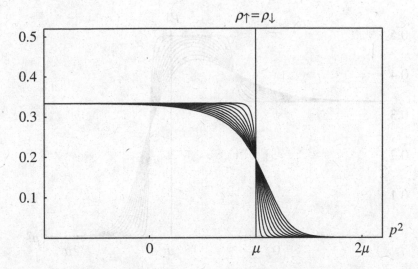

Fig. 3.4. The distribution function $\rho_\uparrow(p) = \rho_\downarrow(p)$ for the U(2) fermionic Sutherland model with $\lambda = 1$ in the absence of magnetic field. Each curve corresponds to temperatures $T/\mu = 0.01, 0.02, \ldots, 0.1$.

in an intriguing way. For energy lower than the chemical potential for the minor component, $\rho_\uparrow(p) \sim 1/(1+2\lambda)$ for sufficiently low temperatures. For energy $\mu_\downarrow \ll \epsilon \ll \mu_\uparrow$, $\rho_\uparrow(p) \sim 1/(1+\lambda)$, the exclusion effect on $\rho_\uparrow$ in an energy region becomes weak when the particles of the minor component are absent in that energy region. For sufficiently low temperatures, the distributions near $\mu_\uparrow$ and $\mu_\downarrow$ do not affect each other. For those temperature regions, a single-component free-particle description, which is different from (3.143), gives a good approximation of the thermodynamics of the U(2) Sutherland model [100].

In Section 2.4, we have seen that there is a duality between the particles and the holes in the single-component anyons. In order to see the duality in thermodynamics explicitly, we write (3.142) in the two-component vector form and multiply

$$\mathbf{g}_{\mathrm{h}} \equiv \begin{pmatrix} g_{\mathrm{h}}^{\uparrow\uparrow} & g_{\mathrm{h}}^{\uparrow\downarrow} \\ g_{\mathrm{h}}^{\downarrow\uparrow} & g_{\mathrm{h}}^{\downarrow\downarrow} \end{pmatrix} = \mathbf{g}_{\mathrm{p}}^{-1} = \frac{1}{2\lambda+1} \begin{pmatrix} \lambda+1 & -\lambda \\ -\lambda & \lambda+1 \end{pmatrix}. \qquad (3.144)$$

Then we obtain

$$\epsilon_{\mathrm{h}\sigma}(p)/T = \ln\left(1 + w_{\mathrm{h}\sigma}\right) - \sum_{\sigma'=\uparrow,\downarrow} g_{\mathrm{h}}^{\sigma\sigma'} \ln\left(1 + w_{\mathrm{h}\sigma'}^{-1}\right), \qquad (3.145)$$

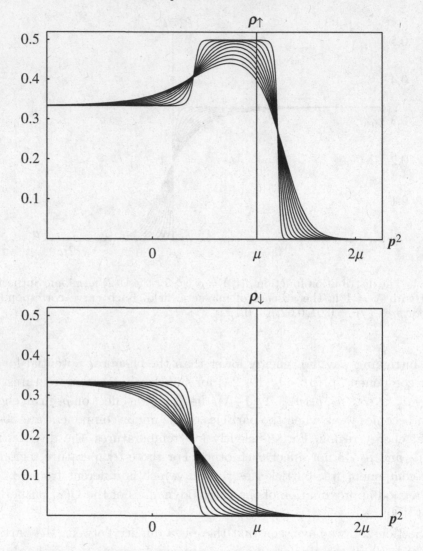

Fig. 3.5. The distribution functions $\rho_\uparrow(p)$ (upper panel) and $\rho_\downarrow(p)$ (lower panel) for the U(2) fermionic Sutherland model with $\lambda = 1$ in the presence of a magnetic field $h/\mu = 0.2$. The horizontal axis is $\epsilon/\mu - 1$. Each curve corresponds to temperature $T/\mu = 0.01, 0.02, \ldots, 0.1$.

with

$$\epsilon_{\mathrm{h}\sigma}(p) = \frac{\mu - p^2}{2\lambda + 1} + \sigma h, \quad w_{\mathrm{h}\sigma} = w_\sigma^{-1}. \tag{3.146}$$

The above relation shows that the duality relation of exclusion statistics leads to the duality of thermodynamics of the U(2) Sutherland model with

Table 3.1. *Elementary excitations of the U(2) Sutherland model with*
$\sigma = \pm 1$. *The quasi-particle is written as* qp, *and the quasi-hole as* qh.

Species	Charge	Spin	Energy	Momentum	Statistics
qp	1	σ	$p^2 - \mu - \sigma h$	p	$\mathbf{g}_p$
qh	$-1/(2\lambda+1)$	σ	$(\mu - p^2)/(2\lambda+1) + \sigma h$	$-p/(2\lambda+1)$	$\mathbf{g}_h$

coupling parameter λ and $-\lambda/(1+2\lambda)$. Similar duality is available to multi-component Sutherland models in general.

Now we consider excitations from the ground state in the unpolarized case ($h = 0$). At $T = 0$, the distributions of particles and holes are given by

$$\rho_\sigma(p) = 1/(2\lambda+1), \quad \rho_\sigma^*(p) = 0, \quad \text{for} \quad |p| < p_F \equiv \mu^{1/2},$$
$$\rho_\sigma(p) = 0, \quad \rho_\sigma^*(p) = 1, \quad \text{for} \quad |p| > p_F. \tag{3.147}$$

The pseudo-Fermi momentum p_F is given as

$$p_F = \pi d (2\lambda+1)/2,$$

by the condition

$$\int_{-\infty}^{\infty} \frac{dp}{2\pi} \sum_{\sigma=\uparrow,\downarrow} \rho_\sigma(p) = \frac{N}{L} \equiv d. \tag{3.148}$$

For $|p| > p_F$, excitations are particle-like; quasi-particles with energy $\epsilon_{p\sigma}$, charge $e_p = +1$, spin $\sigma_p = \sigma$, and statistics $\mathbf{g}_p$. For $|p| < p_F$, on the other hand, excitations are hole-like; quasi-holes with energy $\epsilon_{h\sigma}$, charge $e_h = -1/(2\lambda+1)$, spin $\sigma_h = \sigma$, and statistics $\mathbf{g}_h$. The charge of quasi-holes is renormalized to be fractional, while spin remains the same as that of quasi-particles. Table 3.1 summarizes these results on the elementary excitations.

3.5.3 Generalization to U(K) symmetry

We can straightforwardly generalize the results to the U(K) Sutherland model. The thermodynamic potential of the U(K) model is given by

$$\Omega = -T \int_{-\infty}^{\infty} \frac{dp}{2\pi} \sum_{\sigma=1}^{K} \ln\left(1 + w_\sigma^{-1}\right), \tag{3.149}$$

where w_σ is the real solution of the following equations:

$$\epsilon_{p\sigma}/T \equiv \left(p^2 - \mu_\sigma\right)/T = \ln\left(1 + w_\sigma\right) - \sum_{\sigma'=1}^{K} g_p^{\sigma,\sigma'} \ln\left(1 + w_{\sigma'}^{-1}\right), \qquad (3.150)$$

for $\sigma = 1, 2, \ldots, K$. Here μ_σ represents the chemical potential of each species. The statistical interaction is given by

$$g_p^{\sigma\sigma'} = \delta_{\sigma,\sigma'} + \lambda, \qquad (3.151)$$

which can be summarized by a $K \times K$ matrix: $\mathbf{g}_p = (g_p^{\sigma\sigma'})$.

The quasi-hole spectrum is described as

$$\epsilon_{h\sigma} \equiv -\frac{p^2}{2(K\lambda + 1)} - \mu_{h\sigma}$$

$$= T\ln\left(1 + w_\sigma^{-1}\right) - T\sum_{\sigma'=1}^{K} g_h^{\sigma,\sigma'} \ln\left(1 + w_{\sigma'}\right), \qquad (3.152)$$

$$\mu_{h\sigma} = -\mu_\sigma + \frac{\lambda}{K\lambda + 1}\sum_{\sigma'=1}^{K} \mu_{\sigma'}, \qquad (3.153)$$

$$g_h^{\sigma\sigma'} = \delta_{\sigma,\sigma'} - \lambda/(K\lambda + 1), \qquad (3.154)$$

where the statistical interaction for holes $g_h^{\sigma\sigma'}$ makes the $K \times K$ matrix $\mathbf{g}_h = \mathbf{g}_p^{-1}$. If each species takes the same chemical potential $\mu_\sigma = \mu$, the rapidity distribution functions at $T = 0$ are given by

$$\begin{aligned} \rho_\sigma(p) &= 1/(K\lambda + 1), & \rho_\sigma^*(p) &= 0, & \text{for} \quad |p| < p_F \equiv (2\mu)^{1/2}, \\ \rho_\sigma(p) &= 0, & \rho_\sigma^*(p) &= 1, & \text{for} \quad |p| > p_F, \end{aligned} \qquad (3.155)$$

where the pseudo-Fermi momentum is given by

$$p_F = \pi d\left(K\lambda + 1\right)/K.$$

For $|p| > p_F$, excitations are particle-like; quasi-particles with energy $\epsilon_{p\sigma}$, charge $+1$, spin (color) σ, and statistics $\mathbf{g}_p$. For $|p| < p_F$, on the other hand, excitations are quasi-holes with energy $\epsilon_{h\sigma}$, charge $-1/(K\lambda + 1)$, color σ, and statistics $\mathbf{g}_h$. Table 3.2 summarizes these elementary excitations.

Table 3.2. *Elementary excitations of the $U(K)$ Sutherland model.*

Species	Charge	Energy	Statistics
qp	1	$p^2 - \mu_\sigma$	$\mathbf{g}_p$
qh	$-1/(K\lambda + 1)$	$p^2/(K\lambda + 1) - \mu_{h\sigma}$	$\mathbf{g}_h$

3.6 Eigenfunctions

3.6.1 Non-symmetric Jack polynomials

The non-symmetric eigenfunction Φ (3.25) of $\mathcal{H}$ (3.14) is useful to derive the ground-state wave function of the multi-component Sutherland model. The wave functions Φ_η and $\Phi_{\eta'}$ are orthogonal with respect to the integral norm $\langle \cdot, \cdot \rangle_0$, which has been defined in (2.189) for $\eta^+ \neq (\eta')^+$. These two wave functions are, however, not necessarily orthogonal when η is a rearrangement of η'. For example, we consider a composition

$$\eta^{i1} = (\underbrace{0 \cdots 0}_{i-1}, 1, \underbrace{0 \cdots 0}_{N-i}) \tag{3.156}$$

for $i \in [1, N]$. Then $\Phi_{\eta^{i1}} = z_i$ is obviously an eigenfunction of (3.16) because there are no ways to squeeze η. However, these functions are not orthogonal:

$$\langle z_i, z_j \rangle_0 \neq 0.$$

Non-orthogonality of Φ_η is due to the degeneracy of eigenvalues; the eigenvalue of Φ_η for (3.16) is the same as that of $\Phi_{\eta'}$ when η' is a rearrangement of η. For derivation of dynamical correlation functions, orthogonality of the basis is desirable. In the following, we discuss an orthogonal basis of non-symmetric eigenfunctions of (3.16).

First we introduce an ordering of $\eta, \eta' \in \mathbf{Z}^N$ satisfying $\eta^+ = (\eta')^+ \in \mathcal{L}_N^+$ as $\eta' \prec \eta$ when $\sum_{i=1}^k \eta_i' \leq \sum_{i=1}^k \eta_i$ for all $k = 1, \ldots, N$. An example is given as

$$(0,1,2) \prec (0,2,1) \prec (1,0,2) \prec (1,2,0) \prec (2,0,1) \prec (2,1,0) \tag{3.157}$$

for $N = 3$. We then consider the eigenfunction of (3.16) satisfying

$$E_\eta = \Phi_\eta + \sum_{\eta'(\prec \eta),(\eta')^+ = \eta^+} c_{\eta'} \Phi_{\eta'} \tag{3.158}$$

$$\langle E_\eta, E_{\eta'} \rangle_0 = 0 \text{ when } \eta \neq \eta'. \tag{3.159}$$

These two conditions uniquely determine the orthogonal set $\{E_\eta\}$ resulting from $\{\Phi_\eta\}$. We generalize the definition of the ordering $\prec$ as $\nu \prec \eta$ if $\nu^+ < \eta^+$ or if $\nu^+ = \eta^+$ and $\sum_{i=1}^k \nu_i \leq \sum_{i=1}^k \eta_i$ for all $k = 1, \ldots, N$. An example is given as

$$(1,1,1) \prec (0,2,1) \prec (1,2,0) \prec (0,3,0) \tag{3.160}$$

for $N = 3$.

From the property (3.25) of Φ and (3.158), the wave function E_η can be defined alternatively as the function satisfying

$$E_\eta = \phi_\eta + \sum_{\eta'(\prec\eta)} c'_{\eta'}\phi_{\eta'} \tag{3.161}$$

$$\langle E_\eta, E_{\eta'}\rangle_0 = 0 \text{ when } \eta \neq \eta'.$$

Note that the summation in (3.161) is not restricted to η' satisfying $(\eta')^+ = \eta^+$. When η is a composition, i.e., $\eta_i \geq 0$ for $i \in [1, N]$, the wave function E_η coincides with the so-called non-symmetric Jack polynomial [127, 147]. Even when η is not a composition, the wave function E_η is written as $(z_1 \cdots z_N)^J E_{\eta'}$ with an integer J and a composition η'. We can thus concentrate on E_η with composition η without loss of generality.

To define the non-symmetric Jack polynomials, we introduce a set of operators $\{\hat{d}_1, \ldots, \hat{d}_N\}$ satisfying

(i)   mutual commutativity $[\hat{d}_i, \hat{d}_j] = 0$;
(ii) non-degeneracy of the set of the spectrum of $\{\hat{d}_1, \ldots, \hat{d}_N\}$;
(iii) self-adjointness with respect to $\langle\cdot, \cdot\rangle_0$;
(iv) commutativity $[\hat{d}_i, \mathcal{H}] = 0$ with Hamiltonian (3.16);
(v)   triangularity $\hat{d}_i\phi_\eta = \bar{\eta}_i\phi_\eta + \sum_{\eta'\prec\eta} c_{\eta'}\phi_{\eta'}$ with $\bar{\eta}_i$ defined below.

The following set of operators satisfies the above conditions:

$$\hat{d}_i = \frac{z_i}{\lambda}\frac{\partial}{\partial z_i} + \sum_{j=1}^{i-1} \frac{z_i}{z_i - z_j}(1 - K_{ij})$$

$$+ \sum_{j=i+1}^{N} \frac{z_j}{z_i - z_j}(1 - K_{ij}) - i + 1, \tag{3.162}$$

for $1 \leq i \leq N$. The action of $\hat{d}_i$ on the constant gives $-i + 1$, which ranges from 0 to $-N + 1$. These operators are called Cherednik–Dunkl operators [38, 45], and their properties will be discussed at length in Chapter 7. A more compact expression is obtained in terms of the step function $\theta(i - j) = [1 + \text{sgn}(i - j)]/2$, together with a similarity transformation generated by $\mathcal{O} = \prod_{i<j}|z_i - z_j|^\lambda$. Namely, we obtain

$$\mathcal{O}\hat{d}_i\mathcal{O}^{-1} = \frac{z_i}{\lambda}\frac{\partial}{\partial z_i} + \sum_{j(\neq i)} [\theta(j - i) - \theta_{ij}] K_{ij} + N - 1, \tag{3.163}$$

where $\theta_{ij} = z_i/(z_i - z_j)$. The latter form will be used in Chapter 9.

The property (i) guarantees the existence of simultaneous eigenfunctions of $\{\hat{d}_1, \ldots, \hat{d}_N\}$ and those eigenfunctions are uniquely determined owing to (ii). Those eigenfunctions being polynomials are called non-symmetric Jack polynomials. It follows from (iii) that non-symmetric Jack polynomials are mutually orthogonal with respect to $\langle \cdot, \cdot \rangle_0$. From (iv), non-symmetric Jack polynomials are eigenfunctions of the Hamiltonian (3.16). From (v), the non-symmetric Jack polynomials have the form of (3.161) up to overall factor. Thus we can identify E_η as non-symmetric Jack polynomials. The property (iv) comes from the relation

$$\mathcal{H} = \lambda^2 \sum_{i=1}^{N} \left[\left(\hat{d}_i + \frac{N-1}{2} \right)^2 - \frac{(N-2i+1)^2}{4} \right]. \tag{3.164}$$

The division of summation in (3.162) leads to the triangularity (v) of the matrix representation of $\hat{d}_i$. We have already encountered such triangularity for the three-particle Hamiltonian in (3.23). The eigenvalue of $\hat{d}_i$ is given by

$$\bar{\eta}_i \equiv \frac{\eta_i}{\lambda} - \left(k_i' + k_i'' \right), \tag{3.165}$$

where k_i' and k_i'' have been defined in (3.26). From (3.164) and (3.165), we can confirm that the eigenvalue $\mathcal{E}[\eta]$ of E_η for $\mathcal{H}$ coincides with (3.27).

The non-symmetric Jack polynomials $E_\eta(z_1, \ldots, z_N)$, can be generated successively from the initial condition

$$E_{0 \cdots 0} = 1.$$

We introduce the following generating operators [114]:

(i) The operator Θ is defined for any polynomial $f(z_1, \ldots, z_N)$ and any composition η by

$$\Theta f(z_1, \ldots, z_N) \equiv z_N f(z_N, z_1, \ldots, z_{N-1}), \tag{3.166}$$
$$\Theta \eta = (\eta_2, \ldots, \eta_N, \eta_1 + 1). \tag{3.167}$$

Then we obtain

$$E_{\Theta \eta} = \Theta E_\eta. \tag{3.168}$$

(ii) Another generating operator is the coordinate exchange operator $K_i \equiv K_{i,i+1}$. This operator acts on E_η as

$$K_i E_\eta = \begin{cases} \xi_i E_\eta + \left(1 - \xi_i^2 \right) E_{K_i \eta}, & \eta_i > \eta_{i+1}, \\ E_\eta, & \eta_i = \eta_{i+1}, \\ \xi_i E_\eta + E_{K_i \eta}, & \eta_i < \eta_{i+1}, \end{cases} \tag{3.169}$$

where

$$\xi_i = 1/\left(\bar{\eta}_i - \bar{\eta}_{i+1}\right). \tag{3.170}$$

The details on generators will be given in Section 7.1.5. Let us derive some simplest cases of E_η explicitly in the following.

Examples:

(i) Using (3.168) for $\eta = 0 \cdots 0$, we obtain

$$E_{0 \cdots 01} = z_N. \tag{3.171}$$

(ii) Using (3.171) and (3.169) for $\eta = 0 \cdots 01$, we obtain

$$E_{0 \cdots 010} = z_{N-1} + \frac{\lambda}{1 + (N-1)\lambda} z_N. \tag{3.172}$$

(iii) Using (3.171) and successive use of (3.169), we obtain

$$E_{\eta^{i1}} = z_i + \frac{\lambda}{1 + \lambda i} \left(z_{i+1} + \cdots + z_N\right), \tag{3.173}$$

where η^{i1} has been defined in (3.156).

(iv) With use of (3.173) for $i = 1$, i.e., $\eta = (10 \cdots 0)$, and (3.168), we obtain

$$E_{0 \cdots 02} = z_N^2 + \frac{\lambda}{\lambda + 1} z_N \sum_{i=1}^{N-1} z_i. $$

The non-symmetric Jack polynomials E_η have non-negative integers for all η_i [127,147]. The symmetric Jack polynomials J_κ are obtained from E_η with $\eta^+ = \kappa$ by symmetrization. For example, we can easily confirm that

$$\text{Sym } E_{0 \cdots 02} = m_2 + \frac{2\lambda}{1 + \lambda} m_{11} = J_2.$$

Further, the mathematical formulae given below can be derived with use of the theory of non-symmetric Jack polynomials, as discussed fully in Chapter 7. From the above examples, we see that E_η contains the monomial ϕ_η with a coefficient of unity. This property of the polynomial is called "monic". Further, the eigenvalue of E_η for (3.16) is the same as that of Φ_η. These two facts will be used to construct an orthogonal basis of eigenfunctions of the multi-component Sutherland model.

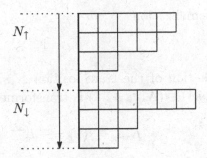

Fig. 3.6. The diagram of $(\kappa^\uparrow, \kappa^\downarrow) = (5, 4, 2, 1, 6, 3, 2)$ for $N = 7$, $N_\uparrow = 4$, $N_\downarrow = 3$.

3.6.2 Jack polynomials with U(2) symmetry

Using E_η, we reconstruct eigenfunctions of (3.16) related to the U(2) fermionic Sutherland model with $\zeta = 1$ and the U(2) bosonic Sutherland model with $\zeta = -1$. Let N_σ be the number of particles having spin $\sigma(=\uparrow, \downarrow)$. As we have seen in Section 3.2.1, each eigenfunction of the U(2) fermionic Sutherland model is specified by $(N_\uparrow, N_\downarrow, \kappa)$ with $\kappa = (\kappa^\uparrow, \kappa^\downarrow) \in \mathcal{L}_{N_\uparrow}^{+>} \otimes \mathcal{L}_{N_\downarrow}^{+>}$. As the wave function of the state $\kappa = (\kappa^\uparrow, \kappa^\downarrow)$, we take

$$\Phi = \text{Asym } E_\kappa v_\alpha, \quad \alpha = (\overbrace{1, \ldots, 1}^{N_\uparrow}, \overbrace{2, \ldots, 2}^{N_\downarrow}) \qquad (3.174)$$

instead of (3.45). It is often useful to have the wave function with spin configuration fixed. Let σ_0 be the spin configuration $(1^{N_\uparrow}, 2^{N_\downarrow})$. The wave function (3.174) reduces to

$$\Phi(\{z_i\}, \{\sigma_i\} = \sigma_0) = \text{Asym}_\uparrow \text{Asym}_\downarrow E_\kappa, \qquad (3.175)$$

where we introduce $\text{Asym}_\uparrow$ as the antisymmetrization operator with respect to the variables z_i in $[1, N_\uparrow]$, and $\text{Asym}_\downarrow$ to the variables z_i in $[N_\uparrow + 1, N]$.

Now we define $\Lambda_N^{+>}$ as

$$\Lambda_N^{+>} = \{\kappa | \kappa = (\kappa_1, \ldots, \kappa_N) \in \mathbf{Z}_{\geq 0}; \kappa_1 > \cdots > \kappa_N > 0\}, \qquad (3.176)$$

where $\mathbf{Z}_{\geq 0}$ is the set of non-negative integers. Figure 3.6 shows a typical example of a composition $(\kappa^\uparrow, \kappa^\downarrow) \in (\Lambda_{N_\uparrow}^{+>}, \Lambda_{N_\downarrow}^{+>})$. When $\kappa \in \Lambda_{N_\uparrow}^{+>} \otimes \Lambda_{N_\downarrow}^{+>}$, the RHS of (3.175) is proportional to the U(2) Jack polynomial $J_\kappa^{(--)}$ defined by the following two conditions [18, 46, 105]:

(i) The polynomial $J_\kappa^{(--)}$ for $\kappa = (\kappa^\uparrow, \kappa^\downarrow) \in (\Lambda_{N_\uparrow}^{+>}, \Lambda_{N_\downarrow}^{+>})$ has the form

$$J_\kappa^{(--)}(z) = \prod_{\sigma=\uparrow, \downarrow} \sum_{(\eta^\sigma)^+ = \kappa^\sigma} a_{\eta^\uparrow, \eta^\downarrow}^{(--)} E_{\eta^\uparrow, \eta^\downarrow}(z), \qquad (3.177)$$

with the normalization

$$a_\kappa = 1.$$

(ii) Under the action of the transposition K_i, the polynomial $J_\kappa^{(--)}(z)$ for $\kappa = (\kappa^\uparrow, \kappa^\downarrow) \in (\Lambda_{N_\uparrow}^{+>}, \Lambda_{N_\downarrow}^{+>})$ is transformed as

$$K_i J_\kappa^{(--)}(z) = -J_\kappa^{(--)}(z), \qquad (3.178)$$

for $i \in [1, N_\uparrow - 1]$ or $i \in [N_\uparrow + 1, N - 1]$.

Every term on the RHS of (3.178) is an eigenfunction of (3.16) with common eigenenergy. Hence the function $J_\kappa^{(--)}$ is also an eigenfunction of (3.16). Owing to the partial antisymmetric property (ii), the function $J_\kappa^{(--)}$ is related to an eigenfunction of the U(2) fermionic Sutherland model. Thus we call it the U(2) Jack polynomial. We use $J_\kappa^{(--)}$ to derive the exact expression for the hole propagator of the U(2) Sutherland model [102, 105].

Let us consider two simple cases: First, when $\kappa^\uparrow = \delta(N_\uparrow) \equiv (N_\uparrow - 1, N_\uparrow - 2, \ldots, 1, 0)$ and $\kappa^\downarrow = \delta(N_\downarrow) \equiv (N_\downarrow - 1, N_\downarrow - 2, \ldots, 1, 0)$, the function $J_{\delta(N_\uparrow), \delta(N_\downarrow)}^{(--)}$ becomes the product of the Vandermonde determinants (2.188)

$$J_{\delta(N_\uparrow), \delta(N_\downarrow)}^{(--)} = \Delta(z_1, \ldots, z_{N_\uparrow}) \Delta(z_{N_\uparrow + 1}, \ldots, z_N). \qquad (3.179)$$

Second, we consider the case $\lambda = 0$ where the function E_η reduces to the monomial ϕ_η. Then $J_{\kappa^\uparrow, \kappa^\downarrow}^{(--)}$ becomes a product of the free fermion functions

$$s_{\tilde\kappa^\uparrow}(z_1, \ldots, z_{N_\uparrow}) s_{\tilde\kappa^\uparrow}(z_{N_\uparrow + 1}, \ldots, z_N) \Delta(z_1, \ldots, z_{N_\uparrow}) \Delta(z_{N_\uparrow + 1}, \ldots, z_N),$$

where $s_{\tilde\kappa^\sigma}$ is the Schur function. The latter has been defined in (2.173) for a partition

$$\tilde\kappa^\sigma \equiv \kappa^\sigma - \delta(N_\sigma).$$

For two general compositions $(\kappa^\uparrow, \kappa^\downarrow), (\mu^\uparrow, \mu^\downarrow) \in (\Lambda_{N_\uparrow}^{+>}, \Lambda_{N_\downarrow}^{+>})$, the orthogonality

$$\langle J_{\kappa^\uparrow, \kappa^\downarrow}^{(--)}, J_{\mu^\uparrow, \mu^\downarrow}^{(--)} \rangle_0 \propto \delta_{(\kappa^\uparrow, \kappa^\downarrow), (\mu^\uparrow, \mu^\downarrow)} \qquad (3.180)$$

holds owing to the orthogonality of E_η. The norm of the U(2) Jack polynomial is given in (7.356) as a product of combinatorial quantities. Furthermore, the expansion formula (called the binomial formula) corresponding to

(2.226) is given as

$$\prod_{i=1}^{N_\uparrow} (1-z_i)^{r-N_\uparrow} \prod_{i=N_\uparrow+1}^{N} (1-z_i)^{r-N_\downarrow} \Delta(z_1,\ldots,z_{N_\uparrow})\Delta(z_{N_\uparrow+1},\ldots,z_N)$$

$$= \sum_{\kappa^\uparrow \in \Lambda_{N_\uparrow}^{+>}} \sum_{\kappa^\downarrow \in \Lambda_{N_\downarrow}^{+>}} b_{\kappa^\uparrow,\kappa^\downarrow}^{(--)}(r) J_{\kappa^\uparrow,\kappa^\downarrow}^{(--)}, \qquad\qquad (3.181)$$

which will be derived in Section 7.4.6. The formula for the expansion coefficient $b_{\kappa^\uparrow,\kappa^\downarrow}^{(--)}$ also has a product-type expression. The LHS of (3.181) appears as a result of annihilating a particle in the ground state of the U(2) Sutherland model. Therefore, the function $J_{\kappa^\uparrow,\kappa^\downarrow}^{(--)}$ plays a central role in the U(2) Sutherland model, as do the symmetric Jack polynomials in the single-component model.

3.7 Dynamics of U(2) Sutherland model

The results on the thermodynamics suggest anyonic elementary excitations in the multi-component Sutherland model. It is interesting to see how anyonic elementary excitations obeying exclusion statistics are reflected in the dynamics in the multi-component model. Exact results on the dynamics of the multi-component Sutherland model are important also in the sense that some of these results can be converted to dynamical correlation functions of the SU(K) Haldane–Shastry model and the supersymmetric t–J model with long-range interactions.

For the U(2) Sutherland model, the expression for the hole propagator was conjectured in [102] relying on finite-size calculations. Uglov derived the exact expression for dynamical density and spin density correlation functions by developing a novel isomorphism between the Yangian Gelfand–Zetlin basis and a degenerate limit of Macdonald symmetric polynomials [189]. The resultant expressions in the thermodynamic limit were naturally interpreted in terms of anyonic elementary excitations [103]. The conjecture in [102] for the hole propagator was proved to be exact in [105]. In [198,199], dynamical density and color correlation functions of the Sutherland model with SU(K) internal symmetry were derived exactly on the basis of the theory of [189]. Dynamical density correlation functions for the U(1,1) Sutherland model [7,8,10,11] were obtained in the course of calculation of dynamical correlation functions of the supersymmetric t–J model with long-range interactions.

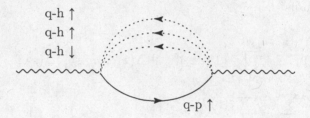

Fig. 3.7. Schematic description of minimum bubble for the U(2) Sutherland model with $\lambda = 1$.

3.7.1 *Hole propagator* $\langle \hat{\psi}_\downarrow^\dagger(x,t)\hat{\psi}_\downarrow(0,0)\rangle$

For simplicity, we consider the fermionic U(2) model for $\zeta = 1$ and even λ, and the bosonic U(2) model for $\zeta = -1$ and odd λ. The boundary condition is taken to be periodic. We take N to be even, and assume $N/2 = M$ to be odd (even) for odd (even) λ, so that the N-particle ground state is non-degenerate. The hole propagator for the singlet ground state in the U(2) Sutherland model is independent of the spin species, and is given by

$$
\begin{aligned}
G^-(x,t) &= \frac{\langle g, N|\hat{\psi}_\downarrow^\dagger(x,t)\hat{\psi}_\downarrow(0,0)|g,N\rangle}{\langle g,N|g,N\rangle} \\
&= \sum_\mu \frac{|\langle \mu, N-1|\hat{\psi}_\downarrow(0,0)|g,N\rangle|^2}{\langle \mu, N-1|\kappa, N-1\rangle\langle g,N|g,N\rangle} e^{-i(E[\mu]-E_0)t+i(P_\mu-P_0)x}.
\end{aligned}
$$

$$(3.182)$$

Here $|\mu, N-1\rangle$ denotes the excited state specified by $\mu \in (\mathcal{L}_{N/2}^{+>}, \mathcal{L}_{N/2-1}^{+>})$. $E[\mu]$ and P_μ denote the energy and momentum, respectively, associated with the state μ, while E_0 and P_0 denote those in the ground state.

First we discuss the excitation contents of excited states μ. Only those excited states with $N-1$ particles and $S_z = 1$ are relevant to (3.182). From results on thermodynamics, we expect that each excited state is described as multiple excitations of quasi-holes and quasi-particles. The charge of quasi-holes is $-1/(2\lambda+1)$ of that for original particles, while the spin of a quasi-hole is the same as that of an original particle. Namely, the following excitation:

$$
\begin{cases}
\text{one quasi-particle with } \sigma_\mathrm{p} = \sigma, \\
\lambda + 1 \text{ quasi-holes with } \sigma_\mathrm{h} = -\sigma, \\
\lambda \text{ quasi-holes with } \sigma_\mathrm{h} = \sigma
\end{cases}
$$

$$(3.183)$$

is the minimum excitation with zero-charge and zero-spin. This set of excitations is called the minimum bubble, as illustrated in Fig. 3.7.

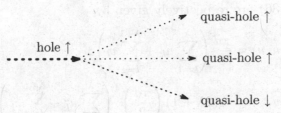

Fig. 3.8. Schematic description of the process of one hole → three quasi-holes in the U(2) Sutherland model with $\lambda = 1$.

The charge and spin neutralities are verified by

$$e_p + (\lambda + 1)\, e_h + \lambda e_h = 0 \tag{3.184}$$

and

$$\sigma + (\lambda + 1)\,(-\sigma) + \lambda\sigma = 0, \tag{3.185}$$

respectively. The relevant excited states should therefore be in the form of

$$(2\lambda + 1)\,\text{quasi-holes} + \text{multiple excitation of minimum bubbles}.$$

An explicit exact calculation, however, reveals that only $2\lambda + 1$ quasi-holes without additional bubbles are relevant to the hole propagator. This fact suggests an underlying strong selection rule. Figure 3.8 illustrates the situation.

The explicit expression for the hole propagator in the thermodynamic limit is given by [102, 103, 105]

$$G^-(x,t) = c(\lambda) \prod_{k=1}^{\lambda} \int_{-1}^{1} \mathrm{d}u_k \prod_{l=1}^{\lambda+1} \int_{-1}^{1} \mathrm{d}v_l$$
$$\times |F(u,v)|^2 \exp\left[-\mathrm{i}\,(E(u,v)t - Q(u,v)x)\right]. \tag{3.186}$$

Here $u = (u_1, \ldots, u_\lambda)$, $v = (v_1, \ldots, v_{\lambda+1})$ represent normalized velocities of quasi-holes with up and down spins, respectively. In (3.186), the form factor F is given, as will be derived in the next section, by

$$F(u,v) =$$
$$\frac{\prod_{1 \le k < l \le \lambda} (u_k - u_l)^{g_d} \prod_{1 \le k < l \le \lambda+1} (v_k - v_l)^{g_d} \prod_{k=1}^{\lambda} \prod_{l=1}^{\lambda+1} (u_k - v_l)^{g_f}}{\prod_{k=1}^{\lambda} \left(1 - u_k^2\right)^{(1-g_d)/2} \prod_{l=1}^{\lambda+1} \left(1 - v_l^2\right)^{(1-g_d)/2}},$$
$$\tag{3.187}$$

where $g_d = (\lambda + 1)/(2\lambda + 1)$ and $g_f = -\lambda/(2\lambda + 1)$. The momentum Q and energy E in (3.186) are respectively given by

$$Q(u, v) = \frac{\pi d}{2} \left(\sum_{k=1}^{\lambda} u_k + \sum_{l=1}^{\lambda+1} v_l \right), \tag{3.188}$$

$$E(u, v) = -(2\lambda + 1) \left(\frac{\pi d}{2} \right)^2 \left(\sum_{k=1}^{\lambda} u_k^2 + \sum_{l=1}^{\lambda+1} v_l^2 \right), \tag{3.189}$$

where $d = N/L$ is the density of particles. Finally, the constant $c(\lambda)$ in (3.186) is given by

$$c(\lambda) = \frac{d}{4 \left(2\lambda + 1 \right)^\lambda \Gamma \left(\lambda + 2 \right)} \prod_{k=1}^{2\lambda+1} \frac{\Gamma \left((\lambda + 1) / (2\lambda + 1) \right)}{\Gamma \left(k/(2\lambda + 1) \right)^2}. \tag{3.190}$$

The support of the spectral function, which has been defined by (2.202), is similar to that of the single-component Sutherland model with replacement of the coupling constant $\lambda \to 1 + 2\lambda$. For example, the support for the U(2) model with $\lambda = 0$ has the same form as that shown in Fig. 2.14. The support for the U(2) model with $\lambda = 2$ has the same form as that shown in Fig. 2.16. From the shape of the support, we easily see that only $2\lambda + 1$ quasi-hole states contribute to the hole propagator for the U(2) model.

We should, however, note that there is no correspondence between the weights of the two spectral functions $A_{2,\lambda}(\omega, P)$ for the U(2) model and $A_{1+2\lambda}(\omega, P)$ for the single-component model. The singularity of the spectral function is determined by the joint density of states and the singularity of the form factor $F(u, v)$. The latter is determined by the exponents g_d and g_f. These two exponents have already appeared in thermodynamics as describing statistical interactions between quasi-holes with the same spins and opposite spins, respectively. Quasi-holes with exclusion statistics behave as free particles in the dynamics and thermodynamics. However, the statistical interaction between them yields the nontrivial singularity of the spectral function, and nontrivial entropy in thermodynamics. In this way, anyonic elementary excitations yield a natural interpretation of physical properties of the U(2) Sutherland model.

3.7.2 Unified description of correlation functions

In addition to the hole propagator, we shall now present and interpret the exact expression for the density $\hat{\rho}(x, t)$ and the spin density $\hat{S}_z(x, t)$ correlation functions

$$\langle \hat{\rho}(x, t) \hat{\rho}(0, 0) \rangle, \quad \langle \hat{S}_z(x, t) \hat{S}_z(0, 0) \rangle \tag{3.191}$$

for the U(2) model. The outline of derivation of $\langle \hat{\rho}(x,t)\hat{\rho}(0,0)\rangle$ is given in the next section. The derivation of $\langle \hat{\rho}(x,t)\hat{\rho}(0,0)\rangle$ and $\langle \hat{S}_z(x,t)\hat{S}_z(0,0)\rangle$ will be given via Uglov's method in Chapter 10. We shall first see that in these correlation functions, only the minimal bubble contributes. Second, the form factors are similar in these correlation functions, and also in the hole propagator. Third, the singularity of the form factor is determined by the statistical interactions between elementary excitations.

Scaling the pseudo-Fermi momenta to unity, the spectra of a quasi-particle and a quasi-hole are given by

$$\epsilon_p(y) = (2\lambda + 1)^2 \left(\frac{\pi d}{2}\right)^2 (y^2 - 1), \tag{3.192}$$

$$\epsilon_h(y) = (2\lambda + 1) \left(\frac{\pi d}{2}\right)^2 (1 - y^2). \tag{3.193}$$

For a compact description of the dynamics, we introduce the following notation [197] with non-negative integers a, b and c:

$$\mathcal{E}(u, v, w; a, b, c) = \sum_{i=1}^{a} \epsilon_p(u_i) + \sum_{j=1}^{b} \epsilon_h(v_j) + \sum_{k=1}^{c} \epsilon_h(w_k), \tag{3.194}$$

$$\mathcal{P}(u, v, w; a, b, c) = \frac{\pi d}{2}\left[-(2\lambda + 1)\sum_{i=1}^{a} u_i + \sum_{j=1}^{b} v_j + \sum_{k=1}^{c} w_k\right], \tag{3.195}$$

$$I(a, b, c)[*] = \prod_{i=1}^{a}\int_1^{\infty} du_i \prod_{j=1}^{b}\int_{-1}^{1} dv_j \prod_{k=1}^{c}\int_{-1}^{1} dw_k(*)|F_\lambda(u, v, w; a, b, c)|^2, \tag{3.196}$$

where d is the density of particles, and $[*]$ is a certain function. The variables $u = (u_1, \ldots, u_a)$, $v = (v_1, \ldots, v_b)$, and $w = (w_1, \ldots, w_c)$ represent the normalized momenta of a quasi-particle with spin σ, and quasi-holes with spin $-\sigma$ and σ, respectively ($\sigma = \pm 1$). The hole propagator in the previous subsection, for example, corresponds to the case $I(0, \lambda+1, \lambda)$ in (3.196). The most important quantity for correlation functions is the form factor, which has the following general form:

$$F_\lambda(u, v, w; a, b, c) = \frac{\prod_{1 \le i < j \le b}(v_i - v_j)^{g_d}}{\prod_{i=1}^{a}\prod_{j=1}^{b}(u_i - v_j)\prod_{i=1}^{a}(u_i^2 - 1)^{(1-g_d^p)/2}}$$

$$\times \frac{\prod_{1 \le i < j \le c}(w_i - w_j)^{g_d}\prod_{i=1}^{b}\prod_{j=1}^{c}(v_i - w_j)^{g_f}}{\prod_{j=1}^{b}(1 - v_j^2)^{(1-g_d)/2}\prod_{k=1}^{c}(1 - w_k^2)^{(1-g_d)/2}}, \tag{3.197}$$

Table 3.3. *Exponents for combinations of rapidities in the form factor.*

	Quasi-hole $-\sigma$	Quasi-hole σ	Pseudo-Fermi points
Quasi-particle σ	-1	$*$	$(g_{\mathrm{d}}^{\mathrm{p}} - 1)/2$
Quasi-hole $-\sigma$	g_{d}	g_{f}	$(g_{\mathrm{d}} - 1)/2$
Quasi-hole σ	g_{f}	g_{d}	$(g_{\mathrm{d}} - 1)/2$

where we use the convention $\prod_{i=1}^{0}(*)=1$ for any quantity $(*)$, and

$$g_{\mathrm{d}} = (\lambda+1)/(2\lambda+1), \ \ g_{\mathrm{f}} = -\lambda/(2\lambda+1), \ \ g_{\mathrm{d}}^{\mathrm{p}} = \lambda+1. \tag{3.198}$$

These exponents are precisely the same as the statistical interaction parameters in thermodynamics. In fact, comparison with (3.94) and (3.95) shows that

- g_{d}: quasi-hole interaction with the same spin,
- g_{f}: quasi-hole interaction with different spins,
- $g_{\mathrm{d}}^{\mathrm{p}}$: quasi-particle interaction with the same spin.

With this identification, the exponents in F_{λ} can also be interpreted intuitively. Namely, two quasi-hole factors with the same spin have g_{d} as in $(v_i - v_j)^{g_{\mathrm{d}}}$, while those with different spins have g_{f} as in $(v_i - w_j)^{g_{\mathrm{f}}}$. Furthermore, quasi-hole and quasi-particle factors with different spins have -1 as in $(u_i - v_j)^{-1}$, while those with the same spin do not appear. We proceed to interpret the factor $v_i^2 - 1 = (v_i - 1)(v_i + 1)$ as the interaction of the rapidity of a quasi-particle and pseudo-Fermi momenta ± 1. Then the exponent $(g_{\mathrm{d}} - 1)/2$ follows because the particle at the pseudo-Fermi momentum is the average of a quasi-particle and a quasi-hole. Similarly, the factor $u_i^2 - 1 = (u_i - 1)(u_i + 1)$, involving the quasi-particle rapidity u_i, has the exponent $(g_{\mathrm{d}}^{\mathrm{p}} - 1)/2$ by interaction with the pseudo-Fermi momenta. Table 3.3 summarizes the exponents of these factors.

Using the notation (3.196) introduced above, the hole propagator as well as the density and spin–density correlation functions can be described in a unified manner [197]. First, the hole propagator, given by (3.187), is rewritten as

$$\langle \psi_{\sigma}^{\dagger}(x,t)\psi_{\sigma}(0,0)\rangle$$
$$= A(\lambda)I(0,\lambda+1,\lambda)[(\pi d/2)\mathrm{e}^{\mathrm{i}(\mathcal{P}(0,\lambda+1,\lambda)x - (\mathcal{E}(0,\lambda+1,\lambda)-\mu)t)}], \tag{3.199}$$

where $\mu = ((2\lambda+1)\pi d)^2$ is the chemical potential. The proportionality constant is written as

$$A(\lambda) = \frac{1}{\pi(2\lambda+1)^{\lambda}} D(\lambda), \tag{3.200}$$

$$D(\lambda) = \frac{1}{\Gamma(\lambda+2)} \prod_{j=1}^{2\lambda+1} \frac{\Gamma((\lambda+1)/(2\lambda+1))}{\Gamma(j/(2\lambda+1))^2}. \tag{3.201}$$

We now quote the results for the density correlation function:

$$\langle \hat{\rho}(x,t)\hat{\rho}(0,0) \rangle$$
$$= B(\lambda)I(1,\lambda+1,\lambda)[\mathcal{P}(1,\lambda+1,\lambda)^2 \cos(\mathcal{P}(1,\lambda+1,\lambda)x)e^{-i\mathcal{E}(1,\lambda+1,\lambda)t}], \tag{3.202}$$

$$B(\lambda) = \frac{1}{\pi^2(2\lambda+1)^{\lambda+1}} D(\lambda). \tag{3.203}$$

The intensity factor $I(1,\lambda+1,\lambda)$ in (3.202) shows that relevant excited states correspond to the minimal bubble in (3.183).

Finally we quote the exact result for the spin–density correlation function. The z-component of the spin density operator $\hat{S}_z(x)$ is given by

$$\hat{S}_z(x) = \frac{1}{2}\left[\sum_{i=1}^{N_\uparrow} \delta(x-x_i) - \sum_{i=N_\uparrow+1}^{N} \delta(x-x_i) \right]. \tag{3.204}$$

Since we are dealing with the singlet ground state of the U(2) Sutherland model, the correlation function is independent of the spin component. The spin correlation function is given by [197]

$$\langle \hat{S}_z(x,t)\hat{S}_z(0,0) \rangle$$
$$= C_{\mathrm{I}}(\lambda)I(1,\lambda,\lambda+1)[(\pi d/2)^2 \cos(\mathcal{P}(1,\lambda,\lambda+1)x)e^{-i\mathcal{E}(1,\lambda,\lambda+1)t}]$$
$$+ C_{\mathrm{II}}(\lambda)I(1,\lambda+2,\lambda-1)[(\pi d/2)^2 \cos(\mathcal{P}(1,\lambda+2,\lambda-1)x)e^{-i\mathcal{E}(1,\lambda+2,\lambda-1)t}], \tag{3.205}$$

where

$$C_{\mathrm{I}}(\lambda) = \frac{1}{4\pi^2(2\lambda+1)^{\lambda-1}} D(\lambda), \tag{3.206}$$

$$C_{\mathrm{II}}(\lambda) = \frac{1}{4\pi^2(2\lambda+1)^{\lambda-1}} \frac{\lambda}{\lambda+2} D(\lambda). \tag{3.207}$$

We shall discuss the derivation of $\langle \hat{S}_z(x,t)\hat{S}_z(0,0) \rangle$ in Section 10.5. Note that in (3.205), $I(1,\lambda,\lambda+1)$ corresponds to the minimal bubble, while $I(1,\lambda+2,\lambda-1)$ contains a spin flip in the quasi-particle and quasi-holes. Then we have $S_z = 0$ also in excited states.

3.8 Derivation of dynamics for finite-sized systems

3.8.1 Hole propagator

In deriving the hole propagator, we first consider the fermion systems and take $\zeta = 1$ in (3.6). We take the periodic boundary condition with N even and $N/2$ odd, so that the ground state is non-degenerate. Then we obtain $\Psi_{g,F}$ in (3.59) as the ground state. If λ is even, the absolute ground state Ψ_0^B involved becomes the same as $\Psi_{0,N}$ given by (2.18). Then we seek a general eigenfunction Ψ in the form of

$$\Psi(x_1, \sigma_1, \ldots, x_N, \sigma_N) = \Phi(x_1, \sigma_1, \ldots, x_N, \sigma_N)\Psi_{0,N} \qquad (3.208)$$

where Φ satisfies the fermionic Fock condition

$$K_{i,j}\Phi = -P_{i,j}\Phi. \qquad (3.209)$$

We do not consider the case of odd λ in the fermion case, where the absolute ground state is not a Galilean-shifted polynomial. As a result, the exact result for the hole propagator has not yet been derived.

In the case of bosons as original particles, we take odd λ with $\zeta = -1$ for an N-particle system with N being twice an even integer. Then the absolute ground state again becomes $\Psi_{0,N}$ given by (2.18). Here Φ is chosen to obey the antiperiodic boundary condition so that $\Psi(x_1, \sigma_1, \ldots, x_N, \sigma_N)$ satisfies the periodic boundary condition. Except for this change of the boundary condition, we have the same conditions (3.208) and (3.209) for general eigenfunctions.

Let us begin with the norm $\langle g, N | g, N \rangle$ in (3.182), which is expressed as

$$\langle g, N | g, N \rangle = \sum_{\sigma_1 = \pm 1} \cdots \sum_{\sigma_N = \pm 1} \int_0^L dx_1 \cdots \int_0^L dx_N |\Psi_g(\{x_i\}, \{\sigma_i\})|^2. \qquad (3.210)$$

The number of spin configurations that contribute to the summation is $N!/((N/2)!)^2$. Since the contribution from each spin configuration is the same, we can fix on any spin configuration and multiply the summand by $N!/((N/2)!)^2$. If we fix the spin variables as

$$\{\sigma_0\} \equiv (\sigma_1 = \cdots = \sigma_{N/2} = 1, \quad \sigma_{N/2+1} = \cdots = \sigma_N = -1), \qquad (3.211)$$

we obtain

$$\Phi_{g,F}(\{x_i\}, \{\sigma_0\}) = \prod_{1 \le i < j \le N/2} (z_i - z_j) \prod_{N/2+1 \le l < m \le N} (z_l - z_m), \qquad (3.212)$$

apart from a phase factor. As mentioned in (3.179) and fully discussed in Section 3.6.2, this function corresponds to a U(2) Jack polynomial $J_\mu^{(--)}$ with the particular case of $\mu = \mu_0$ with

$$\mu_0 = (\delta(N/2), \delta(N/2))$$
$$= (N/2 - 1, N/2 - 2, \ldots, 0, N/2 - 1, N/2 - 2, \ldots, 0).$$

Consequently, the norm of the ground-state wave function is given by

$$\langle g, N | g, N \rangle = \frac{N! L^N}{[(N/2)!]^2} \langle J_{\mu_0}^{(--)}, J_{\mu_0}^{(--)} \rangle_0. \tag{3.213}$$

Next we consider the matrix element in (3.182). When the annihilation operator $\hat{\psi}_\downarrow(0,0)$ acts on the wave function $\Psi(\{x_i\}, \{\sigma_i\})$ of an N-particle state $|\Psi\rangle$, the result is given by

$$\langle \{x_i\}, \{\sigma_i\} | \hat{\psi}_\downarrow(0,0) | g, N \rangle = (-1)^{N-1} \sqrt{N} \Psi(\{x_i\}, \{\sigma_i\})|_{x_N = 0, \sigma_N = -1}. \tag{3.214}$$

Namely, the annihilation fixes the spatial and spin coordinates of the Nth particle. Then we obtain

$$\langle g, N | \hat{\psi}_\downarrow^\dagger(x, t) \hat{\psi}_\downarrow(0,0) | g, N \rangle = N \sum_{\sigma_1 = \pm 1} \cdots \sum_{\sigma_{N-1} = \pm 1} \int_0^L \mathrm{d}x_1 \cdots \int_0^L \mathrm{d}x_{N-1}$$

$$\times \Psi_{g,N}^*(\{x_i\}, \{\sigma_i\})|_{x_N = 0, \sigma_N = -1} \exp\left[-\mathrm{i}\left(H_K - E_{g,N}\right) t + \mathrm{i}Px\right]$$

$$\times \Psi_{g,N}(\{x_i\}, \{\sigma_i\})|_{x_N = 0, \sigma_N = -1}, \tag{3.215}$$

where we have replaced H_P in (3.5) by H_K since we can then fix the spin variables to a configuration and multiply $(N-1)!/[(N/2)!(N/2-1)!]$. When the spin variables are fixed as σ_0 given by (3.211), we obtain

$$\Psi_{g,N}(\{x_i\}, \{\sigma_i\})|_{x_N = 0, \sigma_N = -1} = \mathrm{e}^{\mathrm{i}\alpha} \tilde{\Phi}(z_1, \ldots, z_{N-1}) \Psi_{0,N-1} \tag{3.216}$$

with a phase factor $\mathrm{e}^{\mathrm{i}\alpha}$. Here the function $\tilde{\Phi}$ is defined by

$$\tilde{\Phi} = \prod_{i=1}^{N-1} z_i^J \prod_{i=1}^{N/2} (z_i - 1)^{\lambda+1} \prod_{i=N/2+1}^{N-1} (z_i - 1)^\lambda$$

$$\times \prod_{1 \leq i < j \leq N/2} (z_i - z_j) \prod_{N/2+1 \leq i < j \leq N-1} (z_i - z_j), \tag{3.217}$$

with $J = -\lambda/2 - (N/2 - 1)/2$ and the function $\Psi_{0,N-1}$ in (3.216) obtained from (2.19) by replacement $N \to N - 1$. We can expand $\tilde{\Phi}$ as a linear combination of the eigenfunctions of H_K (3.6). Using the binomial formula (3.181) for U(2) Jack polynomials, we obtain

$$\tilde{\Phi} = \left(\prod_{i=1}^{N-1} z_i^J \right) \sum_{\mu \in (\Lambda_{N/2}^{+>}, \Lambda_{N/2-1}^{+>})} b_\mu^{(--)} \left(\lambda + \frac{N}{2} \right) J_\mu^{(--)}. \tag{3.218}$$

Substituting (3.218) into (3.215), we obtain

$$\langle g, N | \hat{\psi}_\downarrow^\dagger(x,t) \hat{\psi}_\downarrow(0,0) | g, N \rangle = \frac{N! L^{N-1}}{(N/2)!(N/2-1)!}$$

$$\times \sum_{\mu \in (\Lambda_{N/2}^{+>}, \Lambda_{N/2-1}^{+>})} \langle J_\mu^{(--)}, J_\mu^{(--)} \rangle_{0,N-1} |b_\mu^{(--)}(\lambda + N/2)|^2 \exp\left(-i\omega_\mu t + i P_\mu x \right),$$

$$\tag{3.219}$$

where the excitation energy and momentum are given by

$$\omega_\mu = E[\mu + J] - E_{g,N}$$

$$= \left(\frac{2\pi}{L} \right)^2 \sum_{i=1}^{N-1} \left((\mu^+)_i + \frac{\lambda}{2}(N - 2i) + J \right)^2 - E_{g,N}, \tag{3.220}$$

$$P_\mu = \frac{2\pi}{L} \left(|\mu| + (N-1)J \right). \tag{3.221}$$

From (3.219) and (3.213), we obtain the hole propagator in a finite size:

$$G^-(x,t) = \frac{d}{2} \sum_{\mu \in (\Lambda_{N/2}^{+>}, \Lambda_{N/2-1}^{+>})} \frac{\langle J_\mu^{(--)}, J_\mu^{(--)} \rangle_{0,N-1}}{\langle J_{\mu_0}, J_{\mu_0} \rangle_0} |b_\mu^{(--)} \left(\lambda + \frac{N}{2} \right)|^2$$

$$\times \exp\left(-i\omega_\mu t + i P_\mu x \right). \tag{3.222}$$

All factors on the RHS are available in Chapter 7; the norm of the U(2) Jack polynomial is given in (7.356) and the coefficients $b_\mu^{(--)}(\lambda + N/2)$ of the binomial formula are given in (7.384).

We can derive (3.186) from (3.222) in the same way as in Section 2.7.1. First we find the condition that an excited state μ has non-vanishing contribution (3.222). According to (7.384), the matrix element $b_\mu^{(--)}(\lambda + N/2)$ contains the factor $(\lambda + N/2 - (\mu^+)_1)$. Therefore the Young diagram of μ^+ containing the cell $(i,j) = (1, \lambda + N/2)$ does not contribute to the sum in (3.222). Figure 3.9 shows for $\lambda = 1$ a typical Young diagram of $\mu \in (\Lambda_{N/2}^{+>}, \Lambda_{N/2-1}^{+>})$ which contributes to the sum in (3.222). With $\lambda = 1$, the compositions μ relevant to (3.222) can be parameterized by (p_1, q_1, q_2). These three numbers are (dimensionless) momenta of quasi-holes.

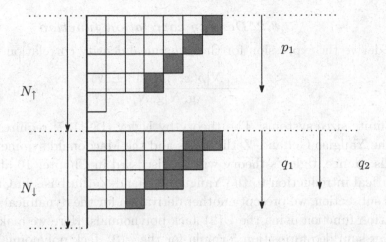

Fig. 3.9. A typical Young diagram of $\mu \in (\Lambda_{N/2}^{+>}, \Lambda_{N/2-1}^{+>})$, which contributes to the sum in (3.222). Here we take $\lambda = 1$ and $N = 12$. Unshaded squares represent the diagram of $(\delta(N/2), \delta(N/2 - 1))$ corresponding to the ground state. Shaded squares are parameterized by three non-negative integers p_1, q_1, and q_2, which are proportional to the momenta of quasi-holes.

For a general non-negative integer λ, the compositions μ relevant to (3.222) can be parameterized by

$$(p_1, \ldots, p_\lambda, q_1, \ldots, q_{\lambda+1}). \tag{3.223}$$

Here $p_1, \ldots, p_\lambda$ are momenta of quasi-holes with spin $\downarrow$ and $q_1, \ldots, q_{\lambda+1}$ with spin $\uparrow$. This interpretation is consistent with the selection rule in Section 3.7.1. The set of parameters (3.223) corresponds to $(\kappa_1', \ldots, \kappa_\lambda')$ in Section 2.6.1. The RHS can be expressed in terms of (3.223) and N, L, λ.

Now we consider the thermodynamic limit with $d = N/L$ fixed. In this limit, it turns out that only the configurations with p_k, q_l, $|p_k - p_l|$, $|q_k - q_l|$ and $|p_k - q_l| \sim \mathcal{O}(N)$ give finite contributions to the hole propagator. Introducing the velocities of quasi-holes

$$u_k = \lim_{N \to \infty} 1 - \frac{4p_k}{N}, \quad \text{for } k = 1, \ldots, \lambda, \tag{3.224}$$

$$v_l = \lim_{N \to \infty} 1 - \frac{4q_l}{N}, \quad \text{for } l = 1, \ldots, \lambda + 1 \tag{3.225}$$

and using the Stirling formula, for $|z| \to \infty$, we arrive at the final expression (3.186) for the hole propagator in the thermodynamic limit.

3.8.2 Density correlation function

We derive the expression for the dynamical density correlation function

$$\frac{\langle g, N | \hat{\rho}(x,t)\hat{\rho}(0,0) | g, N \rangle}{\langle g, N | g, N \rangle} \tag{3.226}$$

for finite-sized systems. The theory by Uglov [189, 197] requires knowledge of the Yangian Gelfand–Zetlin basis and the Macdonald symmetric polynomials. Hence, Uglov's theory will be discussed in Chapter 10 after a mathematical introduction to the Yangian Gelfand–Zetlin basis in Chapter 8. In this subsection, we present another derivation for the dynamical density correlation function using the U(2) Jack polynomials. Here we make use of the power-sum decomposition formula for the U(2) Jack polynomials [12].

We set λ to be a non-negative integer. The original particles are taken to be fermions when $\zeta = 1$. In this case, the absolute ground state is symmetric with respect to particle exchange and obeys the periodic boundary condition. Hence Φ is antisymmetric and obeys the periodic boundary condition. The particle number N is taken to be twice an odd integer so that the ground state has no degeneracy.

We remark that the following derivation is essentially applicable to the case where the original particles are taken as bosons when λ is an integer, $\zeta = -1$ and N is twice an even integer.

Since $\hat{\rho}(0,0)$ does not act on the spin variables, we can fix the spin variables to $\{\sigma_0\}$ given by (3.211). We first rewrite $\Psi_{g,N}$ as

$$i^{N^2/4} \prod_{i=1}^{N} z_i^{J'} \prod_{1 \leq i < j \leq N/2} (z_i - z_j) \prod_{N/2+1 \leq i < j \leq N} (z_i - z_j) \, \Psi_{0,N}, \tag{3.227}$$

with $J' = -(N/2 - 1)/2$. According to (2.307), the action of the density operator $\hat{\rho}(0,0)$ on the wave function is equivalent to multiplying the sum of the power-sum symmetric functions $p_n(z_1, \ldots, z_N)$.

Further, using (7.404), we can decompose the power-sum symmetric polynomials into the linear combination of $\tilde{J}^{(--)}(z_1, \ldots, z_N)$. Combining (3.227), (2.307), and (7.404), we obtain

$$\hat{\rho}(0,0)\Psi_{g,N}(\{x_i\}, \{\sigma_i\})\Big|_{\sigma_1 = \cdots = \sigma_{N/2} = 1, \quad \sigma_{N/2+1} = \cdots = \sigma_N = -1}$$

$$= i^{N^2/4}\Psi_{0,N} \sum_{\mu \in (\Lambda_{N/2}^{+>}, \Lambda_{N/2}^{+>})} \frac{|\mu| c_\mu}{\lambda L} \left\{ J_{\mu+J'}^{(--)}(z_1, \ldots, z_N) + \text{c.c.} \right\}, \tag{3.228}$$

where we denote, by $\mu + J'$,

$$\mu + J' = (\mu_1 + J', \ldots, \mu_N + J'). \tag{3.229}$$

When P acts on (3.228), the two terms

$$J^{(--)}_{\mu+J'}(z_1, \ldots, z_N)\tilde{\Psi}_{0,N}, \quad J^{(--)}_{\mu+J'}(\bar{z}_1, \ldots, \bar{z}_N)\tilde{\Psi}_{0,N} \tag{3.230}$$

in (3.228) yield the opposite eigenvalues

$$2\pi(|\mu| + NJ')/L, \quad -2\pi(|\mu| + NJ')/L$$

for P. The two terms (3.230) in (3.228), on the other hand, are eigenfunctions H' with a common eigenenergy; the expression for eigenenergy is given below. Using the similarity transformation (3.16), we can rewrite (3.228) as

$$\langle g, N|\hat{\rho}(x,t)\hat{\rho}(0,0)|g,N\rangle \tag{3.231}$$

$$= \frac{N!L^N}{((N/2)!)^2} \sum_{\mu \in (\Lambda^{+>}_{N/2}, \Lambda^{+>}_{N/2})} 2\left(\frac{|\mu|c_\mu}{\lambda L}\right)^2 \langle J^{(--)}_{\mu+J'}, J^{(--)}_{\mu+J'}\rangle_0$$

$$\times \cos\left(2\pi(|\mu| + NJ')x/L\right)\exp\left(-\mathrm{i}\omega_\mu t\right). \tag{3.232}$$

Here ω_μ is given by

$$\omega_\mu = E[\mu + J'] - E_{\mathrm{g},N}, \tag{3.233}$$

with

$$E[\mu + J'] = \left(\frac{2\pi}{L}\right)^2 \sum_{i=1}^{N} \left((\mu^+)_i + J' + \frac{\lambda}{2}(N+1-2i)\right)^2 \tag{3.234}$$

and the ground-state energy $E_{\mathrm{g},N}$ given in (3.115). With this result and (3.213), we obtain the expression for the dynamical density correlation function in finite-sized systems:

$$\frac{\langle g, N|\hat{\rho}(x,t)\hat{\rho}(0,0)|g,N\rangle}{\langle g,N|g,N\rangle}$$

$$= \sum_{\mu \in (\Lambda^{+>}_{N/2}, \Lambda^{+>}_{N/2})} 2\left(\frac{|\mu|c_\mu}{\lambda L}\right)^2 \frac{\langle J^{(--)}_\mu, J^{(--)}_\mu\rangle_0}{\langle J^{(--)}_{\mu_0}, J^{(--)}_{\mu_0}\rangle_0}$$

$$\times \exp\left(-\mathrm{i}\omega_\mu t\right)\cos\left(2\pi(|\mu| + NJ')x/L\right). \tag{3.235}$$

The norms of the Jack polynomials appearing in (3.235) are available in Section 7.4.2. The matrix element c_μ can be given by (7.405). With this knowledge, we can obtain the expression (3.202) for the thermodynamic limit. The procedure is the same as that used in Section 2.6 for the single-component model and the hole propagator in Section 3.8.1. The summation with respect to μ in (3.235) can be simplified; the number of parameters is that of the excited quasi-particles and quasi-holes as shown below. The expression for the matrix element c_μ contains the factor

$$b_\mu^{(--)} = \prod_{s'(\neq(1,N/2))\in D(\mu^+)\setminus D(\mu_0^+)} \left\{ \frac{a'(s)+1-N/2}{\lambda} - l'(s) \right\},$$

which is derived in Section 7.4.7. The expression in $\{\cdot\}$ becomes zero when $s = (2, \lambda + N/2)$ and nonzero otherwise. The condition for μ to contribute to the sum in (3.235) is thus $s = (2, \lambda + N/2) \notin D(\mu^+)$, which is followed by

$$(\mu^+)_2 \leq \lambda + N/2 \tag{3.236}$$

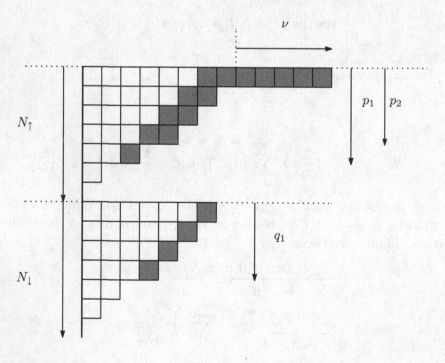

Fig. 3.10. A typical Young diagram of $\mu \in (\Lambda_{N/2}^{+>}, \Lambda_{N/2}^{+>})$ contributes to the sum in (3.235). Here we take $\lambda = 1$ and $N = 14$. Unshaded squares represent the diagram of μ_0 corresponding to the ground state. Shaded squares are parameterized by four non-negative integers ν, p_1, p_2, and q_1. These are, respectively, proportional to the momentum of a quasi-particle or a quasi-hole.

while $(\mu^+)_1$ is unbounded. $(\mu^+)_1$ comes from either $\mu_1^\uparrow$ or $\mu_1^\downarrow$ in

$$\mu = (\mu^\uparrow, \mu^\downarrow), \quad \mu^\sigma \in \Lambda_{N/2}^{+>}, \quad \sigma = \uparrow, \downarrow.$$

We consider the case $(\mu^+)_1 = \mu_1^\uparrow$; the other case can be considered in the same way. $(\mu^+)_2$ comes from either $\mu_2^\uparrow$ or $\mu_1^\downarrow$. The condition (3.236) then becomes

$$\max(\mu_2^\uparrow, \mu_1^\downarrow) \le \lambda + N/2. \tag{3.237}$$

A typical diagram of μ^+ for $\mu \in (\Lambda_{N/2}^{+>}, \Lambda_{N/2}^{+>})$ for $N = 14$ and $\lambda = 1$ is shown in Fig. 3.10. We see that when a quasi-particle with spin $\uparrow$ is excited, two quasi-holes with spin $\downarrow$ and a quasi-hole with spin $\uparrow$ are accompanied. In this figure, μ, p_1, p_2, q_1 are, respectively, integers proportional to the momenta of quasi-particle or quasi-holes. For general non-negative integers λ, μ in (3.235) can be parameterized by

$$(\mu, p_1, \ldots, p_{\lambda+1}, q_1, \ldots, q_\lambda). \tag{3.238}$$

This result describes the selection rule in the density correlation function for the U(2) model. By the action of $\hat{\rho}$ on the singlet ground state, a quasi-particle with spin σ, $\lambda + 1$ quasi-holes with spin $-\sigma$, and $\lambda + 1$ quasi-holes with spin σ are excited. In terms of (3.238), the expression (3.235) can be rewritten. With use of the parameterization (3.225) and

$$\frac{\nu}{(2\lambda + 1)N} = \frac{w - 1}{4}, \tag{3.239}$$

and the Stirling formula, we arrive at the expression (3.202) for the dynamical density correlation in the thermodynamic limit.

4

Spin chain with $1/r^2$ interactions

In this chapter, we discuss the static and dynamic properties of the spin chain which is often referred to as the Haldane–Shastry model [77, 161]. In contrast with more familiar spin chains with the nearest-neighbor exchange interaction, the Haldane–Shastry model has a particular form of the long-range exchange interaction. In spite of its peculiar shape, it has turned out that the Haldane–Shastry model is the most fundamental one-dimensional spin system. We shall start with a discussion of how the model is solved for the ground state in Sections 4.1 to 4.3, and proceed to static correlation functions in Sections 4.4 and 4.5. The excitation spectrum is interpreted by magnons in Section 4.6, and by spinons in Section 4.7. The spinon picture gives the correct degeneracy of the energy levels in Section 4.8, and leads to the derivation of thermodynamics in Section 2.4, and the dynamical correlation function in Section 4.11. Most results have been obtained in an analytic form without any approximation.

There is a close connection between the spectrum of the spin chain and that of the Sutherland model. The most remarkable fact is that the spin chain has a correspondence with two different values of the coupling parameter λ in the Sutherland model. Namely, the spectrum is mapped to the case with either $\lambda = 2$ or $\lambda = \infty$. In the latter case, particles crystallize with equal spacing, and a small oscillation from the equilibrium, as well as exchange interactions, make up the spectrum. In both cases of λ, however, the degeneracy of each level in the spin chain is not reproduced. Hence, the mapping as it stands cannot be used for thermodynamics. We discuss how the degeneracy is related to the basic symmetry in the model, which is identified with the Yangian as in the multi-component Sutherland model. In the present case, however, the relevant Yangian is not $Y(gl_2)$, but $Y(sl_2)$ without the U(1) degrees of freedom.

4.1 Mapping to hard-core bosons

Let us consider a spin chain

$$H_{\mathrm{HS}} = \sum_{i<j} J_{ij} \boldsymbol{S}_i \cdot \boldsymbol{S}_j, \qquad (4.1)$$

where $\boldsymbol{S}_i$ denotes the spin operator with $S = 1/2$ at site i, and the exchange interaction J_{ij} decays as the inverse square of the distance. We consider the N-site system with unit lattice spacing. With the periodic boundary condition, the distance is replaced by the cord distance $|D(i-j)|$ of a ring, where

$$D(i-j) = \frac{N}{\pi} \sin\left[\frac{\pi}{N}(i-j)\right].$$

Namely, J_{ij} is given by

$$J_{ij} = J D(i-j)^{-2}.$$

The Hamiltonian H_{HS} is often called the Haldane–Shastry model [77, 161]. Unless otherwise stated, we consider the case of even number N for the lattice site.

Let us take the hard-core boson representation, which is generally valid for spin chains, and is especially suitable to derive eigenstates of the Haldane–Shastry model. We introduce a creation operator of a hard-core boson:

$$b_i^\dagger = c_{i\downarrow}^\dagger c_{i\uparrow} = S_i^-.$$

It is evident from the properties of spin operators that

$$b_i^\dagger b_i + b_i b_i^\dagger = 1.$$

Namely, the hard-core boson obeys the fermion commutation rule for the same site, and the boson commutation rule for different sites. The hard-core boson may be called a magnon since it describes a spin flip from the fully polarized reference state. The z-component of the spin operator is given by

$$S_i^z = \frac{1}{2} - n_i,$$

where $n_i = b_i^\dagger b_i$ is the number operator of the boson at site i. In terms of hard-core boson (or magnon) operators, H_{HS} is written as

$$H_{\mathrm{HS}} = \frac{1}{2} \sum_{i \neq j} J_{ij}(b_i^\dagger b_j + n_i n_j) + \frac{J(0)}{2} \sum_{i=1}^{N} \left(\frac{1}{4} - n_i\right), \qquad (4.2)$$

with $J(0)$ being the $q = 0$ component of the Fourier transform $J(q)$ of J_{ij}.

In order to make explicit calculations using the specific property of the model, it is most convenient to take the complex coordinate representation. We introduce $z_j = \exp(2\pi i x_j/N)$ with $x_j \in [1, N]$, and $Z_j = \exp(2\pi i j/N)$. Note that we distinguish between Z_j and z_j. The exchange interaction is then represented as

$$J_{ij} = J\left(\frac{2\pi}{N}\right)^2 \frac{1}{|Z_i - Z_j|^2} = J\left(\frac{2\pi}{N}\right)^2 \frac{-Z_i Z_j}{(Z_i - Z_j)^2}, \tag{4.3}$$

where the Fourier transform is given by [176]

$$J(q) = \sum_{j=2}^{N} J_{1j} \exp[iq(j-1)] = \frac{J}{2}(q-\pi)^2 - \frac{\pi^2 J}{6}\left(1 + \frac{2}{N^2}\right), \tag{4.4}$$

where $0 \leq q < 2\pi$ with the spacing of the adjacent q being $2\pi/N$. The constant term makes the sum of $J(q)$ vanish, which is related to the absence of the diagonal term J_{ii}. For reference purposes, we quote the case $q = 0$:

$$J(0) = \sum_j J_{ij}(1 - \delta_{ij}) = \frac{\pi^2 J}{3}\left(1 - \frac{1}{N^2}\right). \tag{4.5}$$

We represent the wave function $\Psi(\{x\})$ of the system, taking the fully up-polarized state as the reference. Let us first consider the one-body wave function. Since z_i^n with $n = 0, 1, \ldots, N-1$ spans a complete set for one-body states, one may restrict the degree of the polynomial to less than N for each z_i. For the case of $\Psi(x) = z^n$, the boson-transfer term acts as

$$\langle x| \sum_{i \neq j} J_{ij} b_i^\dagger b_j |\Psi\rangle = \sum_{j \neq x} J_{xj} Z_j^n$$

$$= \Psi(x) \sum_{j \neq x} J_{xj} \exp[2\pi i n(j-x)/N] = J\left(-i\frac{\partial}{\partial x}\right)\Psi(x).$$

Here we have replaced $q = 2\pi n/N$ in $J(q)$ by $-i\partial/\partial x$ when acting on z^n. A remarkable feature, as apparent in (4.4), is that $J(q)$ is a quadratic function of q [176]. In other words, the spectrum of the magnon has a strong resemblance to that of free particles in continuum space. The difference appears first as the presence of the cut-off in the momentum, and second as the shift π of the origin of q.

4.2 Gutzwiller–Jastrow wave function

4.2.1 Hole representation of lattice fermions

We shall discuss the coordinate representation of lattice particles in more detail. As the simplest example, we first take the case of non-interacting spinless fermions. Let M be the number of fermions, and N the total number of lattice sites. With lattice constant unity, N also represents the length of the system. We take N to be even and impose the periodic boundary condition.

Let us first consider the wave function of the ground state:

$$|0\rangle = \sum_{\{x\}} \Psi_0(x_1, \ldots, x_M) c^\dagger(x_1) c^\dagger(x_2) \cdots c^\dagger(x_M) |\text{vacuum}\rangle, \qquad (4.6)$$

where $c^\dagger(x_i)$ creates an electron at position x_i. We assume that the spectrum ϵ_k is a monotonically increasing function of $|k| \leq \pi$. Then the electrons occupy the states with consecutive momentum k_i with $i = 1, \ldots, M$ given by

$$k_i = -k_\mathrm{F} + 2(i-1)\pi/N.$$

We take M odd, and obtain the minimum $k_1 = -k_\mathrm{F}$ and the maximum

$$k_M = k_\mathrm{F} = 2\pi N_\mathrm{F}/N$$

of momenta where k_F is the Fermi momentum with $M = 2N_\mathrm{F} + 1$. The wave function corresponding to the Slater determinant is given by

$$\Psi_0(x_1, \ldots, x_M) = \det\{\exp(\mathrm{i}k_i x_j)\} = \prod_{i=1}^{M} z_i^{-N_\mathrm{F}} \prod_{i<j}(z_i - z_j), \qquad (4.7)$$

where we have introduced the notation $z_j = \exp(2\pi\mathrm{i}x_j/N)$, and used the property of the Vandermonde determinant. Note that $z_i^N = 1$ for any integer x_i. By using the relation

$$z_i - z_j = \exp\left[\frac{\mathrm{i}}{2}(\theta_i + \theta_j)\right] 2\mathrm{i}\sin\left[\frac{1}{2}(\theta_i - \theta_j)\right]$$

with $\theta_i = 2\pi x_i/N$, the wave function is also written as

$$\Psi_0(x_1, \ldots, x_M) = (2\mathrm{i})^{M(M-1)/2} \prod_{i<j} \sin\left[\frac{1}{2}(\theta_i - \theta_j)\right]. \qquad (4.8)$$

The wave function carries no current or momentum, since it is essentially real except for the constant factor.

In later discussions it is necessary to work with an alternative representation which starts from the fully occupied state:

$$|\mathrm{F}\rangle = c_1^\dagger c_2^\dagger \cdots c_N^\dagger |\text{vacuum}\rangle,$$

where $c_1^\dagger$ is the same as $c^\dagger(x_1 = 1)$. In the momentum representation, $|F\rangle$ has all momentum in the Brillouin zone filled. The consecutive momenta k_j of holes in the ground state are given as follows:

$$k_j = k_F + 2\pi j/N,$$

where $j = 1, 2, \ldots, Q$ with $Q = N - M$. The crystal momentum k is equivalent to $k - 2\pi$. Hence, if k_j is larger than the right boundary of the Brillouin zone π, such k_j is considered to belong to the negative side $(-\pi, 0]$ of the first Brillouin zone. Then the ground state can also be written, apart from a normalization factor, as

$$|0\rangle = \sum_{\{y\}} \Phi_0(y_1, \ldots, y_Q) c(y_Q) c(y_{Q-1}) \cdots c(y_1) |F\rangle. \tag{4.9}$$

We refer to the wave function $\Phi_0(y_1, \ldots, y_Q)$ as the hole representation. In accordance with the anticommuting property of fermion operators $c(y_i)$, the wave function $\Phi_0(y_1, \ldots, y_Q)$ should be antisymmetric against exchange of two coordinates. By the same procedure as that for $\Psi_0(\{x\})$ one obtains

$$\Phi_0(y_1, \ldots, y_Q) = \det\{\exp(\mathrm{i}k_l y_m)\} = \prod_i z_i^{(N_F+1)} \prod_{i<j} (z_i - z_j), \tag{4.10}$$

where $z_j = \exp(2\pi \mathrm{i} y_j/N)$. Alternatively, it is also written as

$$\det\{\exp(\mathrm{i}k_l y_m)\} = \prod_i z_i^{N/2} \prod_{i<j} 2\mathrm{i} \sin\left[\frac{1}{2}(\phi_i - \phi_j)\right] \tag{4.11}$$

with $\phi_i = 2\pi y_i/N$.

With $N/2$ being odd, the state (4.11) may appear to have a finite momentum because of the complex part $z_i^{N/2}$. In other words, the total momentum associated with $\Phi_0(\{y\})$ may appear to be π. However, the total crystal momentum of the state should vanish, to be consistent with (4.8). The momentum π associated with $\Phi_0(\{y\})$ is actually relative to the reference state $|F\rangle$, which also has the momentum π. Hence there is no contradiction. It is useful to remember that the center of momentum distribution of holes is at the edge of the Brillouin zone.

If the electron number M is even, the ground state of the Fermi gas is two-fold degenerate, and the maximum of the occupied momentum $k_F = 2\pi N_F/N$ is characterized by

$$N_F = (M - 1)/2 \pm 1/2.$$

The ground state has a finite momentum in this case.

4.2.2 Gutzwiller wave function in Jastrow form

Now we include spin of electrons. Let us consider the situation where the hard-core repulsion between the electrons prohibits the occupancy of any site by two electrons, but otherwise the wave function is the same as for free electrons. This wave function is represented by a limiting case of the Gutzwiller wave function, which in general allows for finite double occupation. In the particle representation, the limiting case is represented by

$$|\Psi_G\rangle = \sum_{\{x\uparrow\}} \sum_{\{x\downarrow\}}' \Psi_0(x_{1\uparrow},\ldots,x_{N\uparrow}) \Psi_0(x_{1\downarrow},\ldots,x_{N\downarrow})$$

$$\times \prod_{i\in\{x\uparrow\}} c_{i\uparrow}^\dagger \prod_{j\in\{x\downarrow\}} c_{j\downarrow}^\dagger |\text{vacuum}\rangle, \qquad (4.12)$$

where the prime ($'$) in the summation over $\{x\downarrow\}$ means that none of the coordinates in $\{x\uparrow\}$ and $\{x\downarrow\}$ should coincide. It is understood that the creation operators are ordered so that i and j increase from the left.

By using the hole representation, one can neatly avoid the hard-core restriction on the site summation [4]. Namely, one starts from the fully up-polarized state $|\text{F}\uparrow\rangle$ and uses the hole representation for up spins. The hard-core constraint is satisfied automatically provided one creates down-spin electrons only among the hole sites. In the case of spin chain, the number of electrons $N = N_\uparrow + N_\downarrow$ is the same as the number L of lattice sites. Then the hole sites $\{y\}$ for the up spin and the electron sites $\{x\downarrow\}$ for the down spin are identical to each other. Thus we represent both sites by $\{x\}$ and obtain

$$|\Psi_G\rangle = \sum_{\{x\}} \Psi_0(\{x\})\Phi_0(\{x\}) \prod_{i\in\{x\}} c_{i\downarrow}^\dagger c_{i\uparrow} |\text{F}\uparrow\rangle.$$

Here $\{x\}$ denotes the set of coordinates for $M = N_\downarrow$ down-spin electrons. Let us first take the singlet case $M = N/2$. The wave function $\Psi_G(\{x\}) = \Psi_0(\{x\})\Phi_0(\{x\})$ is given by

$$\Psi_G(\{x\}) = \prod_i z_i^{N/2} \prod_{i<j}(-4)\sin^2\left[\frac{1}{2}(\theta_i - \theta_j)\right] = \prod_{i=1}^{N/2} z_i \prod_{i<j}(z_i - z_j)^2. \quad (4.13)$$

It appears that $\Psi_G(\{x\})$ has a momentum π with $N/2$ odd because of the factor $z_i^{N/2}$ in (4.13). As discussed in (4.11), it cancels with the momentum of the reference state, and $|\Psi_G\rangle$ itself carries no momentum. For reference we also quote the Gutzwiller wave function for general $M \leq N/2$ including the case of finite magnetization:

$$\Psi_{\mathrm{G}}(\{x\}; M) = \prod_{i=1}^{M} z_i^{N/2-M+1} \prod_{i<j} (z_i - z_j)^2. \tag{4.14}$$

One may ask whether it is possible to represent the Gutzwiller wave function without using the hole representation. The exclusion of double occupation can be dealt with by a limiting procedure [71], which we now explain. We take a fermionic wave function, which has been discussed in Section 3.2.2, as follows:

$$\Psi_0(\{z_\uparrow, z_\downarrow\}; \lambda) = \prod_{i\sigma} z_i^{-(N_\sigma-1)/2} \prod_{ij} |z_i - z_j|^\lambda (z_i - z_j)^{\delta(\sigma_i, \sigma_j)}, \tag{4.15}$$

where the parameter λ corresponds to the coupling constant in the two-component Sutherland model. With $\lambda = 0$, the wave function gives the ground state of free fermions. On the other hand, double occupation is prohibited with any finite λ. The Gutzwiller projection can then be imposed by considering the limit $\lambda \to 0$, provided the calculation can be done for arbitrary λ.

4.3 Projection to the Sutherland model

Utilizing the quadratic dispersion, we can conveniently work with the first-quantized representation. Within the many-body Hilbert space spanned by polynomials of z_i, we may replace q in (4.4) by $-i\partial/\partial x_i$. The two-body term $J_{ij}n_i n_j$ is easily represented by the first-quantized Hamiltonian $H_{1\mathrm{st}}$. For simplicity, the unit of energy is chosen as $J = 2$. Then $H_{1\mathrm{st}}$ is given by

$$H_{1\mathrm{st}} = \frac{1}{2} \sum_{i=1}^{M} \left(-i\frac{\partial}{\partial x_i} - \pi \right)^2 + 2\sum_{i<j} D(x_i - x_j)^{-2} + E_{\mathrm{c}}, \tag{4.16}$$

where

$$E_{\mathrm{c}} = \frac{J(\pi)}{2} M + \frac{J(0)}{2}\left(\frac{N}{4} - M\right) = \frac{\pi^2 N}{12}\left(1 - \frac{6M}{N} - \frac{1}{N^2}\right).$$

We find that $H_{1\mathrm{st}}$ takes the same form as the Sutherland model with the repulsion parameter $\lambda = 2$. Since the Gutzwiller wave function $\Psi_{\mathrm{G}}(\{x\})$ is a polynomial of z_i, one can use $H_{1\mathrm{st}}$ to act on $\Psi_{\mathrm{G}}(\{x\})$ as the spin-chain Hamiltonian. Then the factor z_i for $i \in [1, M]$ in (4.13) absorbs the shift of the momentum in $H_{1\mathrm{st}}$. The remaining part $\prod_{i<j} \sin^2[(\theta_i - \theta_j)/2]$ in $\Psi_{\mathrm{G}}(\{x\})$ is known to be the ground-state wave function of the Sutherland model.

Hence, $|\Psi_G\rangle$ proves to be an eigenfunction of H_{HS}. The corresponding energy $E_0(M)$ is given by

$$E_0(M) = E_c + \frac{1}{6}\left(\frac{2\pi}{N}\right)^2 M(M^2 - 1), \tag{4.17}$$

where the second term gives the ground-state energy of the Sutherland model with M particles. The minimum of $E_0(M)$ occurs at $M = N/2$, which corresponds to $S^z = 0$. We shall prove in Section 4.8.1 that $\Psi_G(\{x\})$ indeed gives the ground state of H_{HS}. The ground-state energy is given by

$$E_0\left(\frac{N}{2}\right) = -\frac{\pi^2}{12}\left(N + \frac{5}{N}\right). \tag{4.18}$$

We notice here that the polynomial wave function does not span the complete set of the many-boson states. For example, consider the case of $N = 4, M = 2$, and $S = 2$. Such a state $|\Psi_{2,0}\rangle$ can be constructed from $|F\uparrow\rangle$, apart from normalization, by

$$|\Psi_{2,0}\rangle = \left(\sum_{i=1}^{4} S_i^-\right)^2 |F\uparrow\rangle = \sum_{i \neq j} b_i^\dagger b_j^\dagger |F\uparrow\rangle. \tag{4.19}$$

The amplitude $\langle x_1, x_2|\Psi_{2,0}\rangle = \Psi_{2,0}(x_1, x_2)$ becomes constant for $x_1 \neq x_2$, but vanishes for $x_1 = x_2$. Such behavior of the amplitude cannot be described by polynomials of z_1 and z_2. Thus the first-quantized Hamiltonian H_{1st} can act only on a subset of the whole spin states. This subset coincides with the Yangian highest-weight states, as will be discussed in Section 4.8.1.

4.4 Static structure factors

The wave function of the ground state determines the equal-time correlation function $C(x)$ in the real space, or the static structure factor $S(q)$ in the momentum space. If we allow for finite magnetization $m = 1 - 2M/N > 0$ in the presence of a magnetic field, the rapidity k_i of magnons is occupied for $|k| < \pi(1 - m) \equiv k_m$ and is empty near the edge. Here we have defined the critical momentum k_m which plays the role of the Fermi momentum of magnons. Figure 4.3 in Section 4.7.3 illustrates the spectrum, as will be explained together with spinons. Since there is no longer SU(2) symmetry in the magnetic field, we have two independent components for the correlation function, as defined by

$$C^{zz}(i - j) = \langle S_i^z S_j^z \rangle - \langle S_i^z \rangle \langle S_j^z \rangle = \langle b_i^\dagger b_i b_j^\dagger b_j \rangle - \langle b_i^\dagger b_i \rangle \langle b_j^\dagger b_j \rangle, \tag{4.20}$$

$$C^\perp(i - j) = \langle S_i^x S_j^x \rangle = \langle S_i^y S_j^y \rangle = \frac{1}{2}\left(\langle b_i^\dagger b_j \rangle + \langle b_i b_j^\dagger \rangle\right). \tag{4.21}$$

Although the coordinates i, j are integers, we can use the results for the Sutherland model in the continuum space. In other words, when taking the average over $\Psi_G(\{x\})$, we may replace the summation over lattice sites by integration over the continuum space:

$$\sum_{\{x\}} f(\{x\})|\Psi_G(\{x\})|^2 = \int_0^N dx_1 \ldots \int_0^N dx_M f(\{x\})|\Psi_G(\{x\})|^2. \quad (4.22)$$

To check the validity of the above, let us consider the expansion of the summand in terms of $z_1 = \exp(2\pi i x_1/L)$. Then among terms of the form z_1^n, only $n = 0$ survives summation over x_1, and the summation gives N for this term. The same happens in the integration over x_1 on the RHS. Therefore, the replacement by an integral is exact for these polynomials.

Haldane noticed [77] that $C^{zz}(i-j)$ and $C^{\perp}(i-j)$ in the thermodynamic limit are related to the correlation function and the density matrix respectively in the Sutherland model with the repulsion parameter $\lambda = 2$, which have been derived by Sutherland [173]. In a different context, results for $C^{zz}(i-j)$ have been derived by Mehta and Mehta [129] for an arbitrary size of lattice system. These results rely on sophisticated techniques of the random matrix theory [48, 49, 130], and are not easy to access. Therefore we first present the results of correlation functions, deferring the details of derivation to a later part of the section and to Section 4.5.

In the real space the results are given with $k_m/\pi = 2M/N = 1 - m$ by

$$C^{zz}(x) = -\left(\frac{\sin k_m x}{2\pi x}\right)^2 + \left[\frac{k_m \cos k_m x}{4\pi^2 x} - \frac{\sin k_m x}{(2\pi x)^2}\right] \mathrm{Si}(k_m x), \quad (4.23)$$

$$C^{\perp}(x) = \frac{1}{2} m \delta_{x,0} + \frac{(-1)^x}{4\pi x} \mathrm{Si}(k_m x), \quad (4.24)$$

where $\mathrm{Si}(y)$ is the sine integral defined by

$$\mathrm{Si}(y) = \int_0^y dt \frac{\sin t}{t}. \quad (4.25)$$

Note that $C^{\perp}(x)$ has an oscillating factor $(-1)^x$ which comes from the shift π of the momentum in mapping to the Sutherland model. Hence this factor is absent in the density matrix in the Sutherland model. Without magnetization, we obtain $k_m = \pi$ and $\sin \pi x = 0$. Then the correlation function becomes

$$C^{\perp}(x) = C^{zz}(x) = \frac{1}{4\pi x} \mathrm{Si}(\pi x) \cos \pi x \to \frac{(-1)^x}{8|x|}, \quad (|x| \gg 1), \quad (4.26)$$

which is isotropic, thus recovering the SU(2) symmetry.

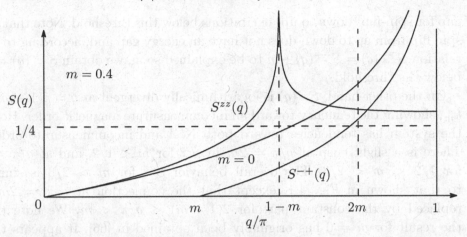

Fig. 4.1. Static structure factors $S^{zz}(q)$ and $S^{-+}(q)$ of the Haldane–Shastry spin chain with and without magnetization m.

The structure factors in the momentum space are given for $0 < q < \pi$ by

$$S^{\perp}(q) = \frac{m}{2} - \frac{1}{4}\theta(q - \pi m)\ln\frac{1 - q/\pi}{1 - m}, \qquad (4.27)$$

for the transverse component. The longitudinal one $S^{zz}(q)$ for $m < 2/3$ is given by

$$
\begin{aligned}
S^{zz}(q) = {} & \theta(2\pi m - q)\frac{q}{4\pi}\left(1 - \frac{1}{2}\ln\left|1 - \frac{q}{k_m}\right|\right) \\
& + \theta(q - 2\pi m)\left(\frac{m}{2} - \frac{1}{4}\ln\left|1 - \frac{2\pi - q}{k_m}\right| + \frac{q}{8\pi}\ln\left|\frac{2\pi - k_m - q}{k_m - q}\right|\right).
\end{aligned}
\qquad (4.28)
$$

For $2/3 < m < 1$, it is given by

$$
\begin{aligned}
S^{zz}(q) = {} & \theta(2k_m - q)\frac{q}{4\pi}\left(1 - \frac{1}{2}\ln\left|1 - \frac{q}{k_m}\right|\right) \\
& + \theta(q - 2k_m)\theta(\pi m - q)\frac{1}{2}(1 - m) \\
& + \theta(q - \pi m)\left(\frac{m}{2} - \frac{1}{4}\ln\left|1 - \frac{2\pi - q}{k_m}\right| + \frac{q}{8\pi}\ln\left|\frac{2\pi - k_m - q}{k_m - q}\right|\right).
\end{aligned}
\qquad (4.29)
$$

Figure 4.1 illustrates the structure factors including the case of $m \neq 0$. The transverse structure factor $S^{\perp}(q)$ is divergent logarithmically at $q = \pi$. The singularity corresponds to a tendency to the Néel order. There is a threshold $q/\pi = m$ below which $S^{-+}(q) = 0$. This indicates the presence of an energy gap for spin-flip (down to up) excitations below this threshold. Note that the spin flip from up to down does not have an energy gap and, according to the relation $S^{+-}(q) = S^{-+}(q) + m$ to be explained soon, we obtain $S^{+-}(q) = m$ below the threshold.

On the other hand, $S^{zz}(q)$ is logarithmically divergent at $q = \pi(1-m) = k_m$, showing the tendency toward an incommensurate magnetic order. Hence the system has tendencies toward both Néel and incommensurate orders. There is a slight cusp at $q/\pi = 2m$ in S^{zz} for $m < 1/2$, and at $q/\pi = 1$ for $1/2 < m < 2/3$. The overall behavior S^{zz} for $m > 2/3$ is similar to that shown in Fig. 4.1, except that the connection at $q/\pi = 2m$ is replaced by the constant part for $2(1-m) < q/\pi < m$. We note that the result for $m = 0$ has originally been obtained in [66]. It appears that $S^{zz}(q)$ and $S^{\perp}(q)$ in the general case $m \neq 0$ have not been reported in the literature.

Now we proceed to the derivation of these results. We begin with the transversal correlation function which can be derived in an elementary manner. The density matrix of the M-magnon system is given by

$$\langle b_l^\dagger b_m \rangle = \frac{1}{C_M N} \int_0^N \frac{\mathrm{d}x_1}{N} \cdots \int_0^N \frac{\mathrm{d}x_{M-1}}{N} \Psi_{\mathrm{G}}(l, \{x\}_{-1}) \Psi_{\mathrm{G}}(m, \{x\}_{-1}), \quad (4.30)$$

where $\{x\}_{-1}$ denotes the set $x_1, \ldots, x_{M-1}$, and $C_M = (2M)!/2^M$ is the norm of the wave function with M magnons. The norm is derived in Section 4.5 in an elementary fashion. The general method for calculation will be presented in Section 7.2.2. In (4.30) we have emphasized that the wave functions are real. By using the complex coordinates for the real wave functions, we can write the integrand as a determinant [60]:

$$\Psi_{\mathrm{G}}(z, \ldots) \Psi_{\mathrm{G}}(\zeta, \ldots)$$

$$= \frac{e^{i\alpha}}{z - \zeta} \det \begin{pmatrix} 1 & z & z^2 & \cdots & z^{2M-1} \\ 1 & \zeta & \zeta^2 & \cdots & \zeta^{2M-1} \\ 1 & z_1 & z_1^2 & \cdots & z_1^{2M-1} \\ 0 & 1 & 2z_1 & \cdots & (2M-1)z_1^{2M-2} \\ 1 & z_2 & z_2^2 & \cdots & z_2^{2M-1} \\ 0 & 1 & 2z_2 & \cdots & (2M-1)z_2^{2M-2} \\ \vdots & \vdots & \vdots & \ddots & \vdots \\ 1 & z_{M-1} & z_{M-1}^2 & \cdots & z_{M-1}^{2M-1} \\ 0 & 1 & 2z_{M-1} & \cdots & (2M-1)z_{M-1}^{2M-2} \end{pmatrix}, \quad (4.31)$$

where $z = \exp(2\pi i l/L)$ and $\zeta = \exp(2\pi i m/L)$. The phase factor $e^{i\alpha}$ is fixed so that the RHS is real. The determinant on the RHS is a generalization of the Vandermonde determinant, and is called the confluent alternant. The simplest example of this type of determinant is given by

$$(z_1 - z_2)^4 = \det \begin{pmatrix} 1 & z_1 & z_1^2 & z_1^3 \\ 0 & 1 & 2z_1 & 3z_1^2 \\ 1 & z_2 & z_2^2 & z_2^3 \\ 0 & 1 & 2z_2 & 3z_2^2 \end{pmatrix}. \tag{4.32}$$

The general case is explained in Section 4.5.

Without loss of generality, we consider the case $l = -m$, which makes it easy to deal with the phase factor. Namely, with $\zeta = z^*$, (4.31) may be rewritten as

$$\Psi_G(z, \ldots)\Psi_G(z^*, \ldots) = (z - z^*)^{-1}$$
$$\times \det_{2M}\left(z^p, z^{-p}, z_1^p, pz_1^p, \ldots, z_{M-1}^p, pz_{M-1}^p\right)_{p=-M+1/2,\ldots,M-1/2}, \tag{4.33}$$

where we have manipulated the determinant as explained in Section 4.5. In expansion of the determinant, the only terms that survive summation or integration over x_i are those that include the product $z_i^a \times bz_i^b$ with $a + b = 0$. This type of term also appears in the calculation of the norm. After integration over $x_1, \ldots, x_{M-1}$, we are left with terms $z^p z^{-q}$ with $p + q = 0$. Then we obtain

$$\frac{1}{N}\int_0^N \frac{dx_1}{N} \cdots \int_0^N \frac{dx_{M-1}}{N}\Psi_G(z, \ldots)\Psi_G(z^*, \ldots)$$
$$= \frac{C_M}{N}\sum_{p=1/2}^{M-1/2} \frac{z^{2p} - z^{-2p}}{2p(z - z^*)}, \tag{4.34}$$

where $2p$ in the denominator comes from correction for the missing factor in producing the norm C_M. In the thermodynamic limit, the summation over p is replaced by the integral for $0 < p < M$. We then obtain the density matrix $C^{-+}(x) = \langle b_x^\dagger b_0 \rangle$ for integer x as

$$C^{-+}(x)(-1)^x = \frac{1}{N}\int_0^M dp \frac{\sin 2\pi px/N}{2p\sin \pi x/N} \to \frac{1}{4\pi x}\mathrm{Si}(k_m x). \tag{4.35}$$

The last expression in (4.35) is valid in the limit $|x|/N \ll 1$. The correlation function is given by $C(x)^\perp = [C(x)^{+-} + C(x)^{-+}]/2$, where $C^{+-}(x) = \langle b_x b_0^\dagger \rangle$ can be obtained from the commutation rule

$$b_i b_j^\dagger = (1 - \delta_{ij})b_j^\dagger b_i + \delta_{ij}(1 - b_i^\dagger b_i) \tag{4.36}$$

of spin operators. Namely, we obtain

$$C^{+-}(x) = C^{-+}(-x) + m\delta_{x,0}, \tag{4.37}$$

where $m = 1 - 2M/N$ and $C^{-+}(-x)$ is in fact an even function of x. In order to obtain the Fourier transform $S^{-+}(q)$, we use the form

$$NC^{-+}(x)(-1)^x = \Re \sum_{p=1/2}^{M-1/2} \frac{z^{2p} - 1}{2p(z-1)} = \sum_{p=1/2}^{M-1/2} \frac{1}{2p} \Re \sum_{m=0}^{2p-1} z^m, \qquad (4.38)$$

where $\Re$ represents the real part. Rearranging the series in ascending powers of z, we identify the coefficient of z^{2p-1} as the sum $\sum_{q>p} 1/(2q)$. Namely, we obtain

$$NC^{-+}(x)(-1)^x = \sum_{p=1/2}^{M-1/2} \frac{1}{2p} + \Re \sum_{l=1}^{M-1} (z^{2l-1} + z^{2l}) \sum_{j=l}^{M-1} \frac{1}{2j+1}. \qquad (4.39)$$

The Fourier component $S^{\perp}(k - \pi)$ with $k = 4\pi l/N > 0$ can easily be read off from the coefficient of z^{2l-1} or z^{2l} in (4.39). Here the momentum shift π comes from the factor $(-1)^x$. In the thermodynamic limit we obtain, with $2M/N = 1 - m$,

$$S^{-+}(k - \pi) = \int_l^M \mathrm{d}j \frac{1}{4j} = \frac{1}{4} \ln \frac{M}{l} = \frac{1}{4} \ln \frac{k_m}{k}, \qquad (4.40)$$

for $0 < k < k_m \equiv \pi(1 - m)$. The Fourier component is 0 for larger k. We put $q = \pi - k > 0$ to obtain

$$S^{\perp}(q) = \frac{1}{2}[S^{-+}(q) + S^{+-}(q)] = \frac{m}{2} - \frac{1}{4}\theta(q - \pi m) \ln \frac{1 - q/\pi}{1 - m}. \qquad (4.41)$$

Calculation of $C^{zz}(x)$ can be done in a similar manner by fixing two coordinates in $\Psi_G(\{z\})$ and integrating over other coordinates. Alternatively, one can derive $C^{zz}(x)$ using a more systematic approach developed in the random matrix theory. A self-contained account is given in Section 4.5. In order to derive $S^{zz}(q)$, we use the formula

$$S^{zz}(q) = \rho \sum_x [g(x) - \rho] \exp(-iqx) = C_q - C_0, \qquad (4.42)$$

where $g(x)$ is the two-body distribution function defined by (4.64) in Section 4.5, and C_q is the Fourier transform of $C^{zz}(x)$. The momentum q is restricted to the Brillouin zone $|q| < \pi$. Because of the restriction, $S^{zz}(q)$ shows a behavior rather different from the structure factor in the corresponding Sutherland model, although $C^{zz}(x)$ is precisely the same in both models. It is convenient to use the form

$$C^{zz}(x) = -\frac{1}{4} \left[s(x)^2 + Ds(x)Is(-x) \right], \qquad (4.43)$$

where the functions on the RHS are defined in (4.84). The Fourier transforms of constituents without the assumption of $|x| \ll N$ are given from (4.79) by

$$s_k = \theta(k_m - |k|), \quad Ds_k = iks_k, \quad Is_k = (ik)^{-1}s_k. \tag{4.44}$$

In contrast with the continuum space, the Fourier components $q - 2n\pi$ with integer n play the same role as q. In the case of $0 < q < \pi$, there can also be a nonzero contribution for $q - 2\pi \equiv q_-$. Then we have to consider both q and q_- in $S^{zz}(q)$. Namely, we obtain the form $S^{zz}(q) = I(q) + I(q_-)$, where

$$I(q) \doteq \frac{1}{2N} \sum_k [s_k(s_{q+k} - s_k) + Is_k(Ds_{q+k} - Ds_k)]. \tag{4.45}$$

Contribution to the summation over k comes from the region with $s_{k+q} = 0$ and $s_k = 1$. For small enough q, the integral region is from $k = k_m - q$ to k_m. In this region we have $I(q_-) = 0$, and obtain

$$S^{zz}(q) = \frac{q}{4\pi} \left(1 - \frac{1}{2} \ln \left|1 - \frac{q}{k_m}\right|\right). \tag{4.46}$$

The upper limit of q for (4.46) is given by $q_c = \min\{2\pi m, 2k_m\}$. Namely, we have $q_c = 2\pi m$ for $m < 2/3$, and $q_c = 2k_m$ for $m > 2/3$. In the latter case, we obtain for $2k_m < q < \pi - k_m$:

$$S^{zz}(q) = \frac{1}{2}(1 - m), \tag{4.47}$$

which does not depend on q. On the other hand, the contribution $I(q_-)$ also becomes effective for $\pi - k_m < q < \pi$. For this range of q, addition of $I(q)$ and $I(q_-)$ gives

$$S^{zz}(q) = \frac{m}{2} - \frac{1}{4} \ln \left|1 - \frac{2\pi - q}{k_m}\right| + \frac{q}{8\pi} \ln \left|\frac{2\pi - k_m - q}{k_m - q}\right|. \tag{4.48}$$

In the case of no magnetization with $k_m = \pi$, $S^{zz}(q)$ is given by (4.48) for all positive q, and is reduced to

$$S^{zz}(q) = -\frac{1}{4} \ln \left|1 - \frac{q}{\pi}\right|. \tag{4.49}$$

In this way, we obtain the results shown in Fig. 4.1.

4.5 *Derivation of static correlation functions

In this section, we shall provide a self-contained account of the derivation of the correlation function $C^{zz}(x)$, which gives $S^{zz}(q)$ by Fourier transform. For this purpose we first summarize some convenient formulae originally developed in the random matrix theory. We begin with non-interacting fermions for which correlation functions can be obtained in terms of Slater determinants. Although the second quantization formalism is best suited for this purpose, we use the first quantization in order to prepare for the case of the spin chain.

The wave function $\Psi_F(x_1, \ldots, x_M)$ of M non-interacting fermions can be written in the form of the Slater determinant with orthonormalized function $\phi_j(x)$ with $j = 1, 2, \ldots, M$. Namely, we have

$$\Psi_F(x_1, \ldots, x_M) = \sum_P \operatorname{sgn}(p)\phi_{P1}(x_1) \cdots \phi_{PM}(x_M). \tag{4.50}$$

To obtain the norm we write the square as

$$|\Psi_F(x_1, \ldots, x_M)|^2 = \sum_P \sum_Q \operatorname{sgn}(PQ)\phi_{P1}(x_1)^* \phi_{Q1}(x_1) \cdots \phi_{PM}(x_M)^*$$
$$\phi_{QM}(x_M)$$
$$= M! \sum_P \operatorname{sgn}(p)\phi_1(x_1)^* \phi_{P1}(x_1) \cdots \phi_M(x_M)^* \phi_{PM}(x_M)$$
$$= M! \det\left(\phi_i(x_i)^* \phi_j(x_i)\right)_{i,j=1,\ldots,M}, \tag{4.51}$$

where we have used the convention of representing the row i and column j of the matrix by subscripts. Because of the orthonormality

$$\int \mathrm{d}x \phi_i(x)^* \phi_j(x) = \delta_{ij}, \tag{4.52}$$

the norm for $\Psi_F(x_1, \ldots, x_M)$ becomes $M!$. It is possible to write the square as another determinant

$$|\Psi_F(x_1, \ldots, x_M)|^2 = \det\left(\sum_i \phi_i(x_\alpha)\phi_i(x_\beta)^*\right)_{\alpha,\beta=1,\ldots,M}. \tag{4.53}$$

By introducing the function

$$K(x, y) = \sum_j \phi_j(x)\phi_j(y)^* = \sum_j \langle x|\phi_j\rangle\langle\phi_j|y\rangle, \tag{4.54}$$

we can write the norm in the form

$$\int \mathrm{d}x_1 \cdots \int \mathrm{d}x_M \det\left(K(x_\alpha, x_\beta)\right)_{\alpha,\beta=1,\ldots,M} = M!, \tag{4.55}$$

which gives a less trivial result for the integral on the LHS. (4.54) shows that $K(x, y)$ plays the role of the particle propagator from y to x. Because of the orthonormality, the propagator has the property

$$\sum_y K(x, y)K(y, z) = K(x, z). \tag{4.56}$$

In order to derive the correlation function, we attach an auxiliary function $1 + f(x_i)$ to each coordinate x_α. Namely, we consider

$$\det \left(\int \phi_j^*(x)\, \phi_k(x)\, (1 + f(x))\, \mathrm{d}x \right)_{j,k=1,\ldots,M}$$

$$= \det \left(\delta_{j,k} + \int \phi_j^*(x)\, \phi_k(x)\, f(x)\, \mathrm{d}x \right)_{j,k=1,\ldots,M}. \tag{4.57}$$

Obviously we recover the norm by setting $f(x) = 0$. The coordinate x can be regarded as discrete to represent the lattice system with N sites. The polynomial wave functions defined in this chapter give the same results either by integration over $0 < x < N$ or summation over $x = 1, 2, \ldots, N$. We prefer the lattice summation scheme since the discreteness makes it easier to understand the matrix structure. Namely, the matrix inside the determinant of (4.57) is of the form $1 + AB$, where A is an $M \times N$ matrix and B is an $N \times M$ one. The structure is clearly seen in the form

$$\langle \phi_j | AB | \phi_k \rangle = \sum_x \langle \phi_j | x \rangle f(x) \langle x | \phi_k \rangle, \tag{4.58}$$

where the summation is over the discrete lattice coordinates x. Now we use the relation $\det_M (I + AB) = \det_N (I + BA)$ to work with the $N \times N$ matrix BA. Namely, we obtain

$$\langle x | BA | y \rangle = \sum_j \langle x | \phi_j \rangle \langle \phi_j | y \rangle f(y) = K(x, y) f(y), \tag{4.59}$$

where the propagator $K(x, y)$ has been defined by (4.54). Therefore the determinant given by (4.57) is equal to $\det(I + K f) \equiv \exp(\mathcal{L})$, where we introduce the quantity $\mathcal{L}$. Here, in analogy to classical statistical mechanics, we relate the norm of the wave function to the partition function, and its logarithm $\mathcal{L}$ to the free energy. Then $f(x)$ is a site-dependent external field. As in the free energy, variation of external fields gives the response function. We recall that the response function is proportional to the irreducible correlation function, or cumulants, in classical theory.

We make the following expansion:

$$\mathcal{L} = \mathrm{Tr}_N \ln(1 + Kf) = \sum_x K(x, x) f(x)$$

$$- \frac{1}{2} \sum_{x,y} K(x, y) f(y) K(y, x) f(x) + \cdots, \tag{4.60}$$

which immediately gives the correlation functions. Namely, we take the variational derivative of $\mathcal{L}$ with respect to infinitesimal $f(x)$. This amounts to fixing one of the particle coordinates to x in (4.55), and gives the density $\langle \rho(x) \rangle$. Namely, we have

$$K(x, x) = \langle \rho(x) \rangle. \tag{4.61}$$

From now on we always consider the system with translational invariance. Then density is independent of x, and is written as $\rho = M/N$. A convenient feature of this formulation for correlation functions is that the norm of the wave function with $f(x) = 0$ need not be considered explicitly, since the norm is just a constant factor for the partition function. Similarly, the second derivative gives

$$\frac{\delta^2 \mathcal{L}}{\delta f(x) \delta f(y)} = -K(x, y) K(y, x) = \langle \rho(x) \rho(y) \rangle - \rho^2. \tag{4.62}$$

In a similar manner, the n-point correlation functions can be obtained as $K(x_1, x_2) K(x_2, x_3) \cdots K(x_n, x_1)$. These n-point functions constitute irreducible or cumulant pieces. The (reducible) correlation functions, which include the lower-order irreducible functions, are constructed from cumulants. For example, we obtain by writing $K(x_1, x_2) = K_{12}$,

$$\langle \rho(x_1) \rho(x_2) \rangle = K_{12} K_{21} + K_{11} K_{22},$$
$$\langle \rho(x_1) \rho(x_2) \rho(x_3) \rangle = K_{12} K_{23} K_{31} + K_{12} K_{21} K_{33} + \cdots . \tag{4.63}$$

The two-point correlation function $C(x - y) = \langle \rho(x) \rho(y) \rangle - \rho^2$ depends only on the difference of coordinates, and can be written as

$$C(x) = \frac{1}{N} \sum_y [\langle \rho(x + y) \rho(y) \rangle - \rho^2] = \frac{1}{N} \sum_{ij} \langle \delta(x - x_i + x_j) \rangle - \rho^2$$
$$= \rho \delta(x) + \frac{1}{N} \sum_{i \neq j} \langle \delta(x - x_i + x_j) \rangle - \rho^2$$
$$\equiv \rho \delta(x) + \rho[g(x) - \rho], \tag{4.64}$$

where the two-body distribution function $g(x)$ has been introduced. Here the delta function should be interpreted as the Kronecker delta symbol for integer variables. The quantity $g(x)$ measures the distribution of particles under the condition that a particle is present at the origin. The sum rules for correlation functions can be obtained from (4.56) and (4.64) as

$$\sum_x C(x) = \rho, \quad \sum_x [g(x) - \rho] = 0. \tag{4.65}$$

It follows from the above that $g(x) \to \rho$ as $|x| \gg 1$, which means complete screening. On the other hand, by definition in (4.64) we have $g(0) = 0$.

In the case of a circular system of our interest, we take

$$\phi_l(\theta) = \frac{1}{\sqrt{2\pi}} \exp(\mathrm{i}p_l\theta) = \frac{1}{\sqrt{2\pi}} \exp(\mathrm{i}k_l x) \qquad (4.66)$$

where we have introduced the coordinate θ in the range of $|\theta| < \pi$. We take the *antiperiodic* boundary condition with N even. Then the p_l are half-integers given by $p_l = l - N/2 - 1/2$, and the distribution of p_l is symmetric about 0. The relation to integer lattice coordinates x and the wave number k_l is given by

$$x = N\theta/(2\pi), \quad k_l = 2\pi p_l/N.$$

The normalization factor in (4.66) is relevant to integration over θ, but should be replaced by $1/\sqrt{N}$ for summation over x. Then we obtain explicitly the propagator

$$K(\theta, 0) = \frac{1}{2\pi} \sum_{l=1}^{N} \exp\left[\mathrm{i}\left(l - \frac{N+1}{2}\right)\theta\right] = \frac{1}{2\pi} \frac{\sin N\theta/2}{\sin \theta/2}. \qquad (4.67)$$

Thus the two-body correlation function of free fermions is given with proper rescaling by

$$\rho g(x) = -\left(\frac{\sin N\theta/2}{N \sin \theta/2}\right)^2 \sim -\left(\frac{\sin \pi x}{\pi x}\right)^2, \qquad (4.68)$$

where the last expression is valid in the limit of $|x| \ll N$.

Now we consider a general wave function where $P_N(x_1, \ldots, x_M)$ is its absolute square. We seek a $K(x, y)$ that satisfies

$$\ln\left\{\sum_{x_i} P_N(x_1, \ldots, x_M) \prod_l [1 + f(x_l)]\right\} \propto \ln \det (I + Kf), \qquad (4.69)$$

for the general function f. As before, K denotes the operator that has N^2 matrix elements $K(x, y)$. Our problem now is to find K for the case of the Gutzwiller–Jastrow function for the spin chain. The integral is of the form

$$\int_{-\pi}^{\pi} \frac{\mathrm{d}\theta_1}{2\pi} \cdots \int_{-\pi}^{\pi} \frac{\mathrm{d}\theta_M}{2\pi} \prod_{l<k} |e^{\mathrm{i}\theta_l} - e^{\mathrm{i}\theta_k}|^4 \prod_l (1 + f(\theta_l)). \qquad (4.70)$$

This form describes the symplectic circular ensemble in the random matrix theory where an elegant formulation using quaternions is available [48, 130]. Here we take an alternative approach following the line explained in [187].

Let us first derive the norm with $f(\theta_l) = 0$ in (4.70). The integrand can be represented in the form of a determinant called the confluent alternant:

$$\prod_{l<k}(z_l - z_k)^4 = \det\left(z_k^l, \ lz_k^{l-1}\right)_{l=0,\dots,2M-1; \ k=1,\dots,M}. \tag{4.71}$$

This is seen by writing the product representation of the Vandermonde determinant

$$\det\left(z_k^l, \ \zeta_k^l\right)_{l=0,\dots,2M-1; \ k=1,\dots,M}, \tag{4.72}$$

then differentiating with respect to each ζ_k and setting $\zeta_k = z_k$.

If we replace each z_k by $e^{i\theta_k}$ and use the relation

$$|e^{i\theta_l} - e^{i\theta_k}|^4 = \exp\left[-2i(\theta_l + \theta_k)\right](e^{i\theta_l} - e^{i\theta_k})^4, \tag{4.73}$$

we see that

$$\prod_{l<k}|e^{i\theta_l} - e^{i\theta_k}|^4 = e^{-2i(M-1)\sum \theta_l}\prod_{l<k}(e^{i\theta_l} - e^{i\theta_k})^4.$$

By using (4.71) this is transformed into

$$e^{-2i(M-1)\sum \theta_l}\det\left(e^{il\theta_k}, \ le^{i(l-1)\theta_k}\right)_{l=0,\dots,2M-1; \ k=1,\dots,M}$$
$$= \det\left(e^{i(l-M+\frac{1}{2})\theta_k}, \ le^{i(l-M+\frac{1}{2})\theta_k}\right)$$
$$= \det\left(e^{i(l-M+\frac{1}{2})\theta_k}, \ (l-M+\tfrac{1}{2})e^{i(l-M+\frac{1}{2})\theta_k}\right)$$
$$= \det\left(e^{ip_l\theta_k}, \ p_le^{ip_l\theta_k}\right). \tag{4.74}$$

In the last determinant we have $k = 1,\dots,M$, and p_l runs through the half integers $-M+\frac{1}{2}, -M+\frac{3}{2}, \dots, M-\frac{1}{2}$. We now expand the determinant as

$$\prod_{i<j}|z_i - z_j|^4 = \sum_P \epsilon_P(P2 - P1)e^{i(P1+P2)\theta_1}\dots$$
$$\times (P(2M) - P(2M-1))e^{i(P(2M-1)+P(2M))\theta_M}. \tag{4.75}$$

Upon integration over θ_1, only such terms that satisfy $P1 + P2 = 0$ survive. The same restriction holds for each θ_i. Thus $|P(2i) - P(2i-1)|$ ranges over $1, 3, \dots, 2M-1$, and we obtain

$$\int_{-\pi}^{\pi}\frac{d\theta_1}{2\pi}\cdots\int_{-\pi}^{\pi}\frac{d\theta_M}{2\pi}\prod_{i<j}|z_i - z_j|^4$$
$$= M!(2M-1)(2M-2)\cdots 2 = 2^{-M}(2M)!. \tag{4.76}$$

which was conjectured by Dyson [48]. This result gives the norm of the Gutzwiller–Jastrow wave function. As we shall see in (7.163) of Chapter 7, there are other (more sophisticated) ways to derive the norm.

Now we use the formula for Pfaffians:

$$\int \cdots \int \det(\phi_j(x_k) \ \psi_j(x_k))_{j=1,\ldots,2M; \ k=1,\ldots,M} \, \mathrm{d}x_1 \cdots \mathrm{d}x_M$$

$$= M! \, \mathrm{Pf} \left(\int (\phi_j(x) \, \psi_k(x) - \phi_k(x) \, \psi_j(x)) \, \mathrm{d}x \right)_{j,k=1,\ldots,2M}, \qquad (4.77)$$

where the Pfaffian "Pf" is the square root of the determinant of the antisymmetric matrix. Then the square of (4.70) is given in terms of the following $2M \times 2M$ determinant:

$$\det \left(\int \frac{\mathrm{d}\theta}{2\pi} (q - p) \, \mathrm{e}^{\mathrm{i}(p+q)\theta} \, (1 + f(\theta)) \right),$$

where both indices p and q run over the half-integers $-M + \frac{1}{2}, \ldots, M - \frac{1}{2}$. If we reverse the order of the rows and divide each column by its index q, we see that this determinant is proportional to another determinant

$$\det \left(\int \frac{\mathrm{d}\theta}{4\pi} \left(1 + \frac{p}{q} \right) \mathrm{e}^{\mathrm{i}(-p+q)\theta} \, (1 + f(\theta)) \, \mathrm{d}\theta \right)$$

$$= \det \left(\delta_{p,q} + \int \frac{\mathrm{d}\theta}{4\pi} \left(1 + \frac{p}{q} \right) \mathrm{e}^{\mathrm{i}(-p+q)\theta} \, f(\theta) \right). \qquad (4.78)$$

We rewrite the factor of $f(\theta)$ as a matrix product

$$\mathrm{e}^{-\mathrm{i}p\theta} \, \mathrm{e}^{\mathrm{i}q\theta} + \frac{p}{q} \mathrm{e}^{-\mathrm{i}p\theta} \, \mathrm{e}^{\mathrm{i}q\theta} = \left(\mathrm{e}^{-\mathrm{i}p\theta}, \ \mathrm{i}p\mathrm{e}^{-\mathrm{i}p\theta} \right) \begin{pmatrix} \mathrm{e}^{\mathrm{i}q\theta} \\ (\mathrm{i}q)^{-1} \, \mathrm{e}^{\mathrm{i}q\theta} \end{pmatrix}.$$

By identifying the matrix structure as

$$\langle p|A|\theta \rangle = \frac{1}{4\pi} f(\theta) \left(\mathrm{e}^{-\mathrm{i}p\theta}, \ \mathrm{i}p\mathrm{e}^{-\mathrm{i}p\theta} \right), \quad \langle \theta|B|q \rangle = \begin{pmatrix} \mathrm{e}^{\mathrm{i}q\theta} \\ (\mathrm{i}q)^{-1} \, \mathrm{e}^{\mathrm{i}q\theta} \end{pmatrix},$$

then the above matrix is of the form $I + AB$. If we identify the integral $\int \mathrm{d}\theta/(2\pi)$ as equivalent to the lattice summation $N^{-1} \sum_x$ with $\theta = 2\pi x/N$, BA is regarded as the $2N \times 2N$ matrix with the propagator $K(\theta, \theta')$ given by

$$K(\theta, \theta') = \frac{1}{4\pi} \sum_p \begin{pmatrix} \mathrm{e}^{\mathrm{i}p(\theta-\theta')}, & \mathrm{i}p\,\mathrm{e}^{\mathrm{i}p(\theta-\theta')} \\ (\mathrm{i}p)^{-1} \, \mathrm{e}^{\mathrm{i}p(\theta-\theta')}, & \mathrm{e}^{\mathrm{i}p(\theta-\theta')} \end{pmatrix}. \qquad (4.79)$$

If we write

$$S_M(\theta) = \frac{1}{2\pi} \sum_p e^{ip\theta} = \frac{1}{2\pi} \frac{\sin \frac{1}{2} M\theta}{\sin \frac{1}{2}\theta},$$

$$DS_M(\theta) = \frac{\mathrm{d}}{\mathrm{d}\theta} S_M(\theta), \quad IS_M(\theta) = \int_0^\theta S_M(\theta')\mathrm{d}\theta', \qquad (4.80)$$

then we can represent the propagator as

$$K(\theta, \theta') = \frac{1}{2} \begin{pmatrix} S_M(\theta - \theta'), & DS_M(\theta - \theta') \\ IS_M(\theta - \theta'), & S_M(\theta - \theta') \end{pmatrix}. \qquad (4.81)$$

The remarkable property of this matrix propagator K is that (4.56) is satisfied as the 2×2 matrix equation. Evidently the transpose of K also serves as the propagator.

Recall that the determinant we are working with now is not the norm of the many-body wave function, but is proportional to its square. Taking the square root amounts to multiplying by a factor $1/2$ in the exponentiated quantity $\mathcal{L}$. Namely, the correlation function can be read off from the expansion

$$\mathcal{L} = \ln \left[\det(1 + Kf)\right]^{1/2} = \frac{1}{2} \int \mathrm{d}\theta f(\theta) \operatorname{Tr} K(\theta, \theta)$$

$$- \frac{1}{4} \int \mathrm{d}\theta f(\theta) \int \mathrm{d}\theta' f(\theta') \operatorname{Tr} K(\theta, \theta') K(\theta', \theta) + \cdots . \qquad (4.82)$$

By noting $2\pi x = N\theta$, we obtain the correlation function $C(x)$ as

$$C(x) = -\frac{1}{2} \left(\frac{N}{2\pi}\right)^2 \operatorname{Tr} K(\theta, 0) K(0, \theta)$$

$$= -\frac{1}{4} \left(\frac{N}{2\pi}\right)^2 \left[S_M(\theta)^2 - IS_M(\theta) DS_M(\theta)\right], \qquad (4.83)$$

where we have used the fact that $S_M(\theta)$ is an even function, and $DS_M(\theta)$, $IS_M(\theta)$ are odd. In the thermodynamic limit with $|x| \ll N$, we define the following functions from (4.80) by setting $\gamma = 2M/N$:

$$s(x) = \frac{\sin k_m x}{\pi x}, \quad Is(x) = \frac{1}{\pi}\operatorname{Si}(k_m x),$$

$$Ds(x) = \frac{\gamma \cos \pi x}{x} - \frac{\sin k_m x}{\pi x^2}. \qquad (4.84)$$

Then the correlation function is obtained explicitly as

$$C^{zz}(x) = -\left(\frac{\sin k_m x}{2\pi x}\right)^2 + \left[\frac{\gamma \cos k_m x}{4\pi x} - \frac{\sin k_m x}{(2\pi x)^2}\right] \mathrm{Si}(k_m x). \qquad (4.85)$$

In the case of no magnetization with $\gamma = 1$, a drastic simplification emerges. Because of the integer lattice coordinate x, we actually have $\sin \pi x = 0$ and $\cos \pi x = (-1)^x$. Then we obtain the form

$$C^{zz}(x) = \frac{\cos \pi x}{4\pi x} \mathrm{Si}(\pi x), \qquad (4.86)$$

which gives $C^{zz}(x) = C^{\perp}(x)$ in (4.21), reflecting the SU(2) symmetry.

4.6 Spectrum of magnons

The spectrum of H_{HS}, or equivalently H_{1st}, is most easily described by the set of momenta of down spins. These particles can be regarded as magnons created from the fully polarized reference state. Since the reference state is not the ground state, the magnons can take negative energy. By using the mapping to the Sutherland model as given by (4.16), we can represent the energy in terms of the rapidity (or renormalized momentum) p_i as discussed in Chapter 2. Then we obtain the energy in terms of the rapidity

$$E = \frac{1}{2}\sum_{i=1}^{M} p_i(p_i - 2\pi) + \frac{N}{8}J(0), \qquad (4.87)$$

where p_i is arranged in the descending order $p_1 > p_2 > \cdots > p_M$.

As the fermionic description (Section 2.2.2) in the Sutherland model with $\lambda = 2$, the rapidity p_i is determined by

$$\frac{N}{2\pi}p_i = I_i + \frac{1}{2}\sum_{j} \mathrm{sgn}(p_i - p_j) = I_i + \frac{1}{2}(M + 1 - 2i) \equiv \kappa_i, \qquad (4.88)$$

where $\{I_i\}$ are distinct integers with $I_1 > I_2 > \cdots > I_M$. In the ground state, we have $I_i = I_{i0} = (N + M + 1)/2 - i$. Hence the corresponding rapidity p_{i0} is given by

$$p_{i0} - \pi = \frac{2\pi}{N}(M + 1 - 2i). \qquad (4.89)$$

The total crystal momentum P of the M-magnon state is given by

$$P = -\pi + \sum_{i} p_i, \qquad (4.90)$$

where $-\pi$ comes from the reference state $|\mathrm{F}\rangle$.

The form of (4.88) is reminiscent of the Bethe ansatz theory with a constant phase shift. The phase shift in the present case is 2π, which means that the minimum separation of occupied momenta is $4\pi/N$. Since the $1/r^2$ interaction eventually decays at long distance, (4.88) is sometimes called the asymptotic Bethe ansatz [107, 177]. An important difference from the Sutherland model is that p_i is restricted to inside the Brillouin zone, where all magnons have negative energies. Since the magnons repel each other in the momentum space, the number of states for magnons cannot exceed $N/2$. All these states are occupied in the ground state.

The distribution of the rapidities $\{p_i\}$ is specified by occupation or vacancy for all momentum states in the Brillouin zone. We associate 1 for an occupied state, and 0 for the vacant state. Then the N-sequence of 0 and 1 determines the energy. Such a sequence of binary digits is called the *motif* [78], which has already appeared in Section 2.3.2. We deal with polynomial wave functions without the constant term. Hence, the first digit in the motif is always 0. It is customary to add an extra 0 at the end, and make the sequence $N + 1$ digits. Then the beginning and the end become equivalent, which corresponds to both boundaries of the Brillouin zone. In the case of a singlet ground state, the motif is given by $0101\ldots010$ with M 1's. Let us represent the κth element $(\kappa = 0, 1, \ldots, N)$ of the motif by d_κ. Then $d_\kappa = 1$ means the occupation of the momentum state. The total crystal momentum P and the energy E are given by

$$P = \frac{2\pi}{N} \sum_{\kappa=0}^{N} \kappa d_\kappa - \pi,$$

$$E = \frac{2\pi}{N} \sum_{\kappa=0}^{N} \kappa(N - \kappa)d_\kappa + \frac{N}{8}J(0), \tag{4.91}$$

in accordance with (4.90) and (4.87).

It is also possible to describe magnons as interacting bosons. The bosonic particles are characterized by a set of integers $\tilde{I}_i$, defined by

$$\tilde{I}_i = I_i - I_{i0}.$$

Namely, $\tilde{I}_i$ gives the deviation from the ground-state quantum numbers I_{i0}. We have the relation $\tilde{I}_1 \geq \tilde{I}_2 \geq \cdots \geq \tilde{I}_M$. In analogy with (4.88), we obtain

$$\frac{N}{2\pi}p_i = \tilde{I}_i + \sum_j \mathrm{sgn}(p_i - p_j) = \tilde{I}_i + (M + 1 - 2i). \tag{4.92}$$

The factor 1 in front of the summation over j is to be compared with another factor $1/2$ in (4.88). Then the energy of the system is written as

$$\left(\frac{N}{2\pi}\right)^2 [E - E_0(M)] = \frac{1}{2} \sum_{i=1}^{M} \tilde{I}_i^2 + \sum_{i<j} |\tilde{I}_i - \tilde{I}_j|, \qquad (4.93)$$

where $E_0(M)$ is given by (4.17). By comparing (4.87) and (4.93), we understand that the free-particle description with renormalized momentum, i.e., rapidity p, is equivalent to the bosonic description with repulsive interaction $|p_i - p_j|$ in the momentum space. We shall use (4.93) in discussing the spinon excitations later. For completeness, we now remark on the case of N being *odd*. The allowed crystal momentum is given by $p_j = 2\pi j/N$ with $j = 1, 2, \ldots, N$. The set of momenta does not include π with N odd. In the mapped Sutherland model, the allowed set of momenta does not include zero, hence the set of momenta corresponds to those under the *antiperiodic* boundary condition. The lattice wave function itself satisfies the periodic boundary condition because of the half-integer power $N/2$ in z_i for magnons. Then the ground state is non-degenerate for M even and doubly degenerate for M odd. The energy of the system is given by (4.87) also with N odd. In the non-degenerate ground state with M even, we obtain the same expression for p_{i0} as (4.89). The corresponding κ_i and I_{i0} are now half integers. The quantum numbers $\tilde{I}_i$ describing bosonic excitations remain integers.

4.7 Spinons

4.7.1 Localized spinons

Elementary excitations from the singlet spin liquid differ from those from magnetically ordered states. The former excitations are called spinons, in contrast with magnons in the latter, and can be viewed as a kind of moving domain wall. Although crude, the real-space picture is the most useful to get an image of spinons. Consider a Néel state with a periodic boundary condition. If the number N of the lattice site is *odd* with $S^z = 1/2$, the up and down sequence of spins must have a part where two up spins are next to each other. This domain wall is the localized representation of a spinon. For N even with $S^z = 0$, on the other hand, the domain wall excitation must occur in pairs. A localized magnon, which is a flip of a spin resulting in $S^z = 1$, has three neighboring up spins. This state is considered as two spinons at closest distance.

For a more detailed inspection including identification of the quantum number of a spinon, we consider the quantum liquid of spins since the Néel state does not have a definite quantum number for the total spin. Let us consider, as in the Néel state, the *odd* number N of the lattice site. Then the ground state has the minimum total spin $S = 1/2$. In the case of the $1/r^2$ exchange model, the wave function is explicitly given by the Gutzwiller-type one as in (4.14). The coordinate index i representing a down spin runs from 1 to $M = (N-1)/2$. Then there is a surplus of $S^z = 1/2$. We now introduce an extra coordinate $z_0 = \exp(2\pi i x_0) = 1$, which corresponds to lattice site $x_0 = 0$, and consider a wave function

$$\Psi_{1s}(\{z\}; z_0) = \prod_{i=1}^{M}(z_i - z_0)\Psi_G(\{x\}) = \prod_{i=1}^{M} z_i(z_i - z_0)\prod_{i<j}(z_i - z_j)^2. \quad (4.94)$$

The amplitude becomes zero if any of the down-spin particles comes to z_0, and tends to that of the singlet liquid as z_i goes away from z_0. Hence $\Psi_s(\{z\}; z_0)$ represents a defect of down spins. Since all sites are occupied by either up or down spins, the site z_0 is occupied by an up spin. We interpret this defect as a localized spinon with spin up [79]. The state given by $\Psi_{1s}(\{z\}; z_0)$ has $S = S^z = 1/2$. This highest-weight property of SU(2) applies generally for polynomial wave functions without the constant term, and will be explained in Section 4.8.1. Note that the maximum power of z_i is $2M(=N-1)$ in (4.94), and the minimum power is one. Namely, the wave function does not contain the constant term.

Let us now consider the case where the total number of lattice sites N is *even*. Then the ground state is a spin singlet, which will be proven formally in Section 4.8.1. It is not possible to have a single spinon as an excited state, since $S = 1/2$ is not allowed for N even. On the other hand, the state with two spinons localized at z_0 and at another arbitrary site z_0' is possible. The corresponding wave function is given by

$$\Psi_{2s}(\{z\}; z_0, z_0') = \prod_{i=1}^{M} z_i(z_i - z_0)(z_i - z_0')\prod_{i<j}(z_i - z_j)^2. \quad (4.95)$$

This wave function corresponds to a state with $N/2 + 1$ up spins and $M = N/2 - 1$ down spins. It is obvious that for any nonzero separation, coordinates z_0 and z_0' are both occupied by up spins, and the total spin of $\Psi_{2s}(\{z\}; z_0, z_0')$ is unity. We interpret this state as a triplet of two spinons. The state Ψ_{2s} has the SU(2) highest weight $S = S^z = 1$. The highest power of z_i in (4.95) is $N - 1$, which is again the maximum allowed within the first Brillouin zone. The case $z_0 = z_0'$ represents a localized magnon as shown above. Thus we obtain the appealing interpretation that a magnon is nothing but two spinons sitting on the same site.

4.7.2 Spectrum of spinons

These localized spinons, however, cannot be eigenstates of the Haldane–Shastry model or, in general, of a spin liquid. One has to go to the momentum space representation of spinons for constructing eigenstates of the translationally invariant system. We consider first the case of N and $M = (N-1)/2$ both being odd. The polynomial ground state is doubly degenerate, as given by

$$\Psi_{G}^{(\pm)}(\{z\}) = \prod_{i=1}^{M} z_{i}^{3/2\pm1/2} \prod_{i<j}(z_{i} - z_{j})^{2}. \qquad (4.96)$$

We take $\Psi_{G}^{(-)}(\{x\})$ and consider an excited state as follows. By expanding the first factor (4.94) in terms of z_{i}, we find terms such as

$$z_{0}^{M-j} z_{1} z_{2} \cdots z_{j} \prod_{i=1}^{M} z_{i}$$

which range from $z_{0}^{M} \prod_{i=1}^{M} z_{i}$ to $\prod_{i=1}^{M} z_{i}^{2}$. Each additional power of z_{i} represents the momentum of a spinon. Then the first contribution is proportional to $\Psi_{G}^{(-)}(\{z\})$ and may be regarded as a spinon excitation with zero momentum. The case $\prod_{i} z_{i}^{2}$ with maximum power is represented, in terms of the Young diagram, by M squares arranged vertically. The polynomial still keeps the maximum power of each z_{i} less than N. Namely, the resultant wave function is within the first Brillouin zone without the constant term.

For a general momentum, a spinon excitation is described by the elementary symmetric polynomial

$$e_{\zeta} = \sum_{i_{1}<\cdots<i_{\zeta}} z_{i_{1}} \cdots z_{i_{\zeta}},$$

which corresponds to the number ζ of vertically arranged squares. The case with $\zeta = M$ corresponds to the Galilean boost by unit momentum, and gives the other degenerate ground states. We interpret the case of $\zeta = 0$ as containing a spinon of zero momentum, and the case with $\zeta = M$ of the maximum momentum. Let us, for example, consider a case of $M = 4$ as shown by Fig. 4.2. As has been shown in (4.87), a state with ζ

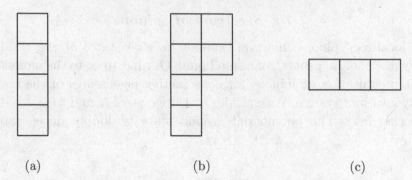

(a) (b) (c)

Fig. 4.2. Young diagrams for spin excitations in the hard-core boson representation of magnons with $M = 4$. The ground state does not have any square since it is taken as the reference. (a) One spinon with $\zeta = 4$ as momentum in units of $2\pi/N$, (b) two spinons with $\zeta_1 = 4$ and $\zeta_2 = 2$, and (c) three spinons with momenta $\zeta_1 = \zeta_2 = \zeta_3 = 1$, or equivalently, one antispinon with particle momentum $\kappa = 3$. The number N ($\gg 1$) of the lattice sites is odd for (a) and (c), and even for (b).

has the energy

$$\epsilon_s = \frac{1}{2}\left(\frac{2\pi}{N}\right)^2 \sum_{i=1}^{\zeta}\left[(\kappa_i + 1)^2 - \kappa_i^2\right] = \left(\frac{2\pi}{N}\right)^2 \zeta(M - \zeta), \qquad (4.97)$$

where we have used $\kappa_i = M + 1/2 - 2i$ for the ground state. Note that the unit of energy is taken as $J/2$. As expected, we obtain $\epsilon_s = 0$ for both $\zeta = 0$ and $\zeta = M$.

In the case of N odd and M even, the polynomial ground state is non-degenerate with $\kappa_i = M + 1 - 2i$. The ground state has number $N - 2M$ of spinons whose spectrum is given by

$$\epsilon_s = \left(\frac{2\pi}{N}\right)^2 \zeta(M - \zeta + 1/2). \qquad (4.98)$$

In the thermodynamic limit, the energy of a spinon is written in terms of $k = 2\pi\zeta/N$ for both even and odd M as

$$\epsilon_s(k) = k(v_m - k), \qquad (4.99)$$

where $v_m = 2\pi M/N$ and $k \geq 0$. We obtain $v_m = \pi(1 - m)$ with m the magnetization per site as defined by $mN = N - 2M$. The spinon velocity v_m determines the maximum of k as $k \leq v_m$. The independent number of one spinon state is obtained as $M + 1$ including the ground state. Namely, the allowed range of k is half the original Brillouin zone with $m = 0$, and becomes smaller as magnetization m increases. The quantity k has the meaning of crystal momentum with $m = 0$. However, in the case of $m \neq 0$, physical momentum q has to be shifted from k by $\pi m/2$, as will be shown later in Fig. 4.3.

We consider how the spin-flip can be interpreted in terms of propagating spinons. This serves also to determine the spinon Brillouin zone. We assume the ground state with N to be even and $M = N/2$ to be odd, and obtain the two-spinon state localized at $z_0' = z_0$ from (4.95):

$$\langle \{z\}|(S_0^x + iS_0^y)\Psi_G\rangle = \Psi_{2s}(\{z\}; z_0, z_0) = z_0 \prod_{i=1}^{M-1} (z_i - z_0)^2 \Psi_{G-}, \qquad (4.100)$$

where Ψ_{G-} is the Gutzwiller wave function with $M - 1$ down spins. The spin-flip described by $S_0^x + iS_0^y$ changes the meaning of z_0 from a coordinate of a magnon to a constant. The momentum associated with Ψ_{G-} is $\pm\pi$, since $M-1$ is even. The extra momentum associated with the spin-flip ranges from 0 to $2\pi(1 - 2/N)$, as can be seen by expansion of $\prod_{i=1}^{M-1}(z_i - z_0)^2$. Then the total crystal momentum of the spin-flip is in the range $[-\pi, \pi)$, as it should be. Since a spin-flip (creating a localized magnon) can be regarded as two spinons sitting on the same site, each spinon should carry a momentum either in the range $[-\pi/2, \pi/2)$ or $[0, \pi)$. The latter range agrees with that obtained above for the spectrum of a single spinon. Since the spinons are created only in pairs with N even, the physical momentum of a spinon has a periodicity π, instead of 2π. Namely, the right half of the spinon branch with $\pi < q < 2\pi$ can be shifted to the negative side of q.

In the hard-core boson representation, a spinon corresponds to a hole in the Sutherland model with the repulsion parameter $\lambda = 2$. Then it follows that spinons with the same spin have the exclusion statistics with $1/\lambda = 1/2$. Moreover, the spinons with opposite spins also have the statistical parameter $1/2$, as will be shown shortly. The particles with $1/2$ statistics are often referred to as semions. We use the terminology "fractional" exclusion statistics for the case where the statistical parameter is between 0 and 1, as in the case of semions. The part of the eigenstates having two spinons is represented by Jack polynomials, by analogy with the Sutherland model. For example, the Young diagram of two spinons with $\zeta = 2, 4$ is shown in Fig. 4.2(b). Note, however, that the two spinons with $S = 1$ but $S^z \neq 1$ *cannot* be represented by a symmetric function of $\{z_i\}$. In general, only the SU(2) highest-weight states with $S = S^z$ of $2S$ spinons are represented by Jack polynomials. These states are called the fully polarized spin gas (FPSG) by Haldane [78]. As we shall explain later, the FPSG turns out to be the SU(2) Yangian highest-weight states (YHWS). We shall also see later that for calculation of the dynamical correlation function without a magnetic field, only these YHWS are relevant. The excited state shown by Fig. 4.2(c) can either be interpreted as three spinons with the same momentum, or an antispinon. This ambiguity is specific to the SU(2) symmetry where the conjugate representation $\bar{\mathbf{2}}$ is isomorphic with the fundamental representation $\mathbf{2}$. In the SU(K) spin chain, a spinon transforms as $\bar{\mathbf{K}}$, as discussed in Chapter 5.

4.7.3 Polarized ground state

In the case of $M < N/2$, we obtain $S^z > 0$ in the corresponding ground state, which is stabilized in the presence of magnetic field $h > 0$. The magnon spectrum is given by

$$\epsilon_{\mathrm{m}}(q) = \frac{1}{2}q(q - 2\pi) + 2h, \tag{4.101}$$

where q is the crystal momentum in the Brillouin zone $[0, 2\pi]$. The spectrum for $-2\pi < q < 0$ is given by replacing q in (4.101) by $|q|$. Figure 4.3(a) shows the spectrum for $q < 0$. As we have identified $\Psi_{2\mathrm{s}}(\{z\}; z_0, z_0)$ as a two-spinon wave function in the real space, we may construct spinon wave functions in

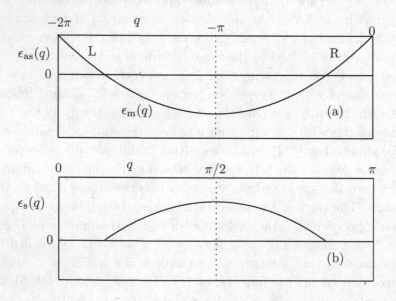

Fig. 4.3. (a) Spectrum of a magnon in a magnetic field. The regions $-\pi m < q < 0$ and $-2\pi < q < -(2 - m)\pi$ with positive energy give the spectrum $\epsilon_{\mathrm{as}}(q)$ of an antispinon with right (R) and left (L) branches. (b) Spectrum of a spinon in a magnetic field. The allowed range of momentum q is given by $m\pi/2 < q < \pi - m\pi/2$.

the momentum space. Then a hole of magnons with negative momentum is interpreted as two spinons with positive momentum. In the ground state with $h > 0$, magnon states are occupied for such q as

$$q_m \equiv \pi m < q < 2\pi - q_m. \tag{4.102}$$

Here $m = 1 - 2M/N$ is the magnetization per site, and is related to h by $\epsilon_m(\pi m) = 0$. The positive part of $\epsilon_m(q)$, which is unoccupied by magnons, is identified as the spectrum of antispinons, i.e., $\epsilon_{as}(q) = \epsilon_m(q)$. Thus antispinons can also be viewed as excited magnons from the polarized ground state. The allowed range of momentum shrinks as m becomes small, and disappears in the singlet state. By restriction of $M < N/2$, antispinons can have only down spin. This means physically the spin-flip excitations from the condensate of up spins.

An antispinon is represented by a row with length κ_a in the Young diagram. Figure 4.2(c) shows an example with $\kappa_a = 3$. The maximum of κ_a is given by $\kappa_a = N/2 - M - 2$. The energy ϵ_{as} of an antispinon is also derived from (4.93) by setting $\tilde{I}_1 > 0$, and $\tilde{I}_j = 0$ with $2 \le j \le M$. Here $\tilde{I}_1$ cannot exceed $N/2 - M = Nm/2$. We obtain

$$\epsilon_{as} = \left(\frac{2\pi}{N}\right)^2 \left[\frac{1}{2}\tilde{I}_1^2 + (M-1)\tilde{I}_1\right] = \frac{1}{2}k^2 + v_m k, \tag{4.103}$$

with $k = 2\pi\tilde{I}_1/N$ and $v_m = 2\pi(M-1)/N$. With the condition $q = k + \pi m$, the spectra given by (4.101) and (4.103) for $k > 0$ become identical. The difference between k and q is that k is measured from the lowest-energy antispinon state, which already has the finite momentum πm.

In terms of the motif introduced in Section 4.6, a spinon corresponds to an extra 0 in the sequence such as ...01010010101... The location of 00 determines the momentum of the spinon. An antispinon with rapidity $k = 2\pi\zeta/N$ corresponds to the motif where the rightmost 1 has the number $\zeta + 1$ of 0's to the left. The successive 0's in the interior of the motif represent multiple spinons with the same momentum. Two spinons are created by annihilating a magnon represented by 1 in the motif of the ground state. The annihilation results in 000 in the motif, which represents two spinons. In the thermodynamic limit, the spectrum of up spinons $\epsilon_s(q)$ is derived most easily from the filled part of the magnons by the particle–hole relation $-\epsilon_m(-2q) = 2\epsilon_s(q)$ (> 0). Namely, we obtain

$$\epsilon_s(q) = q(\pi - q) - h, \tag{4.104}$$

where the crystal momentum q has meaning for $m/2 < q/\pi < 1 - m/2$ with $\epsilon_s(q) > 0$. Figure 4.3(b) shows the spectrum of up spinons. Comparison with (4.99) shows that the crystal momentum q is related to k as

$$q = \begin{cases} k + \pi m/2, & (k > 0), \\ k + \pi(1 - m/2), & (k < 0). \end{cases}$$

Note that the origin of k has been taken to be the spinon states with the lowest energy with momentum $q = \pm \pi m/2$.

4.8 Energy levels and their degeneracy

4.8.1 Degeneracy beyond SU(2) symmetry

By mapping to the Sutherland model, we have found that some eigenfunctions $\Psi_\kappa(\{x\})$ of H_{HS} can be written in the form

$$\Psi_\kappa(\{x\}) = \prod_{i=1}^{M} z_i^{N/2-M+1} \prod_{i<j}(z_i - z_j)^2 J_\kappa(\{z\}), \qquad (4.105)$$

where N is even and $J_\kappa(\{z\})$ is a symmetric Jack polynomial. As long as the degree of $J_\kappa(\{z\})$ for each z_i is less than $N/2 - M + 1$, the polynomial $\Psi_\kappa(\{x\})$ is within the Brillouin zone, and describes a spin state in H_{HS}. Moreover, there is no constant term in (4.105).

The wave function $\Psi_\kappa(\{x\})$ belongs to the highest-weight states of the SU(2) symmetry, namely $S = S^z = N - 2M$. To show this we operate $S^+ = \sum_{j=1}^{N} S_j^+$ to $\Psi_\kappa(\{x\})$. The spin-flip operator S_j^+ gives a null result if the site j is not occupied by a down spin. If the site is occupied, the spin is reversed. In the first quantization, the spin reversal means that one of the coordinates disappears from $\Psi_\kappa(\{x\})$. Namely, one has

$$\langle x_2, \ldots, x_M | S^+ \Psi_\kappa \rangle = \sum_x \Psi_\kappa(x, x_2, \ldots, x_M), \qquad (4.106)$$

where x runs from 1 to N. In the case of the polynomial wave function as described by (4.105), one has the expansion with $z = \exp(2\pi x \text{i}/N)$

$$\Psi_\kappa(x, x_2, \ldots, x_M) = \sum_{p=1}^{N-1} z^p \psi_p(x_2, \ldots, x_M). \qquad (4.107)$$

Because of the oscillatory property of z^p, the summation over x gives $S^+|\Psi_\kappa\rangle = 0$. Thus it is confirmed that the Sutherland-type wave function $\Psi_\kappa(\{x\})$ belongs to the highest-weight states of the SU(2). In the special case of $M = N/2$, we obtain $S^z = 0$, and hence $S = 0$. Thus the Gutzwiller wave

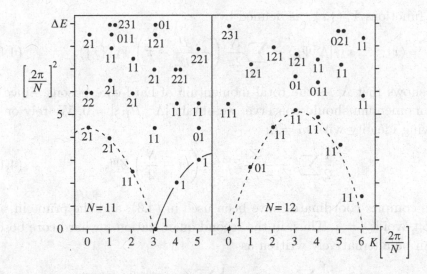

Fig. 4.4. Low-lying levels of the Haldane–Shastry model with sizes $N = 11$ and $N = 12$. Each filled circle indicates a state with a pseudo-momentum in units of $2\pi/N$. Numbers attached to levels show the degenerate structure: "231" means $(S = 1/2)^2 \otimes (S = 3/2)^3 \otimes (S = 5/2)^1$ for $N = 11$ (odd), and $(S = 0)^2 \otimes (S = 1)^3 \otimes (S = 2)^1$ for $N = 12$ (even). Broken and solid lines are a visual guide showing the bottom of the excitation continuum in the thermodynamic limit [77]. Copyright (1988) by the American Physical Society. Reproduced with permission.

function $\Psi_G(\{x\})$ with $M = N/2$ is an SU(2) singlet. It is crucial here that the constant term with $p = 0$ is absent in (4.107). In other words, a polynomial wave function with the constant term does not belong to the SU(2) highest-weight states, and will be excluded in the following analysis.

It was found by Haldane by numerical calculation [77] that the spectrum of the Haldane–Shastry model has an enormous degeneracy. Figure 4.4 shows the results. This degeneracy with different values of the total spin shows the presence of a higher symmetry in the model. By the global SU(2) symmetry, the total spin operator $S = \sum_i S_i$ conserves. Hence the set of states $(S^-)^p|\Psi_\kappa\rangle$ with $S^z = S - p$ has the same energy as that of $|\Psi_\kappa\rangle$, as long as $p \leq 2S$. Hence most of the eigenfunctions of $H_{\rm HS}$ have a form that *cannot* be written as a polynomial like (4.105).

The degeneracy among different values of the total spin is called the supermultiplet, and comes from a symmetry called the Yangian. We now construct heuristically some generators called the "level one" [80] of the Yangian algebra. Let us introduce an operator Λ^z which acts on polynomial wave functions $\Psi_\kappa(\{x\})$ as defined by

$$\langle x_1, \ldots, x_M | \Lambda^z \Psi_\kappa \rangle = \sum_i \frac{N}{2\pi} \left(-i \frac{\partial}{\partial x_i} - \pi \right) \Psi_\kappa(\{x\}). \qquad (4.108)$$

This shows that Λ^z is the total momentum of hard-core bosons. Since the total momentum should conserve, we obtain $[\Lambda^z, H_{HS}] = 0$. We rely on the following identity with $m \neq 0$:

$$\frac{1}{2} \sum_j' \frac{Z_i + Z_j}{Z_i - Z_j} Z_j^m = \left(m - \frac{N}{2} \right) Z_i^m, \qquad (4.109)$$

where complex coordinates have been used in (4.3), and the prime in summation avoids $j = i$. Thus, in the second quantization for hard-core bosons, Λ^z can be alternatively written as

$$\Lambda^z = \frac{1}{2} \sum_{i \neq j} w_{ij} b_i^\dagger b_j = \frac{i}{2} \sum_{i \neq j} w_{ij} (S_i^x S_j^y - S_i^y S_j^x), \qquad (4.110)$$

where we have introduced the notation

$$w_{ij} = \frac{Z_i + Z_j}{Z_i - Z_j} = -i \cot \frac{1}{2} (\theta_i - \theta_j). \qquad (4.111)$$

Note that the total momentum operator given by (4.110) is not restricted to the subspace spanned by polynomial wave functions.

From Λ^z, we obtain a general component α ($= x, y, z$) in the following fashion [80, 162]:

$$\Lambda^\alpha = \frac{i}{2} \sum_{i \neq j} w_{ij} \sum_{\beta\gamma} \epsilon_{\alpha\beta\gamma} S_i^\beta S_j^\gamma = \frac{i}{2} \sum_{i \neq j} w_{ij} (\boldsymbol{S}_i \times \boldsymbol{S}_j)_\alpha, \qquad (4.112)$$

where $\epsilon_{\alpha\beta\gamma}$ is the completely antisymmetric unit tensor. Since the Haldane–Shastry model has the global SU(2) symmetry, the conservation law for Λ^z means

$$[\Lambda^\alpha, H_{HS}] = 0 \qquad (4.113)$$

for $\alpha = x, y, z$. Physically this can be interpreted as conservation of the spin current $\boldsymbol{\Lambda}$ [25, 123]. On the other hand, straightforward algebra using (4.112) gives

$$\left[\sum_i S_i^\alpha, \Lambda^\beta \right] = i \sum_\gamma \epsilon_{\alpha\beta\gamma} \Lambda^\gamma. \qquad (4.114)$$

Thus $\boldsymbol{\Lambda}$ does *not* commute with the total spin. This means that if $|\Psi\rangle$ is an eigenstate of H_{HS}, $\Lambda^\alpha |\Psi\rangle$ is another eigenstate with different total spin, and they are degenerate. Therefore the degeneracy *extends out of the global SU(2) symmetry.*

In order to understand the nature of the degeneracy, we make the following linear combination:

$$\Lambda^x + i\Lambda^y = -\sum_{i \neq j} w_{ij} S_i^z S_j^+. \tag{4.115}$$

By using the up-spin reference state, we put $S_j^z = 1/2 - \delta_{j \in \{x\}}$ where the Kronecker delta becomes unity if Z_j belongs to the set of magnon coordinates. We show now that $\Lambda^x + i\Lambda^y$ annihilates the FPSG state where the corresponding $\Psi_\kappa(\{x\})$ is a symmetric function of magnon coordinates. By applying (4.115) to $\Psi_\kappa(\{x\})$ with M magnons, we obtain

$$\langle x_1, \ldots, x_{M-1} | \Lambda^x + i\Lambda^y | \Psi_\kappa \rangle = -\sum_{x=1}^{N} \sum_{j=1}^{M-1} w_{xj} \Psi(x, x_1, \ldots, x_{M-1}) = 0,$$
$$\tag{4.116}$$

where the summation index j means that of x_j, and we have used the relation $S_j^z = -n_j + 1/2$. The summand in (4.116) is antisymmetric against interchange of x and x_j. Moreover, w_{xj} has the translational invariance $(x, j) \to (x - n, j - n)$. Thus choosing $n = j$, we obtain the last equality by summation over x. On the other hand, successive application of $\Lambda^x - i\Lambda^y$ on $|\Psi_\kappa\rangle$ generates different states with the same energy. The generation stops for certain n where we have $(\Lambda^x - i\Lambda^y)^n |\Psi_\kappa\rangle = 0$. Some simple examples will be discussed later, as illustrated in Fig. 4.7.

This sequence of generation is analogous to the application of raising and lowering operators of angular momentum to an SU(2) highest-weight state. Hence we call such a state that satisfies (4.116) a Yangian highest-weight state (YHWS). All the polynomial eigenstates, i.e., the FPSG states, in the Haldane–Shastry model thus belong not only to the SU(2) highest-weight states but to the YHWS. We shall provide a more formal discussion of Yangian symmetry in Part II.

4.8.2 *Local current operators*

In addition to the conserving current Λ derived above, the Haldane–Shastry model has a set of local operators which annihilate the ground state [162]. We first observe the identity for each variable $\theta_i = 2\pi x_i / N$:

$$\left(-i\frac{\partial}{\partial \theta_i} - \frac{N}{2} - {\sum_j}' w_{ij} \right) \Psi_G(\{x\}) = 0, \tag{4.117}$$

which is analogous to (2.6) in the Sutherland model, and follows from the explicit product form of $\Psi_G(\{x\})$ given by (4.13). In the magnon and spin representations the identity is rewritten as

$$\frac{1}{2}\sum_j{}'w_{ij}(b_i^\dagger b_j - 2n_i n_j)\Psi_G$$

$$= \frac{1}{2}\sum_j{}'w_{ij}\left[S_i^- S_i^+ - 2\left(S_i^z - \frac{1}{2}\right)\left(S_j^z - \frac{1}{2}\right)\right]\Psi_G = 0, \qquad (4.118)$$

where $n_i = b_i^\dagger b_i = S_i^z - 1/2$. Since the singlet Ψ_G remains the same if one reverses the direction of all spins, we also obtain

$$\frac{1}{2}\sum_j{}'w_{ij}\left[S_i^+ S_i^- - 2\left(S_i^z + \frac{1}{2}\right)\left(S_j^z + \frac{1}{2}\right)\right]\Psi_G = 0. \qquad (4.119)$$

By subtracting (4.119) from (4.118), we see that the following operator:

$$\Lambda_i^z = \frac{1}{2}\sum_j{}'w_{ij}\left[\mathrm{i}\left(\boldsymbol{S}_i \times \boldsymbol{S}_j\right) + \boldsymbol{S}_j\right]^z, \qquad (4.120)$$

annihilates Ψ_G. In addition to the z-component Λ_i^z, the vanishing property also holds for the other components Λ_i^x and Λ_i^y because of the singlet nature of Ψ_G. Namely, a vector operator $\boldsymbol{\Lambda}_i$ defined by

$$\boldsymbol{\Lambda}_i = \frac{1}{2}\sum_j{}'w_{ij}\left[\mathrm{i}\left(\boldsymbol{S}_i \times \boldsymbol{S}_j\right) + \boldsymbol{S}_j\right] \qquad (4.121)$$

annihilates Ψ_G [162]. On the other hand, the spin current $\boldsymbol{\Lambda}$ is given by

$$\boldsymbol{\Lambda} = \sum_i \boldsymbol{\Lambda}_i, \qquad (4.122)$$

which justifies the name of the local current operator for $\boldsymbol{\Lambda}_i$. Note that $\boldsymbol{\Lambda}$ is Hermitian, while $\boldsymbol{\Lambda}_i$ is not. It has been shown by straightforward calculation [25, 162] that

$$H_{\mathrm{HS}} = J\left(\frac{2\pi}{N}\right)^2\left[\frac{2}{9}\sum_i \boldsymbol{\Lambda}_i^\dagger \cdot \boldsymbol{\Lambda}_i + \frac{N+1}{12}\boldsymbol{S}^2 - \frac{N(N^2+5)}{48}\right], \qquad (4.123)$$

where $\boldsymbol{S}$ is the total spin of the system. This form confirms that Ψ_G indeed gives the ground state, since $\boldsymbol{\Lambda}_i^\dagger \cdot \boldsymbol{\Lambda}_i$ and $\boldsymbol{S}^2$ are non-negative operators.

In fact there are five more local operators which annihilate the ground state [162]. To derive these, we now add (4.119) to (4.118), and find that the operator

$$Q_i^{zz} = \frac{1}{2} \sideset{}{'}\sum_j w_{ij} \left(S_i^x S_j^x + S_i^y S_j^y - 2 S_i^z S_j^z \right) \tag{4.124}$$

annihilates Ψ_G. It is evident that Q_i^{zz} takes the form of a second-rank tensor, or a quadrupole formed by two spins. By the invariance of Ψ_G against SU(2) rotation, we then find that any component

$$Q_i^{\alpha\beta} = -\frac{1}{2} \sideset{}{'}\sum_j w_{ij} \left[\frac{3}{2} \left(S_i^\alpha S_j^\beta + S_i^\beta S_j^\alpha \right) - \delta_{\alpha\beta} \boldsymbol{S}_i \cdot \boldsymbol{S}_j \right] \tag{4.125}$$

of the tensor also annihilates Ψ_G. There are five independent components of $Q_i^{\alpha\beta}$ since they are symmetric and traceless.

It still remains to clarify what kind of physical roles these local operators play, in addition to annihilating the ground state. In this connection we note that the local currents do not commute with H_{HS}, nor with each other. In order to construct a commuting set of local current operators, a limiting procedure from the Sutherland model is available [183]. In the latter model, Cherednik–Dunkl operators constitute the commuting set, as explained in Section 3.6. The lattice version of Cherednik–Dunkl operators will be discussed in Chapter 9.

4.8.3 Freezing trick

As we have seen in previous sections, the mapping of certain sets of states (YHWS) of the Haldane–Shastry model to eigenstates of the Sutherland model is very powerful in deriving the spectrum of the model. The main advantage is that one can then use the highly developed mathematical technique for orthogonal polynomials in the continuum space. On the other hand, most states in the lattice model cannot be described by orthogonal polynomials. These states do not have the Yangian highest weight, and are degenerate with polynomial wave functions. In order to derive thermodynamics, we have to know not only the energy levels but also their degeneracies.

A nice way to derive all states in a lattice model is to regard the lattice model as the limiting case of the Sutherland model in the continuum space [149, 177]. Namely, we start from the U(2) Sutherland model, and take the limit of $\lambda \to \infty$. Then each particle tries to avoid the others as much as possible, and crystallizes with equal distance $2\pi/N$. The resultant dynamics is the combination of uniform translation, lattice vibration, and exchange of SU(2) components. The last is nothing but the spin degrees of freedom in the Haldane–Shastry spin chain. Provided one can separate unwanted excited states associated with lattice vibrations, one recovers all the excitations in the spin chain.

We start from the U(2) Sutherland model for N-particle systems:

$$H_\lambda = -\sum_{i=1}^{N} \frac{\partial^2}{\partial x_i^2} + \frac{2\pi^2}{L^2} \sum_{i<j} \frac{\lambda(\lambda - P_{ij})}{\sin^2[\pi(x_i - x_j)/L]}. \qquad (4.126)$$

The spin permutation operator P_{ij} is in fact an SU(2) specialization of the SU(K) internal symmetry. We have seen in Chapter 3 that the energy spectrum E_λ of (4.126) has been obtained [22, 99, 177] as follows:

$$\left(\frac{N}{2\pi}\right)^2 (E_\lambda - E_{0,N}) = \sum_{\kappa=-\infty}^{\infty} \kappa^2 \nu(\kappa) + \frac{\lambda}{2} \sum_{\kappa=\infty}^{\infty} \sum_{\kappa'=-\infty}^{\infty} |\kappa - \kappa'| \nu(\kappa)\nu(\kappa'),$$

$$(4.127)$$

where $E_{0,N} = (\pi\lambda/L)^2 N(N^2 - 1)/3$ is the energy of the absolute ground state given by (2.12), and κ runs over integers describing momentum k by the relation $k = 2\pi\kappa/L$. The distribution function consists of

$$\nu(\kappa) = \nu_\uparrow(\kappa) + \nu_\downarrow(\kappa),$$

where $\nu_\sigma(\kappa)$ is the momentum distribution function of fermions with spin σ.

Let us consider the strong coupling limit $\lambda \to \infty$. In this limit, particles localize with a lattice spacing L/N, which is taken to be unity. Up to $\mathcal{O}(\lambda)$, there are two kinds of degrees of freedom: one is the vibration around the lattice points and the other is the exchange of particle species between the two lattice points. The former corresponds to phonons, while the latter describes the spin exchange. We obtain as the Hamiltonian [101, 149, 177]

$$H_{\text{tot}} \equiv \lim_{\lambda\to\infty} \frac{1}{\lambda} (H_\lambda - E_0) = H_{\text{ph}} + 2H'_{\text{HS}}, \qquad (4.128)$$

where H_{ph} corresponds to the limiting form of (4.126) with $P_{ij} = 1$, and describes phonons. On the other hand, H'_{HS} describes the spin dynamics and takes the form

$$H'_{\text{HS}} = \sum_{i<j} D_{ij}^{-2}[P_{ij} - n_i n_j], \qquad (4.129)$$

where $D_{ij} \equiv (N/\pi)\sin[\pi(i-j)/N]$. The original particles in the Sutherland model can be taken as fermions. In the large λ limit, each site is singly occupied by a fermion with either up or down spin. Thus the system is equivalent to a set of interacting quantum spins. By using the identity

$$P_{ij} = 2\boldsymbol{S}_i \cdot \boldsymbol{S}_j + n_i n_j/2, \qquad (4.130)$$

for hard-core fermions with the occupation number $n_i = 1$, we recover from H_{HS} the Haldane–Shastry model with $J = 2$ minus the constant term $NJ(0)$. By construction the eigenvalue of the fully polarized state is zero in H'_{HS}, since we have $P_{ij} = n_i n_j = 1$ for this state. The same result follows if we start from bosons in the Sutherland model.

Let us consider a route to construct thermodynamics for H'_{HS}: first we consider thermodynamics for H_{tot} and subsequently subtract the phonon contribution. Using (4.127) and the relation (4.128), we obtain the expression for the energy spectrum of H_{tot} [101, 177]

$$E_{tot} = \frac{\pi^2}{N^2} \sum_{\kappa=\infty}^{\infty} \sum_{\kappa'=-\infty}^{\infty} |\kappa - \kappa'| \nu(\kappa)\nu(\kappa'). \qquad (4.131)$$

The phonon contribution can be derived by regarding $\nu(\kappa)$ as a single component. The ground state corresponds to $\kappa_i^{(0)} = N - i$ with $i = 1, 2, \ldots, N$, or equivalent ones with a uniform shift. The Young diagram for the difference $\Delta\kappa_i = \kappa_i - \kappa_i^{(0)}$ takes the form for bosons. The uniform increment of $\Delta\kappa_i$ describes the Galilean boost of the whole particles, and does not affect E_{tot} in (4.131). On the other hand, an excitation with $\Delta\kappa_i = 1$ for $1 \leq i \leq m$ and $\Delta\kappa_i = 0$ for $m > i$ corresponds to a phonon with momentum $q = 2\pi m/N$. The excitation energy is derived from (4.131) as

$$\omega_q = \frac{2\pi^2}{N^2} \sum_{i=1}^{m} \sum_{j=m+1}^{N} (\Delta\kappa_i - \Delta\kappa_j) = \frac{2\pi^2}{N^2} m(N - m) = \frac{1}{2}q(2\pi - q). \quad (4.132)$$

If we take $\Delta\kappa_i = n$ for $1 \leq i \leq m$ and $\Delta\kappa_i = 0$ for $m > i$, the energy is given by $n\omega_q$. By definition the excitation given by (4.132) obeys the boson statistics and should properly be called a phonon. It is seen from the above derivation that the phonon in the present system is harmonic for any magnitude of the displacement. Namely, a larger displacement simply increases the number of harmonic phonons.

We now turn to the two-component case of (4.131) with N even and $M = N/2$ odd. In the ground state, we obtain for each component $\kappa_{i\sigma}^{(0)} = (M + 1)/2 - i$ for $1 \leq i \leq M$. Since we have derived all the eigenvalues of the model given by (4.128), and all eigenvalues of the phonon part, we can easily check that the Gutzwiller state gives the lowest eigenvalue for the spin part, and there is no degeneracy with $N_\uparrow = N_\downarrow = N/2$ odd. Therefore, we have obtained another proof, in addition to (4.123), that the Gutzwiller wave function (4.13) indeed gives the ground state.

4.9 From Young diagrams to ribbons

4.9.1 Removal of phonons

To make physical excitations in the spin chain explicit in the freezing trick, one has to remove the U(1) component, i.e., the phonons, and retain only the SU(2) components in counting the available states. We now discuss explicitly such a procedure in which the correspondence between a Young diagram and another diagram called a ribbon diagram, or simply a ribbon, is relevant. Figure 4.5 shows partitions of two-component particles giving the ground state and two kinds of excited states in (4.131) with $N = 6$. Unlike ordinary Young diagrams, each row in a partition diagram has a spin index, and is tentatively called U(2) partitions here. Although antisymmetrization of different rows has not been performed, we keep only such diagrams that survive the antisymmetrization. Hence at most two rows have the same length.

In state (b), the spin indices x and y can either be up or down, leading to four possible combinations of x and y. The four different spin states can be reorganized into one singlet and one triplet. On the other hand, the state (c) is an SU(2) singlet where both up and down spins have the same set of momenta. The state (c) represents a phonon excitation. In general, a totally white column (without any shaded square) represents a phonon excitation.

Let us proceed to systematic classification of U(1) and SU(2) excitations. It is clear that the location of the rightmost squares in each row, which are shaded in Fig. 4.5, carries all the information on the states. Then we rearrange the diagram, keeping only the shaded squares. Namely, some shaded squares are shifted vertically so that the adjacent shaded squares

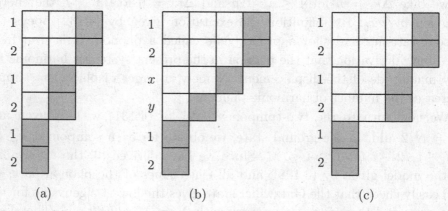

(a)   (b)   (c)

Fig. 4.5. U(2) partitions for (a) the ground state, (b) an excited state with two spinons, and (c) an excited state with a phonon. The index 1 represents the up spin, and 2 the down spin. Shaded squares show the momentum of each row.

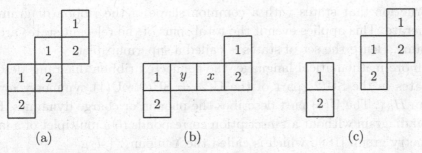

(a)   (b)   (c)

Fig. 4.6. Skew Young diagrams for (a) the ground state and (b) an excited state with two spinons. Both (a) and (b) are called ribbon diagrams. A disconnected skew Young diagram like (c) represents an excited state with a phonon, and does not belong to ribbon diagrams.

share a common side if possible. Figure 4.6 shows the arrangement of six squares keeping the horizontal position. The spin state is inscribed as a number 1 or 2 in each square.

Since the shaded squares are separated in Fig. 4.5(c), the rearrangement leads to disconnected pieces as shown in Fig. 4.6(c). Then by keeping only the connected diagrams after vertical shifts, one can exclude phonon or U(1) excitations. These connected diagrams are called ribbon diagrams [111], or simply ribbons.

We note that ribbon diagrams are a particular subclass of skew Young diagrams, which are constructed from a partition λ and its subset μ to represent λ/μ. Namely, ribbon diagrams do not have 2×2 or larger blocks of connected squares for λ/μ.

A phonon-free excited state in the freezing limit of the Sutherland model has one-to-one correspondence to a ribbon diagram with inscription. It is understood in the ribbon diagram that a symmetrization of spin states has been made for each row. In order to avoid overcounting, we require $y \leq x$ in Fig. 4.6(b). Thus, of four states in Fig. 4.5(b), only the triplet states survive the symmetrization. The triplet excitation is interpreted as two spinons. Note that two spinons can also form a singlet provided they are separated by one or more vertical pairs in a ribbon diagram.

The energy of each state can easily be read off by returning to the corresponding partition diagram, since we have never made a horizontal move of the square. Then the energy remains the same as given by (4.131), provided each square in the ribbon diagram represents a single-particle state. The energy of the whole state depends solely on the momentum distribution $\nu(\kappa)$, and is independent of spin structures in $\nu(\kappa)$. Moreover, the energy does not depend on the total momentum, or the Galilean boost. Hence, it is clear by construction that states with a common shape of the ribbon diagram are degenerate. This applies even if the total spins of the relevant states are not the same. Hence the set of states is called a supermultiplet.

In more mathematical language, the connected ribbon diagrams describe all states in the SU(2) part of the U(2) = SU(2)⊗U(1) symmetry for the system H_{tot}. The U(1) part describes the phonon or charge dynamics. Each ribbon diagram without an inscription corresponds to a multiplet of a larger symmetry group [188], which is called the Yangian $Y(sl_2)$.

4.9.2 Completeness of spinon basis

The spinons span the complete basis of the Hilbert space of N spins [78]. Before proving the completeness, we illustrate the simplest case of $N = 2$ where the four states are classified into one singlet and one triplet. The latter is interpreted as two spinon states with $S^z = \pm 1, 0$. Hence the states with zero and two spinons indeed span the complete set. In the next simplest case of $N = 3$, the $2^3 = 8$ states are classified into two doublets and one quartet with $S = 3/2$.

To visualize the classification in the general case we use ribbon diagrams. We arrange the squares, each of which has an inscription of either 1 (spin up) or 2 (spin down), as shown in Fig. 4.7. The sequence of inscribed numbers is called a semistandard tableau. Although squares can represent any basis set, it is most convenient to take the momentum basis so that the momentum increases by $2\pi/N$ by going one square to the right. A horizontal array implies symmetrization and a vertical array antisymmetrization, as in the standard Young diagrams. To avoid overcounting we require that the numbers inscribed should increase on going down the array, and should

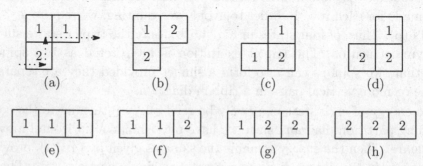

Fig. 4.7. Skew Young diagrams, or ribbons, for $N = 3$. The dotted lines in (a) illustrate the path leading to the motif 0100 as explained in the text. The Yangian lowering operator $\Lambda^x - i\Lambda^y$ of (4.133) changes (a) to (b), and (c) to (d). Successive application of $\Lambda^x - i\Lambda^y$ changes (e) to (h) through (f) and (g).

not decrease on going right. A vertical pair of squares represents the singlet state of two spins. In the general case, we need a limiting procedure to obtain the wave function as explained in Section 4.8.3. This is in contrast with the FPSG case, where the actual form of the wave functions is given by Jack polynomials. A ribbon diagram without inscribed numbers represents a supermultiplet which contains degenerate states with different spin configurations.

It is possible to identify spinons in any ribbon diagram from its shape. Namely, a spinon is represented by a square which is not connected to a vertical neighbor. Equivalently, one can associate with any ribbon diagram a sequence of 0 and 1, which is called the *motif*. We have already introduced the motif in Section 4.6 in an apparently different way. In the present case, the motif is constructed by tracing the N squares from the left bottom to the right top. If one crosses an edge of squares by a horizontal move, one associates the crossing with the digit 0. On the other hand, a crossing of an edge by a vertical move is recorded by the digit 1. The first and last crossings of an edge are always done horizontally. For example, the motif of (a) and (b) in Fig. 4.7 is given by 0100, while (c) and (d) by 0010. Now a spinon is identified as a successive pair of 0's in the motif. The states shown as (a)–(d) in Fig. 4.7 contain a spinon. The momentum of a spinon differs between the multiplet containing (a), (b) and the other containing (c), (d). Similarly the multiplet containing (e)–(h) is characterized by the motif 0000 with three spinons. Thus we see that a motif is in one-to-one correspondence with a ribbon diagram, and hence specifies a supermultiplet.

We now show that the present definition of the motif is equivalent to that determined by the occupation pattern of rapidities, as introduced in Section 4.6. We regard the sequence of N squares from left bottom as all momentum states in the Brillouin zone. Let us restrict ourselves to the YHWS in the ribbon diagram. Then the horizontal crossing of an edge is always from a square of spin up, since the down spin can only appear at the right end of each row. Thus we make a correspondence from the digit 0 to the spin up. In the polynomial wave function, the spin up is not counted as a particle. Hence the rapidity at this position is vacant. On the other hand, a vertical move is always from a down spin by the ordering convention of spins. Thus the digit 1 for the move is in correspondence with the occupation of this momentum state by a down spin. In this way, we see the equivalence between the shape of a ribbon diagram and the occupation pattern of rapidities in the YHWS. Both definitions give the same results for the motif. The energy associated with the motif is given either by (4.87) or by (4.131). It is remarkable that such different expressions give equivalent results.

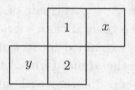

Fig. 4.8. An example of a supermultiplet straddling $S = 0$ and $S = 1$ with $N = 4$. The values of x and y can be either 1 or 2.

For the simple cases as shown in Fig. 4.7, each multiplet is characterized by the total spin of the system. A new situation arises for larger N, as shown in Fig. 4.8 with $N = 4$. The state with both x and y being 1 has $S = 1$ and $S^z = 1$, while that with $x = y = 2$ has $S = 1$ and $S^z = -1$. In the case of $x = 1$ and $y = 2$, the resultant state has $S^z = 0$, but may contain both $S = 0$ and 1 components. Namely it is *not* an eigenstate of S. The same applies to the case where $x = 2$ and $y = 1$. Thus different values of the total spin have the same Young diagram and the same energy as the Haldane–Shastry model. Hence one uses the name of the supermultiplet. This degeneracy, larger than that from the SU(2) symmetry, is due to the Yangian symmetry. We have shown in Section 4.8.1 that the Yangian lowering operator

$$\Lambda^x - i\Lambda^y = -\sum_{i \neq j} w_{ij} S_i^z S_j^- \tag{4.133}$$

commutes with H_{HS} but not with the total spin. In terms of ribbon diagrams, the action of $\Lambda^x - i\Lambda^y$ leaves the shape of the diagram intact but changes one of the inscriptions from 1 to 2 because of S_j^-. It is possible to choose the convention such that the change applies to the rightmost 1 in each row, provided the resultant state is not null. Thus all members of the supermultiplet are generated by successive application of $\Lambda^x - i\Lambda^y$ starting from the YHWS. Some examples are illustrated in Fig. 4.7.

We now show that the spinon basis spans the complete set of 2^N spin states. There are a few alternative methods for the proof. In Section 6.5.2, we shall give a general proof which is applicable to any internal symmetry. Here we use the most direct counting of states. Let N_{pair} denote the number of vertical (singlet) pairs. Then the number N_{sp} of spinons with arbitrary composition of up and down spins is given by $N_{\mathrm{sp}} = N - 2N_{\mathrm{pair}}$. The spinons correspond to squares between the pairs and outside the first and last pairs. In the case of FPSG [78], we have $N_{\mathrm{pair}} = M$. Obviously the number of possible positions (y-coordinates) along the vertical direction is $N_{\mathrm{pair}} + 1$. We identify each y-coordinate as an orbital of spinons, in direct analogy to the momentum of a hole in the Young diagram for the

Sutherland model. Each orbital can accommodate any number of spinons with any spin direction. If there are no singlet pairs as in Fig. 4.7(e)–(h), all spinons occupy the same orbital. This corresponds to the bosonic character of spinons in the space of orbitals, but the statistics of spinons in the physical Hilbert space is not bosonic, as shown later.

Let $N_{sp} = N_\uparrow + N_\downarrow$ be the total number of up and down spinons. The total number $D_{sp} = D_\uparrow + D_\downarrow$ of available single-spinon states is given by $2(N_{pair} + 1)$, where the factor 2 accounts for the spin factor and $(N_{pair} + 1)$ the orbital factor. Since the N_{sp} spinons can choose any of these states, the total number $W(N, N_{sp})$ of N_{sp} spinon states is given by [79]

$$W(N, N_{sp}) = {}_{D_{sp}}H_{N_{sp}} = \binom{(N - N_{sp}) + 2 + N_{sp} - 1}{N_{sp}}$$

$$= \frac{(N+1)!}{N_{sp}!(N + 1 - N_{sp})!} = {}_{N+1}C_{N_{sp}}, \qquad (4.134)$$

where

$$ {}_nH_m = {}_{n+m-1}C_m \qquad (4.135) $$

denotes the number of ways of choosing m elements out of n objects with allowance of duplication. For example, we obtain $W(3,3) = {}_2H_3 = 4$, and all states are contained in the motif 0000 shown by (e)–(h) in Fig. 4.7. In general there are plural ribbon diagrams for given N_{sp} and N. For example, Fig. 4.7(b) and (c) both have $N = 3$ and $N_{sp} = 1$. The total number of spinon states for an N-site system is then given by

$$\sum_{N_{sp}=0}^{N} \frac{1}{2} \left[1 + (-1)^{N - N_{sp}}\right] W(N, N_{sp}), \qquad (4.136)$$

where the factor in front of $W(N, N_{sp})$ picks up only even integers for $N - N_{sp}$, and the summation over N_{sp} without this restriction corresponds to a binomial expansion of $(1 \pm 1)^{N+1}$. Thus the total number of states in the spinon basis exhausts all the states in the N-site spin chain. In other words, the spinons form the complete set.

4.9.3 Semionic statistics of spinons

We proceed to the derivation of the statistical parameters of spinons. We first show $g_{\uparrow\uparrow} = 1/2$ by considering the YHWS states. Then we argue on symmetry grounds that $g_{\sigma\sigma'} = g_{\uparrow\uparrow}$ for arbitrary components σ and σ' of spins. Let us consider the number $W(M)$ of magnon states for given numbers of N and M. This $W(M)$ is equivalent to the number of supermultiplets, which in turn is equal to the number of ribbon diagrams, or the number of motifs with $N_\uparrow = N - 2M$ of up spinons. Since a ribbon diagram with N squares consists of singlet pairs and spinons, a supermultiplet is specified by the location of singlets.

A convenient way of counting the supermultiplets is to first set aside M squares, and construct ribbon diagrams. A magnon appears as a square forming a corner, which specifies the location of magnon momenta among the $N - M$ squares. In terms of a motif, the location of 1's in the sequence of $N - M$ digits is to be specified first. The number of 0's is given by $N - 2M + 1$, one of which is placed to the leftmost. To recover the supermultiplet in the motif, we put a 0 (zero) to the right of each 1 representing a momentum of a magnon. In terms of the ribbon diagram, putting a 0 is equivalent to placing a square on top of each square for magnons, which recovers the N squares representing a supermultiplet. The number of ways of choosing M 1's is given by

$$W(M) = \frac{(N - M)!}{M!(N - 2M)!}. \tag{4.137}$$

We rewrite $W(M)$ using the relation $M = (N - N_\uparrow)/2$ as

$$W(M) = {}_{(N+N_\uparrow)/2}C_{N_\uparrow} = {}_{D_\uparrow}H_{N_\uparrow}. \tag{4.138}$$

Here $D_\uparrow = 1 + (N - N_\uparrow)/2$ gives the number of available orbitals for up spinons. We then obtain the relation $\Delta D_\uparrow = -\Delta N_\uparrow/2$, which means $g_{\uparrow\uparrow} = 1/2$. As we have seen, the reduction of available states depends only on the total number of spinons. Hence we obtain

$$\Delta D_\sigma / \Delta N_{\sigma'} = -1/2. \tag{4.139}$$

In this way, we have derived the statistical parameter $g_{\sigma\sigma'} = 1/2$, which is independent of spin species.

The statistical parameter can be generalized for elementary excitations with the $SU(K)$ with $K > 2$, as we shall discuss in Chapter 5. The concept of spinons can naturally be extended for general K. Surprisingly, however, the statistical parameters for $SU(K \neq 2)$ spinons are now negative; $g_{\alpha\beta} = -1/K$ for any $SU(K)$ indices α and β. These statistical parameters also show up in dynamical correlation functions.

4.9.4 Variants of Young diagrams

The case of $x = y = 1$ in Fig. 4.6(b) is an example of the FPSG explained in Section 4.9.2. In this case a ribbon diagram can be reduced to a Young

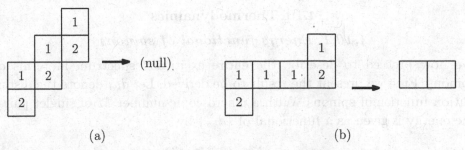

(a)   (b)

Fig. 4.9. Reduction from ribbon diagrams shown in Fig. 4.6 to Young diagrams in the magnon representation: (a) the ground state; (b) an excited state with two spinons.

diagram of hard-core bosons, or magnons. We keep only the component $2\,(=\downarrow)$ and erase the squares of the other component $1\,(=\uparrow)$. The latter is not regarded as a particle but contributes to the fully polarized reference state. The Young diagrams for magnons represent a part of the wave functions multiplied by the ground state. Examples of reduction are shown in Fig. 4.9 with (a) $M = 3$ and (b) $M = 2$. For the number M of magnons, in general, the reduction proceeds first as leftward shifts of each square by i $(=1,\ldots,M+1)$ in the $(M+2-i)$th row. Those squares to the right of the bottom square represent excitations, and the corresponding Young diagram is constructed by making each row up to these squares. The ground state for any M corresponds to a null Young diagram. It is also possible to construct a ribbon diagram from a given magnon diagram.

Let us summarize the three kinds of Young diagrams, and clarify the relationship between them as follows:

(i) U(2) partition diagrams such as Fig. 4.5 for N particles have N rows with either up or down spin, and the length of each row corresponds to the momentum of the particle. The wave function is antisymmetric against interchange of any two particles.

(ii) Ribbon diagrams such as Fig. 4.6 filter out phonons present in U(2) Young diagrams, and give all states in the SU(2) spin chain. There are N squares in a ribbon diagram.

(iii) A Young diagram in the magnon representation such as Fig. 4.2 is obtained from the corresponding ribbon diagram if and only if the state belongs to the FPSG.

4.10 Thermodynamics

4.10.1 Energy functional of spinons

We now proceed to describe the entire excitation spectrum in terms of spinons. First we present the result to be derived. Let $d_{k\sigma}$ denote the distribution function of spinons. With a macroscopic number M of singlet pairs, the energy is given as a functional of $\{d_{k\sigma}\}$ by

$$\Delta E(M) = E(M) - E_0(M) - \frac{\pi^2}{2} M \left(1 - \frac{4M^2}{N^2} \right)$$

$$= \sum_{k\sigma} \left(k_0^2 - k^2 \right) d_{k\sigma} + \frac{\pi}{2N} \sum_{k\sigma} \sum_{k'\sigma'} \left(k_0 - |k - k'| \right) d_{k\sigma} d_{k'\sigma'}, \quad (4.140)$$

with $E_0(M)$ given by (4.17). The sum over k runs from $-k_0$ to $k_0 = \pi M/N \leq \pi/2$. In the case of $M = N/2$, namely $S^z = 0$, the range of k is half the Brillouin zone. The finite magnetization at the ground state corresponds to Bose condensation of spinons at $k = \pm k_0$. The momentum dependence of the energy functional is very similar to that in the U(2) Sutherland model as given by (4.127), except for the reversal of signs. This duality reflects the Yangian symmetry encompassing both Sutherland and Haldane–Shastry models.

We rely on two important observations to derive (4.140):

(i) The spectrum of the FPSG is parameterized by a set κ of dimensionless momentum.

(ii) With a given set of κ_i, the energy of the system does not depend on the spin configuration within κ.

A Young diagram for the FPSG completely specifies the energy of the corresponding state. We make a Galilean transformation to obtain the partition μ by

$$\mu_i = \kappa_i - \kappa_M, \quad (4.141)$$

which leads to $\mu_M = 0$. We will derive the spinon description of excited states for arbitrary shape of Young diagrams. The key concept is the particle–hole duality explained in Chapter 2. Following the method developed in Chapter 2, we convert the representation in terms of particles (magnons or antispinons) to that of holes (spinons). The quantity $\mathcal{E}$ proportional to the energy is converted from magnon to spinon representation by adopting the hole momentum ζ instead of the particle momentum μ by setting $\zeta_i = \mu'_i$ in the conjugate diagram. Then we obtain, using (2.90)

$$\mathcal{E} \equiv \sum_i \mu_i^2 + \lambda \sum_{i<j}(\mu_i - \mu_j)$$

$$= \sum_{j=1}^{\mu_1}\left[-\lambda\zeta_j^2 + (\lambda M + 2j - 1)\zeta_j\right]$$

$$= \lambda \sum_j \zeta_j(M - \zeta_j) - \sum_{i<j}(\zeta_i - \zeta_j) + \mu_1\sum_j \zeta_j, \qquad (4.142)$$

where we need the specific case $\lambda = 2$ in the following.

Since any Young diagram represents only non-negative values of ζ_i, we have to make another Galilean transformation to obtain the physical momentum of spinons. Let us derive the excitation energy for N_{sp} spinons. With $\mu_M = 0$, we have $N_{\mathrm{sp}} = \mu_1$ by construction. Then we determine the proper Galilean shift by requiring the same energy and the same absolute value of the momentum for the following states:

(a) $\mu_i = 0$ for $i \geq 2$, and
(b) $\mu_i = N_{\mathrm{sp}}$ for $i \neq M$.

The physical momentum of (b) should have the opposite sign to that of (a). Defining the physical momentum by $\mu_i - N_{\mathrm{sp}}/2$, we obtain the same kinetic energies of (a) and (b), as can be checked easily.

For a general distribution of μ_i, we obtain the physical kinetic energy by using the same Galilean shift as

$$\mathcal{E}_{\mathrm{kin}} \equiv \sum_i\left(\mu_i - \frac{1}{2}N_{\mathrm{sp}}\right)^2 = \sum_i \mu_i^2 - N_{\mathrm{sp}}\Pi + \frac{1}{4}MN_{\mathrm{sp}}^2, \qquad (4.143)$$

where $\Pi = \sum_i \mu_i = \sum_j \zeta_j$. We take the difference $\mathcal{E}_{\mathrm{kin}} - \sum_i \mu_i^2$ also in (4.142) to obtain the physical energy in the spinon representation. The result is given by

$$\mathcal{E}_{\mathrm{phys}} \equiv \mathcal{E} - N_{\mathrm{sp}}\Pi + \frac{1}{4}MN_{\mathrm{sp}}^2$$

$$= \lambda\sum_j \zeta_j(M - \zeta_j) - \sum_{i<j}(\zeta_i - \zeta_j) + \frac{1}{4}MN_{\mathrm{sp}}^2. \qquad (4.144)$$

Note that the last term in (4.142) is cancelled by the Galilean term $-N_{\mathrm{sp}}\Pi$. With N and $N_{\mathrm{sp}} = N - 2M$ being fixed, the last term in (4.144) does not depend on the distribution of spinons.

The forms given by (4.142) and (4.144) conform to the particle–hole duality in the Sutherland model discussed in Section 2.3.3. Namely, if particles have the repulsion λ, holes have the repulsion $1/\lambda$. One needs to multiply the unit of energy by λ to see the repulsion $1/\lambda$ explicitly. The following remarks are in order concerning (4.144):

(i) The kinetic energy is parabolic, but the coefficient is negative.
(ii) The sign of the interaction is negative.
(iii) By the features (i) and (ii), the statistical interaction between spinons becomes repulsive, i.e., $1/\lambda = 1/2$.

To understand the last remark, we note that $-\mathcal{E}_{\mathrm{phys}}/\lambda$ takes the same form as the energy of the Sutherland model, and that the sign reversal of energy does not change the allowed distribution of momenta. If we had a positive sign for the parabolic spectrum, the negative interaction of course would give attractive statistical interaction. In (4.144), we shift the origin of ζ_j to $M/2$ and absorb the last term in the interaction term as follows:

$$\mathcal{E}_{\mathrm{phys}} = \lambda \sum_j \left(\frac{1}{4}M^2 - \zeta_j^2 \right) + \frac{1}{2} \sum_{i,j} \left[\frac{1}{2}M - (\zeta_i - \zeta_j) \right]. \tag{4.145}$$

So far we have dealt with FPSG states. By the Yangian symmetry, we know that replacement of an up spinon with a down spinon with the same momentum does not change the energy without magnetic field. Hence we obtain the energy E of the system with arbitrary distribution $d_{k\sigma}$ of bosonic spinons with spin σ. Then we obtain (4.140).

By exploiting the Yangian symmetry, we have thus been able to derive the energy as a functional of up and down spinon distributions. The result of (4.140) was first obtained by Haldane [78] by analysis of numerical results. In the two-spinon case, the energy was derived by straightforward calculation in [25]. Since there is no restriction to the distribution of $d_{k\sigma}$, we may regard the spinon as obeying bosonic statistics with repulsive interaction. In strong contrast to ordinary interaction, the interaction between bosonic spinons merely serves to modify the occupation pattern of the rapidity from that of bosons to semions, and constitutes the statistical interaction.

We shall now make the free-semion nature more explicit. For this purpose we return to the YHWS and introduce the rapidity p_i of a spinon by the relation

$$p_i = k_i + \frac{\pi}{2N} \sum_j \mathrm{sgn}(p_i - p_j) = k_i + \frac{\pi}{2N}(N_{\mathrm{sp}} + 1 - 2i), \tag{4.146}$$

where $k_i = 2\pi\zeta_i/N$ is the bosonic momentum. For any magnetization $N - 2M > 0$ in the thermodynamic limit, the maximum of p_i is given by $\pi/2$, which corresponds to $i = 1$ and $\zeta_1 = M/2$. Likewise, the minimum of p_i

is given by $-\pi/2$ with $i = N_{\mathrm{sp}}$ and $\zeta_{N-2M} = -M/2$. The spinon becomes gapless at these extremal rapidities $p_i = \pm\pi/2$. Then we obtain from (4.145) and (4.140)

$$E(M) = E_2(M) + \sum_i \left[\left(\frac{\pi}{2}\right)^2 - p_i^2 \right], \qquad (4.147)$$

where $E_2(M)$ is an M-dependent reference energy. It is then clear that the spinons with rapidity p_i behave as free particles. They obey fractional exclusion statistics with the repulsion parameter $1/2$, as seen from (4.146).

In analogy with the bosonic spinons, we now derive the energy with inclusion of non-YHWS. It is convenient to introduce the distribution function $\rho_\sigma(p)$ of spinons with spin σ by requiring $d_{k\sigma}dk = \rho_\sigma(p)dp$ in the thermodynamic limit. By including down spinons for determining the rapidity in (4.146), we obtain the relationship

$$\mathrm{d}p - \mathrm{d}k = \frac{1}{2}d_k\mathrm{d}k = \frac{1}{2}\rho(p)\mathrm{d}p, \qquad (4.148)$$

where $d_k = d_{k\uparrow} + d_{k\downarrow}$ and $\rho(p) = \rho_\uparrow(p) + \rho_\downarrow(p)$. Then the bosonic and semionic distributions are related as

$$\rho_\sigma(p) = d_{k\sigma} \left(1 + \frac{1}{2}d_k \right)^{-1} = \left[1 - \frac{1}{2}\rho(p) \right] d_{k\sigma}. \qquad (4.149)$$

In the ground state with no magnetization, we have $\rho_\sigma(p) = 0$. In this case the energy is reduced to $E_0(N/2)$, defined by (4.18). We can represent the energy for arbitrary states in terms of $\rho_\sigma(p)$. In the language of semionic spinons, the positive magnetization appears as $\rho_\uparrow(p) > \rho_\downarrow(p)$ for a finite range of p. The energy with inclusion of the Zeeman term is represented by

$$U(h) = E - h(N_\uparrow - N_\downarrow) = E_0\left(\frac{N}{2}\right) + \sum_{p\sigma} \left[\left(\frac{\pi}{2}\right)^2 - p^2 - \sigma h \right] \rho_\sigma(p), \quad (4.150)$$

which applies to any value of magnetization. From this energy functional, we can derive the thermodynamics of the system. Note that the origin of the spinon momentum is shifted by π in (4.150). Such a shift does not influence the thermodynamics, but should be restored properly in considering the dynamical response, as will be discussed later. Before going on to the case of finite temperature T, we derive the magnetization at $T = 0$ as a function of $h > 0$. Let us denote by $\pm p_{\mathrm{s}}$ the critical rapidities where the up spinon takes zero energy. The condition is given by

$$\epsilon_\uparrow(p_{\mathrm{s}}) = (\pi/2)^2 - p_{\mathrm{s}}^2 - h = 0. \qquad (4.151)$$

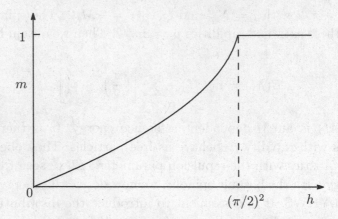

Fig. 4.10. The magnetization of the Haldane–Shastry model at $T = 0$.

The negative energy states are occupied at $T = 0$ with the semionic distribution $\rho_\uparrow(p) = 2$. The volume of regions of $\epsilon_\uparrow(p) < 0$ determines the magnetization $mN \equiv N_\uparrow - N_\downarrow$. We obtain

$$m = 1 - \frac{2}{\pi}p_{\mathrm{s}} = 1 - \sqrt{1 - \left(\frac{2}{\pi}\right)^2 h} \tag{4.152}$$

for $0 < h < h_{\mathrm{c}} = (\pi/2)^2$, and $m = 1$ for $h > h_{\mathrm{c}}$. Figure 4.10 shows the magnetization curve. From (4.152) we obtain the relation

$$2p_{\mathrm{s}} = \pi(1 - m) = v_m. \tag{4.153}$$

The zero-field susceptibility is given by

$$\chi_m = \lim_{h \to 0} \partial m/\partial h = 2/\pi^2 = 2/(\pi v_m), \tag{4.154}$$

where the spinon velocity v_m is reduced to π at $m = 0$. If one uses $v_m = \pi(1 - m)$ for finite magnetization, the differential susceptibility is given by (4.154) for arbitrary magnetization. We note that the differential susceptibility at $h = h_{\mathrm{c}}$ is divergent.

4.10.2 Thermodynamic potential of spinons

For derivation of the thermodynamics of the spin chain, the spinon picture with semionic statistics provides the most economical framework. Once we know the statistical parameters, we can construct the thermodynamics according to the general scheme of the fractional exclusion statistics. We start with the entropy obtained in the bosonic picture as

$$S = \sum_{k\sigma} [(d_{k\sigma} + 1) \ln(d_{k\sigma} + 1) - d_{k\sigma} \ln d_{k\sigma}]. \qquad (4.155)$$

It is possible to rewrite (4.155) as the entropy of semionic spinons. For this purpose we use the distribution function $\rho_\sigma(p)$ that describes the semionic spinons. Complementary to $\rho_\sigma(p)$, we introduce a function $\rho_\sigma^*(p)$ which gives the distribution of dual particles of spinons, namely antispinons or magnons. In the case of finite magnetization, $\rho_\uparrow^*(p)$ describes excitation from the condensate of spinons. By definition of the available orbitals in the narrow range Δp or Δk we obtain

$$\frac{\Delta D_\sigma}{\Delta N_\sigma} = \frac{\rho_\sigma^*(p)\Delta p}{\rho_\sigma(p)\Delta p} = \frac{\Delta k}{d_{k\sigma}\Delta k}, \qquad (4.156)$$

where the last equality follows from the fact that the available states of bosons remain constant independent of the particle occupation. Then we obtain another form of the entropy

$$S = \sum_{p\sigma} [(\rho_\sigma + \rho_\sigma^*) \ln(\rho_\sigma + \rho_\sigma^*) - \rho_\sigma \ln \rho_\sigma - \rho_\sigma^* \ln \rho_\sigma^*], \qquad (4.157)$$

where we have used $d_{k\sigma} dk = \rho_\sigma(p) dp$. On the other hand, using (4.149) and (4.156) we obtain

$$\rho_\sigma^*(p) = 1 - \frac{1}{2} \sum_\alpha \rho_\alpha(p), \qquad (4.158)$$

which corresponds to the semionic exclusion statistics with the exclusion parameter $g_{\sigma\alpha} = 1/2$ for all combinations σ, α of spins. This is consistent with the state counting in terms of spinons that leads to (4.139). While the state counting was performed for all the states with various momenta, the same exclusion parameter follows for a range of states with given momentum in the thermodynamic limit. The form of (4.157) confirms the interpretation that the entropy is given by particles obeying the fractional exclusion statistics.

We write the spectrum of a spinon in a magnetic field as

$$\epsilon_\sigma(p) = (\pi/2)^2 - p^2 - \sigma h = \epsilon_0(p) - \sigma h, \qquad (4.159)$$

where it is understood that $\sigma = \uparrow, \downarrow$ for the suffix, and $\sigma = \pm 1$ on the RHS. Without a magnetic field, $\epsilon_0(p)$ describes the spinon spectrum. While p runs from $-\pi/2$ to $\pi/2$ without a magnetic field, only the part with $\epsilon_\uparrow(p) > 0$ is meaningful with $h > 0$. Therefore, the spinons can exist only for $|p| < p_s n = \pi(1-m)/2$, where m is the magnetization given by (4.152). On the other hand, the spectrum $\epsilon_\downarrow(p)$ has a gap $2h$ at $p = p_s$. Note that the down spinons have the same restriction for the momentum range because of the Yangian symmetry. The up–down symmetry in the spinon energy is most clearly seen in (4.131).

According to the free-particle picture with semionic statistics, the energy of the system with inclusion of the Zeeman term is given by

$$U(h) = \sum_p \left[\epsilon_\uparrow(p)\rho_\uparrow(p) + \epsilon_\downarrow(p)\rho_\downarrow(p) \right]. \tag{4.160}$$

Then the thermodynamic potential is given by $\Omega(T, h) = U(h) - TS$. The distribution function is determined by the variational condition $\delta\Omega/\delta\rho_\sigma(p) = 0$. As a result, we obtain for each p the relation

$$\beta\epsilon_\sigma = \ln\left(1 + w_\sigma\right) - \frac{1}{2}\sum_\nu \ln\left(1 + w_\nu^{-1}\right) \tag{4.161}$$

with $\beta = 1/T$ and $w_\sigma \equiv \rho_\sigma^*/\rho_\sigma$.

Then the thermodynamic potential takes the simple form

$$\Omega = -T\sum_{p\sigma} \ln\left[1 + w_\sigma(p)^{-1}\right]. \tag{4.162}$$

Without a magnetic field, we obviously have $\rho_\uparrow(p) = \rho_\downarrow(p)$. Then (4.158) is reduced to

$$\rho_\sigma^*(p) = 1 - \rho_\sigma(p), \tag{4.163}$$

which gives nothing but the Fermi statistics. Thus we obtain $w_\sigma(p) = \exp[\epsilon_0(p)/T]$ for $h = 0$, and the free-fermion form for Ω from (4.162). The reason why the free fermions appear without h is that a pair of semions decreases the available number of states by just one, and hence they act as fermions. It is instructive to compare the situation with the XY chain where the thermodynamics is also described by free fermions with the spectrum $\epsilon_k = J\cos k$. The energy ϵ_k takes both positive and negative values, while $\epsilon_0(p)$ can only be positive. The positiveness means that the number of fermions goes to zero as the temperature approaches absolute zero. Hence there is no Fermi sea in the Haldane–Shastry spin chain.

At $h = 0$ the entropy is written in the form

$$S = 2\sum_p [\ln\left(2\cosh x_p\right) - x_p \tanh x_p], \tag{4.164}$$

where $x_p = \beta J\epsilon_0(p)/4$ with $J = 2$. Note that the entropy is an even function of x_p, which is a general property of ideal fermions. In our case, this means an unexpected symmetry that the ferromagnetic Haldane–Shastry model has the same entropy as that of the antiferromagnetic case. With $J < 0$ in the Haldane–Shastry model, the ground state is completely polarized. In counting the entropy, however, we should consider all possible polarizations in the ground state as well as excited states. Then we have zero polarization on average, even in the ferromagnetic system at $T \neq 0$. Magnons can then be described as antispinon excitations out of the condensate of up and down spinons with equal numbers. Because of (4.163), the magnons also obey the fermionic exclusion statistics as dual particles of spinons. In fact, the spectrum of ferromagnetic magnons in the p-space is the same as that of spinons in the antiferromagnetic system. We emphasize again that the symmetry between ferro- and antiferromagnetic systems applies only with $h = 0$.

4.10.3 Susceptibility and specific heat

We now derive the susceptibility and specific heat at low temperatures. By inserting an explicit expression for ϵ_σ in (4.161), we obtain

$$w_\sigma(p) = e^{-\beta[\epsilon_0(p)+\sigma h]} \left[\sqrt{\gamma(p)^2 + 1} + \sigma \gamma(p) \right], \qquad (4.165)$$

where $\gamma(p) = \exp[-\beta\epsilon_0(p)] \sinh \beta h$. Then we can derive $\rho_\sigma(p)$ by using (4.149). The magnetization $m = n_\uparrow - n_\downarrow$ per site is given by

$$m = \int_{-\pi/2}^{\pi/2} \frac{\mathrm{d}p}{2\pi} \left[\rho_\uparrow(p) - \rho_\downarrow(p) \right]. \qquad (4.166)$$

For small $h > 0$, $\epsilon_\uparrow(p)$ becomes negative near $p = \pm\pi/2$. This implies condensation of up spinons, and leads to finite magnetization. With $h > 0$, a gap develops for creating a down spinon, and an excitation from the condensate constitutes an antiparticle. In Section 4.7.2, we have called this antiparticle an antispinon [121].

Let us derive the magnetization at low temperature T. We can safely set $\rho_\downarrow(p) = 0$ for all p. The other component $\rho_\uparrow(p)$ is derived using (4.149) as

$$\rho_\uparrow(p) = \frac{2}{\sqrt{4 \exp[2\beta\epsilon_\uparrow(p)] + 1}}, \qquad (4.167)$$

which tends to 2 for $\epsilon_\uparrow < 0$ and 0 otherwise. By comparing this with (2.159) in Chapter 2, we see that the distribution corresponds to the single-component fractional exclusion statistics with the statistical parameter $1/2$. The differential susceptibility is derived by expanding $\rho_\uparrow$ near the zero energy. For this purpose we introduce the Sommerfeld-type expansion with coefficients

$$I_n \equiv \int_{-\infty}^{\infty} d\epsilon \left(-\frac{\partial \rho_\uparrow}{\partial \epsilon}\right) \epsilon^n. \tag{4.168}$$

These are calculated to be

$$I_0 = 1, \ I_1 = 0, \ I_2 = 2\pi^2 T^2/3, \ I_3 = 12\zeta(3)T^3,$$

with $\zeta(x)$ being the zeta function. We make an expansion near $\epsilon_\uparrow(p_s) = 0$ as

$$p(\epsilon) \simeq p_s + \epsilon/p_s - \epsilon^2/(2p_s^3) + \epsilon^3/(2p_s^5). \tag{4.169}$$

Then we obtain

$$\frac{\partial m}{\partial h} \equiv \chi_m(m) = \frac{2}{\pi^2(1-m)} \left[1 + \frac{2T^2}{3\pi^2(1-m)^4}\right] + O(T^3). \tag{4.170}$$

If we take the limit of zero magnetic field first, i.e., $h/T \ll 1$, we obtain

$$\rho_\uparrow(p) - \rho_\downarrow(p) = 4\beta h \exp\left[-\beta\epsilon_0(p)\right], \tag{4.171}$$

from (4.165) and (4.149). Alternatively, we start from the bosonic distribution $d_{k\sigma}$, and take the limit $h \to 0$ to obtain

$$\chi_m = \frac{1}{N} \sum_{k\sigma} \sigma \frac{\partial}{\partial h} d_{k\sigma} = 4\beta \int_{-\pi/2}^{\pi/2} \frac{dp}{2\pi} \frac{\rho_\sigma(p)}{1 - \rho_\sigma(p)}, \tag{4.172}$$

where we have used $\partial d_{k\sigma}/\partial h = \beta\sigma d_{k\sigma}(1 + d_{k\sigma})$, and (4.149). Since $\rho_\sigma(p)$ at $h = 0$ obeys the fermionic distribution, we obtain $\rho_\sigma(p)/[1 - \rho_\sigma(p)] = \exp\left[-\beta\epsilon_0(p)\right]$ in consistency with (4.171). The susceptibility is derived as

$$\chi_m = \frac{2}{\pi^2}\left(1 + \frac{2}{\pi^2}T\right) + O(T^2), \tag{4.173}$$

which has the leading correction of $O(T)$ in contrast to (4.170). The difference comes from the p-linear spinon spectrum near $p = \pm\pi/2$. The susceptibility in the limit $T \to 0$ is consistent with (4.154).

The entropy or specific heat at low temperature can be derived for arbitrary magnetization. We start with the entropy given by (4.157), and observe that $\rho_\uparrow$ is given by (4.167) and $\rho_\downarrow = 0$. Then changing the integration variable to $\epsilon_\uparrow$, we obtain the low-temperature expansion by the Sommerfeld-type formula. The leading contribution is given by $S/N \simeq \pi T/(6p_s)$, with $p_s = \pi(1-m)/2$ as given by (4.152). We can rewrite the result as

$$\frac{S}{N} \simeq \frac{T}{3(1-m)} = \frac{\pi^2}{6}T\chi_m \equiv \gamma T, \tag{4.174}$$

which is consistent with the specific heat of a Tomonaga–Luttinger liquid. In the opposite limit of high T, we may neglect the p-dependence of $\rho_\alpha(p)$. Then for fixed n_σ we recover the result corresponding to local moments,

$$S = -N \sum_\sigma n_\sigma \ln n_\sigma.$$

The numerical results on specific heat and susceptibility for a general temperature will be presented in Chapter 6, together with the case of a finite number of holes.

The thermodynamic quantities thus derived are functions of temperature without logarithmic singularity. This feature is in contrast to those in the Heisenberg model with the nearest-neighbor interaction. It is known that the susceptibility increases in a singular manner as the temperature becomes nonzero [54, 179]. The simpler behavior of the $1/r^2$ exchange model reflects the fact that the model represents the fixed point of the Tomonaga–Luttinger liquid.

In Section 4.10.4, we discuss an alternative method to derive the thermodynamics by using the freezing trick.

4.10.4 *Thermodynamics by freezing trick*

It is instructive to derive the thermodynamics by an alternative picture using the freezing trick. We work with H_{tot} in (4.128). In order to extract the physical results, we have to remove the phonon contribution. We introduce a variable $k = 2\pi\kappa/N$ that corresponds to a wave number. In the thermodynamic limit $N \to \infty$, the energy is rewritten as

$$\frac{1}{N} E_{\text{tot}} = \frac{1}{8\pi} \int_{-\infty}^{\infty} dk \int_{-\infty}^{\infty} dk' \, |k - k'| \, \nu[k]\nu[k'], \qquad (4.175)$$

with $\nu[k] = \nu(\kappa)$. Since each microscopic eigenstate is uniquely characterized by the momentum distribution functions $\{\nu_\sigma(\kappa)\}$, the entropy s_{tot} per site has the same form as that of the U(2) free fermion system:

$$s_{\text{tot}} = -\frac{1}{2\pi} \int_{-\infty}^{\infty} dk \sum_\sigma \left[(1 - \nu_\sigma) \ln (1 - \nu_\sigma) + \nu_\sigma \ln \nu_\sigma \right].$$

The density of each spin component is given by

$$n_\sigma = \frac{1}{2\pi} \int_{-\infty}^{\infty} dk \nu_\sigma[k]. \qquad (4.176)$$

From (4.175), (4.176), and (4.176), we obtain the thermodynamic potential

$$\Omega_{\text{tot}}(\{\nu\}) = E_{\text{tot}}(\{\nu\}) - NT s_{\text{tot}}(\{\nu\}) - N \sum_{\sigma} \mu_{\sigma} n_{\sigma}(\{\nu\}). \qquad (4.177)$$

The equilibrium conditions $\delta\Omega_{\text{tot}}/\delta\nu_{\sigma} = 0$ yield the equilibrium momentum distribution functions

$$\nu_{\sigma}[k] = \frac{1}{\exp\left[(\epsilon(k) - \mu_{\sigma})/T\right] + 1}, \qquad (4.178)$$

where we have introduced the one-particle energy $\epsilon(k)$ defined by

$$\epsilon(k) = \frac{2\pi}{N}\frac{\delta E_{\text{tot}}(\{\nu_{\kappa}\})}{\delta\nu(\kappa)} = \frac{1}{2}\int_{-\infty}^{\infty} dk' \, |k - k'| \, \nu[k']. \qquad (4.179)$$

(4.179) with (4.178) gives a functional equation for $\epsilon(k)$. Substituting the resultant expression for $\epsilon(k)$ into (4.175)–(4.177), we can obtain thermodynamic quantities for H_{tot}.

In order to derive the thermodynamics of the spin chain described by H_{HS}, we have to subtract the phonon contribution. For this purpose we introduce a new variable $p(k)$ defined as

$$p(k) \equiv \frac{\partial\epsilon(k)}{\partial k} = \frac{1}{2}\int_{-\infty}^{\infty} dk' \text{sgn}\,(k - k')\,\nu[k'], \qquad (4.180)$$

which corresponds to rapidity of holes. The boundary condition is given by

$$p(k = \pm\infty) = \pm\pi, \qquad (4.181)$$

which is obtained from (4.180) by noting that

$$\int_{-\infty}^{\infty} dk\,\nu[k] = 2\pi. \qquad (4.182)$$

The range of p being $[-\pi, \pi]$ shows its character as the crystal momentum. It could be taken as $[0, 2\pi]$ if one adopted an alternative boundary condition: $p(k = -\infty) = 0$. Thus the origin of p can be shifted freely for convenience, while the range 2π is fixed.

We shall now rewrite the thermodynamic quantities in terms of p. First we shall obtain $\epsilon(k)$ as a function of p, namely $\epsilon(k(p)) = \epsilon_p$. Differentiation of (4.180) with respect to k gives

$$\frac{\partial p(k)}{\partial k} = \nu[k] = \nu_p, \qquad (4.183)$$

where we have introduced the notation ν_p to emphasize the dependence on p. Hence we can use the relation $dk = dp/\nu_p$ to change the integration variable from k to p.

By multiplying both sides of (4.183) by $p(k)$ and integrating from k to ∞, we obtain the following relation:

$$(\pi^2 - p^2)/2 \equiv \omega_0(p) = -T \sum_\sigma \ln(1 - \nu_{p\sigma}), \qquad (4.184)$$

where $\nu_{p\sigma}$ is equal to the Fermi function given by (4.178), and we have used the property $pdk = d\epsilon(k)$. We can obtain $\nu_{p\sigma}$ as a function of p by using (4.184). Let us first study the simplest case $\mu_\uparrow = \mu_\downarrow$, i.e., without a magnetic field. We obtain from (4.184)

$$\nu_{p\sigma} = \nu_p/2 = 1 - \exp[-\beta\omega_0(p)/2]. \qquad (4.185)$$

The phonon part of the distribution function is obtained by regarding ν_p as a single component. We obtain $\nu_p = 1 - \exp(-\beta\omega_0(p))$ from (4.184) by omitting the σ summation. The corresponding entropy is given by

$$S_{\mathrm{ph}} = N \int_{-\pi}^{\pi} \frac{dp}{2\pi} \left[\frac{\beta\omega_0(p)}{\exp(\beta\omega_0(p)) - 1} - \ln(1 - e^{-\beta\omega_0(p)}) \right], \qquad (4.186)$$

where we have used (4.183) to change the integral variable from k to p in the spinless version of (4.176). By using the variable p we deal with excitations described by $\omega_0(p)$, instead of dealing with N particles. Namely, we deal with hole excitations with the range of momentum limited by the size of the system. This entropy takes the form of a difference between the internal energy and the thermodynamic potential of free bosons with energy $\omega_0(p)$ and chemical potential zero. By shifting the origin of p by π in $\omega_0(p)$, we interpret S_{ph} as the entropy of phonons. More explicitly, we obtain

$$\Omega_{\mathrm{ph}}(T) = -N \int_0^T dT' \, S_{\mathrm{ph}}(T') = NT \int_{-\pi}^{\pi} \frac{dp}{2\pi} \ln[1 - \exp(-\beta\omega_0(p))], \qquad (4.187)$$

which corresponds to the thermodynamic potential of free bosons with the spectrum $\omega_0(p)$. The thermodynamic potential Ω_{tot} can be derived in a similar manner. Without a magnetic field, $\mu_\uparrow = \mu_\downarrow$, the only difference from the phonon case is that $\omega_0(p)$ is replaced by $\omega_0(p)/2$ as a result of spin summation in (4.184). The result is given by

$$\Omega_{\mathrm{tot}}(T) = NT \int_{-\pi}^{\pi} \frac{dp}{2\pi} \ln[1 - \exp(-\beta\omega_0(p)/2)]. \qquad (4.188)$$

Hence we obtain the desired quantity Ω_S of the spin chain:

$$\Omega_S(T) = \Omega_{\mathrm{tot}}(T) - \Omega_{\mathrm{ph}}(T) = -NT \int_{-\pi}^{\pi} \frac{dp}{2\pi} \ln[1 + \exp(-\beta\omega_0(p)/2)]. \qquad (4.189)$$

It can easily be seen that the result of $\Omega_S(T)$ is equivalent to (4.162) with $h = 0$ by the change $p \to p/2$ of the momentum scale.

With a finite magnetic field, the distribution function becomes more complicated. Using the relation

$$\nu_{p\uparrow}^{-1} - 1 = \exp(-2\beta h)[\nu_{p\downarrow}^{-1} - 1] \tag{4.190}$$

for the Fermi function, we can eliminate $\nu_{p\uparrow}$ in (4.184) to obtain the equation for $\nu_{p\downarrow}$. The resultant quadratic equation is solved to give

$$\nu_{p\sigma} = 1 - e^{-\beta(\omega_0(p)/2 + \sigma h)}\left(\sqrt{\gamma^2 + 1} + \sigma\gamma\right), \tag{4.191}$$

where $\gamma = \exp[-\beta\omega_0(p)/2]\sinh\beta h$. Given the explicit form of the distribution function, we can obtain thermodynamic quantities. For example, magnetization is derived by the formula

$$m = \frac{1}{N}\sum_k (\nu_\uparrow[k] - \nu_\downarrow[k]) = \frac{1}{N}\sum_p \frac{\nu_{p\uparrow} - \nu_{p\downarrow}}{\nu_{p\uparrow} + \nu_{p\downarrow}}, \tag{4.192}$$

which leads to the same result as that using the semionic statistics. In the present treatment, $\nu_{p\sigma}$ describes hole excitations with U(2) symmetry.

4.11 Dynamical structure factor

4.11.1 Brief survey on dynamical theory

The dynamics of spin chains has been studied for a long time, and much attention has been paid to the antiferromagnetic Heisenberg model, or its variants. The quantity of central importance is the dynamical structure factor. At zero temperature it is defined with the spin component $\alpha, \beta = x, y, z$ by

$$S^{\alpha\beta}(q, \omega) = \sum_\nu \langle 0|S_q^\alpha|\nu\rangle\langle\nu|S_{-q}^\beta|0\rangle\delta(\omega - E_\nu + E_0), \tag{4.193}$$

where $|\nu\rangle$ denotes an eigenstate of the Hamiltonian with energy E_ν (E_0 being the ground energy). The Fourier transform S_q^α is given by

$$S_q^\alpha = \frac{1}{\sqrt{N}}\sum_l S_l^\alpha e^{-iql}, \tag{4.194}$$

in terms of the spin at site l.

An important contribution was made by des Cloizeaux and Pearson [42], who derived the lowest branch of the spin 1 excitation from the singlet ground state. The dispersion relation is derived as

$$\omega_q = \frac{\pi}{2} J \sin q, \tag{4.195}$$

where J is the nearest-neighbor exchange, and the lattice constant is taken to be unity. They used the Bethe ansatz theory to derive the results. At that time the difference from the spin-wave spectrum [3], which arises from the Néel ordered state, was emphasized. However, the continuum nature of excitations was not noticed.

It was then recognized [196] that the des Cloizeaux–Pearson mode is the lower bound of the continuum spin excitations. This feature is shared with the spectrum of the XY chain, which permits an exact solution to be obtained easily by the Jordan–Wigner transformation. Let us briefly review the solution. We introduce fermion creation and annihilation operators $\psi_i^\dagger$ and ψ_i at each site i as

$$S_i^z = \psi_i^\dagger \psi_i - 1/2 = n_i - 1/2, \tag{4.196}$$

$$S_i^- = S_i^x - \mathrm{i}S_i^y = \psi_i \exp\left(\mathrm{i}\pi \sum_{j=1}^{i-1} n_j\right), \tag{4.197}$$

where the site index begins at $i = 1$ on the leftmost side of the system, and ends at $i = N$ on the rightmost. Let us consider the Hamiltonian called the XXZ model

$$H = J \sum_i \left(S_i^x S_{i+1}^x + S_i^y S_{i+1}^y + \Delta S_i^z S_{i+1}^z \right), \tag{4.198}$$

with the anisotropy parameter Δ. The limit $\Delta = 0$ is called the one-dimensional XX model or, less precisely, the XY chain. The Heisenberg model corresponds to $\Delta = 1$, and the Ising model to $\Delta = \infty$. In terms of the Jordan–Wigner fermions, (4.198) is written as

$$H = \frac{J}{2} \sum_i \left[\psi_i^\dagger \psi_{i+1} + \psi_{i+1}^\dagger \psi_i + 2\Delta \left(n_i - \frac{1}{2} \right) \left(n_{i+1} - \frac{1}{2} \right) \right]. \tag{4.199}$$

The XY chain with $\Delta = 0$ is thus equivalent to spinless free fermions with the spectrum $\epsilon_k = J \cos k$ for $-\pi < k \leq \pi$. The ground state has $N/2$ fermions with the Fermi wave number $k_F = \pi/2$. The z-component of the dynamical spin structure factor of the XY chain is obtained as the charge structure factor of the Jordan–Wigner fermions [106, 145]. The result is given by

$$S^{zz}(q, \omega) = \sum_k [f(\epsilon_k) - f(\epsilon_{k+q})]\delta(\omega + \epsilon_k - \epsilon_{k+q}), \tag{4.200}$$

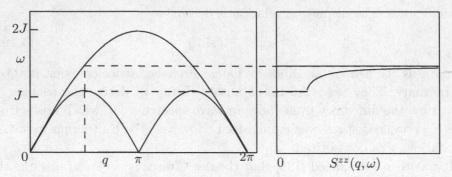

Fig. 4.11. The finite spectral region (support) of $S^{zz}(q,\omega)$ in the XY chain, and a schematic view of the spectral intensity at $q = \pi/2$.

where $f(\epsilon_k)$ is the Fermi function. This result is actually exact even at finite temperatures. (4.200) shows that the spin excitation spectrum involves two elementary excitations. This is in strong contrast to the Néel state, which involves only a single mode in the spin wave theory. We obtain from (4.200) at zero temperature

$$S^{zz}(q,\omega) = \frac{\theta(\omega - \omega_{\mathrm{L}}(q))\theta(\omega_{\mathrm{U}}(q) - \omega)}{2\pi\sqrt{\omega_{\mathrm{U}}(q)^2 - \omega^2}}, \qquad (4.201)$$

where $\theta(x)$ is the step function. The lower bound $\omega_{\mathrm{L}}(q)$ and the upper bound $\omega_{\mathrm{U}}(q)$ are given by

$$\omega_{\mathrm{L}}(q) = J|\sin q|, \quad \omega_{\mathrm{U}}(q) = 2J\sin(q/2), \qquad (4.202)$$

for $0 \leq q \leq 2\pi$.

The lower bound $\omega_{\mathrm{L}}(q)$ happens to be the same as the spin-wave result for the Heisenberg chain [3]. Figure 4.11 illustrates the spectrum. With fixed q, the spectral intensity $S^{zz}(q,\omega)$ as a function of ω rises stepwise from the lower bound, and diverges with the exponent $1/2$ in the upper bound. This feature differs very much from the isotropic Heisenberg chain.

For $\Delta \neq 0$, it is difficult to derive the dynamics of the XXZ chain. Theoretical development up to 1981 was summarized by Müller *et al.* [136]. Performing numerical diagonalization of the Heisenberg chain ($\Delta = 1$) for a finite size, and using the Bethe ansatz results, Müller *et al.* proposed an approximate formula [136]

$$S(q,\omega) = \frac{A\theta(\omega - \omega_{\mathrm{L}}(q))\theta(\omega_{\mathrm{U}}(q) - \omega)}{\sqrt{\omega^2 - \omega_{\mathrm{L}}(q)^2}}, \qquad (4.203)$$

where A is a constant and the component index has been removed since $S(q, \omega)$ is a scalar. According to exact information on $S(q, \omega)$ available now, the intensity vanishes continuously at the upper threshold with the exponent $1/2$ [30]. We shall show later that the above formula, though approximate for the Heisenberg chain, is in fact exact for the Haldane–Shastry model provided one uses modified functions for the thresholds $\omega_L(q)$ and $\omega_U(q)$.

The nature of the elementary excitations in the Heisenberg chain was studied by Faddeev and Takhtajan [55], who showed that the spin $1/2$ excitation, which is now called the spinon, plays a fundamental role in dynamics. The exact derivation of the dynamical structure factor for the Heisenberg model is a difficult task, partly because the two-spinon contribution does not exhaust the whole intensity. The two-spinon part was derived in [30], and more recently in [96, 112] in a magnetic field. In the Haldane–Shastry model, on the other hand, the spectrum is exhausted by the two-spinon contribution. Haldane and Zirnbauer [81, 200] derived the first exact dynamical results for lattice spin systems by using a sophisticated method. We shall derive the dynamics of the spin chain in an elementary fashion, as presented below.

4.11.2 Exact analytic results

Because of the rotational invariance, $S^{\alpha\beta}(q, \omega)$ is diagonal in the component indices α, β and does not depend on them. Then it is most convenient to work with the spin-flip component generated by

$$S_q^+ = \frac{1}{\sqrt{N}} \sum_{l=1}^{N} \exp(-\mathrm{i}ql)(S_l^x + \mathrm{i}S_l^y) \tag{4.204}$$

and its conjugate. The flip of a down spin corresponds to annihilation of a hard-core boson, as seen in (4.106). In other words, the action of $S_l^x + \mathrm{i}S_l^y$ on the ground state Ψ_G, given by (4.13), is conveniently represented in terms of another ground-state wave function Ψ_{G-} with $N/2 - 1$ down-spin electrons as seen from (4.100). Namely, the spin-flip creates two spinons at the same site z_0:

$$(S_0^x + \mathrm{i}S_0^y)\Psi_G = \Psi_{2s}(\{z\}; z_0, z_0) = z_0 \prod_{i=1}^{N/2-1} (z_i - z_0)^2 \Psi_{G-}. \tag{4.205}$$

We have thus obtained the explicit form of the wave function of the final state after the spin-flip. The form of the final state shows clearly that the two-spinon contribution exhausts the dynamical structure factor.

The final state is a superposition of eigenfunctions of (4.16). Namely, we can represent $\Psi_{2s}(\{z\}; z_0, z_0)$ in terms of the Jack polynomial, which forms the complete eigenbasis for the Sutherland model. Taking $z_0 = 1$ without loss of generality, we perform the following expansion:

$$\prod_{i=1}^{N/2-1} (z_i - 1)^2 \Psi_{G-} = \sum_\mu b_\mu \Psi_\mu, \tag{4.206}$$

where Ψ_μ is an eigenfunction given by

$$\Psi_\mu(z) = J_\mu(z) \Psi_{G-}(z) \exp\left[i\theta_\mu(z)\right], \tag{4.207}$$

in terms of the Jack polynomial $J_\mu(z)$, and a phase factor to account for the Galilean shift of the total momentum. The coefficient b_μ is called the binomial expansion coefficient, and is given by

$$b_\mu = \prod_{s \in \mathcal{D}(\mu)} \frac{-a'(s)/\lambda + l'(s) + 1}{(a(s)+1)/\lambda + l(s)}, \tag{4.208}$$

which is the same as (2.227) with $\lambda = 2$.

We then obtain

$$S^{-+}(q, \omega) = \frac{1}{2} N \sum_\mu{}' \delta(\omega - \omega_\mu) |b_\mu|^2 \frac{\langle \Psi_\mu, \Psi_\mu \rangle_0}{\langle \Psi_G, \Psi_G \rangle_0}, \tag{4.209}$$

where the prime means restriction of the summation over μ so as to satisfy the momentum conservation. Namely, the state μ has the excitation energy ω_ν as given by $\omega_\mu = E_\mu - E_0$, and the momentum q. As discussed in Chapter 2, the inner product $\langle f, g \rangle_0$ is defined as the constant term in the Laurent expansion of $f(z)^* g(z)$.

The above expansion is the same as the one that we found for the hole propagator in the Sutherland model. Therefore we can immediately utilize the mathematical results obtained there. It is important, however, to recognize that the maximum power of z_i should be less than the size N of the lattice system. This is automatically satisfied in the present case since the spin-flip does not increase the power of each z_i. Equivalently, the excited states created by the spin-flip belong to the YHWS with $S = S^z = 1$.

For a finite size of the system, b_ν has a complicated and unappealing form. In the thermodynamic limit, however, the result simplifies dramatically. Following the same procedure as discussed in detail in Section 2.7.1, we obtain the dynamical structure factor $S(q,\omega) = S^{-+}(q,\omega)/2$ as follows:

$$S(q,\omega) = \int_0^\pi dq_1 \int_0^\pi dq_2 \, F_{2s}(q_1,q_2)\delta(q - q_1 - q_2)\delta(\omega - \epsilon_s(q_1) - \epsilon_s(q_2)),$$
(4.210)

where $F_{2s}(q_1,q_2)$ is the square of the form factor given by

$$F_{2s}(q_1,q_2) = \frac{1}{4}|q_1 - q_2|^{2g_s} \prod_{i=1}^2 \epsilon_s(q_i)^{g_s-1},$$
(4.211)

with $g_s = 1/2$. This g_s corresponds to the semionic statistical parameter of spinons. By proper shift of the momentum, the form factor agrees with the form (2.203) obtained for the hole propagator in the Sutherland model, which gives rise to ideal particles with fractional exclusion statistics.

(4.210) involves two spinons, instead of two Jordan–Wigner fermions in the XY chain, each of which has the spectrum

$$\epsilon_s(q) = \frac{1}{2}Jq(\pi - q)$$
(4.212)

with $0 \leq q \leq \pi$. We have taken $J = 2$ in the previous discussion. Note that spinons have the momentum span π which is half the size of the Brillouin zone. For the momentum q outside this range, $\epsilon_s(q)$ is defined by requiring the periodicity of π. Due care must be taken as to the origin of the momentum. In mapping to the Sutherland model, we have shifted the origin by π which gives the bottom of the parabolic spectrum. Creation of a magnon at the bottom corresponds to the creation of two spinons, each of which has the momentum $q = \pi/2$, being consistent with (4.212). On the other hand, the origin of momentum is not important in thermodynamics, and the alternative spectrum $\epsilon_0(p) = (\pi/2)^2 - p^2$ proves to be more convenient for spinons. Both forms are simply related by the shift in the momentum $p = q - \pi/2$.

The integral in (4.210) can be performed analytically. As a result we obtain for $0 \leq q \leq \pi$

$$S(q,\omega) = \frac{\theta(\omega - \epsilon_s(q))\theta(2\epsilon_s(q/2) - \omega)}{4\sqrt{\omega - \epsilon_s(q)}\sqrt{\omega - \epsilon_+(q)}},$$
(4.213)

where we have introduced the notation $\epsilon_+(q) = (2\pi - q)(q - \pi)$. For $\pi < q < 2\pi$, we have the relation $S(q,\omega) = S(2\pi - q,\omega)$. It is seen that (4.213) is rather similar to (4.203). The boundaries of the spectrum are obtained by replacement of the dispersion $\omega_L(q)$ by $\epsilon_s(q)$ for $0 \leq q \leq \pi$, and $\omega_U(q)$ by $2\epsilon_s(q/2)$. The lower threshold comes from two-spinon excitations in which one spinon has momentum q and the other has zero or π. In the upper

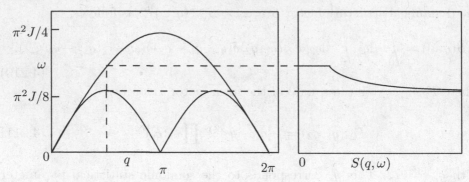

Fig. 4.12. The finite spectral region (support) of $S(q,\omega)$ in the Haldane–Shastry chain, and a schematic view of the spectral intensity at $q = \pi/2$.

threshold, two spinons move with the same momentum $q/2$. For each momentum q, in general two spinons have energies between these two extremes, and constitute the continuum. Figure 4.12 illustrates the feature schematically. The intensity $S(q,\omega)$ given by (4.213) has a stepwise singularity at the upper threshold, and diverges at the lower threshold with the exponent $1/2$.

Let us examine how the divergence of $S(q,\omega)$ in the lower threshold in the Haldane–Shastry model comes about. It is instructive to compare it with $S^{zz}(q,\omega)$ in the XY chain given by (4.201). In the latter case, the divergence is at the upper threshold, and comes from the cosine band spectrum of the Jordan–Wigner fermions. Suppose one introduces the z-component $J\Delta$ of the exchange interaction, and increases Δ from zero in the XY limit to one in the Heisenberg limit. In terms of the Jordan–Wigner fermions, $J\Delta$ describes the repulsive interaction between fermions because of the correspondence $2S_i^z = 1 - 2n_i$, where n_i is the fermion number operator at site i. The repulsive interaction works as an effective attractive interaction for the particle–hole excitation, just like excitons in semiconductors. Hence the large weight near the lower threshold is interpreted as a kind of exciton-like resonance. One can also make an analogy with the X-ray threshold singularity where divergent intensity appears in the presence of the electron–hole attraction [128].

With small enough Δ, one may expect a divergent spectrum at both the upper and lower thresholds: the former comes from the cosine-band spectrum, and the latter from the particle–hole attraction. This feature has indeed been derived [157] using the Bethe ansatz and the so-called determinant representation [112]. In the isotropic limit $\Delta = 1$, the divergence in the upper edge is completely suppressed [98]. In the particle–hole excitation picture for the Jordan–Wigner fermions, the suppression of the singularity is interpreted as a result of effective repulsive interactions. Although the particle–hole interaction always has a negative sign, the effective interaction near the upper end of the spectrum becomes repulsive because of the negative effective masses of Jordan–Wigner fermions in the relevant momentum range.

Now we take the spinon picture to interpret the behavior of $S(q, \omega)$ of the Haldane–Shastry model. In thermodynamics the spinons are regarded as non-interacting particles obeying semionic statistics. However, spinons have a repulsive statistical interaction according to the definition of the statistical interaction where free bosons are taken as the reference. This statistical interaction may be distinguished from the ordinary interaction, which in general leads to a momentum dependence in the phase shift for the two-particle scattering matrix [53].

In dynamics, on the other hand, statistical interaction and the ordinary interaction both cause deviation of the form factor from unity. The semionic repulsion $g_s = 1/2$ between the spinons appears explicitly as the factor $|q_1 - q_2|^{2g_s}$ in (4.211), which suppresses the divergent intensity in the upper threshold of $S(q, \omega)$. On the other hand, the divergence of $S(q, \omega)$ in the lower threshold comes from the factor $\epsilon_s(q_i)^{g_s - 1}$ in (4.211). Namely, the semionic statistics $g_s = 1/2$ together with the gapless spectrum of spinons is responsible for the divergence. Some conflicting viewpoints in the literature [25, 70] about the effective interaction between spinons seem to be resolved by the interpretation described above.

4.11.3 Dynamics in magnetic field

In the presence of a magnetic field, the system loses the SU(2) symmetry. Consequently, dynamical structure factors $S^{\alpha\beta}(q, \omega)$ as defined by (4.193) depend on the components α, β. Among these components, we can derive $S^{-+}(q, \omega)$ analytically since all relevant excited states $|\nu\rangle$ contributing to $\langle \nu | S^{\beta}_{-q} | 0 \rangle$ belong to the YHWS. Other components $S^{+-}(q, \omega)$ and $S^{zz}(q, \omega)$ contain excited states out of the YHWS, and it is difficult to obtain these components analytically, and only numerical results are available [156, 181]. However, the mapping to the Sutherland model gives a part of the intensity coming from the YHWS. The excitation contents involved in the YHWS part helps us to understand the numerical results.

We shall now present the available analytic and numerical results for a finite-sized system. In the numerical study [181] it is argued that the maximum number of quasi-particles is at most four: two spinons and two antispinons.

Two-spinon creation in $S^{-+}(q,\omega)$

Let us begin with the simplest case of $S^{-+}(q,\omega)$. In the presence of $M(<N/2)$ magnons in the ground state, the spin-flip S_i^+ annihilates a magnon with spin down. Equivalently, the spin-flip creates two spinons with spin up. The final state can be represented in the same way as in (4.206), with $N/2$ replaced by M. The result for $S^{-+}(q,\omega)$ in the thermodynamic limit is given for $0 < q < 2\pi$ and $\omega > 0$ by

$$S^{-+}(q,\omega) = \frac{\Theta\left(\epsilon_U(p) - \omega\right)\Theta\left(\omega - \epsilon_{L-}(p)\right)\Theta\left(\omega - \epsilon_{L+}(p)\right)}{2\sqrt{(\omega - \epsilon_{L-}(p))(\omega - \epsilon_{L+}(p))}}, \qquad (4.214)$$

where $\Theta(\omega)$ is the step function to give the bounds of the support. We have used the variable $p = q - \pi$ to make explicit the reflection symmetry about $q = \pi$. The upper bound is given by

$$\epsilon_U(p) = 2p_s^2 - p^2/2 = 2\epsilon_\uparrow(p/2), \qquad (4.215)$$

where the up-spinon energy has appeared. The lower bounds are given by

$$\epsilon_{L\pm}(p) = -p(p \mp q_m), \qquad (4.216)$$

which is just the spectrum of an up spinon with the momentum shifted by $\mp q_m/2 = \mp \pi m/2$. The shift is due to another spinon with zero energy. In the zero-polarization limit $(m = 0)$, $S^{-+}(q,\omega)$ given by (4.214) indeed reduces to $2S(q,\omega)$ given by (4.213). Note that both the upper and lower thresholds are determined by the spectrum of up spinons. Figure 4.13(a) shows the results for a finite-sized system with $N = 16$ and $m = 0.25$. The threshold momentum $q = q_m \equiv m\pi$ corresponds to twice the minimum momentum of up spinons.

Density response of magnons in $S^{zz}(q,\omega)$

The component $S^{zz}(q,\omega)$ gives the dynamical structure factor of magnons. A part of the spectrum can be obtained by mapping to the Sutherland model together with the freezing trick [7]. However, there are no analytical results available for the whole spectrum. Figure 4.13(b) shows numerical results for $N = 16$ and $m = 0.25$ [156]. The upper and lower thresholds of the spectrum are easily identified. Namely, the upper threshold labeled A near $q = 0$ is given by the antispinon $\bar{s}_R$ with the boundary $q(q + v_m)$, where two spinons with zero energy provide the momentum shift $2\pi - \pi m$, which gives the zero energy of $\bar{s}_R$. On the other hand, the lower threshold labeled B is given by the up spinon where the momentum shift πm is compensated by the zero-energy antispinon. The difference from the Sutherland model appears to be due to the periodicity 2π in the antispinon spectrum. Namely, at the

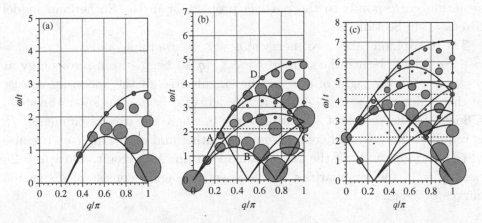

Fig. 4.13. Numerical results of dynamical structure factor (a) $S^{-+}(q,\omega)$, (b) $S^{zz}(q,\omega)$, and (c) $S^{+-}(q,\omega)$ with $N = 16$ and $m = 0.25$ [156]. The intensity is proportional to the area of each circle. The solid lines are determined by dispersion relations of the elementary excitations in the thermodynamic limit.

momentum 0 or 2π of the antispinon as shown in Fig. 4.3, a cusp appears in the spectrum. Correspondingly, new threshold singularities arise in $S^{\alpha\beta}(q,\omega)$ that are absent in the Sutherland model. An example is seen at the meeting point of the curves A and D, where the antispinon $\bar{s}_R$ reaches $q = 2\pi$. Other dispersion curves like C and D in Fig. 4.13(b) are determined by proper combination of up spinons and an antispinon. It is found by numerical work that at most one antispinon and two spinons are involved in the spectrum [181]. This restriction of the excitation contents agrees with the density spectrum of the Sutherland model with the coupling parameter $\lambda = 2$, where an antispinon corresponds to a particle and a spinon to a hole.

Antispinon creation in $S^{+-}(q,\omega)$

The action S_i^- on the ground state corresponds to the creation of a down-spin magnon, i.e., an antispinon. Equivalently, S_i^- annihilates two spinons with spin up in the condensate. The analytic expression of $S^{+-}(q,\omega)$ has not yet been derived. Figure 4.13(c) shows numerical results of $S^{+-}(q,\omega)$ for $N = 16$ and $m = 0.25$ [156]. An analytical solution is possible for such part of the final states as belongs to the YHWS by mapping to the Sutherland model with the coupling parameter $\lambda = 2$. Within the YHWS, the right $(\bar{s}_R)$ or left $(\bar{s}_L)$ antispinon with the spin component $S^z = -1$ corresponds to a particle, and the up spinon to a hole as illustrated in Fig. 4.3. Then the addition of $\bar{s}_{L,R}$ can accompany the excitations of $\bar{s}_{R,L} + 2s$ in the YHWS [160]. The situation corresponds to the particle propagator in the Sutherland model discussed in Section 2.6.2.

The antispinon takes zero energy at $q = q_m = \pi m$ for $\bar{s}_L$ and $q = 2\pi - q_m = \pi + 2p_s$ for $\bar{s}_R$. Hence the support for $S^{+-}(q, \omega)$ begins from zero energy at $q = \pm q_m$. To the right of $q = q_m$ the singularities of the spectrum consist of dispersion curves of $\bar{s}_R + 2s$. To the left of $q = q_m$ the lower threshold follows the dispersion of $\bar{s}_L$ down to $q = 0$, where the excitation energy $2h$ is required. Here the difference from the Sutherland model appears because of the periodicity 2π of the antispinon spectrum. As a result, above $\omega = 2h$ near $q = 0$, the singularities of the spectrum again consist of dispersion curves of $\bar{s}_R + 2s$.

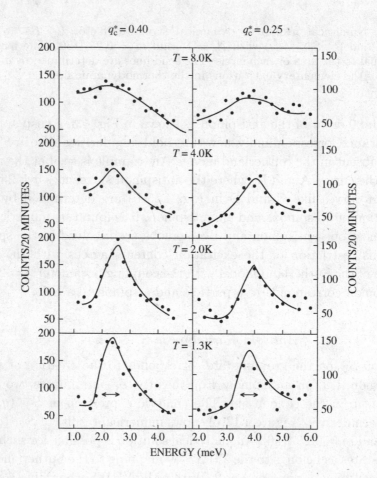

Fig. 4.14. Neutron scattering intensity from $CuCl_2 \cdot 2N(C_5D_5)$ for two values of reduced wave number $q^* = q/(2\pi)$ at various temperatures [52]. Copyright (1974) by the American Physical Society. Reproduced with permission. The exchange J is estimated to be 6.7K.

Table 4.1. *Maximum contents of quasi-particles for $S^{\alpha\beta}(q,\omega)$.*
The subscripts R *and* L *indicate the branches of antispinons.*

$S^{-+}(q,\omega)$	$2s$
$S^{+-}(q,\omega)$	$2s + \bar{s}_{R} + \bar{s}_{L}$
$S^{zz}(q,\omega)$	$2s + \bar{s}_{R}, 2s + \bar{s}_{L}$

Table 4.1 summarizes the excitation contents relevant to the dynamical structure factor in a magnetic field.

4.11.4 Comments on experimental results

The first neutron scattering experiment to probe the dynamics of quasi-one-dimensional spin systems was carried out for $CuCl_2 \cdot 2N(C_5D_5)$ [52]. The spectral function for each q has asymmetric shape, as shown in Fig. 4.14.

The peak corresponds to the lower threshold, the des Cloizeaux Pearson mode, and the asymmetry reflects the continuum above the threshold.

More recently, a detailed excitation spectrum of quasi-one-dimensional spin systems has been measured by many groups. Among others, we mention the result for $KCuF_3$ [185], $CuGeO_3$ [5], Yb_4As_3 [116], and copper pyrazine dinitrate (CuPzN) [170]. For CuPzN, good agreement is found with the two-spinon contribution derived theoretically for the nearest-neighbor Heisenberg chain [30]. The spectrum in a magnetic field has also been measured for CuPzN [170], and compared with numerical results for the Heisenberg chain.

5

SU(K) spin chain

As an extension of the Haldane–Shastry spin chain, this chapter discusses the SU(K) spin chain where K denotes the number of internal degrees of freedom, referred to as *color*. The coordinate representation is also possible in this case, provided we choose the reference state to be fully polarized to one of the components. In Section 5.1 we derive the ground state by mapping to the U($K-1$) Sutherland model. Then the spectrum is discussed by use of the ribbon diagram and the motif, both of which respect the SU(K) Yangian symmetry.

The elementary excitations in the system have a rather subtle feature: they have different statistical parameters in thermodynamics $g_{\alpha\beta}$ and dynamics $\tilde{g}_{\alpha\beta}$, where α and β denote color components. In Section 5.3, we discuss thermodynamics by using the freezing trick, and derive the statistical parameters. It turns out for $K = 2$ that the off-diagonal component acquires extra repulsion when starting from the U(2) Sutherland model with infinite repulsion. As a result we have $g_{\alpha\beta} = 1 - 1/K = 1/2$, which is independent of spin indices, consistent with the argument in Section 4.9.3. In the unpolarized case, the resultant thermodynamic potential can be interpreted as those of parafermions.

In Section 5.4, we present the analytic result for the dynamical structure factor, which includes the SU(K) spinons. They transform as the conjugate representation $\bar{K}$, with the statistical parameter $\tilde{g}_{\alpha\beta} = \delta_{\alpha\beta} - 1/K$. Note that the case $K = 2$ does not reproduce the thermodynamic statistical parameters $g_{\alpha\beta} = 1 - 1/K = 1/2$ for all spin combinations. In the SU(2) dynamics, however, the dynamical structure factor does not involve $g_{\uparrow\downarrow}$. Hence the difference between $\tilde{g}_{\alpha\beta}$ and $g_{\alpha\beta}$ does not show up in the case of $K = 2$.

5.1 Coordinate representation of ground state

Let us consider an $SU(K)$ chain with the number K of internal degrees of freedom. In order to distinguish this from the $SU(2)$ case, we call each component "color", in analogy with the $SU(3)$ color in the quark theory. The exchange of colors is described by the permutation operator P_{ij} given by

$$P_{ij} = \sum_{\alpha\beta} X_i^{\alpha\beta} X_j^{\beta\alpha}, \tag{5.1}$$

where the X-operators are defined as $X_i^{\alpha\beta} = |\alpha\rangle\langle\beta|$ for each site i. The Hamiltonian is given by

$$H_{\mathrm{SU}(K)} = \frac{1}{2} \sum_{i<j} J_{ij}(P_{ij}+1), \tag{5.2}$$

where $J_{ij} = J/D(x_i - x_j)^2$ with $D(x) = (N/\pi)\sin(\pi x/N)$. We have added unity in $P_{ij}+1$ for later convenience in mapping to the Sutherland model.

Starting with the fully polarized reference state $|F\rangle$ in which all sites are occupied by the Kth color component, we can represent any state $|\Psi\rangle$ in terms of the occupied set of coordinates $\{x^\alpha\}$, which specifies the positions of M_α particles with color species $\alpha = 1, \ldots, (K-1)$. The number M_K of the Kth color sites is then given by $M_K = N - \sum_{\alpha=1}^{K-1} M_\alpha$. As a generalization from the $SU(2)$ case, we introduce for each site i the creation operator $b_{i\alpha}^\dagger = |\alpha\rangle\langle K|$ of an αth magnon. The many-magnon state is described by the wave function $\Psi(\{x^\alpha\})$ as

$$|\Psi\rangle = \sum_{\{x^\alpha\}} \Psi(\{x^\alpha\}) \prod_{\alpha=1}^{K-1} \prod_{i\in\{x^\alpha\}} b_{i\alpha}^\dagger |F\rangle, \tag{5.3}$$

where $\Psi(\{x^\alpha\})$ is symmetric against interchange of equal-color coordinates. We separate P_{ij} into two parts:

$$P_{ij} = \sum_{\alpha=1}^{K-1} \left(b_{i\alpha}^\dagger b_{j\alpha} + b_{j\alpha}^\dagger b_{i\alpha} \right) + P_{ij}^{(K-1)}, \tag{5.4}$$

where $P_{ij}^{(K-1)}$ exchanges the $(K-1)$-color degrees of freedom other than $\alpha = K$. Although $\Psi(\{x^\alpha\})$ does not have a particular parity against interchange of coordinates of different components, $\Psi(\{x^\alpha\})$ is invariant against simultaneous interchange of color and spatial coordinates of any pair.

Provided $P_{ij}^{(K-1)}$ acts on this wave function, we may replace it by the operator K_{ij} which exchanges the spatial coordinates in $\Psi(\{x^\alpha\})$. Namely, we have

$$P_{ij}^{(K-1)}|\Psi\rangle = \sum_{\{x^\alpha\}} K_{ij}\Psi(\{x^\alpha\}) \prod_{\alpha=1}^{K-1} \prod_{i\in\{x^\alpha\}} b_{i\alpha}^\dagger |F\rangle. \qquad (5.5)$$

In analogy with cases of the SU(2) chain, we can work with the first-quantized form of the Hamiltonian for a subset of states which is given by a polynomial of $z_i = \exp(2\pi i x_i/N)$ with degrees ranging from 1 to $N-1$ for each color $\alpha = 1, \ldots, (K-1)$. For such polynomials we obtain the first-quantized Hamiltonian as in the previous cases:

$$H = \frac{1}{2} \sum_{i=1}^{N-M_K} \left(-i\frac{\partial}{\partial x_i} - \pi\right)^2 + \frac{1}{2} \sum_{i\neq j} \frac{1+K_{ij}}{D(x_i-x_j)^2} + \frac{1}{2}J(\pi)(N-M_K), \quad (5.6)$$

where we have put $J = 2$, and the Fourier component $J(q = \pi)$ has been obtained as $J(\pi) = -(\pi^2/3)(1 + 2/N^2)$ in (4.4). The first term results from the permutation involving the Kth species, and $-\partial/\partial x_i$ is the momentum of a magnon of any species $\alpha \neq K$.

The polynomial eigenfunctions are obtained most conveniently by regarding all magnons as distinguishable particles in the first step, and then imposing the proper symmetry. This is precisely the method we discussed in Chapter 3. In the case of all M_α ($\alpha = 1, \ldots, K$) odd, and N being divisible by K, the ground state is non-degenerate. We confine ourselves to this simplest case in the following, since the thermodynamic limit does not depend on these specific properties for N and M_α.

With the hard-core constraint, the absolute ground state, which does not have any constraint on the symmetry of the wave function, is given by a Slater determinant

$$\Psi_{\rm h}(\{z\}) = \prod_{i=1}^{N-M_K} z_i^{N/2-(N-M_K-1)/2} \prod_{i<j}(z_i - z_j), \qquad (5.7)$$

where the factor $z_i^{N/2}$ comes from the momentum shift to the edge π of the Brillouin zone, and another factor $z_i^{-(N-M_K-1)/2}$ makes the zero total momentum from the shifted origin. For the antisymmetric wave function, we obtain $K_{ij} = -1$ for the exchange operator. Then the interaction term in (5.6) vanishes, and the free-fermion wave function $\Psi_{\rm h}(\{z\})$ proves to be an eigenstate.

We now impose proper symmetry for the wave function for the mixture of magnons. We multiply $\Psi_h(\{z\})$ by a polynomial which is antisymmetric against interchange of magnon coordinates with the same color. The lowest-order antisymmetric polynomial together with the proper momentum shift is given by the Slater determinant with M_α coordinates:

$$\Psi_\alpha(\{z\}) = \prod_{i=1}^{M_\alpha} z_i^{-(M_\alpha-1)/2} \prod_{i<j\leq M_\alpha} (z_i - z_j), \tag{5.8}$$

where we have assigned the index $i = 1, \ldots, M_\alpha$ to magnons with color α. Then, following the same argument as in Section 3.2.2, the polynomial wave function

$$\Psi_G(\{z\}) = \Psi_h(\{z\}) \prod_{\alpha=1}^{K-1} \Psi_\alpha(\{z^\alpha\}) \tag{5.9}$$

proves to be an eigenfunction of (5.6). It is easily seen that all other polynomial eigenfunctions, which are obtained from (5.9) by multiplying symmetric polynomials of magnons, have higher energies. Hence, (5.9) is the ground-state wave function within the polynomial family. This wave function corresponds to the Gutzwiller wave function, which can also be obtained by imposing the hard-core condition on the product of Slater determinants of each color $\alpha = 1, \ldots, K$, as discussed in Section 3.2.2.

By using the freezing trick, for example, it can be proved that (5.9) in fact describes the ground state without restriction to the polynomial family. In the non-degenerate ground state, the wave function can be written in the explicitly real form. Namely, in terms of the chord distance $D(x)$, the eigenfunction of (5.9) is represented with a real factor C by

$$\Psi_G(\{z\}) = C \prod_{i>j} D(x_i - x_j) \prod_{\alpha=1}^{K-1} \prod_{l>m} D(x_{l\alpha} - x_{m\alpha}), \tag{5.10}$$

where x_i ($i = 1, \ldots, N - M_K$) denotes all magnon coordinates, and $x_{l\alpha}$ ($l = 1, \ldots, M_\alpha$) denotes those of α-magnons.

5.2 Spectrum and motif

By using the Yangian symmetry, we can enumerate all states including non-polynomial wave functions, whose energies are degenerate with those of polynomial wave functions. A Yangian generator is defined by analogy with (4.110) as

$$\Lambda_K = \frac{1}{2} \sum_{i\neq j} w_{ij} \sum_{\alpha=1}^{K-1} b_{i\alpha}^\dagger b_{j\alpha} = \frac{1}{2} \sum_{i\neq j} w_{ij} \sum_{\alpha=1}^{K-1} X_i^{\alpha K} X_j^{K\alpha}. \tag{5.11}$$

This operator corresponds to the total momentum of magnons, and as such commutes with the Hamiltonian $H_{\mathrm{SU}(K)}$. Then by the global $\mathrm{SU}(K)$ symmetry, we may define a set of $K^2 - 1$ Yangian generators starting from Λ_K. In terms of $\mathrm{SU}(K)$ generators J^α with $\alpha = 1, 2, \ldots, K^2 - 1$, the set of Yangian generators is given by

$$\Lambda^\alpha = \frac{1}{2} \sum_{i \neq j} w_{ij} \sum_{\beta, \gamma} f_{\alpha\beta\gamma} J_i^\beta J_j^\gamma, \tag{5.12}$$

where $f_{\alpha\beta\gamma}$ is the structure constant of the $\mathrm{SU}(K)$ algebra. Each operator Λ^α commutes with $H_{\mathrm{SU}(K)}$ but not with $\sum_i J_i^\beta$, which corresponds to the β-component of the total spin. Hence, application of Λ^α to the YHWS generates a supermultiplet which may have different total spins but has the same energy. Each supermultiplet is pictorially represented by a skew Young diagram, or more concisely by a motif as explained below. In the case of $K = 2$, $f_{\alpha\beta\gamma}$ is reduced to the completely antisymmetric tensor $\epsilon_{\alpha\beta\gamma}$, and J_i^α to the spin operator as given by (4.112).

Following the previous cases of the spin chain, we introduce the rapidity k_j which plays a role of the renormalized (and shifted) momentum of magnons spanning the whole Brillouin zone $[-\pi, \pi)$. Following the same logic as in the Sutherland model, we obtain the energy characterized by the rapidities as follows:

$$E = \frac{1}{2} \sum_{j=1}^{N-M_K} (k_j^2 - \pi^2) + J(0)(N - M_K), \tag{5.13}$$

where the Fourier component $J(q = 0)$ has been obtained in (4.5) as $J(0) = (\pi^2/3)(1 - 1/N^2)$.

The energy looks like that of free particles with the parabolic spectrum. The total crystal momentum P of the system is given by

$$P = \sum_{j=1}^{N-M_K} (k_j \pm \pi) \mp \pi = \sum_{j=1}^{N-M_K} k_j \pm (N - M_K - 1)\pi, \tag{5.14}$$

where the upper sign corresponds to the Brillouin zone $[0, 2\pi]$ and the lower sign to $[-2\pi, 0]$. This is a generalization of (4.90) in Chapter 4.

The distribution of the rapidity is most conveniently represented by the motif as in the previous cases. Namely, we put 0 if a one-body state in the Brillouin zone $[-\pi, \pi)$ is empty, and put 1 if it is filled by a magnon. It is customary to put an extra 0 in the rightmost position. This position corresponds to π in the Brillouin zone, which is equivalent to $-\pi$. This restriction of putting two zeros, one at each end, actually gives the N degrees of freedom for the occupation pattern. A sequence of $N+1$ digits $d_0d_1 \ldots d_N$ with $d_\kappa = 0$ or 1 completely characterizes the distribution of the rapidity, and hence the momentum. This definition of the motif as an occupation pattern of rapidities has been used in the SU(2) spin chain. We obtain

$$P = \frac{2\pi}{N} \sum_{\kappa=0}^{N} \kappa d_\kappa + \pi,$$

$$E = \frac{2\pi}{N} \sum_{\kappa=0}^{N} \kappa(N-\kappa)d_\kappa + \frac{N}{8}J(0). \tag{5.15}$$

Let us consider the case of SU(3). The ground state with $N = 9$, for example, is represented by

$$0110110110. \tag{5.16}$$

Let's choose the color B (blue) for $\alpha = 3$ as the reference state. Then the occupation of momentum is either by R (red) for $\alpha = 1$ or G (green) for $\alpha = 2$ particles. The momentum distribution in the ground state is understood from the wave function given by the product of Ψ_h and Ψ_α with $\alpha = 1, 2$. The part coming from Ψ_h requires unit separation between neighboring momenta, and the part from Ψ_α requires additional unit separation between the neighboring momenta for the same color α. Thus the distribution unit 011 follows, where 11 should be shared by the colors R and G. In the corresponding ribbon diagram, we choose R for the upmost square as a convention when drawing an inscribed Young diagram.

The singlet motif given by (5.16) is equivalent to the ribbon diagram in Fig. 5.1(a). In order to describe excited states, it is convenient to modify the motif as follows [80]:

(i) Replace "0" surrounding a single 1 or successive 1's by ")(".
(ii) Replace the first "0" by "(" and the last "0" by ")".

For example, the motif 01101011 is written as

$$(11)(1)(11), \tag{5.17}$$

which shows the presence of a spinon by "(1)" in the central part. The corresponding ribbon diagram is given in Fig. 5.1(b). On the other hand, a Yangian multiplet shown in Fig. 5.1(c) is described by

$$(11)()(11), \tag{5.18}$$

where the unit "()" is interpreted as a multiplet in which two spinons have the same momentum. In the SU(3) case, any motif consists of a sequence

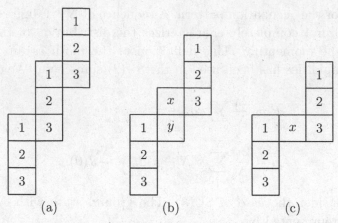

(a)   (b)   (c)

Fig. 5.1. Ribbon diagrams for (a) the singlet ground state of the SU(3) chain with $N = 9$, (b) a one-spinon state characterized by the conjugate representation $\bar{\mathbf{3}}$, and (c) a state with the fundamental representation $\mathbf{3}$, which is interpreted as a two-spinon state with the same momentum.

of elementary motifs (), (1), and (11), where the last one describes the SU(3) singlet.

In the ribbon diagram of Fig. 5.1(c), the color x corresponding to "()" in the motif can be either of R, G, or B. In the case of R or G, we assign a particle in a polynomial wave function. Hence in the SU(3) chain, and more generally in SU(K) chains with $K \geq 3$, a polynomial wave function does not necessarily belong to the YHWS, which is uniquely given for each ribbon diagram. This is in contrast to the Haldane–Shastry model and the SUSY t–J model, which will be discussed in Chapter 6, where all polynomial wave functions belong to the YHWS.

In the general SU(K) case, we have K elementary motifs since the number of 1's inside the brackets ranges from 0 to $K - 1$. The elementary motif with r 1's corresponds to the fundamental representation Λ_r of SU(K). Alternatively, the motif may be interpreted as a state where $K - r - 1$ spinons have the same momentum. In order to avoid double-counting of spinon states, the one-to-one correspondence should be made between a state in the fundamental representation Λ_r and one of the $K - r - 1$ spinon state. This can be accomplished by antisymmetrization of spinon states with different components. As a result, spinons with the same momentum obey the parafermion statistics of order K [68,69], which allows up to $K - 1$ spinons in the same orbital. This situation generalizes the $K = 2$ case, where symmetric excitation of up and down spinons leads to a fermionic description of thermodynamics. Further study, however, is necessary to clarify this situation completely.

The parafermion description of $SU(K)$ spinons was first obtained by application of the freezing trick [101, 121, 150]. The same results have been obtained by a recurrence relation using the Yangian symmetry [158]. On the other hand, straightforward application of the fractional exclusion statistics does not work for $K \geq 3$, since thermodynamics needs proper subtraction to remove overcounting.

If the spectrum of spinons is replaced by a linear one, which is relevant in the low-energy limit, simplification occurs for the thermodynamics in that the partition function can be obtained in terms of certain polynomials. These polynomials are a multi-variable version of the q-deformed Hermite polynomials, which are called the Rogers–Szegö polynomials [87, 150]. We refer readers for a detailed discussion of the q-deformation to the general theory of the quantum group [43, 44], but explain an example called Macdonald polynomials in Chapter 10. If one neglects the q-deformation, the partition function is reduced to its high-temperature limit K^N. Detailed analysis has been performed in relation to the conformal field theory [32, 33], including a supersymmetric generalization [88]. Ordinary fractional exclusion statistics is applicable if none of the spinons share the same momentum. This situation is relevant to the dynamical response of the systems with any K as described below.

The dispersion relations of elementary excitations can be derived from those of magnons by proper reinterpretation. In order to describe the single spinon state for finite size, it is most convenient to take M odd and $N = 3M - 1$ as in Fig. 5.1(b). Here the ground state has a deficit of color α, which is neither x nor y. In the case of $x = 1$, $y = 2$, for example, the spinon component is written as $\bar{3}$, which means a deficit of 3. The set $\{\bar{1}, \bar{2}, \bar{3}\}$ constitutes the conjugate representation of $SU(3)$, which is written as $\bar{\mathbf{3}}$. For general K, the spinons form the conjugate representation $\bar{\mathbf{K}}$. On the other hand, a state shown in Fig. 5.1(c) can have either $x = 1, 2$, or 3. The set $\{1, 2, 3\}$ constitutes the fundamental representation of $SU(3)$, which is written as $\mathbf{3}$.

Let us take the $SU(3)$ case, and start from a state in which a spinon is at the rightmost end of the Brillouin zone. In terms of the modified motif, the state is described by $(11) \dots (11)(1)$. Now we consider the change of energy as the elementary motif "(1)" shifts to the left as in $(11) \dots (11)(1)(11)$. The change of momentum when "(1)" reaches the leftmost position is $2\pi/3$ in the case of $N \gg 1$. This result is understood when one compares the motifs

$$011011011 \dots 011010$$
$$010110111 \dots 110110,$$

where every third momentum in the motif has been shifted to the right in the lower motif. Namely, the Brillouin zone of a spinon in the SU(3) chain is one-third of the original one. For general K, the size is shrunk to $2\pi/K$. Now we derive the energy of a spinon with rapidity $2\pi\zeta/N$, which is characterized by the shift $\kappa_j \to \kappa_j + 1$ for $j = 1, \ldots, \zeta \le M$ in the momentum distribution of the ground state. Here the maximum of j is $M = N/K$ in the case of an SU(K) chain, as shown below. The spectrum of a spinon for finite size can be derived by calculating the increment of energy associated with the momentum shift. For $0 < \zeta \le M$ we obtain [70]

$$\epsilon_s = \frac{1}{2}\left(\frac{2\pi}{N}\right)^2 \sum_{i=1}^{\zeta} \left[(\kappa_i + 1)^2 - \kappa_i^2\right] = \frac{3}{2}q\left(\frac{2\pi}{3} - q\right), \qquad (5.19)$$

where $\kappa_i = N/2 + 1 - 3i$, and we have introduced $q = 2\pi\zeta/N$.

In the thermodynamic limit, the spinon spectrum can be derived very simply. Let us consider a spin-flip which corresponds to annihilation of a magnon. In terms of a motif for the SU(3) chain, we consider a change

$$\ldots(11)(11)(11)\ldots \Rightarrow \ldots(11)(1)()(11)\ldots, \text{ or}$$

$$\ldots(11)()(1)(11)\ldots, \qquad (5.20)$$

either of which describes the creation of three spinons with neighboring momenta. In the thermodynamic limit, the difference between these momenta is negligible. Thus we realize, for general K, that the magnon annihilation with momentum $-Kq = k_j - \pi$ corresponds to the creation of K spinons with the same momentum q, i.e., $\epsilon_m(-Kq) = -K\epsilon_s(q)$. We immediately obtain the spectrum by taking $0 < q < 2\pi/K$ as

$$\epsilon_s(q) = \frac{1}{2}Kq\left(\frac{2\pi}{K} - q\right), \qquad (5.21)$$

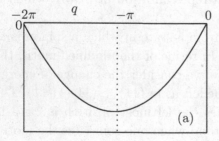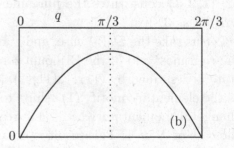

Fig. 5.2. Comparison of (a) the magnon spectrum and (b) the spinon spectrum in the SU(3) case.

which is consistent with (5.19) with $K = 3$. Figure 5.2 shows the relation between the spectra of magnons and spinons.

5.3 Statistical parameters via freezing trick

In terms of the distribution function $\nu(\kappa)$, which is the sum of K components, we obtain the energy

$$E_{\text{tot}} = \frac{\pi^2}{N^2} \sum_{\kappa=\infty}^{\infty} \sum_{\kappa'=-\infty}^{\infty} |\kappa - \kappa'| \nu(\kappa)\nu(\kappa'). \tag{5.22}$$

Note that E_{tot} is independent of spin components making up $\nu(\kappa)$. The independence reflects the $U(K)$ Yangian symmetry, which includes the charge degrees of freedom. In order to apply to the $SU(K)$ chain, we have to remove the phonon contribution from $\nu(\kappa)$, which can be accomplished by restriction to the ribbon diagrams for the momentum distribution. Each ribbon diagram represents a supermultiplet, which constitutes an irreducible representation of the $SU(K)$ Yangian, which is written as $Y(sl_K)$. A ribbon diagram with allowed inscription of integers $1, \dots, K$ is called a semistandard tableau (SST).

The statistical interaction among spinons can be obtained by the following consideration. Let us return to the K-component Sutherland model with the repulsion parameter λ. The exclusion statistics of particles is described by (2.78) with the parameters

$$G_{\alpha\beta} = \delta_{\alpha\beta} + \lambda, \tag{5.23}$$

with $\alpha, \beta = 1, \dots, K$. Note that we have changed the notation to a capital letter. The set of statistical parameters is regarded as a $K \times K$ matrix $\mathbf{G}$. In addition to particle excitations, there arise hole excitations from the many-particle ground state. The statistical parameters for the hole excitations are given by the inverse matrix $\mathbf{g}^{-1} \equiv \mathbf{G}$. We obtain

$$g_{\alpha\beta} = \delta_{\alpha\beta} - \frac{\lambda}{1 + \lambda K}, \tag{5.24}$$

which can be verified by direct calculation.

The energy spectrum of the $SU(K)$ chain is most easily obtained by the freezing trick, as discussed in Section 4.8.3. In the limit of $\lambda \to \infty$, we obtain from (5.24)

$$g_{\alpha\beta} = \delta_{\alpha\beta} - 1/K. \tag{5.25}$$

This result is relevant to hole excitations in the $U(K)$ chain, which includes phonons as the $U(1)$ component. For description of the $SU(K)$ chain in terms of the exclusion statistics, we have to remove the $U(1)$ component from (5.25). This removal is achieved graphically by restricting to the ribbon diagrams for relevant $SU(K)$ states. As we shall discuss later in Section 5.4, the dynamical structure factor does not involve the $U(1)$ component as final states. Hence the statistical parameters for the $U(K)$ system remain the same in the $SU(K)$ chain. This statement applies to all values of K.

In the case of $K = 2$, on the other hand, two holes with up and down spins can make a singlet. This singlet should be excluded in the $SU(2)$ spinon picture, where the neighboring spinons always make a triplet. For thermodynamics, the singlet state with the same momentum of two particles is excluded by adding unity in $g_{\sigma\sigma'}$ with $\sigma \neq \sigma'$ in (5.25). Namely, we move from $U(2)$ thermodynamics to $SU(2)$ thermodynamics by the replacement

$$\begin{pmatrix} g_{\uparrow\uparrow} & g_{\uparrow\downarrow} \\ g_{\downarrow\uparrow} & g_{\downarrow\downarrow} \end{pmatrix} = \begin{pmatrix} 1/2 & -1/2 \\ -1/2 & 1/2 \end{pmatrix} \Rightarrow \begin{pmatrix} 1/2 & 1/2 \\ 1/2 & 1/2 \end{pmatrix}. \quad (5.26)$$

Hence we obtain the statistical parameter $g_{\sigma\sigma'} = 1/2$ for the thermodynamics of spinons, as obtained for the Haldane–Shastry chain. Note that the off-diagonal element $g_{\sigma\sigma'}$ with $\sigma \neq \sigma'$ does not appear as a dynamical structure factor in the $SU(2)$ case. It should be emphasized that the statistics of spinons is rather different in the case of $K = 2$ and $K > 2$.

A merit of the description based on the freezing trick appears in thermodynamics. We consider the case without a magnetic field and proceed just as in the $SU(2)$ case discussed in Chapter 4. As a result, the distribution function is obtained as

$$\nu_\alpha(p) = 1 - \exp[-\beta\omega_0(p)/K]$$

for $\alpha = 1, \ldots, K$, which generalizes (4.185). Then the thermodynamic potential Ω_{tot} takes the same form as the phonon part Ω_{ph}, except for the replacement $\omega_0 \to \omega_0/K$ in the former. By subtracting the phonon part, the thermodynamic potential $\Omega_{\text{SU}(K)} = \Omega_{\text{tot}} - \Omega_{\text{ph}}$ is written as

$$\Omega_{\text{SU}(K)} = -T \int_{-\pi}^{\pi} \frac{dp}{2\pi} \ln[1+e^{-\beta\omega_0/K}+e^{-2\beta\omega_0/K}+\cdots+e^{-(K-1)\beta\omega_0/K}], \quad (5.27)$$

for any temperature T. It is natural to interpret $\Omega_{\text{SU}(K)}$ as that of ideal parafermions of order K [68, 69], for which up to $K - 1$ particles can take the same quantum number p. The parafermion is reduced to the fermion with $K = 2$.

5.4 Dynamical structure factor

Let us define the following operator in the momentum space by

$$X^{\alpha\beta}(q) = \frac{1}{\sqrt{N}} \sum_j X_j^{\alpha\beta} \exp(-iqx_j), \qquad (5.28)$$

which describes a color flip for $\alpha \neq \beta$, or color projection for $\alpha = \beta$. Then we consider the dynamical structure factor defined by

$$S^{(\alpha\beta)}(q,\omega) = \sum_\nu |\langle \nu | X^{\beta\alpha}(-q) | 0 \rangle|^2 \delta(\omega - E_\nu + E_0), \qquad (5.29)$$

where $|\nu\rangle$ denotes an eigenstate of the Hamiltonian with energy E_ν, and E_0 is the ground-state energy. Because of the SU(K) symmetry, the dynamical structure factor does not depend on the color components α, β as long as they are different from each other. Then we may regard $S^{(\alpha\beta)}(q,\omega) \to S(q,\omega)$ as a scalar, and choose $\beta = K$ together with the coordinate representation of wave functions. Since the operator $X^{K\alpha}$ is equal to the magnon annihilation operator b_α, the dynamical susceptibility is equivalent to the magnon propagator in this representation.

The excited state after the magnon annihilation is written in polynomial form. In order to simplify the notation, we work with the SU(3) symmetry ($K = 3$) in the following. The coordinates $z_1, \ldots, z_M$ denote magnons with color $\alpha = 1$, and $\zeta_1, \ldots, \zeta_M$ denote magnons with $\alpha = 2$, where $N = 3M$ with the singlet ground state. The excited state after annihilating the magnon with $\alpha = 1$ at $z = 1$ (or $x = 0$) is given by

$$\langle z_1, \ldots, z_{M-1}; \zeta_1, \ldots, \zeta_M | X_0^{31} \Psi_G \rangle = \Psi_G(1, z_1, \ldots, z_{M-1}; \zeta_1, \ldots, \zeta_M). \qquad (5.30)$$

Namely, the final states are represented by the Gutzwiller wave function where one of the complex coordinates is fixed. Because of its polynomial form, the final state can also be mapped to a wave function in the Sutherland model with the repulsion parameter $\lambda = 1$. In the case of $K = 3$, the relevant internal symmetry is SU(2) in the Sutherland model [13]. The magnon propagator in the SU(3) chain is given by the hole propagator in the SU(2) chain in the mapped Sutherland model, for which the exact results are obtained in Section 3.7.1. The excited states consist of three spinons, two of which have the component $\bar{1}$ and the other $\bar{2}$.

For a general value of K, we can obtain $S(q,\omega)$ in the singlet ground state from the hole propagator in the U($K - 1$) Sutherland model. Alternatively, the freezing trick discussed in Section 4.8.3 can be utilized together with Uglov polynomials explained in Chapter 10. Very detailed calculation along the latter route has been performed by Yamamoto *et al.* [199]. The result is given by

$$S(q,\omega) = A_K \sum_{1 \leq a < b \leq K} \prod_{i=1}^{K} \int_{-1}^{1} dk_i |F_{ab}^{(K)}(k)|^2 \delta(q - \pi - p(k))\delta(\omega - \epsilon(k)),$$

(5.31)

where

$$\epsilon(k) = \frac{\pi v_s}{2K} \sum_{i=1}^{K}(1 - k_i^2), \quad p(k) = \frac{\pi}{K} \sum_{i=1}^{K} k_i,$$

(5.32)

with $v_s = \pi/2$. Here we have taken the unit $J = 2$, and A_K is a normalization constant given by

$$A_K = \frac{\pi}{K^3(K-1)} \prod_{j=1}^{K} \frac{\Gamma((K-1)/K)}{\Gamma(j/K)^2},$$

(5.33)

which has been corrected by Arikawa [13] so as to remove an erroneous numerical factor 2^K for A_K in the original result [198, 199]. The form factor is given by [199]

$$F_{ab}^{(K)}(k) = \frac{|k_a - k_b|^{g_K} \prod_{1 \leq i < j \leq K, (i,j) \neq (a,b)} |k_i - k_j|^{g'_K}}{\prod_{i=1}^{K}(1 - k_i^2)^{(1-g_K)/2}},$$

(5.34)

with $g_K = (K-1)/K$ and $g'_K = -1/K$. For $K = 2$, the formula reproduces the result explained in Chapter 4. Notice that the second product in the numerator of (5.34) is absent in the SU(2) case.

6

Supersymmetric t–J model with $1/r^2$ interaction

A one-dimensional electron has both spin and charge degrees of freedom. In the limit of strong on-site repulsion, double occupation of a site can be neglected since it has a large energy cost. If the number N_e of electrons is less than the number N of lattice sites, vacant sites appear and the system acquires both spin and charge degrees of freedom. The simplest model to incorporate these degrees of freedom is called the t–J model, which is discussed in this chapter.

We begin by reviewing the SU(2,1) supersymmetry (SUSY) as an extension of the SU(2) symmetry in the spin chain. With the supersymmetry, the hopping and exchange in the t–J model can be treated in a unified way as the graded permutation. Then the ground state of the SUSY t–J model is derived in Section 6.2 by the coordinate representation of wave functions by using the fully polarized state as the reference. The static structure factor at zero temperature is completely determined by the wave function at the ground state. The spin and charge components are derived using the determinant and its generalizations in Section 6.3.

In Section 6.4, we proceed to derive the spectrum of the SUSY t–J model. The elementary excitations consist of spinons, holons, and their antiparticles. It is proved in Section 6.5 that these excitations span the complete set. The argument relies on a generalization of Young diagrams, which are called ribbon diagrams. At the same time, the degeneracy of energy levels beyond the global SU(2) symmetry is ascribed to the presence of a Yangian supersymmetry $Y(sl_{2|1})$.

With the knowledge of energy levels and their degeneracies thus acquired, we derive the thermodynamics of the SUSY t–J model in Section 6.6. Physical quantities such as magnetic susceptibility, charge susceptibility, entropy, and specific heat are derived explicitly. The results are interpreted naturally in terms of elementary excitations obeying fractional exclusion statistics.

Finally, as the most advanced result of the one-dimensional electron theory, we present exact dynamical correlation functions and the single-particle spectral functions in Section 6.7.

6.1 Global supersymmetry in $t-J$ model

The t–J model is given by

$$H_{tJ} = \mathcal{P} \sum_{i<j} \left[-t_{ij} \sum_{\sigma=\uparrow,\downarrow} \left(c_{i\sigma}^\dagger c_{j\sigma} + \text{h.c.} \right) + J_{ij} \left(\boldsymbol{S}_i \cdot \boldsymbol{S}_j - \frac{1}{4} n_i n_j \right) \right] \mathcal{P}, \quad (6.1)$$

where $c_{i\sigma}$ is the annihilation operator of an electron with spin σ at site i, n_i is the number operator, and $\mathcal{P}$ is the projection operator to exclude double occupation at a given site. We have introduced in Section 1.5 the X-operators $X^{\alpha\beta} = |\alpha\rangle\langle\beta|$, which describe transition from one state β to another state α at each site. If the transfer energy t_{ij} and exchange energy J_{ij} satisfy the relation $t_{ij} = J_{ij}/2$, we can combine the spin permutation and the hopping using the graded permutation operator given by (1.41). The X-operators can also be represented by a fictitious boson annihilation operator $B_{i\sigma}$ and a fermion creation operator $F_i^\dagger$ as

$$X_i^{0\sigma} = F_i^\dagger B_{i\sigma}, \quad X_i^{\downarrow\uparrow} = B_{i\downarrow}^\dagger B_{i\uparrow}, \quad X_i^{00} = F_i^\dagger F_i. \quad (6.2)$$

Because of the constraint, $F_i^\dagger F_i + \sum_\sigma B_{i\sigma}^\dagger B_{i\sigma} = 1$, these fermions and bosons do not obey the standard commutation rule. The graded permutation operator introduced by (1.41) is then written as

$$\tilde{P}_{ij} = \sum_{\alpha\beta} p(\beta) A_{i\alpha}^\dagger A_{i\beta} A_{j\beta}^\dagger A_{j\alpha} = \sum_{\alpha\beta} A_{i\alpha}^\dagger A_{j\beta}^\dagger A_{i\beta} A_{j\alpha}, \quad (6.3)$$

where $A_{i\alpha}$ denotes either $B_{i\sigma}$ or F_i. Since the vacant and singly occupied states appear to be on an equal footing, the model possesses a corresponding symmetry. This symmetry is called the supersymmetry, which in its most general sense connects fermionic and bosonic degrees of freedom.

In terms of the graded permutation operator $\tilde{P}_{ij}$, the supersymmetric (SUSY) t–J model is written concisely as

$$H_{tJ} = \sum_{i<j} t_{ij}(\tilde{P}_{ij} - 1 + X_i^{00} + X_j^{00}). \quad (6.4)$$

The term $X_i^{00} + X_j^{00}$ serves to associate the zero energy with the vacant state $N_e = 0$, and plays the role of the chemical potential. This term breaks the supersymmetry by distinguishing states with different N_e. We note that the energy is also zero in the completely polarized case with $N_e = N$. We obtain $\tilde{P}_{ij} - 1 = 0 = X_i^{00}$ for this case.

Removing the chemical potential and constant terms, let us consider a system

$$H_{SUSY} = \sum_{i<j} t_{ij} \tilde{P}_{ij}, \qquad (6.5)$$

which is invariant under the global supersymmetry operation. Namely, we obtain by direct calculation the N-site version of (1.44):

$$\left[\tilde{P}_{ij}, \sum_i X_i^{\alpha\beta} \right] = 0, \qquad (6.6)$$

for any combination of α, β. This means that the eigenvalues of H_{SUSY} are degenerate not only for such states as have the same total spin, but also for certain states with different number of holes. The latter set of states are generated by application of $\sum_i X_i^{0\sigma}$ and $\sum_i X_i^{\sigma 0}$, as is evident from (1.44).

Note that the global supersymmetry demonstrated above does not depend on the form of t_{ij}. It has been shown [175] for $t_{ij} = J_{ij}/2 = t\delta_{i,j\pm1}$ that the model can be solved exactly by the Bethe ansatz. Another solvable case is given by the inverse-square interaction [119]

$$t_{ij} = J_{ij}/2 = tD_{ij}^{-2}, \qquad (6.7)$$

where $D_{ij} = (N/\pi)\sin[\pi(i-j)/N]$. Henceforth we take t as the unit of energy, and concentrate on the case specified by (6.7).

6.2 Mapping to U(1,1) Sutherland model

As in the case of the spin chain, it is useful to start with the fully up-polarized state as the reference. We introduce for each site the hole creation operator $h_i^\dagger = F_i^\dagger B_{i\uparrow}$. In addition, we identify S_i^- as the creation operator $b_i^\dagger$ of the hard-core boson as in the case of the Haldane–Shastry spin chain. Hereafter we regard $b_i^\dagger$ as the creation operator of a magnon. The kinetic energy term of up-spin electrons is regarded as hole hopping:

$$-t_{ij}c_{i\uparrow}^\dagger c_{j\uparrow} = t_{ij}h_j^\dagger h_i, \qquad (6.8)$$

where the anticommuting property of the hole operators is used. On the other hand, annihilation of a down-spin electron at each site can be represented by $c_{i\downarrow} = F_i^\dagger B_{i\downarrow} = h_i^\dagger b_i$. Hence the kinetic energy term of down-spin electrons takes the form of a graded permutation:

$$-t_{ij}c_{i\downarrow}^\dagger c_{j\downarrow} = -t_{ij}b_j^\dagger h_j h_i^\dagger b_i = t_{ij}b_j^\dagger h_i^\dagger h_j b_i. \qquad (6.9)$$

Instead of the $U(2,1)$ supersymmetry with spin degrees of freedom for bosons, the new particles created by $h_i^\dagger$ (holes) and $b_i^\dagger$ (magnons) have the $U(1,1)$ supersymmetry with no spin degrees of freedom. Now each site is not necessarily filled by the new particles. Hence the system gets the $U(1)$ charge degrees of freedom in addition to the $SU(1,1)$ internal symmetry. Thus the total number of states can remain 3^N in the new representation.

In this way we take the up-spin site as a vacuum, and obtain the $U(1,1)$ representation of the model as

$$\mathcal{H}_{tJ} = \sum_{i<j} t_{ij} \left[b_i^\dagger b_j + h_i^\dagger h_j + \frac{1}{2}(\tilde{P}_{ij} + \tilde{n}_i \tilde{n}_j) \right] - M t(0), \qquad (6.10)$$

where M is the number of down spins (magnons), $\tilde{n}_i = b_i^\dagger b_i + h_i^\dagger h_i$, and $t(0) = \sum_j t_{ij} = \pi^2(1 - N^{-2})/3$. Here the Fourier transform $t(q)$ of t_{ij} is given in analogy with (4.4) by

$$t(q) = \sum_{j=2}^{N} t_{1j} \exp[iq(j-1)] = \frac{t}{2}(q-\pi)^2 - \frac{\pi^2 t}{6}\left(1 + \frac{2}{N^2}\right), \qquad (6.11)$$

where $t = 1$ in our unit. The graded permutation operator $\tilde{P}_{ij}$ acts now in the space of magnons and holes. The chemical potential term for magnons breaks the $U(1,1)$ supersymmetry, and gives rise to the last term in (6.10).

We can represent any state in terms of the wave function $\Psi(\{x^h\}, \{x^s\})$ as [4]

$$|\Psi\rangle = \sum_{\{x^h\},\{x^s\}} \Psi(\{x^h\}, \{x^s\}) \prod_{i\in\{x^s\}} S_i^- \prod_{j\in\{x^h\}} h_j^\dagger |F\rangle, \qquad (6.12)$$

where the set of coordinates $\{x^s\}$ specifies the positions of M magnons, and $\{x^h\}$ specifies those of Q holes. By definition $\Psi(\{x^h\}, \{x^s\})$ is symmetric against interchange of down-spin coordinates, and antisymmetric against hole coordinates. This is related to the commutation rule $S_i^- S_j^- = S_j^- S_i^-$ and the anticommutation rule $h_i^\dagger h_j^\dagger = -h_j^\dagger h_i^\dagger$. These relations allow us to replace the operator $\tilde{P}_{ij}$, which permutes the internal degrees of freedom, by the operator K_{ij}, which exchanges the spatial coordinates in $\Psi(\{x^h\}, \{x^s\})$. Namely, we have

$$\tilde{P}_{ij}|\Psi\rangle = \sum_{\{x^h\},\{x^s\}} K_{ij}\Psi(\{x^h\}, \{x^s\}) \prod_{i\in\{x^s\}} S_i^- \prod_{j\in\{x^h\}} h_j^\dagger |F\rangle. \qquad (6.13)$$

Note that the only constraint in using K_{ij} is that we have $K_{ij}\tilde{P}_{ij} = 1$ against interchange of a pair of particles. We do not impose any commutation rule between F and B operators in (6.2). Then $\Psi(\{x^h\}, \{x^s\})$ do not in general have a definite parity against interchange of hole and magnon coordinates. Hence we regard a hole and a magnon as distinguishable.

In analogy with the case of the Haldane–Shastry spin chain, we can work with the first-quantized form of the Hamiltonian for a subset of states which is given by a polynomial of $z_i = \exp(2\pi i x_i/N)$ with degrees ranging from 1 to $N-1$ for each variable. For such polynomials we proceed as in the spin chain to obtain

$$H_{t-J}^{(\text{poly})} = \frac{1}{2} \sum_{i=1}^{M+Q} \left(-i\frac{\partial}{\partial x_i} - \pi \right)^2 + \frac{1}{2} \sum_{i \neq j} \frac{1 + K_{ij}}{D(x_i - x_j)^2} + E_M, \qquad (6.14)$$

where $-\partial/\partial x_i$ is the momentum of a hole or a down-spin boson, and $D(x) = (N/\pi)\sin(\pi x/N)$. The constant term E_{MQ} is given by

$$E_{MQ} = (M + Q)t(\pi) - Mt(0). \qquad (6.15)$$

The polynomial eigenfunctions are obtained most conveniently by first regarding holes and magnons as distinguishable particles, and then imposing the proper symmetry. This is precisely the method we discussed in Chapter 3. With the hard-core constraint, the absolute ground state is given by the Slater determinant. In the case of N even and $Q + M$ odd, the absolute ground state is non-degenerate and is given by

$$\Psi_{\text{h}}(\{z\}) = \prod_{i=1}^{Q+M} z_i^{N/2 - (Q+M-1)/2} \prod_{i<j}(z_i - z_j), \qquad (6.16)$$

which has the proper antisymmetry if there are no magnons, i.e., $M = 0$. The factor $z_i^{N/2}$ comes from the momentum shift to the edge π of the Brillouin zone, and another factor $z_i^{-(Q+M-1)/2}$ makes the total momentum zero from the shifted origin. With N odd, the mapped Sutherland model should be solved with the *antiperiodic* boundary condition, as discussed in Chapter 4. Then the absolute ground state is degenerate with $Q + M$ odd. One should make the replacement of exponent

$$(Q + M - 1)/2 \rightarrow (Q + M - 1 \pm 1)/2 \qquad (6.17)$$

for the degenerate ground states. Note that $\Psi_{\text{h}}(\{z\})$ remains a polynomial with integer powers even in the case of N odd. For $Q + M$ even and N odd, (6.16) describes the non-degenerate absolute ground state.

For the antisymmetric wave function, we obtain $K_{ij} = -1$ for the coordinate exchange operator. Then the interaction term in (6.14) vanishes, and the free-fermion wave function $\Psi_{\text{h}}(\{z\})$ proves to be an eigenstate. In the opposite case of $Q = 0$, all particles are magnons and we should have the symmetric wave function. The lowest-order polynomial is obtained by taking the square of $\Psi_h(\{z\})$ except for the factor $z_i^{N/2}$. The result is nothing but the eigenfunction of the Haldane–Shastry model. The interaction term becomes the same as the single-component Sutherland model with $\lambda = 2$, since we obtain $K_{ij} = 1$ in this case.

For the mixture of Q holes and M magnons, we obtain the proper symmetry by multiplying $\Psi_h(\{z\})$ by an antisymmetric polynomial containing only magnon variables. The lowest-order antisymmetric polynomial together with the proper momentum shift is given by the Slater determinant. With M odd, we obtain

$$\Psi_m(\{z\}) = \prod_{i=1}^{M} z_i^{-(M-1)/2} \prod_{i<j\leq M} (z_i - z_j), \qquad (6.18)$$

where we have assigned the index $i = 1, \ldots, M$ to magnons. In the case of M even, we should make the replacement

$$z_i^{-(M-1)/2} \rightarrow z_i^{-(M-1\mp1)/2} \qquad (6.19)$$

in (6.18). In either case, we obtain $\Psi_m(\{z\})$ with an integer exponent for magnon variables. It is convenient to regard $\Psi_m(\{z\})$ as including the hole variables, but independent of them. In Section 2.1.2 we showed that the Sutherland model after similarity transformation is given by a triangular matrix. Then the polynomial wave function

$$\Psi_G(\{z\}) = \Psi_h(\{z\})\Psi_m(\{z\}) \qquad (6.20)$$

proves to be an eigenfunction of (6.14) for all cases of even and odd numbers of M, Q, and N. It is easily seen that all other polynomial eigenfunctions, which are obtained from (6.20) by multiplying symmetric polynomials of hole or magnon, have higher energies. Hence, (6.20) is the ground-state wave function within the polynomial family. This wave function corresponds to the Gutzwiller wave function with a finite number of holes. It will be proved in Section 6.5.3 that (6.20) in fact describes the ground state without restriction to the polynomial family. The degeneracy of the ground state is at most two.

In the non-degenerate case with N, Q even and M odd, the wave function can be written in the explicitly real form, provided the momentum is measured from the boundary π of the Brillouin zone. Namely, in terms of the chord distance $D(x)$, the eigenfunction of (6.20) is represented with a numerical factor C by

$$\Psi_G(\{x\},\{y\}) = C \prod_{j=1}^{Q+M} z_j^{N/2} \prod_{i>j} D(x_i - x_j)^2 \prod_{l>m} D(y_l - y_m) \prod_{i,l} D(x_i - y_l),$$

(6.21)

where x_i ($i = 1, \ldots, M$) denotes the magnon coordinates, and y_l ($l = 1, \ldots, Q$) denotes those of holes.

6.3 Static structure factors

The static correlation functions of the t–J model are completely determined by the ground-state wave function. In Section 4.4, we have already presented the results of correlation functions for the spin chain, and described the method of calculation. Here we describe the static correlation functions in the presence of holes. It is instructive to compare these with the spin chain, and identify the effect of holes. Let us first present the results in the real space, and then explain the derivation. We modify the form first derived in [60] so as to make explicit the relations to magnon and hole distribution functions. The magnon–magnon correlation function is written as $C^{bb}(x)$, the hole–hole one as $C^{hh}(x)$, and the magnon–hole correlation function as $C^{bh}(x)$.

Then the results are given for arbitrary density n and magnetization m in the form

$$C^{bb}(x) = -\frac{1}{4}[s_-(x)^2 - Ds_-(x)Is_-(x)],$$

(6.22)

$$C^{hh}(x) = -s_h(x)^2 - Ds_h(x)Is_h(x),$$

(6.23)

$$C^{bh}(x) = \frac{1}{2}[s_h(x)s_-(x) + Ds_h(x)Is_-(x)],$$

(6.24)

where the basic constituents are defined by

$$s_-(x) = s_b(x) - s_h(x), \quad s_b(x) = \frac{\sin k_m x}{\pi x},$$

$$s_h(x) = \frac{\sin k_c x}{\pi x} = \frac{\sin \pi n x}{\pi x},$$

(6.25)

with $k_m = \pi(1-m)$ and $k_c = \pi(1-n) \leq k_m$. As in the spin chain discussed in Section 4.4, $Ds_\alpha(x)$ and $Is_\alpha(x)$ are the derivative and integral of the constituent function $s_\alpha(x)$ with $\alpha = -$, h, or b. For example we obtain

$$Is_-(x) = \frac{1}{\pi}[\text{Si}(k_m x) - \text{Si}(k_c x)],$$

(6.26)

with $\mathrm{Si}(x)$ being the sine integral. The function $s_-(x)$ constitutes a part of the 2×2 matrix propagator of magnons, while $s_\mathrm{h}(x)$ is the propagator of holes. Because of exclusion of momentum due to holes, the magnon propagator cannot involve the low-momentum component.

The transverse correlation function is given by

$$C^\perp(x) = \frac{1}{2}\left[C^{+-}(x) + C^{-+}(x)\right], \tag{6.27}$$

where $C^{-+}(x) = \langle b_x^\dagger b_0 \rangle$ is the density matrix of magnons. It is given by

$$C^{-+}(x) = \frac{(-1)^x}{x} I s_-(x) = \frac{(-1)^x}{\pi x}[\mathrm{Si}(k_m x) - \mathrm{Si}(k_c x)], \tag{6.28}$$

as will be shown soon. We then obtain $C^\perp(x) = C^{-+}(x) + \frac{1}{2}(1 - n + m)\delta_{x,0}$.

To obtain the longitudinal spin correlation function $C^{zz}(x)$, we use the following relation at each site:

$$S^z = \left(\frac{1}{2} - b^\dagger b\right)(1 - h^\dagger h) = \frac{1}{2} - b^\dagger b - \frac{1}{2}h^\dagger h, \tag{6.29}$$

where we have used the hard-core property of magnons and holes. Then we obtain

$$C^{zz}(x) = C^{\mathrm{bb}}(x) + \frac{1}{4}C^{\mathrm{hh}}(x) + C^{\mathrm{bh}}(x). \tag{6.30}$$

In the singlet case with $k_m = \pi$, we should obtain $C^{zz}(x) = C^\perp(x)$ because of the rotational symmetry. This can be checked by first noting the following cancellation among terms contributing to (6.30):

$$-\frac{1}{4}s_-(x)^2 - \frac{1}{4}s_\mathrm{h}(x)^2 - \frac{1}{2}s_\mathrm{h}(x)s_-(x) = -\frac{1}{4}s_\mathrm{b}(x)^2 = 0, \tag{6.31}$$

because of $\sin \pi x = 0$ for integer x. Therefore with $m = 0$, the first terms of $C^{\alpha\beta}(x)$ in (6.22) to (6.24) combine to zero for $C^{zz}(x)$. On the other hand, the second terms in (6.22) to (6.24) combine to give

$$\left[\frac{1}{4}Ds_-(x) - \frac{1}{4}Ds_\mathrm{h}(x) + \frac{1}{2}Ds_\mathrm{h}(x)\right] I s_-(x) \tag{6.32}$$

$$= \frac{1}{4}Ds_\mathrm{b}(x)Is_-(x) \rightarrow \frac{(-1)^x}{4x} I s_-(x), \tag{6.33}$$

again because of $\sin \pi x = 0$. Hence we indeed recover $C^{zz}(x) = C^\perp(x)$ for the case of $m = 0$.

Now we describe the derivation, taking $C^\perp(x)$ as an example. We consider the system with M magnons and Q holes, and generalize the derivation of the correlation function for the spin chain. We need the norm of Ψ_G which is most conveniently obtained from the expression

$$|\Psi_G(\{x,y\})|^2 = (-1)^R \sum_P \epsilon_P z_{P1}^{-(Q-1)/2} z_{P2}^{-(Q-3)/2} \cdots z_{PQ}^{(Q-1)/2}$$

$$\times \det_{2M+Q} \left(z_1^p, \ldots, z_Q^p, \xi_1^p, p\xi_1^p, \ldots, \xi_M^p, p\xi_M^p \right)_{|p|\leq M+(Q-1)/2}, \qquad (6.34)$$

where the sign factor $(-1)^R$ is derived below as $R = MQ + Q(Q-1)/2$. The RHS of (6.34) is a symmetric function of both magnon (ξ_i) and hole (z_j) coordinates. This form can be obtained by manipulation of the determinant as in the case of the spin chain. Among terms appearing by expansion of the determinant, we first integrate over the real hole coordinates $y_1, \ldots, y_Q$ in $z_j = \exp(2\pi i y_j/N)$. The only term that survives integration over y_i is the product $z_i^p z_i^{Pi}$ with $p + Pi = 0$. Then we obtain the factor $Q!$ coming from the number of permutations. After integration over the hole coordinates, we are left with magnon terms like $(p-q)\zeta^p \zeta^{-q}$ with $p+q = 0$, and $p > (Q-1)/2$. The exclusion of the small value of p is due to the presence of holes. Since $p-q = 2p$ ranges from $Q+1$ to $2M+Q-1$ in two steps, integration over the magnon coordinates gives the factor $M!(2M + Q - 1) \cdots (Q+1)$ where $M!$ comes from the number of permutations for magnon coordinates. Combining these factors, we obtain the norm as [59]

$$C_{M,Q} = Q!M!(2M + Q - 1)(2M + Q - 3) \cdots (Q+1)$$

$$= \begin{cases} \dfrac{(2M + Q)!M!(Q/2)!}{2^M (M + Q/2)!}, & (Q : \text{even}), \\[3mm] \dfrac{M!Q!2^M[M + (Q-1)/2]!}{[(Q-1)/2]!}, & (Q : \text{odd}). \end{cases} \qquad (6.35)$$

The sign factor $(-1)^R$ is obtained by inspecting the sign of a permutation which gives a positive integral. For example, a permutation

$$(P1, P2, \ldots, PQ) = (Q, Q-1, \ldots, 2, 1)$$

in (6.34) gives $\epsilon_P = (-1)^{Q(Q-1)/2}$. Then the positive integral results by a permutation of $2M + Q$ numbers such that $(1, 2, \ldots, Q)$ for hole coordinates are inserted between $(Q+1, Q+2, \ldots, Q+M)$ and $(Q+M+1, Q+M+2, \ldots, Q+2M)$. This permutation gives a sign factor $(-1)^{MQ}$. Another derivation using sophisticated combinatorics is given in Section 7.5.6.

Let us evaluate the density matrix from the product of $\Psi_G^* \Psi_G$ with complex coordinates. We first fix a complex magnon coordinate η in Ψ_G^*, and another coordinate ζ in Ψ_G. Then the rest of the magnon coordinates

$\xi_j = \exp(2\pi i j/N)$ and the hole coordinates $z_l = \exp(2\pi i l/N)$ are to be integrated out. Generalizing the procedure used for the spin chain, we start from the following representation with the confluent alternant:

$$\Psi_G(\eta,\ldots)^*\Psi_G(\zeta,\ldots) = \frac{e^{i\alpha}}{\eta - \zeta}$$
$$\times \det\left(\eta^j,\zeta^j,z_1^j,\ldots,z_Q^j,\xi_1^j,j\xi_1^{j-1},\ldots,\xi_{M-1}^j,j\xi_{M-1}^{j-1}\right)_{j=0,1,\ldots,2M+Q-1}$$
$$\times \prod_{j<k}(z_j - z_k), \tag{6.36}$$

where $\eta = \exp(2\pi i l/N)$ and $\zeta = \exp(2\pi i n/N)$ are specified in terms of integer coordinates l and n. The phase factor $\exp(i\alpha)$ is fixed so that the RHS is real. Without loss of generality, we consider the case $l = -n$, which means $\zeta = \eta^*$ in (6.36). Then we manipulate the determinant as in the case of the norm for the spin chain. We obtain

$$\Psi_G(\eta,\ldots)^*\Psi_G(\eta^*,\ldots) = \exp(i\alpha')(\eta - \eta^*)^{-1}$$
$$\times \det_{2M+Q}\left(\eta^p,\eta^{-p},z_1^p,\ldots,z_Q^p,\xi_1^p,p\xi_1^p,\ldots,\xi_{M-1}^p,p\xi_{M-1}^p\right)_{|p|\leq M+(Q-1)/2}$$
$$\times \det_Q\left(z_1^p,\ldots,z_Q^p\right)_{|p|\leq(Q-1)/2}, \tag{6.37}$$

where $\exp(i\alpha')$ is a phase factor. With the result for the norm, we can now derive the density matrix of magnons in the form

$$\frac{1}{N}\int_0^N \frac{dx_1}{N}\cdots\int_0^N \frac{dx_{M-1}}{N}\int_0^N \frac{dy_1}{N}\cdots\int_0^N \frac{dy_Q}{N}\Psi(z,\ldots)^*\Psi(z^*,\ldots)$$
$$= \frac{C_{M,Q}}{N}\sum_{p=(Q+1)/2}^{M+(Q-1)/2}\frac{z^{2p} - z^{-2p}}{2p(z - z^*)}, \tag{6.38}$$

where $2p$ in the denominator comes from correction for the missing factor in producing the norm $C_{M,Q}$. In the thermodynamic limit, the summation over p is replaced by the integral for $Q/2 < p < M$. We then obtain for integer x,

$$C^{-+}(x)(-1)^x = \frac{1}{2N}\int_{Q/2}^{M+Q/2}dp\frac{\sin 2\pi px/N}{2p\sin \pi x/N} \rightarrow \frac{1}{4\pi x}[\mathrm{Si}(k_m x) - \mathrm{Si}(k_c x)], \tag{6.39}$$

where $k_m/\pi = M/N$ and $k_c/\pi = Q/N$. As in the spin-chain case given by (4.24), the oscillating factor $(-1)^x$ comes from the shift π of the momentum in mapping to the Sutherland model. The last expression in (6.39) is valid in the limit $|x|/N \ll 1$. The other correlation functions can be obtained

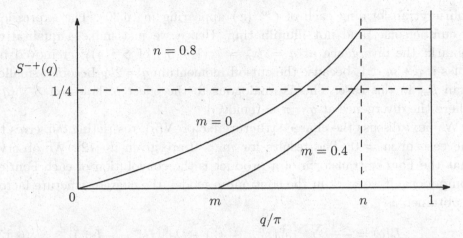

Fig. 6.1. The spin-flip structure factor $S^{-+}(q)$ of the SUSY t–J model with and without magnetization m. The transverse component $S^{\perp}(q) = S^{xx}(q) = S^{yy}(q)$ is given by $S^{\perp}(q) = S^{-+}(q) + (1 - n + m)/2$.

in a similar manner by fixing two coordinates in the wave function and integrating over the others. The results for $C^{hh}(x)$ are written explicitly as

$$C^{hh}(x) = -\left(\frac{\sin \pi n x}{\pi x}\right)^2 + \left(\frac{k_c \cos \pi n x}{\pi x} + \frac{\sin \pi n x}{(\pi x)^2}\right) [\mathrm{Si}(k_m x) - \mathrm{Si}(k_c x)].$$

$$(6.40)$$

We now derive structure factors by Fourier transforming the correlation functions. The simplest is $S^{-+}(q)$ that can be obtained by combination of terms derived already for the spin chain. Namely, we obtain

$$S^{-+}(q) = \frac{1}{4}\theta(q - \pi m)\left[\theta(q - \pi n)\ln\frac{1-m}{1-n} - \theta(\pi n - q)\ln\frac{1 - q/\pi}{1-m}\right],$$

$$(6.41)$$

with $0 < q < \pi$. The result with $m = 0$ has been obtained in [66]. Figure 6.1 illustrates the momentum dependence of $S^{-+}(q)$ with and without magnetization. It is remarkable that the only dependence on n is through the cut-off momentum $q = \pi n = 2k_\mathrm{F}$ for arbitrary m. Thus the charge and spin degrees of freedom behave rather independently in the SUSY t–J model. This mutual independence also appears in thermodynamics and dynamics, as we shall see later. Evidently the tendency to magnetic ordering is suppressed as the hole concentration increases. The longitudinal component $S^{zz}(q)$ in the presence of holes and with nonzero magnetization can be obtained by

Fourier transforming each of $C^{\alpha\beta}(x)$ appearing in (6.30). The expression is cumbersome, and not illuminating. Hence we just make a qualitative remark: the divergence at $q = k_m = \pi(1 - m)$ of $S^{zz}(q)$ is removed by holes if $n + m < 1$, because the cut-off momentum $q = 2k_F$ becomes smaller than k_m in this case. This is analogous to the effect of holes on $S^{-+}(q)$, where the divergence at $q = \pi$ is removed.

We next discuss the charge structure factor $N(q)$, restricting ourselves to the case of $m = 0$. The results for $m \neq 0$ are given in [12]. We observe that the Fourier transform of a product is the convolution of each Fourier component. If we work in the continuum model, the charge structure factor is obtained as

$$I_c(q) = -\frac{1}{N} \sum_k \left[s_k^h(s_{k+q}^h - s_k^h) + D s_k^h(I s_{k+q}^- - I s_k^-) \right]. \qquad (6.42)$$

The calculation of k-summation is analogous to the case of a spin chain, where we have described calculation of $S^{zz}(q)$. In the lattice system, the charge structure factor $N(q)$ becomes a periodic function of q with the Brillouin zone $-\pi < q < \pi$ as the basic period. Because of the inversion symmetry, we may assume $0 < q < \pi$ without loss of generality. In this case we obtain $N(q) = I_c(q) + I_c(q - 2\pi)$, where the second term is specific to the lattice system. This term becomes nonzero only for q beyond a threshold depending on n. The summation range over k is classified according to the value of q. As a result, there appear three regions I, II, III in the momentum space. The momentum at each boundary of neighboring regions is written as $q_{\text{I–II}}$ or $q_{\text{II–III}}$. For example, we obtain $N(q)$ for $n < 1/2$ as

$$N(q) = N_I(q/\pi)\theta_I(q) + N_{IIa}(q/\pi)\theta_{II}(q) + N_{IIIa}(q/\pi)\theta_{III}(q), \qquad (6.43)$$

where $\theta_{II}(q) = \theta(q - q_{\text{I–II}})\theta(q_{\text{II–III}} - q)$ is the step function which is nonzero only for the region II. Other step functions are defined in a similar manner. The constituent functions $N_{IIa}(q/\pi)$, etc. depend on the density n. Namely, we define the following functions:

$$N_I(x) = x - \frac{x}{2} \ln \frac{1 - n + x}{1 - n}, \qquad (6.44)$$

$$N_{IIa}(x) = x - \frac{x}{2} \ln \frac{1 + n - x}{1 - n} + \ln(1 + n - x), \qquad (6.45)$$

$$N_{IIb}(x) = 2 - 2n + \frac{x}{2} \ln \frac{n - 1 + x}{1 - n + x}, \qquad (6.46)$$

$$N_{IIIa} = 2n + \ln(1 - n), \qquad (6.47)$$

$$N_{IIIb}(x) = 2 - 2n + \frac{x}{2} \ln \frac{n - 1 + x}{1 + n - x} + \ln(1 + n - x). \qquad (6.48)$$

Table 6.1. *List of functions relevant to the charge structure factor $N(q)$ for each q range I, II or III and for a given density n.*

Density	I	$q_{\text{I–II}}/\pi$	II	$q_{\text{II–III}}/\pi$	III
$n < 1/2$	N_1	n	N_{IIa}	$2n$	N_{IIIa}
$1/2 < n < 2/3$	N_1	n	N_{IIa}	$2 - 2n$	N_{IIIb}
$n > 2/3$	N_1	$2 - 2n$	N_{IIb}	n	N_{IIIb}

Table 6.1 summarizes the proper components, and the boundary momentum between different ranges of q. They become degenerate, i.e., $q_{\text{I–II}} = q_{\text{II–III}}$ at $n = 1/2$ and $n = 2/3$. At these special values, the region II shrinks to zero. Figure 6.2 shows $N(q)$ for three cases of n. For small n, the behavior is similar to that in free fermions:

$$N_0(q) = \theta(n\pi - q)q/\pi + \theta(q - n\pi)n, \qquad (6.49)$$

which can be seen in the case of $n = 0.3$. On the other hand, $N(q)$ hardly depends on q for $n \sim 1$ and becomes small because the strong repulsion leads to almost local character of electrons. The case $n = 0.9$ strongly reflects this behavior. In the limit of $n = 1$, we obviously have $N(q) = 0$ for all q. The initial slope $1/\pi$ at $q = 0$ is independent of n, and is the same as that for free fermions. Note that the cusp at $q = \pi n = 2k_{\text{F}}$ is much less significant compared with the divergence in $S^{zz}(q)$ at $q = (1-m)\pi$, and that in $S^{-+}(q)$ at $q = \pi$. This shows that the SUSY t–J model does not have a tendency toward the charge density wave.

6.4 Spectrum of elementary excitations

6.4.1 Energy of polynomial wave functions

We take the non-degenerate ground state with M odd and N, Q even, and analyze the structure of polynomial wave functions. In $\Psi_{\text{m}}(\{z\})$, exponents $I_i^{(\text{m})}$ of a magnon coordinate are given in descending order by $I_i^{(\text{m})} = (M+1)/2 - i$, with $i = 1, \ldots, M$. The distribution is symmetric about zero. The exponents $I_l^{(\text{h})}$ of hole coordinates are zero for all $l = 1, \ldots, Q$. We arrange these $M + Q$ exponents in descending order and relabel them by the common index $j = 1, 2, \ldots, M + Q$. Namely, we introduce I_j by

$$I_j = \begin{cases} I_j^{(\text{m})}, & (j = 1, 2, \ldots, (M+1)/2), \\ I_{j-(M+1)/2}^{(\text{h})}, & (j - (M+1)/2 = 1, 2, \ldots, Q), \\ I_{j-Q}^{(\text{m})}, & (j - (M+1)/2 - Q = 1, 2, \ldots, (M-1)/2). \end{cases} \qquad (6.50)$$

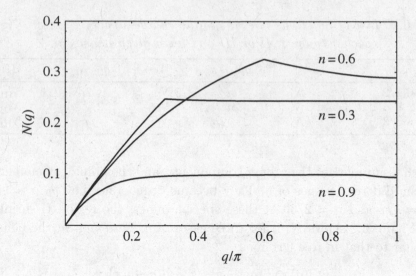

Fig. 6.2. The charge structure factor $N(q)$ for three cases of n. The cusp at $q/\pi = n$ is visible for each n, but becomes faint for $n = 0.9$.

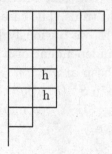

Fig. 6.3. Momentum distributions of magnons ($M = 5$) and holes ($Q = 2$) shown by h. Of the $M + Q = 7$ rows, the lowest one does not have a square.

The corresponding Young diagram can be written in terms of non-negative integers by adding a constant, e.g., $I_j + (M - 1)/2$. We loosely call this set of integers the momentum distribution, although the factor $2\pi/N$ should be attached to the physical momentum. Figure 6.3 shows the momentum distribution in the case of $M = 5$ and $Q = 2$. The ground state is a singlet with $N = 2M + Q = 12$, and has a finite magnetization $N - 12$ for larger N.

Now we take excited states of the system into consideration. Let us rewrite I_j derived above as I_{j0} restricted to the ground state. As in the spin chain, there are three alternative but equivalent ways to describe excited states.

In the first we introduce a set of non-increasing integers $\tilde{I}_1 \geq \tilde{I}_2 \geq \tilde{I}_{M+Q}$, and describe the excitations as bosonic. Namely, an excited state is specified by

$$I_j = I_{j0} + \tilde{I}_j. \tag{6.51}$$

In the second, we deal with I_j as it stands in (6.51). In contrast with the spin chain, hole momenta $I_j^{(\mathrm{h})}$ can take the same value. Hence in the second description, magnon excitations appear as fermions and hole excitations as bosons. For unique indexing of j in the set $\{I_j\}$, we set the convention that a magnon index is smaller than a hole index if they have the same value of I_j. The energy E associated with I_j's is given by

$$\left(\frac{N}{2\pi}\right)^2 (E - E_{\mathrm{c}}) = \frac{1}{2} \sum_{i=1}^{M+Q} I_i^2 + \frac{1}{2} \sum_{i<j} |I_i - I_j|, \tag{6.52}$$

where E_{c} corresponds to the energy of the absolute ground state given by (6.16). In the third description, the distribution of the rapidity is represented by the *motif* in the following way. We put 0 if a one-body state in the Brillouin zone is empty, and 1 if it is filled either by a magnon or a hole. This definition of the motif as an occupation pattern of rapidities is a straightforward generalization of that discussed in Section 4.6 for the spin chain. We deal with integers κ_j with $j = 1, \ldots, M + Q$ defined by

$$\kappa_j = I_j + (M + Q + 1)/2 - j. \tag{6.53}$$

The distribution of κ_j in the ground state is symmetric about zero, with maximum $\kappa_1 = (2M + Q)/2 - 1$ and minimum $\kappa_{Q+M} = (2M + Q)/2 - N$. Therefore in the singlet state with $N = 2M + Q$, κ_j ranges from $-N/2$ to $N/2 - 1$. In terms of the rapidity $k_j = 2\pi\kappa_j/N$, which plays a role of the renormalized momentum, the set spans the whole Brillouin zone $[-\pi, \pi)$. In order to emphasize the equivalence between $k_j = \pm\pi$, it is customary to put an extra 0 in the rightmost position. Then the sequence of $N + 1$ digits completely characterizes the distribution of the rapidity, and hence the momentum. For example, the ground state shown in Fig. 6.3 is represented by

0101011101010.

More generally, in the non-degenerate ground state with finite magnetization $2M < N$, we have Q successive 1's in the center which are sandwiched by $(M-1)/2$ number of 10's both from left and right. In the rest of the $N+1$ positions we have $N/2-M+1$ number of 0's on the left, and $N/2-M$ number of 0's on the right. It will be shown later that this definition of the motif is equivalent to the one introduced in Chapter 4. A pair k_j and k_{j+1} with the minimum separation $2\pi/N$ corresponds to either a pair of holes or a hole–magnon pair. According to the construction above, the smaller one, k_{j+1}, is associated with a hole. Following the same logic as in the Sutherland model, we obtain the energy characterized by the rapidities as follows:

$$E = \frac{1}{2} \sum_{j=1}^{M+Q} (k_j^2 - \pi^2) + \frac{\pi^2}{3} Q \left(1 - \frac{1}{N^2}\right), \tag{6.54}$$

for both ground and excited states. The energy looks like that of free particles with the parabolic spectrum. The total crystal momentum P of the system is given by

$$P = \sum_{j=1}^{M+Q} (k_j + \pi) - \pi = (M + Q - 1)\pi + \sum_{j=1}^{M+Q} k_j, \tag{6.55}$$

which is a generalization of (4.90) in Chapter 4. This method of description in terms of free magnons and holes is the most convenient to use to derive the spectrum of elementary excitations in the thermodynamic limit, as discussed in Section 6.4.2.

In any of three methods of description, the energy depends only on the distribution of momenta, but not on the particle species, i.e., magnon or hole for each momentum. This is a consequence of the U(1,1) supersymmetry in the mapped Hamiltonian given by (6.14). In order to classify spin and charge excitations, it is necessary to specify either magnon or hole for each occupied momentum. As an alternative to $I_i^{(m)}$ and $I_l^{(h)}$ given earlier for such a specification, we now introduce an auxiliary set q_i $(i = 1, \ldots, Q)$ of descending half-integers as follows. If a κ_j with certain j is the lth momentum of holes we set

$$q_l = \kappa_j + 1/2. \tag{6.56}$$

Thus the largest hole momentum is given by $q_1 - 1/2$, and the smallest one by $q_Q - 1/2$. Since the number q_i merely specifies the location of a hole, they have some arbitrariness. Namely, instead of 1/2 in (6.56), any number between 0

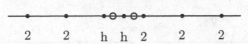

$$2 \qquad 2 \qquad \text{h} \quad \text{h} \quad 2 \qquad 2 \qquad 2$$

Fig. 6.4. Momentum distributions in the ground state with $M = 5, Q = 2$, and $N = 12$, i.e., the same state as shown by Fig. 6.3. The dots indicate occupied momenta which increase from left to right in the line. The circles indicate locations of q_i. With dots and circles given, locations of magnons (down spins) and holes are specified as indicated by 2 and h, respectively.

and 1 can be used for specification. The distribution of κ_j with information on particle species is described by the following equations [107, 193]:

$$\kappa_i = I_i^{(\mathrm{m+h})} + \frac{1}{2} \sum_{j=1}^{M+Q} \mathrm{sgn}(\kappa_i - \kappa_j) - \frac{1}{2} \sum_{l=1}^{Q} \mathrm{sgn}(\kappa_j - q_l), \qquad (6.57)$$

$$I_l^{(\mathrm{c})} = \frac{1}{2} \sum_{j=1}^{M+Q} \mathrm{sgn}(q_l - \kappa_j), \qquad (6.58)$$

where $I_j^{(\mathrm{m+h})}$ and $I_l^{(\mathrm{c})}$ are distinct quantum numbers in descending order. In the non-degenerate ground state, they are given by

$$I_j^{(\mathrm{m+h})} = (M + Q + 1)/2 - j, \ (j = 1, 2, \ldots, M + Q), \qquad (6.59)$$

$$I_l^{(\mathrm{c})} = (Q + 1)/2 - l, \ (l = 1, 2, \ldots, Q), \qquad (6.60)$$

with $I_j^{(\mathrm{m+h})}$ integers and $I_l^{(\mathrm{c})}$ half-integers, both of which are distributed symmetrically about zero. (6.57) and (6.58) are analogous to those for rapidities in the Bethe ansatz theory. The simplifying feature here is that the phase shifts for two-particle scatterings are independent of momenta. The distribution of κ_j, or equivalently k_j, in the ground state is shown in Fig. 6.4. We comment on combinations other than N even and M odd: (i) the case with both N and M even, and (ii) the case of N odd. In both cases, the energy is still given by (6.54). In case (i), the quantum numbers $I_j, \tilde{I}_j, \kappa_j$ remain integers, but their distributions are modified in accordance with the degenerate ground state. In case (ii), the mapped Sutherland model is effectively imposed on the antiperiodic boundary condition. We then have to consider κ_j and I_j as half-integers, as discussed in the Haldane–Shastry model. The quantum numbers $\tilde{I}_j$ for bosonic excitations are still integers.

In the thermodynamic limit, we need not care about degeneracies of the ground state and the situation simplifies. Namely, the holes occupy the rapidities $|k| < \pi(1 - n)$ with $n = 1 - Q/N$ being the electron number per site. The magnons then occupy the outer region $\pi(1 - n) < |k| < 2\pi M/N$.

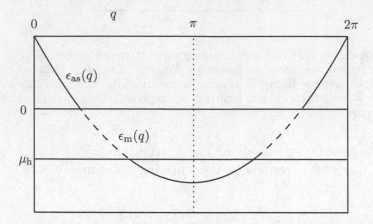

Fig. 6.5. The spectrum of magnons, antispinons, and holes in a magnetic field. The bottom region $n\pi < q < (2-n)\pi$ is occupied by holes whose chemical potential is written as μ_h. The regions $m\pi < q < n\pi$ and $(2-n)\pi < q < (2-m)\pi$ with dashed spectrum are occupied by magnons. The positive energy regions are unoccupied, and correspond to the spectrum of antispinons.

Figure 6.5 illustrates the spectra of holes and magnons as a function of the crystal momentum $q = k + \pi$.

6.4.2 Spinons and antispinons

The dispersion relations of elementary excitations can be derived from those of magnons and holes by proper reinterpretation. Let us begin with spinons following the argument of Section 4.7.3 in Chapter 4. In the thermodynamic limit, annihilation of a down-spin magnon with momentum $-2q$ corresponds to the creation of two spin-up spinons with adjacent momenta, i.e., $\epsilon_m(-2q) = -2\epsilon_s(q)$, which gives the same change of the total momentum of the system. Then we obtain, by taking $0 < q < \pi$

$$\epsilon_s(q) = q(\pi - q) - h, \tag{6.61}$$

where $h = (\pi/2)^2 m(2-m)$ is the magnetic field. Figure 6.6(c) shows the spectrum of spinons. In the presence of holes, spinons can be defined only in the momentum ranges $\pi m/2 < q < \pi n/2$ and $\pi(1 - n/2) < q < \pi(1 - m/2)$. The latter range is also represented as $\pi m/2 < |q| < \pi n/2$ by extending (6.61) to negative q by the replacement $q \to |q|$. The part with positive group velocity is called the right (R) branch, while the one with negative group velocity is called the left (L) branch, as indicated in Fig. 6.6.

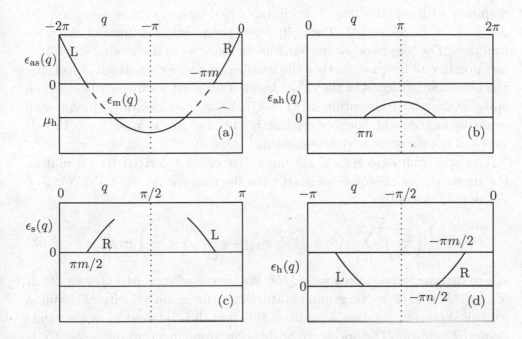

Fig. 6.6. Dispersion relation of (a) antispinon, (b) antiholon, (c) spinon, and (d) holon. The panel (a), which is the same as Fig. 6.5 except for the shift -2π in q, illustrates the regions in the momentum space for the U(1,1) Sutherland model. The spinon and holon are both defined in the momentum regions: $\pi m/2 < |q| < \pi n/2 = k_F$ and $\pi(1 - n/2) < |q| < \pi(1 - m/2)$, but with different signs for q. The right-going (R) and left-going (L) branches are indicated in each case.

On the other hand, the region with $\epsilon_m(q) > 0$ defines the spectrum of antispinons just as in the case of the spin chain. The spectrum of antispinons is given for $0 < q < \pi m$ and $(2 - m)\pi < q < 2\pi$ by

$$\epsilon_{as}(q) = \frac{1}{2}q(q - 2\pi) + 2h, \qquad (6.62)$$

where $h = (\pi/2)^2 m(2 - m)$. The presence of holes does not affect the antispinon. By replacing q by $|q|$ in (6.62), we rewrite the allowed momentum range simply as $|q| < \pi m$. We can define the right (R) and left (L) branches according to the sign of the group velocity.

Now we turn to the case of finite size. It is most convenient to take Q even, M odd, and $N = 2M + Q + 1$. The ground state has spin $S = S_z = 1/2$, and is doubly degenerate with total momentum $\pm\pi(M + Q)/N$. In Fig. 6.4, for example, the ground state with odd $N = 13$ corresponds to $\kappa_4 = \pm 1/2$. Here we regard all occupied κ_j as half-integers. Let us take one of the ground states with total momentum $-\pi(M + Q)/N$. As in the spin chain, a spinon with rapidity $2\pi\zeta/N$ is characterized by the shift $\kappa_j \to \kappa_j + 1$ for $j = 1, \ldots, \zeta \leq (M + 1)/2$. The shift keeps the polynomial form of the wave function. The hole momentum remains the same as in the ground state. It is not possible to increase further the number of magnons without disturbing the hole distribution. On the other hand, if we start with the other ground state with total momentum $\pi(M + Q)/N$, we can create a spinon with negative momentum. Namely, by shifting from κ_{M+Q} to $\kappa_{M+Q+1-\zeta}$ by -1, we can create a spinon with momentum $-2\pi\zeta/N$.

The spectrum of a spinon for finite size can be derived by calculating the increment of energy associated with the momentum shift. For $0 < \zeta \leq (M + 1)/2$, we obtain

$$\epsilon_{\mathrm{s}} = \frac{1}{2} \left(\frac{2\pi}{N} \right)^2 \sum_{i=1}^{\zeta} \left[(\kappa_i + 1)^2 - \kappa_i^2 \right] = q \left[\pi \left(1 - \frac{1}{N} \right) - q \right], \qquad (6.63)$$

where we have introduced $q = 2\pi\zeta/N$. We have used $\kappa_i = M + Q/2 + 1/2 - 2i$ for $i \leq (M + 1)/2$ in the ground state. Since the group velocity of a spinon for this momentum range is positive, this branch is referred to as the right-going (R) spinon. The spectrum with negative momentum $q = -2\pi\zeta/N$ is obtained from the ground state with total momentum $\pi(M + Q)/N$. By counting the increment of energy associated with the negative shift by $\zeta \leq (M - 1)/2$, we obtain

$$\epsilon_{\mathrm{s}} = \frac{1}{2} \left(\frac{2\pi}{N} \right)^2 \sum_{i=1}^{\zeta} \left[(\kappa_{M+Q+1-i} - 1)^2 - \kappa_{M+Q+1-i}^2 \right] = -q \left[\pi \left(1 - \frac{1}{N} \right) + q \right], \tag{6.64}$$

where we have used $\kappa_{M+Q+1-i} = -(M + Q/2 + 1/2) + 2i$ for $i \leq \zeta$ in this ground state. Since the group velocity of a spinon for this momentum range is negative, we refer to this branch as the left-going (L) spinon. Combining the R and L branches of the spectrum, we obtain the spinon dispersion relation from the singlet ground state. In the thermodynamic limit we obtain

$$\epsilon_{\mathrm{s}}(q) = |q|(\pi - |q|), \qquad (6.65)$$

which agrees with (6.61) for $h = 0$. Here the Brillouin zone should be defined as $[-\pi/2, -\pi/2]$ rather than $[0, \pi]$ in (6.61). The maximum of the energy takes place at $|q| = k_{\mathrm{F}} = \pi n/2$, as shown in Fig. 6.6(c). Spinons are forbidden for $|q| > k_{\mathrm{F}}$. Except for this restriction of the momentum range, the spinon spectrum does not depend on the number of holes.

In one dimension, the velocities of spin and charge excitations are in general different. This difference is often referred to as the spin–charge separation. The spin–charge separation is in marked contrast to a free-electron system or a Fermi liquid, where excitation of an electron accompanies both spin and charge excitations with the same Fermi velocity. However, in the Hubbard model [57] or the nearest-neighbor t–J model [19], the spin velocity still depends on the electron density. Since the spin velocity in the supersymmetric t–J model does *not* depend on the electron density, the independence of spin and charge in the supersymmetric t–J model is more complete than in other integrable models. Hence the supersymmetric t–J model is characterized by the *strong* spin–charge separation [121].

The spectrum of spinons in the presence of finite magnetization can be obtained similarly. Spinons with spin down do not belong to the FPSG states, and cannot be represented by polynomial wave functions. As has been discussed in the Haldane–Shastry model, a down spinon has an energy gap $2h$ originating from the Zeeman splitting. The momentum dependence is the same as that of the up spinon due to the Yangian symmetry.

6.4.3 Holons and antiholons

The elementary excitation of charge is called a holon. The holon spectrum is most easily obtained in the thermodynamic limit. Two holons can be created by annihilating a spin singlet pair in the ground state. This event accompanies a magnon annihilation in a polynomial wave function. However, there is no change in the total spin, in contrast with a spin-flip in the spinon pair creation. In the momentum space, an event of magnon annihilation creates two holons with adjacent momenta. Thus the momentum range of magnons is mapped to that of holons, as seen from Fig. 6.5. It is convenient to introduce the quantity p_c by

$$p_c = \pi(1 - n)/2 = \pi/2 - k_F, \tag{6.66}$$

in analogy with $p_s = \pi(1 - m)/2$ defined by (4.152). We obtain the holon spectrum $\epsilon_h(q)$ by adjusting the origin of energy as follows:

$$2[\epsilon_h(q) - \epsilon_h(\pi n/2)] = \epsilon_m(2q) - \epsilon_m(\pi n). \tag{6.67}$$

We then obtain

$$\epsilon_h(q) = q(q - \pi) + (\pi/2)^2 n(2 - n), \tag{6.68}$$

where q is allowed in the range with $\epsilon_h(q) > 0$; $\pi m/2 < q < \pi n/2 = k_F$ and $\pi(1 - n/2) = \pi/2 + p_c < q < \pi(1 - m/2) = \pi/2 + p_s$. There are right (R) and left (L) branches in the holon spectrum according to the sign of the group velocity. It is trivial to obtain the spectrum for $q < 0$ by replacing q by $|q|$ in (6.68). The spectrum for negative q is shown in Fig. 6.6(d).

The ground state with occupation of consecutive momenta by holes can be regarded as a condensate of holons. A defect in this consecutive distribution is regarded as a hole in the holon condensate, i.e., an antiholon. Thus, an antiholon emerges by removing a hole in the ground state. The spectrum of the antiholon is just the minus of the hole relative to μ_{h}. Then the antiholon spectrum is given in the thermodynamic limit by

$$\epsilon_{\mathrm{ah}}(q) = q(2\pi - q)/2 - \pi^2 n(2 - n)/2 = 2p_{\mathrm{c}}^2 - (q - \pi)^2/2. \qquad (6.69)$$

The spectrum is illustrated in Fig. 6.6(b). In contrast to other excitations, antiholons do not have right and left branches.

We now derive the holon spectrum for a finite-sized system. It is convenient to take the spin singlet case $N = 2M + Q$ with M odd. The ground state with N odd is non-degenerate, as shown in Fig. 6.7(a). Consider an excitation where the rightmost hole, which corresponds to $\kappa_5 = 3/2$ in Fig. 6.7(a), moves to the right by a certain step, and correspondingly the magnons in between move to the left by the unit step. The magnons to the right of the moved hole remain the same. Figure 6.7(b) shows the case where two magnons moved to the left. This is referred to as the right-going holon excitation. It is easy to derive the energy ϵ_{h} and the rapidity $k > 0$ associated with the excitation. We obtain

$$\epsilon_{\mathrm{h}} = \frac{1}{2}\left(\frac{2\pi}{N}\right)^2 \sum_{i=1}^{\zeta} \left[(\kappa_{(M+3)/2-i} + 1)^2 - \kappa_{(M+3)/2-i}{}^2\right] = k(v_{\mathrm{c}} + k), \quad (6.70)$$

with $v_{\mathrm{c}} = \pi(1 - n)$ and $k = 2\pi\zeta/N < \pi n/2$. The spectrum is the same as that given by (6.68), with identification $k = q - \pi(1 - n/2)$. Note that the origin of k is the holon state with minimum energy.

On the other hand, the left-going holon is identified as the excitation where the leftmost hole in the ground state moves to the left by a certain step, and the magnons in between move to the right by the unit step. The corresponding spectrum is obtained as

$$\epsilon_{\mathrm{h}} = v_{\mathrm{c}}|k| + k^2. \qquad (6.71)$$

The spectrum is the same as given by (6.68), with identification $k = q - k_{\mathrm{F}}$ with $0 < q < k_{\mathrm{F}}$.

It is not possible to choose, by any combination of N, M, and Q, a ground state where an antiholon is present. This is in contrast with other elementary excitations. Therefore, as the reference we take a state where the rightmost

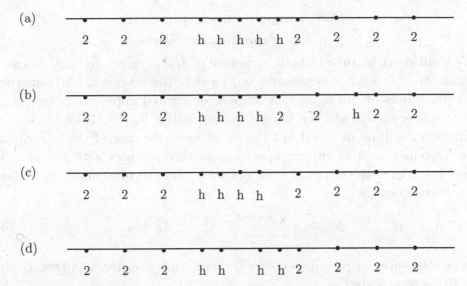

Fig. 6.7. Examples of momentum distributions of holes shown by h, and down spins (magnons) shown by 2: (a) the ground state with $N = 19, Q = 5$, and $M = 7$; (b) holon excitation with $\zeta = 2$; (c) reference state for antiholon excitation with $\zeta = 0$; (d) antiholon excitation with $\zeta = 2$.

hole is removed from the ground state. Figure 6.7(c) illustrates the situation for $N = 19, M = 7$, and $Q = 4$. Note that this state has $S_z = 1/2$, since a hole is replaced by an electron with spin up. If ζ holes shift to the right, the momentum of the antiholon increases by $k = 2\pi\zeta/N$. The increase $\epsilon_{\rm ah}$ of antiholon energy is given by

$$\epsilon_{\rm ah} = \frac{1}{2}\left(\frac{2\pi}{N}\right)^2 \sum_{i=1}^{\zeta} \left[(\kappa_{(M+1)/2+i} + 1)^2 - \kappa_{(M+1)/2+i}{}^2\right] = \frac{1}{2}k(2v_{\rm c} - k),$$

(6.72)

where we have used $\kappa_{(M+1)/2+i} = (Q-1)/2-i$. The rapidity k of an antiholon is related to the crystal momentum q by $k = q - \pi n$, and runs from 0 to $2v_{\rm c}$. Its spectrum is independent of magnetization. In the limit $n \to 0$, the antiholon is reduced to an electron with spin up. This direction of the spin is of course due to the definition of polynomial wave functions.

When a multiple number of elementary excitations is present, there arise statistical interactions which we have already encountered in the spin chain and in the Sutherland model with internal degrees of freedom. We discuss this problem in Section 6.6.1.

6.5 Yangian supersymmetry

6.5.1 Yangian generators

We shall describe the algebraic structure of H_{SUSY} given by (6.5) in more detail. We follow a less systematic, but more heuristic approach to construct the generators of the algebra. A more sophisticated approach, generalizing the transfer matrix and the Yang–Baxter relation [22, 80, 82, 182] to supersymmetry, will be sketched in Chapter 9. As in the case of the spin chain, we introduce a set of current operators which commutes with H_{SUSY}, but does not commute with total spin, nor even with the total number of holes. They are given by

$$\Lambda^{\alpha\beta} = \frac{1}{2} \sum_{ij} \sum_{\gamma} w_{ij} A^{\dagger}_{i\alpha} A^{\dagger}_{j\gamma} A_{i\gamma} A_{j\beta}, \tag{6.73}$$

where the superscript ranges over up, down, and hole states. Alternatively, (6.73) is represented as

$$\Lambda^{\alpha\beta} = \frac{1}{2} \sum_{ij} w_{ij} X^{\alpha\beta}_{j} \tilde{P}_{ij} = \frac{1}{2} \sum_{ij} w_{ij} \tilde{P}_{ij} X^{\alpha\beta}_{j}. \tag{6.74}$$

We now show that $\Lambda^{\alpha\beta}$ commutes with H_{SUSY}. If the superscript does not involve the hole state, $\Lambda^{\alpha\beta}$ commutes with H_{tJ} given by (6.1) as well.

The proof can be done most intuitively by appealing to the momentum conservation. The momentum of a hole is given in the coordinate and second-quantized representations by

$$-\mathrm{i}\frac{\partial}{\partial x_i} - \frac{N}{2} \Rightarrow \frac{1}{2} \sum_{j} w_{ij} h^{\dagger}_i h_j, \tag{6.75}$$

which is analogous to the magnon momentum discussed in Chapter 4. Thus the following operator:

$$\Lambda_t = \frac{1}{2} \sum_{i\neq j} w_{ij} (h^{\dagger}_i h_j + b^{\dagger}_i b_j) \tag{6.76}$$

gives the total momentum of U(1,1) particles, and therefore commutes with the Hamiltonian. In the representation of (6.2), on the other hand, the transfer terms are rewritten in the exchange form:

$$h^{\dagger}_i h_j = F^{\dagger}_i B_{i\uparrow} B^{\dagger}_{j\uparrow} F_j, \quad b^{\dagger}_i b_j = B^{\dagger}_{i\downarrow} B_{i\uparrow} B^{\dagger}_{j\uparrow} B_{j\downarrow}. \tag{6.77}$$

Then we identify $\Lambda_t = \Lambda^{\uparrow\uparrow}$, namely

$$\Lambda_t = \frac{1}{2} \sum_{i\neq j} \sum_{\alpha} w_{ij} A^{\dagger}_{i\uparrow} A^{\dagger}_{j\alpha} A_{i\alpha} A_{j\uparrow} = \frac{1}{2} \sum_{i\neq j} w_{ij} X^{\uparrow\uparrow}_i \tilde{P}_{ij}, \tag{6.78}$$

where α runs over three internal degrees of freedom. In going from (6.75) to (6.78), we have used the odd property $w_{ij} = -w_{ji}$ to include the canceling contribution $\alpha = \uparrow$. Since the total momentum is conserved with any number of holes, we obtain

$$[\Lambda^{\uparrow\uparrow}, H_{t-J}] = 0. \tag{6.79}$$

We now prove that another component $\Lambda^{\uparrow\downarrow}$ also commutes with H_{t-J} without a magnetic field. For this purpose we introduce the notation $X^{\alpha\beta} = \sum_i X_i^{\alpha\beta}$ and write $\Lambda^{\uparrow\downarrow}$ in the form

$$\Lambda^{\uparrow\downarrow} = [\Lambda^{\uparrow\uparrow}, X^{\uparrow\downarrow}] = \lim_{\delta\to 0} \delta^{-1} \left[\exp(-\delta X^{\uparrow\downarrow})\Lambda^{\uparrow\uparrow}\exp(\delta X^{\uparrow\downarrow}) - \Lambda^{\uparrow\uparrow}\right], \tag{6.80}$$

where $\exp(\delta X^{\uparrow\downarrow})$ describes a global SU(2) rotation. Since $X^{\uparrow\downarrow}$, and hence $U(\delta) \equiv \exp(\delta X^{\uparrow\downarrow})$, commute with H_{t-J} without a magnetic field, we can show

$$[\Lambda^{\uparrow\downarrow}, H_{t-J}] = U(-\delta)[\Lambda^{\uparrow\uparrow}, H_{t-J}]U(\delta) = 0. \tag{6.81}$$

Alternatively, one may directly derive (6.81) by combining $[\Lambda^{\uparrow\uparrow}, H_{t-J}] = 0$ and $[X^{\uparrow\downarrow}, H_{t-J}] = 0$.

In a manner similar to (6.80), we can write all components $\Lambda^{\alpha\beta}$ of the current operator as a commutator of a conserving quantity. Thus we have proved

$$[\Lambda^{\alpha\beta}, H_{\text{SUSY}}] = 0, \tag{6.82}$$

for all combinations of three internal quantum numbers α, β.

These current operators, however, do not commute with the total spin, nor with each other. They constitute the supersymmetric Yangian generators. We now show that polynomial wave functions $\Psi(\{x^{\text{h}}\}, \{x^{\text{s}}\})$ are annihilated by $\Lambda^{\uparrow\downarrow}$. In order to understand their action, we represent $\Lambda^{\uparrow\downarrow}$ as

$$\Lambda^{\uparrow\downarrow} = \sum_{i\neq j} w_{ij}\left(S_i^z S_j^+ - \frac{1}{2}X_i^{\uparrow 0}X_j^{0\downarrow}\right). \tag{6.83}$$

In the high-density limit, the fermion transfer term vanishes, and we recover the corresponding result for the spin chain as given by (4.115). For general filling of a site, $\Lambda^{\uparrow\downarrow}$ becomes a conserving quantity only by including the fermion transfer term. We rewrite $\Lambda^{\uparrow\downarrow}$ as

$$\Lambda^{\uparrow\downarrow} = \sum_{i\neq j} w_{ij}\left(-2b_i^\dagger b_i b_j - h_i^\dagger h_i b_j + h_j^\dagger h_i b_j\right), \tag{6.84}$$

using the U(1,1) representation. The part coming from the first term in parentheses annihilates $\Psi(\{x^{\text{h}}\}, \{x^{\text{s}}\})$, as in the case of the spin chain. To see the effect of second and third terms, we consider the simplest case $\Psi(x_h, x_s)$ with only one hole and one magnon. After annihilating the magnon by $\Lambda^{\uparrow\downarrow}$, we end up with a single hole state and obtain

$$\langle x_h| \sum_{i \neq j} w_{ij} \left(h_i^\dagger - h_j^\dagger \right) h_i b_j |\Psi\rangle = \sum_{x=1}^{N} w_{hx} \left[\Psi(x_h, x) + \Psi(x, x_h) \right], \qquad (6.85)$$

where we have used $w_{xh} = -w_{hx}$. The summand is antisymmetric against interchange of x and x_h and, furthermore, w_{xh} is translationally invariant. Thus by the same reason as in (4.116), (6.85) vanishes by summation over x. It is easy to see that the annihilation by $\Lambda^{\uparrow\downarrow}$ holds true for a general case of $\Psi(\{x^h\}, \{x^s\})$ with any numbers of holes and magnons.

We have thus shown that $\Psi(\{x^h\}, \{x^s\})$ belongs to the YHWS, and that $\Lambda^{\uparrow\downarrow}$ is a Yangian raising operator. By successive application of the lowering operator $\Lambda^{\downarrow\uparrow}$ to $\Psi(\{x^h\}, \{x^s\})$, we can construct a set of degenerate states. This set is called a Yangian supermultiplet as in the Haldane–Shastry spin chain. At certain power of n, the series should terminate, i.e., $(\Lambda^{\downarrow\uparrow})^n \Psi(\{x^h\}, \{x^s\}) = 0$.

The supersymmetric Yangian contains those generators that change the number of holes. Let us consider, for example, $\Lambda^{\uparrow 0}$. With a little algebra we obtain

$$\Lambda^{\uparrow 0} = \sum_{i \neq j} w_{ij} X_i^{\uparrow 0} \tilde{P}_{ij} = \sum_{i \neq j} w_{ij} (b_j^\dagger - b_i^\dagger) b_i h_j. \qquad (6.86)$$

In the same way as before, we can prove that $\Lambda^{\uparrow 0}$ annihilates the polynomial wave function $\Psi(\{x^h\}, \{x^s\})$. Thus $\Psi(\{x^h\}, \{x^s\})$ belongs to the supersymmetric YHWS. The lowering operator $\Lambda^{0\uparrow}$, and also $\Lambda^{0\downarrow}$ and $\Lambda^{\downarrow 0}$, generate a Yangian supermultiplet with a different number of holes. The energy of these states is degenerate for the Hamiltonian H_{SUSY}, but not for H_{t-J}. The difference is caused by the chemical potential term in (6.4). Hence we find it convenient to use the terminology of YHWS also in the restricted space with a fixed number of holes.

One may now naturally ask if there are YHWS that do not take a polynomial form. The answer is no: all YHWS are exhausted by polynomial wave functions. We can show that any of the 3^N states allowed in the $t-J$ model is in a supermultiplet to which one of the polynomial wave functions belongs. In other words, any state can be generated by successive operation of Yangian lowering operators starting from one of the polynomial states. The structure of supermultiplets can be made transparent in terms of ribbon diagrams, to be explained below.

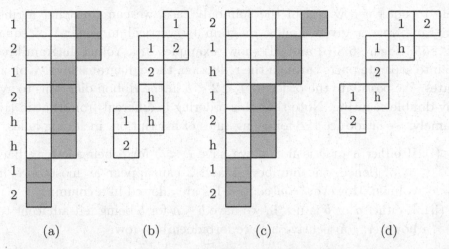

Fig. 6.8. Young diagrams (a), (c) and the corresponding ribbon diagrams (b), (d). Each shaded square represents the momentum of a particle, and only such squares are kept for ribbon diagrams. The ground state with two holes and four electrons is represented by (a) and (b), while an excited state with increased momentum is shown in (c) and (d).

6.5.2 Ribbon diagrams and supermultiplets

We now consider all states in the system without restricting ourselves to the family of polynomial wave functions which is equivalent to the YHWS. We start from the following result for the energy of the $U(2,1)$ Sutherland model in the strong-coupling limit:

$$E_{\text{tot}} = \frac{\pi^2}{N^2} \sum_{\kappa=\infty}^{\infty} \sum_{\kappa'=-\infty}^{\infty} |\kappa - \kappa'| \, \nu(\kappa)\nu(\kappa'). \tag{6.87}$$

Here the distribution function $\nu(\kappa)$ has three components, given by

$$\nu(\kappa) = \nu_\uparrow(\kappa) + \nu_\downarrow(\kappa) + \nu_{\text{h}}(\kappa).$$

The hole occupation number $\nu_{\text{h}}(\kappa)$ can take any non-negative integer because of its bosonic nature. For example, in the ground state with $N = 6$, $Q = 2$, two holes have $\kappa = 2$, and up and down electrons have $\kappa = 1, 2, 3$ for both spins. The distribution of κ is best seen by the Young diagrams in Fig. 6.8. It is evident from (6.87) that the energy does not depend on the total momentum. In other words, there is a large degeneracy related to a Galilean boost.

In Young diagrams, different spin components are always symmetrized in a row, and antisymmetrized in a column. On the other hand, holes in a column are symmetrized. Horizontal antisymmetrization prohibits more than one hole to enter a row. As in the spin-only case, we can construct a ribbon diagram from a given Young diagram by vertical movement of squares [88, 155]. Figure 6.8(b) and (d) shows examples. If a Young diagram breaks up into separate parts through the reduction, these diagrams involve phonon states. We postulate the ordering $1 < 2 < h$ in the ribbon diagrams to avoid any double counting. Note that this ordering is different from that in [155]. Namely, we introduce the following rules of inscription, including holes:

(i) If either a or b is not h, we have $a < b$ for b being lower-adjacent to a. Hence, the numbers 1 and 2 can appear at most once in a column. However, consecutive h's are allowed in a column.

(ii) If either a or b is not h, we have $b \leq a$ for b being left-adjacent to a; however, consecutive h's are forbidden in a row.

For example, the ground state has three singlet pairs of electrons and can be represented in an appealing manner in Fig. 6.8(b). The inscription is unique according to rules (i) and (ii).

Now we consider an excited state where an electron and a hole increase the momentum by one unit, as shown in Fig. 6.8(c). Each row is regarded as an orbital for spinons as in the SU(2) case, while each column is regarded as an orbital of a holon. We use the term *holon* when we emphasize its nature as elementary excitations. Thus we identify two spinons in Fig. 6.8(d), and the spinons occupy the same orbital. A connected ribbon diagram has filtered out the phonon excitation. Hence from each ribbon diagram, the distribution $\nu(\kappa)$ of the supersymmetric t–J model is recovered, which gives the energy of the system according to (6.87). By definition the energy associated with a ribbon diagram does not depend on the inscription. Hence each ribbon diagram represents a supermultiplet of the supersymmetric Yangian $Y(sl_{2,1})$.

We show now that the supersymmetric Yangian multiplets form the complete set in the Hilbert space for the t–J model. In order to prove this, we establish a one-to-one correspondence between any of 3^N states to a Yangian state. We start from a diagram with two boxes arranged as sharing a corner. See Fig. 6.9. There are $3^2 = 9$ possibilities to inscribe either 1, 2, or h in the box. Then we require the ordering $1 < 2 < h$, and move the lower box either up or to the right. Namely, if the lower box is larger according to the ordering, it moves right as shown in (a) and (c). In the opposite case, it moves up as shown in (b) and (d). If they both have h, the lower box moves right as shown in (e). If the two boxes have the same number, either 1 or 2, the lower box moves up as shown in (f). This rule is applicable to a corner-sharing diagram of any size. Namely, we start from the bottom and move the last box either up or right according to the ordering of the last and the

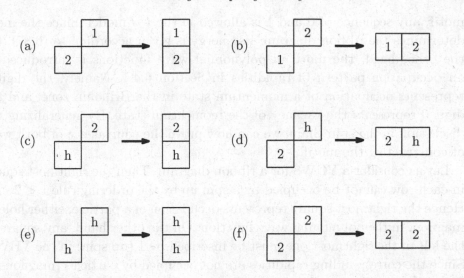

Fig. 6.9. Examples of reduction from a diagonal corner-sharing diagram to a ribbon diagram.

next last inscriptions. Then we compare the inscriptions of the second last and the third last boxes. By following the same rule of ordering, we move two boxes which are arranged either horizontally or vertically. The process continues until we move the $N-1$ boxes by comparing the first and second inscriptions.

It is then obvious that any inscription of the corner-sharing diagram with N boxes leads uniquely to a corresponding ribbon Young diagram. Conversely, starting from any inscribed ribbon diagram, we can move back the boxes to a corner-sharing diagram uniquely. Thus it is proved that the correspondence between an inscribed corner-sharing diagram and an inscribed ribbon diagram is one-to-one. Namely, we obtain the complete set of 3^N states for the t–J model. Since the energy of a state is determined only by the shape of a ribbon diagram without inscription, each shape defines a supermultiplet.

6.5.3 Motif as representation of supermultiplets

The shape of a ribbon diagram is determined completely by the motif as in the case of the spin chain. Namely, the digit 0 represents horizontally adjacent squares, and the digit 1 represents vertically adjacent squares. Since an arbitrary number of holes can be arranged vertically, it is allowed now to have the sequence $\cdots 1111 \cdots$. Except for the 0's at each end of the motif, any sequence of 0 and 1 is allowed in the t–J model. Since the motif determines the ribbon diagram, its energy is given according to (6.87). On the other hand, the motif in polynomial wave functions is introduced as an occupation pattern of rapidities in Section 6.4.1. Namely, the digit 1 represents occupation of a momentum state in the Brillouin zone, and the digit 0 represents the vacancy of the momentum state. By generalizing the discussion in the spin chain, we can now prove the equivalence of both ways of constructing the motif.

Let us consider a YHWS for a ribbon diagram. Then the rightmost square in each row cannot be occupied by a spin up by the ordering rule $1 < 2 < $ h. Hence the rightmost square represents occupation of a particle, either hole or magnon, in the polynomial wave function. On the other hand, any square to the left of the rightmost one must be inscribed by 1 (up spin) in the YHWS. Since the corresponding rapidities are not occupied by particles (magnons or holes) in the polynomial wave function, the digit 0 in the motif is associated with these squares. Thus we have established the equivalence of both definitions of the motif; one from the shape of a ribbon diagram, and the other from the occupation pattern of rapidities. The energy of a supermultiplet is completely determined by the motif, or by the ribbon diagram. With a given motif, the occupation pattern of magnons and holes gives the energy by (6.54). On the other hand, a ribbon diagram uniquely gives an ordinary Young diagram by vertical movement of squares, as shown in Fig. 6.8. The resultant distribution $\nu(\kappa)$ gives the energy by (6.87). It is amazing that two apparently different expressions give identical results with a proper shift of the origin.

It is now trivial to prove that the wave function $\Psi_G(\{z\})$ of (6.20) gives the true ground state of the supersymmetric t–J model. We have shown that the energy of each supermultiplet is given in terms of the YHWS, which can be represented by a polynomial wave function. Among these, $\Psi_G(\{z\})$ gives the lowest energy. Hence the ground state is given by $\Psi_G(\{z\})$ with the motif $0101\ldots111101\ldots01$.

6.6 Thermodynamics

6.6.1 Parameters for exclusion statistics

In the Haldane–Shastry spin chain, we have derived thermodynamics by the free-semion picture of spinons. One may naturally ask if the fractional particle picture can be applied in the presence of holes. The answer is affirmative, as we now discuss in detail. In order to derive the fractional exclusion statistics in the SUSY t–J model, we first count the available number $W(M, Q)$ of states for the case where M magnons and Q holes are present. This number is the same as the number of YHWS for a given number $N_\uparrow = N - Q - 2M$ of up spinons, and Q holons. The number $W(M, Q)$ is also equal to the number of motifs, or the number of ribbon diagrams without inscription.

Of the N squares for the lattice sites, we set aside M squares which will come back when recovering the part of a ribbon diagram for singlet pairs. Equivalently, in terms of a motif, we first consider a sequence of $N - M$ digits of either 0 or 1. The digit 1 represents either a magnon or a hole. Then the number of magnon plus hole states is given by

$$W(M, Q) = \frac{(N - M)!}{(M + Q)!(N - 2M - Q)!} \cdot \frac{(M + Q)!}{Q!M!}, \tag{6.88}$$

where the first combinatorial factor gives the possible ways of placing $M + Q$ 1's among $N - M$ digits. This number in fact gives the number of singlet states. The second combinatorial factor gives the number of hole states for given location of 1's. To recover $N + 1$ digits representing a supermultiplet, a 0 (zero) is inserted to the right of each 1, provided this 1 represents a magnon. In terms of a ribbon diagram, a square out of M reserved ones is placed on top of each magnon square.

In order to obtain the statistical parameter, we rewrite $W(M, Q)$ in terms of number $N_\uparrow$ of up spinons, and Q of holons. Namely (6.88) is written as

$$W(M, Q) = {}_{D_\uparrow}H_{N_\uparrow} \times {}_{D_h}H_Q, \tag{6.89}$$

where the notation ${}_nH_m$ has been defined by (4.135), and

$$D_\uparrow = \frac{1}{2}(N + Q - N_\uparrow) + 1, \quad D_h = \frac{1}{2}(N - Q - N_\uparrow) + 1. \tag{6.90}$$

Here $D_\uparrow$ gives the number of available orbitals for up spinons, and D_h for holons. We then obtain the relation

$$\Delta D_\uparrow = -g_{\uparrow\uparrow}\Delta N_\uparrow - g_{\uparrow h}\Delta Q, \tag{6.91}$$

$$\Delta D_h = -g_{h\uparrow}\Delta N_\uparrow - g_{hh}\Delta Q, \tag{6.92}$$

with

$$\begin{pmatrix} g_{\uparrow\uparrow} & g_{\uparrow h} \\ g_{h\uparrow} & g_{hh} \end{pmatrix} = \begin{pmatrix} 1/2 & -1/2 \\ 1/2 & 1/2 \end{pmatrix}. \tag{6.93}$$

It is remarkable that the matrix is antisymmetric. Namely, spinons have more states in the presence of holons, while holons have less states in the presence of spinons. Another remarkable feature is that holons acquire the statistical parameter 1/2, which is different from that of holes. The semionic holon statistics appears explicitly in the absence of spinons. On the other hand, the fermionic statistics of holes appears explicitly in the absence of the singlet pairs. The latter follows straightforwardly from (6.88) with $M = 0$. Thus we conclude the following: the statistical parameter depends on the reference state in the t–J model.

According to the general framework of exclusion statistics, the statistical matrix $\{g_{\alpha\beta}\}$ gives sufficient information for constructing the thermodynamics of spinons and holons. In order to apply the framework to the t–J model, we have to extend the above results to be applicable to

(i) all states including the non-YHWS, and

(ii) a macroscopic number of states within the narrow momentum range.

We note that the magnon–hole picture applies only to the YHWS state, and cannot meet the requirement (i). On the other hand, the spinon–holon picture can be extended to meet (i), as has been demonstrated in the spin chain. The spinons in the YHWS all have spin up by definition. As in the spin chain, the statistical parameters are independent of the spin direction of spinons and are commonly given by 1/2. Hence we obtain the result including down spinons as follows:

$$\{g_{\alpha\beta}\} = \begin{pmatrix} 1/2 & 1/2 & -1/2 \\ 1/2 & 1/2 & -1/2 \\ 1/2 & 1/2 & 1/2 \end{pmatrix}, \tag{6.94}$$

where the first and second components are for up and down spinons and the third one for holons. We note that the total number 3^N of states is recovered by construction of ribbon diagrams, as explained in Section 6.5.2. Alternatively, the number 3^N comes out if one sums up the number of combinations

$$W(N_\uparrow, N_\downarrow, Q) = {}_{D_h}H_Q \times {}_{D_{sp}}H_{N_\uparrow} \times {}_{D_{sp}}H_{N_\downarrow} \tag{6.95}$$

over all possible combinations of up and down spinon numbers $N_\uparrow, N_\downarrow$, and the holon number Q [101]. Here D_h and D_{sp} are given by (6.90) with the replacement $N_\uparrow \rightarrow N_{sp} = N_\uparrow + N_\downarrow$.

The other necessary extension (ii) can be achieved by the standard technique established by Yang and Yang [195]. Namely, as we have considered in Section 4.10, we take a region with small width $\Delta p \ll 2\pi$ in the Brillouin zone. Under the condition $N\Delta p \gg 2\pi$, the small region has a macroscopic number of states for magnons and holes. We can apply almost the same argument of counting the available states as that used for taking the whole energy range. Since the exhaustion of available states applies to this narrow region as well, we also obtain (6.94) for the range around p, provided the momentum p is common to both spinons and holons. We shall next provide such a description.

6.6.2 Energy and thermodynamic potential

Let us return to the YHWS described by polynomial wave functions. By regarding spinons as antiparticles of magnons, we can easily derive a part of the excitation spectrum of the system by using (6.54). Although the present derivation assumes a pair creation with the same momentum, the restriction has no effect in the thermodynamic limit. Instead of changing the momenta of many magnons as described by (6.64), we create two spinons by moving a magnon with rapidity k to π. Here we assume $M = N/2$, i.e., no spin polarization. The energy change associated with the move is given by

$$\Delta E = -\frac{1}{2}(k^2 - \pi^2) = 2\left[\left(\frac{\pi}{2}\right)^2 - p^2\right] = 2\epsilon_0(p), \qquad (6.96)$$

where $p = -k/2$. Namely, the change in energy and rapidity k is shared by two spinons. The sign of the momentum is reversed because the spinons are holes of magnons. For the fractional exclusion statistics to be described soon, it is convenient to regard p as the rapidity of a spinon. It is related to physical momentum q by the relation $p = q - \pi/2$, as in the case of the spin chain. Then we have $\epsilon_s(q) = \epsilon_0(p)$, and the range of p is $[-\pi/2, \pi/2]$.

In order to describe mutual exclusion of spinons and holons, we should use the common rapidity p for a holon excitation as well. The simplest derivation of the holon spectrum as a function of p is to use (6.54), and consider the creation of a spinon–holon pair. Namely, we choose the rapidity k_j of a magnon, and reinterpret this as the rapidity of a hole with $M \to M - 1$ and $Q \to Q + 1$. This is equivalent to creating a spinon–holon pair with the same rapidity. Owing to the supersymmetry with $M + Q$ kept constant, the change of total energy comes only from the last term in (6.54), which sets the chemical potential to the center of the non-interacting energy band described by $t(k)$. Hence, apart from this chemical potential term $\pi^2/3$, the holon energy should compensate the spinon energy. From this consideration we obtain the holon spectrum as $\pi^2/3 - \epsilon_0(p)$. The physical momentum q is related to p by $p = q - \pi/2$ with vanishing hole concentration.

It is clear in the present derivation that the total energy of the system is given by the sum of elementary excitation energies, and there is no interaction term in the p-space. Although the spectrum derived above is for the YHWS, the result can be generalized to all excited states other than polynomial wave functions because of the Yangian symmetry. Taking the reference energy as the ground-state energy of the Haldane–Shastry model, we obtain the internal energy $U(h)$ with inclusion of the Zeeman term as

$$U(h) = \frac{\pi^2}{3}Q + \sum_p [\epsilon_\uparrow(p)\rho_\uparrow(p) + \epsilon_\downarrow(p)\rho_\downarrow(p) - \epsilon_0(p)\rho_h(p)], \qquad (6.97)$$

where $\epsilon_\sigma(p) = \epsilon_0(p) - \sigma h$ and $\rho_\alpha(p)$ is the distribution function for the component $\alpha =\uparrow, \downarrow$, and h. We define the density of available orbitals for each species α by $\rho_\alpha^*(p)$ by the relation

$$\rho_\alpha^*(p)\Delta p = \Delta D_\alpha, \quad \rho_\alpha(p)\Delta p = \Delta N_\alpha, \qquad (6.98)$$

where N_α denotes the number of particles with species α, i.e., up and down spinons and holons. Then we obtain the entropy

$$S = \sum_{p\alpha} [(\rho_\alpha + \rho_\alpha^*) \ln (\rho_\alpha + \rho_\alpha^*) - \rho_\alpha \ln \rho_\alpha - \rho_\alpha^* \ln \rho_\alpha^*]. \qquad (6.99)$$

The particle and hole distribution functions are related by the statistical matrix $g_{\alpha\beta}$ as

$$\rho_\alpha^*(p) = 1 - \sum_\beta g_{\alpha\beta}\, \rho_\beta(p), \qquad (6.100)$$

where the indices α, β specify either spin up, down, or holon.

In terms of these distribution functions the thermodynamic potential $\Omega = U(h) - TS + (\mu - \pi^2/3)Q$ is obtained. The distribution functions are determined by the stationary condition $\delta\Omega/\delta\rho_\alpha(p) = 0$. As a result, we obtain for each p

$$\beta\epsilon_\alpha = \ln (1 + w_\alpha) - \sum_\gamma g_{\gamma\alpha} \ln (1 + w_\gamma^{-1}), \qquad (6.101)$$

with $\beta = 1/T$ and $w_\alpha \equiv \rho_\alpha^*/\rho_\alpha$. The quasi-particle energy ϵ_α includes the Zeeman term for spinons, and the chemical potential term for holons. Explicitly they are written as

$$\epsilon_\uparrow(p) = p_s^2 - p^2, \quad \epsilon_\downarrow(p) = p_s^2 - p^2 + 2h, \quad \epsilon_h(p) = p^2 - p_c^2, \qquad (6.102)$$
$$p_s^2 = (\pi/2)^2 - h = (\pi/2)^2(1 - m)^2, \qquad (6.103)$$
$$p_c^2 = (\pi/2)^2 - \mu = (\pi/2)^2(1 - n)^2. \qquad (6.104)$$

Note that p_s has already been used in Chapter 4. Provided the stationary condition is satisfied, we obtain the simple form for Ω as

$$\Omega = -T \sum_{p\alpha} \ln\left(1 + w_\alpha^{-1}\right). \tag{6.105}$$

The spectrum of spinons and holons can be defined only for rapidities out of the condensate regions $p_c < |p| < p_s$. On the other hand, antiparticles arise as elementary excitations in the condensate regions. Namely, antispinons can be defined for $p_s < |p| < \pi/2$, and antiholons for $|p| < p_c$.

6.6.3 Fully polarized limit

It is instructive to see the situation in the fully polarized case $m = n$, or $\rho_\downarrow(p) = 0$. In this limit, double occupation of a site is prohibited by the exclusion principle. Hence the hard-core repulsion between electrons becomes irrelevant, and the system behaves as free fermions. Note that this limiting situation applies to any form of the t–J model, and has nothing to do with the supersymmetry. We shall now discuss how the description in terms of semionic spinons and holons with supersymmetry reduces to that of free fermions. Let us put $w_\downarrow^{-1} = 0$ in accordance with $\rho_\downarrow(p) = 0$. This corresponds to $h \to \infty$. On the LHS of (6.101), we consider cases $\alpha = \uparrow$ and h. Adding both sides in these cases we obtain

$$(\mu - h)\beta = \ln\left[\frac{1}{\rho_\uparrow \rho_h}\left(1 - \frac{\rho^2}{4}\right)\right], \tag{6.106}$$

where $\rho = \rho_\uparrow + \rho_h$ with omission of argument p. We obtain the solution $\rho(p) = 2$, which gives $-\infty$ on both sides. We now subtract the case $\alpha = \uparrow$ in (6.101) from the case $\alpha = $ h to obtain

$$[2\epsilon_0(p) - h - \mu]\beta = \ln\left[\left(1 - \frac{1}{4}\tilde{\rho}^2\right) / (\rho_\uparrow \rho_h)\right] = \ln[f(p)^{-1} - 1], \tag{6.107}$$

where $\tilde{\rho} = \rho_\uparrow - \rho_h$ and $f(p) = \rho_\uparrow(p)/2 = 1 - \rho_h(p)/2$. To make a connection with free electrons we interpret $\zeta = \mu + h$ as the chemical potential, and $k = 2p$ as the physical momentum of the electron. Then we obtain from (6.107)

$$f(p) = [\exp\beta(\epsilon_k - \zeta) + 1]^{-1}, \tag{6.108}$$

where $\epsilon_k = (\pi^2 - k^2)/2$. The function $f(p)$ is precisely the Fermi distribution with the band energy ϵ_k. The internal energy of (6.97) is reduced to

$$U(h) = U_0(h) + \sum_k (\epsilon_k - \zeta)f(k/2), \tag{6.109}$$

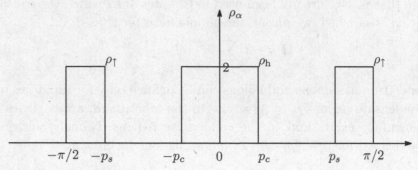

Fig. 6.10. Distribution functions of holons $\rho_h(p)$ and up spinons $\rho_\uparrow(p)$ at zero temperature.

where $U_0(h)$ is a constant.

The entropy of the system is written as

$$S = \sum_p \rho_\alpha [w_\alpha \ln(1 + w_\alpha^{-1}) + \ln(1 + w_\alpha)]. \qquad (6.110)$$

In the fully polarized limit, the finite contribution comes only from $\alpha = \uparrow$ since $w_h = (1 - \rho/2)/\rho_h = 0$ and $\rho_\downarrow = 0$. By putting $w_\uparrow = f^{-1} - 1$ and $\rho_\uparrow = 2f$, we obtain

$$S = -2 \sum_p [f \ln f + (1 - f) \ln(1 - f)], \qquad (6.111)$$

which gives the entropy of free fermions. Note that the factor 2 is absorbed by changing the summation variable to k. Thus we see that all thermodynamic quantities derived from $\Omega = U - TS$ give the free-fermion behavior.

6.6.4 Distribution functions at low temperature

We consider the zero-temperature limit where $\rho_\alpha(p)$ reduces to step functions. For example, we have $\rho_h = 2$ and $\rho_\sigma(p) = 0$ for $|p| < p_c$, where $\epsilon_h(p) < 0$. Figure 6.10 illustrates the distribution functions at zero temperature. We obtain the relation at $T = 0$,

$$\int_{-\pi/2}^{\pi/2} \frac{dp}{2\pi} \rho_h(p) = \frac{2p_c}{\pi} = 1 - n. \qquad (6.112)$$

Note that the halving of the Brillouin zone for holons is compensated by $\rho_h = 2$ in the semionic statistics in giving the hole density. With use of (6.104) we obtain the relation at $T = 0$

$$n = 1 - \sqrt{1 - \left(\frac{2}{\pi}\right)^2 \mu},$$ (6.113)

with $\mu = (\pi/2)^2 - p_c^2$. This is analogous to (4.152) or (6.103) for magnetization. At finite but low temperature T, (6.101) with $\alpha = h$ gives the distribution function

$$\rho_h(p) = \frac{2}{\sqrt{4 \exp[2\beta\epsilon_h(p)] + 1}},$$ (6.114)

which corresponds to a single-component system with the statistical parameter $1/2$ as discussed in Section 2.4.3. Thus holons at low temperature behave as single-component semions. Because of the relation $n \geq m$, we have $p_c < p_s$.

With finite magnetization, we obtain $\epsilon_\uparrow(p) < 0$ for $p_s < |p| < \pi/2$. In this momentum range we have $\rho_h = \rho_\downarrow = 0$ at $T = 0$. Then we obtain

$$\int_{-\pi/2}^{\pi/2} \frac{dp}{2\pi} \rho_\uparrow(p) = 1 - \frac{2p_s}{\pi} = m.$$ (6.115)

Using (4.152) we obtain the magnetization at $T = 0$

$$m = 1 - \sqrt{1 - \left(\frac{2}{\pi}\right)^2 h},$$ (6.116)

for $h < (\pi n/2)^2$. For larger h, we have $m = n$. Remarkably, the magnetization increases in exactly the same way for any hole concentration. The concentration n appears only in the value of saturation. This property is an example of the strong spin–charge separation, as will be discussed in more detail later.

With $|p|$ much beyond p_c, we may keep only $\rho_\uparrow$ at low temperatures, and neglect ρ_h and $\rho_\downarrow$ which remain exponentially small. For $|p|$ much smaller than p_s, on the other hand, we have to keep only ρ_h and neglect ρ_σ. Namely, the overlap of distribution functions of spinons and holons has an exponentially small factor $\exp(-T_{mix}/T)$. Here the characteristic temperature T_{mix} is given by

$$T_{mix} = 2(p_s^2 - p_c^2) = \frac{1}{2}\pi^2(2 - n - m)(n - m).$$ (6.117)

We have seen that holons and spinons at low temperature are separated in the rapidity space, and behave as independent particles. (6.101) gives

$$\rho_\uparrow(p) = \frac{2}{\sqrt{4\exp[2\beta\epsilon_\uparrow(p)] + 1}}, \tag{6.118}$$

which describes thermally excited spinons with $|p| > p_s$, and the condensate of up spinons with $|p| < p_s$. As in the case of holons, up spinons behave as single-component semions at low temperature.

6.6.5 Magnetic susceptibility

Given the thermodynamic potential one can derive the susceptibilities and the specific heat by appropriate thermodynamic derivatives. We shall first derive the magnetic susceptibility at low temperature. In deriving $\rho_\uparrow(p)$ near $p = p_s$, we can set $\rho_h(p) = \rho_\downarrow(p) = 0$. The magnetization m is given by integration of $\rho_\uparrow(p)$ and the differential susceptibility is derived as

$$\frac{\partial m}{\partial h} \equiv \chi_m(m) = \frac{2}{\pi^2(1-m)}\left[1 + \frac{2T^2}{3\pi^2(1-m)^4}\right] + O(T^3), \tag{6.119}$$

which agrees exactly with (4.170) in Chapter 4 up to $O(T^2)$. Accordingly, there is no logarithmic singularity in the temperature dependence for any charge density n. The independence of n is an example of the strong spin–charge separation [101, 121] at low T. If we take the limit of zero magnetic field first, i.e., $h/T \ll 1$, we have

$$\chi_m = \frac{2}{\pi^2}\left(1 + \frac{2}{\pi^2}T\right) + O(T^2), \tag{6.120}$$

which is again the same as that for the spin chain. The spin susceptibility χ_s is related to the magnetic susceptibility χ_m by $\chi_m = 4\chi_s$. For general temperature, it is necessary to include the mixing between spin and charge degrees of freedom. By using explicit values of $g_{\alpha\beta}$ given by (6.94) for fractional exclusion statistics, we obtain

$$w_\sigma = \frac{1}{\rho_\sigma}\left[1 - \frac{1}{2}(\rho_\uparrow + \rho_\downarrow - \rho_h)\right] \equiv \frac{1}{\rho_\sigma}\left[1 - \frac{1}{2}\tilde{\rho}\right], \tag{6.121}$$

$$w_h = \frac{1}{\rho_h}\left[1 - \frac{1}{2}(\rho_\uparrow + \rho_\downarrow + \rho_h)\right] = \frac{1}{\rho_h}\left[1 - \rho_h + \frac{1}{2}\tilde{\rho}\right], \tag{6.122}$$

omitting the obvious argument p. Complementary to $\rho_\alpha(p)$, we now introduce distribution functions $d_h(k)$ of fermionic holons and $d_\sigma(k)$ of bosonic spinons with a variable k by requiring $w_\sigma(p) = 1/d_\sigma(k)$ and

$w_h(p) = [1 - d_h(k)]/d_h(k)$. The variable k is specified by the relation $d_\alpha(k)$ $\mathrm{d}k = \rho_\alpha(p)\mathrm{d}p$. Then differentials $\mathrm{d}p$ and $\mathrm{d}k$ satisfy

$$\mathrm{d}p - \mathrm{d}k = \frac{1}{2}\tilde{\rho}(p)\mathrm{d}k = -\frac{1}{2}\tilde{d}(k)\mathrm{d}p. \tag{6.123}$$

In accordance with fermionic and bosonic distribution functions, we rewrite (6.101) as

$$\epsilon_\sigma(p) = T\ln[1 + w_\sigma(p)] + \frac{1}{2}\Omega_p, \tag{6.124}$$

$$\epsilon_h(p) = T\ln w_h(p) - \frac{1}{2}\Omega_p, \tag{6.125}$$

where

$$\Omega_p = -T\sum_\alpha \ln[1 + w_\alpha(p)^{-1}]. \tag{6.126}$$

Summation over p of Ω_p gives the thermodynamic potential Ω as shown by (6.105). The fermionic nature of holons and the bosonic nature of spinons in d_α become explicit in the form

$$d_h(k) = \left[\exp\beta\left(\epsilon_h + \frac{1}{2}\Omega_p\right) + 1\right]^{-1}, \tag{6.127}$$

$$d_\sigma(k) = \left[\exp\beta\left(\epsilon_\sigma - \frac{1}{2}\Omega_p\right) - 1\right]^{-1}, \tag{6.128}$$

which can be obtained from (6.124) and (6.125). It is obviously possible to regard d_α also as a function of p. We can actually solve for either d_α or ρ_α as follows:

$$d_\sigma = \rho_\sigma\left(1 - \frac{1}{2}\tilde{\rho}\right)^{-1}, \quad d_h = \rho_h\left(1 - \frac{1}{2}\tilde{\rho}\right)^{-1}, \tag{6.129}$$

$$\rho_\sigma = d_\sigma\left(1 + \frac{1}{2}\tilde{d}\right)^{-1}, \quad \rho_h = d_h\left(1 + \frac{1}{2}\tilde{d}\right)^{-1}, \tag{6.130}$$

with $\tilde{d} = d_\uparrow + d_\downarrow - d_h$.

It is now straightforward to derive the magnetic susceptibility for an arbitrary temperature. We obtain in the limit of $h \to 0$

$$\chi_m = \frac{\partial}{\partial h}\sum_{p\sigma}\sigma\rho_\sigma = 2\beta\sum_p \frac{d_\sigma(1 + d_\sigma)}{1 + d_\sigma - d_h/2} = 2\beta\sum_p \frac{\rho_\sigma(1 + \rho_h/2)}{1 - \tilde{\rho}/2}, \tag{6.131}$$

where d_σ actually does not depend on σ, and we have used (6.129). In the case of $d_h = 0$, (6.131) reduces to (4.172) in Chapter 4. Figure 6.11 shows

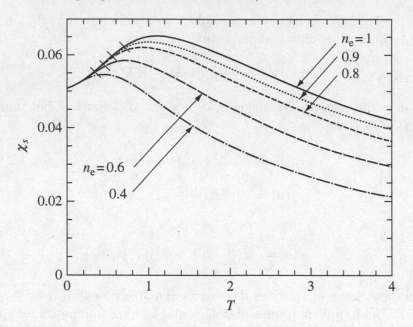

Fig. 6.11. Temperature dependence of the spin susceptibility for various electron concentrations $n = n_e$ [101]. The crosses represent the crossover temperature T_X above which the strong spin–charge separation breaks down. See (6.132) for the definition of T_X.

the numerical results for the spin susceptibility $\chi_s = \chi_m/4$ [101]. Here we have introduced a crossover temperature T_X such that

$$\rho_h(p_s) = 0.01, \qquad (6.132)$$

where the number 0.01 was just chosen as a typical magnitude to characterize the deviation from complete spin–charge separation. The temperatures T_X and T_{mix} defined by (6.117) are related by a numerical factor. At temperatures above T_X the charge degrees of freedom influence the magnetic susceptibility by more than an order of 1%. It is clear that χ_s is independent of $n = n_e$ at low temperature, consistent with (6.120).

6.6.6 Charge susceptibility

Now we turn to the charge susceptibility. The density n at low enough T is given by

$$1 - n = \int_0^{\pi/2} \frac{\mathrm{d}p}{\pi} \rho_h(p) \simeq \frac{1}{\pi} \int_{-\infty}^{\infty} \mathrm{d}\epsilon_c p(\epsilon_c) \left(-\frac{\partial \rho_h}{\partial \epsilon_c} \right), \qquad (6.133)$$

with $\epsilon_c = 2\epsilon_h(p)$. Here we have used the delta-function-like character of $-\partial\rho_h/\partial\epsilon_c$ in extending the range of integration. We make the expansion

$$p(\epsilon_c) \simeq p_c + \epsilon_c/p_c - \epsilon_c^2/(2p_c^3) + \epsilon_c^3/(2p_c^5). \tag{6.134}$$

Then in terms of quantities

$$I_n \equiv \int_{-\infty}^{\infty} d\epsilon_c \left(-\frac{\partial\rho_h}{\partial\epsilon_c}\right) \epsilon_c^n, \tag{6.135}$$

which is actually the same as appeared in (4.168) of Chapter 4; one can perform a low-T expansion of n. The charge susceptibility $\chi_c(n)$ is then given by

$$\frac{\partial n}{\partial\mu} \equiv \chi_c(n) = \frac{2}{\pi^2(1-n)} \left[1 + \frac{2T^2}{3\pi^2(1-n)^4}\right] + O(T^3), \tag{6.136}$$

which is independent of m. The presence of the $O(T^3)$ term in (6.136) results in a difference from the standard Sommerfeld expansion. It should be emphasized that $\chi_c(n)$ has precisely the same functional form as $\chi_m(m)$. That $\chi_c(n)$ is independent of m is another indication of the strong spin–charge separation.

At general temperature, we use bosonic and fermionic distribution functions d_α to derive the convenient formula for χ_c. We observe

$$\frac{\partial\Omega_p}{\partial\mu} = \left(1 + \frac{1}{2}\frac{\partial\Omega_p}{\partial\mu}\right) d_h - \frac{1}{2}\frac{\partial\Omega_p}{\partial\mu}\sum_\sigma d_\sigma = \frac{d_h}{1 + \tilde{d}/2} = \rho_h. \tag{6.137}$$

Then we obtain the derivatives

$$\frac{\partial d_h}{\partial\mu} = -\beta d_h(1 - d_h)\left(1 - \frac{1}{2}\rho_h\right), \tag{6.138}$$

$$\frac{\partial d_\sigma}{\partial\mu} = -\frac{\beta}{2}d_\sigma(1 + d_\sigma)\rho_h. \tag{6.139}$$

These results are substituted for derivatives appearing in

$$\chi_c = \frac{\partial}{\partial\mu}\sum_p \rho_h(p) = \frac{\partial}{\partial\mu}\sum_p \frac{d_h}{1 + \tilde{d}/2}. \tag{6.140}$$

Then we obtain the formula for the charge susceptibility

$$\chi_c = \beta\sum_p \rho_h \left\{(1 - d_h) + \frac{1}{4}\rho_h\left[\sum_\sigma \rho_\sigma(1 + d_\sigma) - \rho_h(1 - d_h)\right]\right\}. \tag{6.141}$$

We obtain ρ_α and d_α as functions of p by solving (6.101). Hence we can evaluate χ_c by integration over p. Figure 6.12 shows the charge susceptibility

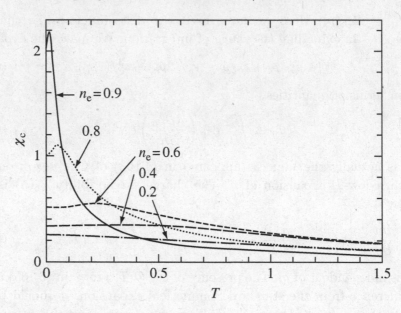

Fig. 6.12. Temperature dependence of the charge susceptibility for various electron concentrations $n = d_e$ [101].

obtained numerically [101]. Consistent with (6.136), χ_c tends to diverge in the high-density limit at $T = 0$.

6.6.7 Entropy and specific heat

At low temperatures, thermodynamics is determined only by the distribution functions near p_c and p_s. Hence they play the role of "Fermi momenta". The mutual statistical parameter is not important as long as $p_c \neq p_s$, which is the basis of the independent semion picture. Let us now consider the entropy S_{tJ} per site of the t–J model which consists of the charge part S_c, and the spin part S_σ for each spin. As in the case of a spin chain, we use the Sommerfeld-type expansion including the semionic distribution function. Then S_σ is not affected by the presence of holes in the low-temperature limit. We obtain $S_\sigma \simeq \pi T/(6p_s)$, as given in Chapter 4. The charge part S_c is given at low temperature by

$$S_c = \frac{\beta}{2\pi} \int_{-\infty}^{\infty} d\epsilon\, p(\epsilon)\epsilon \left(-\frac{\partial \rho_h}{\partial \epsilon} \right) \simeq \frac{\pi T}{6p_c}. \tag{6.142}$$

The total entropy per site is given by

$$S_{tJ} \simeq \frac{T}{3}\left(\frac{1}{1-n} + \frac{1}{1-m}\right) = \frac{\pi^2}{6}T\left(\chi_c + \chi_m\right), \qquad (6.143)$$

which also corresponds to the specific heat γT. The result describes a two-component (spin and charge) Tomonaga–Luttinger liquid.

As the temperature T of the system increases above the characteristic temperature T_{mix} or T_X, spin and charge contributions are no longer decoupled. The entropy can be derived from (6.99) with the solution of w_α from (6.101). To derive the specific heat C_N of the N-site system, one can use either the relation $C_N = T\partial S/\partial T)_n$ or more conveniently the following formula:

$$C_N = \left.\frac{\partial U}{\partial T}\right)_n = \left.\frac{\partial U}{\partial T}\right)_\mu + \left.\frac{\partial U}{\partial \mu}\right)_T \left.\frac{\partial \mu}{\partial T}\right)_n. \qquad (6.144)$$

The first term on the RHS is given by

$$\left.\frac{\partial U}{\partial T}\right)_\mu = \sum_\alpha \epsilon_\alpha \left.\frac{\partial \rho_\alpha}{\partial T}\right)_\mu. \qquad (6.145)$$

As in the case of susceptibility, it is convenient to use the distribution functions d_α in calculating the derivative. We obtain

$$\frac{1}{\rho_\alpha}\left.\frac{\partial \rho_\alpha}{\partial T}\right)_\mu = \frac{1}{d_\alpha}\left.\frac{\partial d_\alpha}{\partial T}\right)_\mu - \frac{1}{2+\tilde{d}}\left.\frac{\partial \tilde{d}}{\partial T}\right)_\mu. \qquad (6.146)$$

The derivatives of fermionic and bosonic distributions are given by

$$\left.\frac{\partial d_h}{\partial T}\right)_\mu = \beta d_h(1 - d_h)\left[\beta(\epsilon_h + \mu) - \frac{1}{2}S_p\right], \qquad (6.147)$$

$$\left.\frac{\partial d_\sigma}{\partial T}\right)_\mu = \beta d_\sigma(1 + d_\sigma)\left[\beta\epsilon_\sigma + \frac{1}{2}S_p\right], \qquad (6.148)$$

where $S_p = -\partial\Omega_p/\partial T$ is the contribution to entropy from p. On the other hand, to obtain the second term of (6.144) we observe

$$\left.\frac{\partial U}{\partial \mu}\right)_T = -Nn, \qquad (6.149)$$

$$\left.\frac{\partial \mu}{\partial T}\right)_n = -\sum_p \left.\frac{\partial \rho_h}{\partial T}\right)_\mu \Big/ \sum_p \left.\frac{\partial \rho_h}{\partial \mu}\right)_T = \frac{1}{\chi_c N}\sum_p \left.\frac{\partial \rho_h}{\partial T}\right)_\mu. \qquad (6.150)$$

In this way, it is possible to derive the specific heat once we have the distribution functions. Figure 6.13 shows the numerical results for the specific heat $C = C_N/N$ per site [101]. At low temperature, C becomes linear in T,

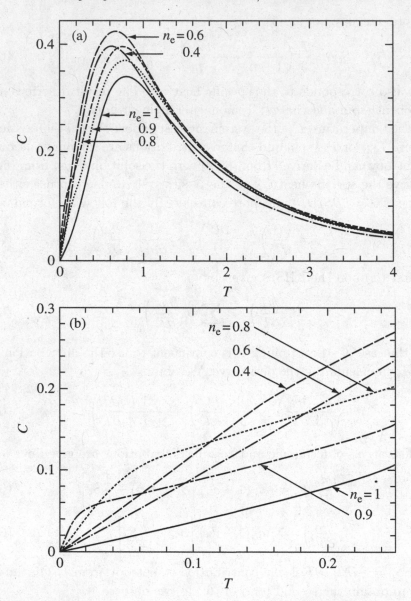

Fig. 6.13. Temperature dependence of the specific heat per site for various electron concentrations $n = n_e$ [101]: (a) overall behavior and (b) low-temperature behavior.

consistent with (6.99). The charge contribution has a larger slope as n comes closer to unity because of the factor $(1-n)^{-1}$. Figure 6.14 shows the decomposition of C into spin and charge contributions. It is remarkable that even at T larger than the hopping $t = 1$, the approximate picture of independent spinons and holons reproduces the exact results fairly well.

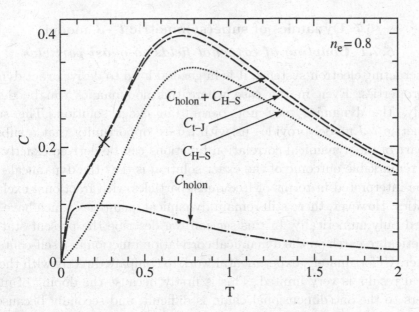

Fig. 6.14. Decomposition of the specific heat C_{t-J} into spin and charge contributions [101]. C_{H-S}: specific heat of the spin chain, C_{holon}: contribution of holons regarded as single-component semions. The cross denotes the crossover temperature T_X defined by (6.132).

In the limit of high T, we may neglect the p-dependence of $\rho_\alpha(p)$. Then for fixed density n_σ of each spin we recover the obvious result

$$S_{tJ} = -\sum_\sigma n_\sigma \ln n_\sigma - (1-n)\ln(1-n). \qquad (6.151)$$

We have shown that the thermodynamics of the supersymmetric $t–J$ model is described exactly in terms of free particles obeying the fractional exclusion statistics. At low T, the effects of mutual exclusion statistics become negligible, and the strong spin–charge separation emerges. At higher T, the multi-component character of fractional exclusion statistics appears explicitly, and the spin–charge separation breaks down. Thermodynamics for the nearest-neighbor SUSY $t–J$ model has been derived by the transfer matrix technique [168]. The temperature dependence derived is similar to the case of the $1/r^2$ model. However, the strong spin–charge separation as seen in the magnetic susceptibility is absent in the nearest-neighbor model.

6.7 Dynamics of supersymmetric $t−J$ model

6.7.1 Coupling of external fields to quasi-particles

In interacting electron systems, it is in general hard to derive exact dynamical properties. Even in the case where thermodynamics can be derived exactly, the dynamics may not permit an exact solution. The supersymmetric $t−J$ model provides us with a rare opportunity that a substantial part of the dynamical correlation functions can be derived exactly. The most remarkable outcome of the exact solution is that the dynamical result can be interpreted in terms of free quasi-particles with fractional exclusion statistics. However, there still remain dynamical quantities which have been derived only numerically. In this section, we describe the present status of theoretical research on the dynamical correlation function in a self-contained manner. Unfortunately, experimental work to compare directly with the theoretical results is very limited. This is firstly because the doping of mobile carriers to the one-dimensional chain is difficult, and secondly because the charge velocity is much larger than the spin velocity in real materials such as TTF-TCNQ [41] where the supersymmetry is strongly broken.

The quasi-particles are characterized by their intrinsic quantum numbers. Table 6.2 summarizes the spin and charge of each quasi-particle including non-interacting electrons and holes. Let us consider the coupling of external fields to quasi-particle excitations. We have already seen in Section 4.11.2 that the magnetic excitation spectrum in the spin chain is exhausted by two spinons. The two spinons provide the minimal coupling to external magnetic fields with integer angular momentum. In the presence of holes, the charge excitation may in principle be coupled. In reality, as long as the wave number does not exceed the Fermi wave number k_F, holons and antiholons are not excited by external magnetic fields, as shown in this chapter. This is another indication of the strong spin–charge separation, which has already been discussed in thermodynamics. Similarly, the minimal coupling of

Table 6.2. *Spin and charge of quasi-particles and their antiparticles.*

Quasi-particle species	Charge	Spin
electron	−1	1/2
hole	1	1/2
spinon	0	1/2
antispinon	0	1
holon	1	0
antiholon	−2	0

external charge is the two holons plus one antiholon. This combination gives the neutral total charge and zero spin. In addition to these excitations, spinons can in principle participate provided their total spin is zero. Actually the strong spin–charge separation holds again for small wave numbers $q < k_F$ where no spinons are excited by external fields.

The relevant variable for photoemission is the removal of an electron. The minimal coupling in this case is the one spinon plus one holon excitation which gives the right quantum numbers of spin and charge for an electron. In this case, however, there appear two spinon excitations accompanying the electron removal for any momentum. Because of this complexity, derivation of the electron removal spectrum still has to rely on numerical work, although the spinon–holon part has been derived analytically.

On the other hand, addition of an electron to the system is relevant to Bremsstrahlung spectroscopy. The process corresponds to the inverse of photoemission. In the electron addition spectrum, the minimal coupling is provided by the creation of an antiholon, a holon, and a spinon. This coupling in fact exhausts the whole spectrum. In deriving the response function, we have to use the physical momentum q for each quasi-particle. The spectrum of quasi-particles has been illustrated in Fig. 6.6. The easiest way to derive the spectrum is to use the YHWS state which can be mapped to eigenstates in the U(1,1) Sutherland model. Let us first consider the case without magnetic field for simplicity.

In the mapping to the U(1,1) Sutherland model, we now take $[-2\pi, 0]$ as the Brillouin zone as shown in Fig. 6.6(a). This choice gives positive momentum for spinons and antiholons, but negative momentum for holons. The size for the Brillouin zone is 2π for antiholons, but π for spinons and holons. In dynamics, however, one *cannot* freely shift the momentum of spinons and holons by π. Namely, the conservation of physical momentum requires that the shift π should be done simultaneously for a pair of semions. Thus holes occupy the lowest part with $2k_F < -q < 2\pi - 2k_F$, and magnons occupy the rest of the Brillouin zone. A hole-like excitation from the ground-state distribution of holes is the antiholon. The spinons are regarded as holes of magnons, and can be defined in the following momentum range: $0 < q < k_F$ which is called the right (R) side, and $\pi - k_F < q < \pi$ which is called the left (L) side. Note that R and L spinons keep each sign of the group velocity after the particle–hole reinterpretation:

$$-\epsilon_m(-2q) = 2\epsilon_s(q), \tag{6.152}$$

which has already been used in Section 6.4. The positive momentum of spinons has a corresponding negative momentum of magnons in the range

$[-2\pi, 0]$, as shown in Fig. 6.6. On the other hand, holons are regarded as particle excitations from the holon condensate and take negative values of q. The allowed momentum range of holons is given by sign reversal of that of spinons.

For convenience of reference, we summarize the spectra in terms of physical momentum q.

spinon: $\qquad\qquad\qquad \epsilon_s(q) = q(\pi - q), \qquad (0 < q < \pi),$

holon: $\qquad\qquad \epsilon_h(q) = (q + \pi/2)^2 - p_c^2, \quad (-\pi < q < -\pi(1 + n/2)$
$\qquad\qquad\qquad\qquad\qquad\qquad\qquad\qquad \text{and} -\pi(1 - n/2) < q < 0),$

antiholon: $\quad \epsilon_{ah}(q) = -(q - \pi)^2/2 + 2p_c^2, \quad (\pi n < q < \pi(2 - n)),$

where $p_c = \pi(1 - n)/2$, and we have taken $t = 1$ for the transfer energy. The spin velocity is given by $v_s = \pi$ without magnetization, and the charge velocity is given by $v_c = \pi(1 - n)$.

We now consider the case with finite magnetization m $(0 < m \leq n)$ per site. Then antispinons arise as antiparticles from the spinon condensate. We have obtained the spectrum in Section 6.4 as

$$\text{antispinon:} \quad \epsilon_{as}(q) = \tfrac{1}{2}(q + \pi)^2 - 2p_s^2, \quad (-2\pi < q < 0), \quad (6.153)$$

where $p_s = \pi(1 - m)/2$, and the allowed range is (R) $0 < -q < \pi m$, and (L) $\pi(2 - m) < -q < 2\pi$. The momentum range of spinons is accordingly reduced: (R) $\pi m/2 < q < k_F$, and (L) $\pi - k_F < q < \pi - \pi m/2$, to avoid the antispinon region. In this case, down spinons have the energy gap

$$2h = (\pi^2/2)m(2 - m), \qquad\qquad (6.154)$$

and the spin velocity is reduced to $v_m = \pi(1 - m)$. The holon spectrum is not affected by finite magnetization.

6.7.2 Dynamical spin structure factor

We begin with the magnetic response from the singlet ground state $|0\rangle$. We define the Fourier transform S_q^+ of the spin operator $S_l^+ = S_l^x + iS_l^y$ at site l as

$$S_q^+ = \frac{1}{\sqrt{N}} \sum_l S_l^+ \exp(-iql). \qquad\qquad (6.155)$$

The dynamical structure factor is given by

$$S(q, \omega) = \frac{1}{2} \sum_\nu |\langle \nu | S_{-q}^+ | 0 \rangle|^2 \delta(\omega - E_\nu + E_0), \qquad\qquad (6.156)$$

where $|\nu\rangle$ denotes an eigenstate of the Hamiltonian (6.1) with energy E_ν. As in the case of a spin chain, which is recovered in the limit of high electron density, $S(q,\omega)$ is equivalently obtained if we use $\sqrt{2}S_l^z$ instead of S_l^+ in the singlet ground state. In the polarized case, however, the transverse response with S_l^+ is different from the longitudinal one with $\sqrt{2}S_l^z$. In the limit of low electron density, the response should reduce to that of free electrons since the many-body effect does not operate in this limit.

In the hard-core boson representation with the completely up-polarized reference state, S_l^+ annihilates a magnon at site l. Given S_l^+ with $l = 0$ acting on the ground state Ψ_G with M magnons and Q holes, we obtain

$$\langle x_1, \ldots, x_{M-1}; y_1, \ldots, y_Q | S_0^+ \Psi_G \rangle$$
$$= \Psi_G(x_1, \ldots, x_{M-1}, x_M = 0; y_1, \ldots, y_Q). \tag{6.157}$$

Namely, the final states are represented by polynomials of $M − 1$ complex magnon coordinates $z_l = \exp(2\pi i x_l / N)$ with $l = 1, \ldots, M − 1$, and Q hole coordinates. Because of its polynomial form, the final state belongs to the YHWS.

It is instructive, and useful for later purposes, to provide an alternative proof of being the YHWS in terms of the Yangian generator introduced in Section 6.5.1. By definition a raising operator $\Lambda^{\uparrow\downarrow}$ of the Yangian, which has been defined by (6.84), annihilates the YHWS. Therefore, $S_i^+ \Psi_G$ belongs to YHWS provided the following holds:

$$\Lambda^{\uparrow\downarrow} S_i^+ \Psi_G = [\Lambda^{\uparrow\downarrow}, S_i^+] \Psi_G + S_i^+ \Lambda^{\uparrow\downarrow} \Psi_G = 0. \tag{6.158}$$

Here we already know that $\Lambda^{\uparrow\downarrow} \Psi_G = 0$. Furthermore, in the magnon–hole representation, we put $S_i^+ = b_i$ and derive the commutator as

$$[\Lambda^{\uparrow\downarrow}, b_i] = 2 \sum_{j(\neq i)} w_{ij} b_i b_j, \tag{6.159}$$

where the summand is antisymmetric against interchange of i and j. Then by using the same reasoning as discussed below (4.116) in Chapter 4, we obtain $[\Lambda^{\uparrow\downarrow}, b_i] \Psi_G = 0$. Hence, (6.158) indeed holds.

We derive the matrix element $c_\nu = \langle \nu | S_q^+ | 0 \rangle$ explicitly by using the coordinate representation. Let us represent the set of complex magnon coordinates by z^s, and the set for holes by z^h. Note that any state vector $|\Psi\rangle$ in the $t−J$ model can be expressed in terms of the wave function $\Psi(z^s, z^h)$ as

$$|\Psi\rangle = \sum_{z^s, z^h} \Psi(z^s, z^h) \prod_{i \in z^h} b_i^\dagger \prod_{j \in z^s} h_j^\dagger |F\rangle, \tag{6.160}$$

where $|F\rangle$ is the fully up-polarized state. Let $z_-^{\rm s}$ represent the complex coordinates of $M-1$ down-spin electrons after the spin-flip at $z_0 = 1$ in $|\Psi\rangle$. We can represent the resultant wave function $\Psi_{\rm flip}(z_-^{\rm s}, z^{\rm h}) = \langle z_-^{\rm s}, z^{\rm h}|S_j^+|0\rangle$ with $x_j = 0$ by

$$\Psi_{\rm flip} = z_0 \prod_{i=1}^{Q}(z_i^{\rm h} - z_0) \prod_{i=1}^{M-1}(z_i^{\rm s} - z_0)^2 \Psi_{\rm Gs-} = \sum_{\mu} b_\mu \Psi_\mu, \qquad (6.161)$$

where $\Psi_{\rm Gs-}$ is the ground state, and Ψ_μ is a general eigenfunction for the system with Q holes and $M-1$ magnons. Note that the state after the spin-flip is given explicitly as the polynomial factors multiplying $\Psi_{\rm Gs-}$. The factor $\prod(z_i^{\rm s} - z_0)^2$ represents two spinons both localized at z_0, and another factor $\prod(z_i^{\rm h} - z_0)$ represents a deficit of holes around z_0. In the momentum space, these factors represent elementary excitations. The actual excitation contents were first identified by a numerical study [72] of a system with a small size. The excitations are composed at most of two spinons, two holons, and an antiholon. Their spin and charge are summarized in Table 6.2.

It is evident from (6.161) that relevant eigenstates contributing to b_μ are given in terms of polynomials of z_i. These states belong to the Yangian highest-weight states (YHWS) related to the U(1,1) supersymmetry, and can be given in terms of U(1,1) Jack polynomials [11] written as $J_\kappa^{(+-)}(z)$. These polynomials are also eigenfunctions of the U(1,1) Sutherland model [99,193], and their basic mathematical properties are summarized in Section 7.5.

From the knowledge of $J_\kappa^{(+-)}(z)$, we can calculate most of b_μ exactly. Following the same procedure as that in the spinless Sutherland model [74,125], we obtain the analytic expressions of $S(q,\omega)$ in the thermodynamic limit. The formula for $S(q,\omega)$ depends on dispersion relations: $\epsilon_{\rm s}(q)$ of spinons, $\epsilon_{\rm h}(q)$ of holons, and $\epsilon_{\rm a}(q)$ of antiholons. Since the necessary mathematics is fairly complicated, we present in this section the results, and discuss their physical implications. The details of derivation are given in Section 6.8.2.

In terms of quasi-particles, $S(q,\omega)$ has three types of excitation contents: (i) two spinons; (ii) two spinons, two holons, and one antiholon; (iii) one spinon, one holon, and one electron. In type (iii), we obtain the spectrum as if an electron is contributing without spin–charge separation. We shall come back to this point later in discussing the electron addition spectrum. According to types from (i) to (iii), we divide $S(q,\omega)$ into three components:

$$S(q,\omega) = S_{2s}(q,\omega) + S_{2s2h\bar{h}}(q,\omega) + S_{she}(q,\omega). \qquad (6.162)$$

Each term on the RHS of (6.162) is further divided into subcomponents $S_X(q,\omega)$, where X specifies the excitation content of quasi-particles. For each X, an integral region, or the support, $D_X = D_X(\{q_i\})$ is defined in terms of the set $\{q_i\}_{i \in X}$ of momenta. We obtain

$$
S_X(q,\omega) = \int_{D_X} \prod_{i \in X} dq_i F_X(\{q_i\}) \delta\left(q + 2n\pi - \sum_{i \in X} q_i\right) \delta\left(\omega - \sum_{i \in X} \epsilon_i(q_i)\right),
$$
(6.163)

where n is an integer to bring q inside the Brillouin zone, and $F_X(\{q_i\})$ gives the intensity, and is given in terms of the square of the matrix element called the form factor.

Contribution from $2s_R$ or $2s_L$

For the two-spinon contribution of (6.162), the set X consists of $2s_R$ or $2s_L$, i.e., $S_{2s}(q,\omega) = S_{2s_R}(q,\omega) + S_{2s_L}(q,\omega)$.

The supports D_{2s_R} and D_{2s_L} are shown in Fig. 6.15. Due to the relation $S_{2s_L}(q,\omega) = S_{2s_R}(2\pi - q, \omega)$, we need only derive the formulae for $S_{2s_R}(q,\omega)$. The integral region is defined by $D_{2s_R}(q_{s1}, q_{s2}) = \{0 \leq q_{s1}, q_{s2} \leq k_F\}$. The corresponding squared form factor $F_{2s}(q_{s1}, q_{s2})$ is given by

$$
F_{2s}(q_1, q_2) = \frac{1}{4}|q_1 - q_2|^{2g_s} \prod_{i=1}^{2} \epsilon_s(q_i)^{g_s - 1},
$$
(6.164)

where $g_s = 1/2$ is the semionic statistical parameter of spinons. The form factor for the region L is the same as that given by (6.164) except for the allowed range which is now given by $D_{2s_L}(q_{s1}, q_{s2}) = \{\pi - k_F \leq q_{s1}, q_{s2} \leq \pi\}$.

It is remarkable that the two-spinon contribution (6.164) is the same as that for the Haldane–Shastry model [81, 198, 199]. Namely, the spectral weight in these integral regions does not depend on the electron density, provided the external momentum is less than the Fermi momentum. The

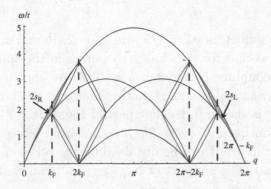

Fig. 6.15. The supports (shaded regions) for $S_{2s_R}(q,\omega)$ and $S_{2s_L}(q,\omega)$ for $n = 0.5$. Note that t is taken as 1 in the text.

only effect of holes is to reduce the integral regions $D_{2s_{R,L}}$. These phenomena of mutual independence of spin and charge dynamics have been referred to as the strong spin–charge separation [101, 121]. We shall show later that the dynamical charge structure factor has the corresponding spin–charge separation in that the intensity does not depend on magnetization.

Contribution from $2s_R 2h_R \bar{h}$, $2s_L 2h_L \bar{h}$, or $s_R s_L h_R h_L \bar{h}$

As the external momentum increases beyond the threshold $q = k_F$, charge excitations also enter the spectrum. The contribution from two spinons, two holons, and one antiholon is given by either $X = 2s_R 2h_R \bar{h}$, $2s_L 2h_L \bar{h}$, or $s_R s_L h_R h_L \bar{h}$. It is found that in all cases the squared form factor factorizes into the spin part F_{2s} and the charge part $F_{2h\bar{h}}$. For example, $F_{2s_R 2h_R \bar{h}}$ is given by

$$F_{2s_R 2h_R \bar{h}}(q_{s1}, q_{s2}; q_{h1}, q_{h2}; q_a) = F_{2s}(q_{s1}, q_{s2}) F_{2h\bar{h}}(q_{h1}, q_{h2}; q_a), \qquad (6.165)$$

where each momentum of spinons or holons runs inside the region R for each particle. Namely, $D_{2s_R 2h_R \bar{h}}$ is given by

$$\{0 < q_{s1} < q_{s2} < k_F, \ -k_F < q_{h1} < q_{h2} < 0, \ q_{s2} + q_{h2} > 0\}. \qquad (6.166)$$

The momentum of the antiholon runs over the full range $2k_F < q_a < 2\pi - 2k_F$, or equivalently $|q_a - \pi| < 2p_c$. The full range for the antiholon is always the case for any support D_X, and its specification may be omitted.

In the product $F_{2s} F_{2h\bar{h}}$, the analytic expression for the spin part F_{2s} is given by (6.164), and the charge part is given by

$$F_{2h\bar{h}}(q_1, q_2; q_a) = \frac{1}{2\pi^2} |q_1 - q_2|^{2g_h} \epsilon_a(q_a)^{g_a - 1} \prod_{i=1}^{2} \epsilon_h(q_i)^{g_h - 1} \left(q_i + \frac{q_a}{2} \right)^{-2},$$
$$(6.167)$$

with statistical parameters $g_h = 1/2$ and $g_a = 2$. Remarkably the function $F_{2h\bar{h}}$ is the same as that for the charge dynamics in the spinless Sutherland model with the coupling parameter $\lambda = 1/2$. As shown later, $F_{2h\bar{h}}$ also appears in the dynamical charge structure factor in the t–J model [6]. The support $D_{2s_R 2h_R \bar{h}}$ is shown in the left part of Fig. 6.16.

The contribution from $X = 2s_L 2h_L \bar{h}$ comes from left-going spinons and holons. Actually the result can be obtained by the relation $S_{2s_L 2h_L \bar{h}}(q, \omega) = S_{2s_R 2h_R \bar{h}}(2\pi - q, \omega)$. The support of $S_{2s_L 2h_L \bar{h}}(q, \omega)$ can also be obtained by making the replacement $q \to 2\pi - q$.

The other contribution $S_{s_R s_L h_R h_L \bar{h}}(q, \omega)$ has the zero energy threshold at $q = 2k_F$ and $2\pi - 2k_F$. The corresponding momentum of each quasi-particle

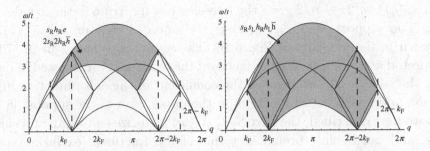

Fig. 6.16. (Left): The support of $S_{2s_R 2h_R \bar{h}}(q,\omega)$ and $S_{s_R h_R e}(q,\omega)$. (Right): The support of $S_{s_R s_L h_R h_L \bar{h}}(q,\omega)$ for $n = 0.5$.

is given with $2p_c = \pi - 2k_F$ by

$$q_{s1} = 0, \ q_{s2} = \pi, \ q_{h1} = -\pi + k_F, \ q_{h2} = -k_F, \ q_a = \pi \pm 2p_c, \qquad (6.168)$$

all of which have zero excitation energy. The momentum range for $s_R s_L h_R h_L \bar{h}$ can be obtained from the case $2s_R 2h_R \bar{h}$ by replacing one of the right spinons and one of the right holons in $S_{2s_R 2h_R \bar{h}}(q,\omega)$ with left ones, i.e.,

$$q_{s_L} = \pi - q_{s1}, \ q_{h_L} = -\pi - q_{h2}.$$

Then we can use the same form of $F_{2s} F_{2h\bar{h}}$ as that for the region R, except for the integral region. The relevant region $D_{s_R s_L h_R h_L \bar{h}}$ is shown in the right part of Fig. 6.16.

Contribution from $s_R h_R e$ or $s_L h_L e$

We proceed to discuss the remaining contribution $S_{she}(q,\omega)$, which consists of R and L components as

$$S_{she}(q,\omega) = S_{s_R h_R e}(q,\omega) + S_{s_L h_L e}(q,\omega)$$

with $S_{s_L h_L e}(q,\omega) = S_{s_R h_R e}(2\pi - q, \omega)$. It is therefore sufficient to derive the R component using the R branch of spinons and holons. The integral region for the R branch is defined by

$$D_{s_R h_R e}(q_s; q_h; q_a) = \{-k_F \le q_h \le 0, \ 0 \le q_s \le \pi - q_h\}, \qquad (6.169)$$

with the full range $|q_a - \pi| < 2p_c$ for antiholons as in other cases. The squared form factor $F_{she}(q_s; q_h; q_a)$ has been derived in the following form [10]:

$$F_{she}(q_s; q_h; q_a) = \frac{1}{2\pi} \frac{q_a/2 - q_s}{q_h + q_a/2} \epsilon_s(q_s)^{g_s - 1} \epsilon_h(q_h)^{g_h - 1} \sqrt{\frac{\epsilon_a(q_a)}{\epsilon_0(q_a)}}, \qquad (6.170)$$

where $\epsilon_0(q) = q(2\pi - q)/2$ gives the free-electron spectrum in the low-density limit. The support of $S_{s_{\mathrm{R}} h_{\mathrm{R}} e}(q, \omega)$ is the same as that of $S_{2s_{\mathrm{R}} 2h_{\mathrm{R}} \bar{h}}(q, \omega)$, as shown in the left part of Fig. 6.16. The square-root factor in (6.170) is interpreted as a renormalization factor of the electron. As discussed in detail later, the same factor appears as the coefficient of the delta function in the electron addition spectral function on the upper edge of the support [8, 11].

As we have described, the part $S_{2s2h\bar{h}}(q, \omega)$ factorizes into spin and charge parts, and can be interpreted in terms of the fractional exclusion statistics [79]. This part of the form factor and $S_{2s}(q, \omega)$ in the SUSY t–J model look very similar to those in the related continuum systems [103]. On the other hand, the part $S_{she}(q, \omega)$ does not factorize into the spin and charge parts, and cannot be interpreted by fractional exclusion statistics straightforwardly. Thus we prefer an alternative interpretation in terms of unfractionalized electrons. Namely, one spinon with spin (S) and charge (C), $(S, C) = (1/2, 0)$, one holon with $(0, +e)$, and one antiholon with $(0, -2e)$ combine to give one electron $(1/2, -e)$. However, this "electron" should not be regarded as an ordinary bound state since there is no binding energy, and the momenta of the holon and the spinon are fixed, i.e., $(q_{\mathrm{sR}}, q_{\mathrm{hR}}) = (0, 0)$ or $(q_{\mathrm{sL}}, q_{\mathrm{hL}}) = (\pi, \pi)$. This spinon–holon pair does not give the momentum shift but increases the energy by $\epsilon_{\mathrm{s}}(k_{\mathrm{F}})$ in the spectrum.

By rewriting a part of the form factor as

$$\frac{q_{\mathrm{a}}/2 - q_{\mathrm{s}}}{q_{\mathrm{h}} + q_{\mathrm{a}}/2} = \frac{(q_{\mathrm{a}}/2 - q_{\mathrm{s}})^{2g_{\mathrm{s}}}(q_{\mathrm{h}} + q_{\mathrm{a}}/2)^{2g_{\mathrm{h}}}}{(q_{\mathrm{h}} + q_{\mathrm{a}}/2)^2}, \tag{6.171}$$

with $g_{\mathrm{h}} = g_{\mathrm{s}} = 1/2$, we can interpret that a fractionalized spinon (q_{s}) or a holon $(-q_{\mathrm{h}})$ interact with each unfractionalized counterpart $(q_{\mathrm{a}}/2)$ making up the electron. In the electron addition spectrum as described later, the unfractionalized electron appears more explicitly as a delta-function peak.

We can obtain analytic results even for a finite size, although the complicated analytic expressions, to be given by (6.220) and (6.221), do not allow any intuition. Figure 6.17 shows the results of $S(q, \omega)$ for a large but finite-size system $(N = 60)$, which is due to Arikawa (unpublished), together with the lines that give thresholds of the spectrum.

From these analytic results, we can derive the behavior of the $S(q, \omega)$ near boundaries of the supports. There are divergent singularities at lower boundaries $\omega = \epsilon_{\mathrm{s}}(q)$ and $\omega = \epsilon_{\mathrm{s}}(2\pi - q)$ $(0 \leq q \leq k_{\mathrm{F}})$. At these boundaries $S(q, \omega)$ diverges by the power law with exponent $-1/2$. At other boundaries, $S(q, \omega)$ has threshold singularities but no divergence. Along the "electron" dispersion which appears in the lower edge of the spectrum in the left part of Fig. 6.16, the threshold singularity is large, as seen in the central region

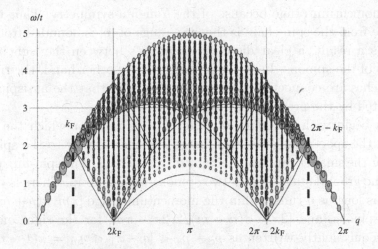

Fig. 6.17. The dynamical spin structure factor $S(q,\omega)$ for $N = 60$, $Q = 30$, and $M = 15$ with Fermi momentum $k_F = \pi/4$ ($n = 0.5$), calculated by Arikawa (unpublished). The intensity, which is derived analytically, is proportional to the area of each oval. The solid lines are determined by dispersion relations of the elementary excitations in the thermodynamic limit.

of $S(q,\omega)$ in Fig. 6.17. This feature clearly distinguishes the dynamics in the presence of holes from that in the spin chain.

6.7.3 Dynamical structure factor in magnetic fields

To understand the effects of magnetic field in the SUSY $t–J$ model, it is instructive to compare it with the dynamics for the spin chain which we have discussed in Chapter 4. As in the spin chain, we can derive $S^{-+}(q,\omega)$ analytically since all relevant excited states $|\nu\rangle$ contributing to $\langle\nu|S^+_{-q}|0\rangle$ belong to the YHWS. Other components $S^{+-}(q,\omega)$ and $S^{zz}(q,\omega)$ contain excited states out of the YHWS. Hence we can derive only a part belonging to the YHWS analytically. Derivation of the whole intensity needs numerical calculation.

In order to simplify the notation of the limiting values of the momentum for each quasi-particle, we use here the variable $p = q - \pi/2$ for spinons and holons, and $p = q - \pi$ for antiholons and antispinons. By this shift all spectra of quasi-particles become symmetric about $p = 0$. The variable p has already been used in discussing thermodynamics. In the presence of holes and magnetic field, the spinons can exist only for $p_c < |p| < p_s$, where $p_c = \pi(1 - n)/2$ and $p_s = \pi(1 - m)/2$ with m being the magnetization per site. As in the spin chain, up and down spinons have the same restriction for the momentum range because of the Yangian symmetry. Thus the only difference from the spin chain is that the range of $|p|$ is bounded from below by p_c. As a result, a clear distinction emerges between the right and left branches of the spinons. In the spin chain, on the other hand, the right and left branches are connected continuously. We note that the antispinons are not affected by the presence of holes.

Let us begin with the simplest case of $S^{-+}(q, \omega)$ for which the YHWS exhausts the spectrum. The form factor is rather similar to the spin chain except for the smaller range of momentum for spinons. The spin-flip operator S^+_{-q} on the ground state creates two spinons with spin up. There is no effect of holes as long as q runs within the momentum range $[\pi m, \pi(n + m)/2]$ for the left spinons, and $[2\pi - \pi(n + m)/2, 2\pi - \pi m]$ for the right ones. The range is equivalently written as $p_c + p_s < |q - \pi| < v_m = \pi(1 - m)$. The result of $S^{-+}(q, \omega)$ for this range is given by

$$S^{-+}(q, \omega) = \frac{\Theta\left(\epsilon_U(p) - \omega\right)\Theta\left(\omega - \epsilon_{L-}(p)\right)\Theta\left(\omega - \epsilon_{L+}(p)\right)}{2\sqrt{(\omega - \epsilon_{L-}(p))(\omega - \epsilon_{L+}(p))}}, \tag{6.172}$$

where $\Theta(\omega)$ is the step function, $p = q - \pi$, and

$$\epsilon_{L\pm}(p) = -p(p \mp q_m) = \epsilon_\uparrow(p \mp p_s), \tag{6.173}$$

$$\epsilon_U(p) = 2p_s^2 - p^2/2 = 2\epsilon_\uparrow(p/2). \tag{6.174}$$

In the high-density limit $(n = 1)$, this expression reduces to (4.214) for the spin chain. The upper and lower thresholds are determined by the spectrum of up spinons. Figure 6.18(a) shows the numerical results with $N = 16$ and $m = 0.25$ [156]. In addition to the two-spinon contribution derived above, there appear contributions $2h\bar{h}$ beyond the threshold $q = \pi(n + m)/2$. The difference from the zero-field case is the finite threshold $q = \pi m$ of the momentum below which there is no intensity.

We proceed to discuss the component $S^{zz}(q, \omega)$. Analytic results relying on the YHWS are exact only for small enough q that does not create charge excitations. In this range $S^{zz}(q, \omega)$ can be interpreted as the density response of magnons as in the spin chain. Then the spectrum agrees with the spinless Sutherland model with the coupling parameter $\lambda = 2$. The excitation contents in the range $0 < q < \pi(n-m)/2$ are simply given by $2s_R\bar{s}_R$. Beyond the threshold momentum, charge excitations with the excitation contents $2h_R\bar{h}$ or $2h_L\bar{h}$ also enter the spectrum. As a result, the support of $S^{zz}(q, \omega)$ becomes rather complicated. Figure 6.18(b) shows numerical results for $N = 16$ and $m = 0.25$ [156]. The upper and lower thresholds of the spectrum are easily identified. The upper threshold near $q = 0$ is determined by the spectrum

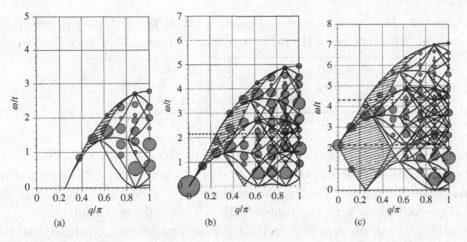

Fig. 6.18. Numerical results of dynamical structure factor: (a) $S^{-+}(q,\omega)$, (b) $S^{zz}(q,\omega)$, and (c) $S^{+-}(q,\omega)$ with $N = 16, Q = 2$, and $m = 0.25$ [156]. The intensity is proportional to the area of each circle. The solid lines are determined by the dispersion relations of the elementary excitations in the thermodynamic limit.

of the antispinon $\bar{s}_R$. The difference $q_m = \pi m$ of the threshold momentum from 2π is compensated by the two spinons with zero energy. On the other hand, the lower threshold near $q = 0$ is determined by the up spinon, whose finite momentum for the threshold is compensated by the zero-energy antispinon. As in the spin chain, the periodicity of the antispinon spectrum makes the difference from the dynamics of the Sutherland model.

Finally we discuss the component $S^{+-}(q,\omega)$. The action S^-_{-q} on the ground state corresponds to the creation of a down-spin magnon, i.e., an antispinon. Equivalently, S^-_{-q} annihilates two spinons with spin up in the condensate. The analytic expression of $S^{+-}(q,\omega)$ within the YHWS is analogous to the particle addition spectrum in the U(1,1) Sutherland model with the coupling parameter $\lambda = 1$. Namely, the maximum excitation contents are given by $2s\bar{s}_R\bar{s}_L 2h\bar{h}$, which has been found numerically [155]. Although the analytic expression for the whole spectrum is not available now, upper and lower thresholds for the support can be derived from the excitation contents. Figure 6.18(c) shows the numerical results of $S^{+-}(q,\omega)$ for $N = 16$ and $m = 0.25$ [156]. The upper and lower thresholds can be interpreted in terms of quasi-particles as in the spin chain. For example, the antispinon $\bar{s}_L$ takes zero energy at $q = q_m = \pi m$, and $\bar{s}_R$ takes zero energy at $q = 2\pi - q_m = \pi + 2p_s$. Hence the support for $S^{+-}(q,\omega)$ begins from zero energy at $q = \pm q_m$. To the right of $q = q_m$ the singularities of the spectrum

Table 6.3. *Maximum contents of quasi-particles for* $S^{\alpha\beta}(q,\omega)$ *in magnetic field. The subscripts* R *and* L *indicate the branches of antispinons.*

$S^{-+}(q,\omega)$	$2s + 2h + \bar{h}$
$S^{+-}(q,\omega)$	$2s + \bar{s}_R + \bar{s}_L + 2h + \bar{h}$
$S^{zz}(q,\omega)$	$2s + \bar{s}_R + 2h + \bar{h},\ 2s + \bar{s}_L + 2h + \bar{h}$

consist of the dispersion curves of $2s\bar{s}_R$. To the left of $q = q_m$ the lower threshold follows the dispersion of $\bar{s}_L$ down to $q = 0$, where the excitation energy $2h$ is required. The support for larger q is very complicated since the charge excitations with the contents $2h + \bar{h}$ enter the spectrum.

Table 6.3 summarizes the excitation contents relevant to the dynamical structure factor of the SUSY t–J model in a magnetic field. The effect of charge degrees of freedom can be seen by comparing it with the corresponding Table 4.1 in Chapter 4. The spinons and holons can take various combinations of R and L. Numerical results show that the number of spinons plus holons for each branch R and L should be even. For example, combinations such as $s_R s_L 2h_L \bar{h}$ are not allowed for $S^{-+}(q,\omega)$.

6.7.4 Dynamical charge structure factor

The charge correlation function is relevant to dielectric and elastic responses of real systems which have quasi-one-dimensional electrons as their active constituent. The density operator is defined by

$$n_q = N^{-1/2} \sum_{\sigma} \sum_{l=1}^{N} X_l^{\sigma\sigma} \exp(-iql), \qquad (6.175)$$

where the X-operators have been defined in (1.39). Because of the constraint of completeness, we can alternatively use

$$n_q = -N^{-1/2} \sum_{l=1}^{N} h_l^{\dagger} h_l \exp(-iql), \qquad (6.176)$$

as long as $q \neq 0$. The dynamical charge structure factor is defined by

$$N(q,\omega) = \sum_{\nu} |\langle \nu | n_{-q} | 0 \rangle|^2 \delta(\omega - E_\nu + E_0), \qquad (6.177)$$

where $|\nu\rangle$ denotes a normalized eigenstate of the Hamiltonian (6.1) with energy E_ν.

Analytic derivation of $N(q,\omega)$ is more difficult than the spin structure factor $S(q,\omega)$. The reason is that the state $n_q|0\rangle$ has a part outside the YHWS, and the polynomial wave functions are not sufficient. To see this, let us look at the commutator with the Yangian generator

$$[h_i^\dagger h_i, \Lambda^{\uparrow\downarrow}] = \sum_{j(\neq i)} w_{ij}(h_i^\dagger h_j - h_j^\dagger h_i)b_j, \qquad (6.178)$$

where the summand is no longer antisymmetric against an interchange of i and j because of the presence of b_j. Thus $h_i^\dagger h_i \Psi_G$ does not belong to the YHWS. However, if the external momentum q is sufficiently small, only charge excitations enter in the dynamical charge response. In this region the final state belongs to the YHWS and we can obtain the analytic solution. The first analytic solution for $N(q,\omega)$ with small q was obtained using the freezing trick [6]. In the following, however, we shall use mapping to the U(1,1) Sutherland model in line with the derivation of other dynamical quantities.

The form factor with excitations $2h + \bar{h}$ is precisely the same as the dynamical correlation function of the single-component Sutherland model with the repulsion parameter $g_h = 1/2$, which properly describes the semionic statistics of holons. One may wonder how the fermionic hole with $\lambda = 1$ acquires the semionic repulsion parameter without spin excitations. By taking the allowed states in the lattice, we have discussed in detail the origin of the semionic exclusion statistics in Section 6.6.1. Similar evolution of fractional exclusion statistics in the multi-component Sutherland model has been discussed [15, 197]. Namely, even though the magnons are not excited in the mapped U(1,1) Sutherland model, their presence in the ground state leads to a renormalization of the coupling parameter for hole excitations as $\lambda \to \lambda/(\lambda + 1)$. In the case of $\lambda = 1$, we obtain $g_h = \lambda/(\lambda + 1) = 1/2$ as the statistical parameter for holons. The mathematical structure leading to this situation will be explained later in Section 7.5.3.

The threshold momentum for spinon excitation is given by $q_{th} = \pi(n - m)/2$, which corresponds to the Fermi momentum of minority-spin electrons. Below the threshold, $N(q,\omega)$ is independent of the magnetization m. This independence corresponds to the strong spin–charge separation. We thus obtain for $q < q_{th}$:

$$N(q,\omega) = \frac{q^2}{\pi} \int_{\pi-2p_c}^{\pi+2p_c} dq' \int_{-\pi n/2}^{-\pi m/2} dq_1 \int_{-\pi n/2}^{-\pi m/2} dq_2\, \delta\left(q - q' - \sum_{i=1}^{2} q_i\right)$$

$$\times \delta\left(\omega - \epsilon_a(q') - \sum_{i=1}^{2} \epsilon_h(q_i)\right) F_{2h\bar{h}}(q_1, q_2; q'), \qquad (6.179)$$

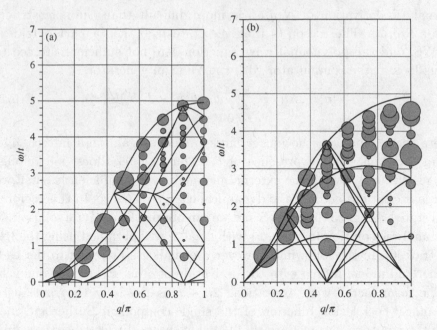

Fig. 6.19. Numerical results of dynamical charge structure factor $N(q,\omega)$ with $N = 16$ and (a) $Q = 2$, (b) $Q = 8$ [155]. The intensity is proportional to the area of each circle. The solid lines are determined by the dispersion relations of the elementary excitations in the thermodynamic limit.

with $q > 0$ and m being the magnetization. The squared form factor $F_{2h\bar{h}}$ is the same as the one given by (6.167).

It should be emphasized that (6.179) is valid only for such momentum q that is too small to excite spinons. In this range, it can be shown that $N(q,\omega)$ diverges as $[\epsilon_h(q - \pi n/2) - \omega]^{-1/2}$ as the frequency approaches the upper edge determined by the holon dispersion [6]. This behavior has been found in the corresponding quantity in the Sutherland model [138], which has been discussed in Section 2.6.3.

For $q > q_{\mathrm{th}}$, we have to rely on numerical calculation. Figure 6.19 shows numerical results for the dynamical charge structure factor [156]. In addition to two holons and an antiholon, spinons enter in the excitation spectrum for $q > k_{\mathrm{F}}$ in the following way:

$$2s_{\mathrm{R}} + 2h_{\mathrm{R}} + \bar{h}, \quad 2s_{\mathrm{L}} + 2h_{\mathrm{L}} + \bar{h}, \quad (s_{\mathrm{R}} + h_{\mathrm{R}}) + (s_{\mathrm{L}} + h_{\mathrm{L}}) + \bar{h}, \qquad (6.180)$$

where the spinons make the singlet and participate in the charge excitation.

6.7.5 Electron addition spectrum

The electron addition spectral function with the ground state $|0\rangle$ is defined by

$$A^+ (k,\omega) = \sum_\nu |\langle \nu; N_e + 1|c_{k\sigma}^\dagger|0; N_e\rangle|^2 \delta \left(\omega - E_\nu + E_0 + \mu\right), \qquad (6.181)$$

where N_e is the total electron number and μ the chemical potential, $c_{k\sigma}^\dagger$ is the electron creation operator with momentum k and spin σ, and $|\nu\rangle$ denotes an eigenstate of the Hamiltonian with energy E_ν. We choose $\sigma =\uparrow$ and work with the representation $c_{i\uparrow}^\dagger = h_i$. The excited states are exhausted by the YHWS since the wave function of the final state is given by a polynomial. Alternatively, the supersymmetric Yangian generator $\Lambda^{\uparrow 0}$ as given by (6.86) has the anticommutation property

$$\{h_i, \Lambda^{\uparrow 0}\} = 0, \qquad (6.182)$$

which leads to $\Lambda^{\uparrow 0} h_i |0\rangle = 0$, since $\Lambda^{\uparrow 0}|0\rangle = 0$. Thus we confirm that $h_i|0\rangle$ belongs to the YHWS.

Let us consider annihilation of a hole from the ground-state wave function $\Psi_{\mathrm{Gh+}}$ with $Q + 1$ holes and M magnons, the explicit form of which is given by (6.18) to (6.20). After the annihilation, the resultant wave function $\Psi_f(z^s, z^h) = \langle z^s, z^h|h_j|\Psi_{\mathrm{Gh+}}\rangle$ with Q holes and M magnons is written as

$$\Psi_f(z^s, z^h) = \prod_{i=1}^{Q}(z_0 - z_i^h) \prod_{i=1}^{M}(z_0 - z_i^s)\Psi_{\mathrm{G}}(z^s, z^h), \qquad (6.183)$$

where z_0 is the complex coordinate of the annihilated hole, and $\Psi_{\mathrm{G}}(z^s, z^h)$ is the wave function of the ground state for the system with Q holes and M magnons. Note that if $Q + M$ is odd with even N, the wave function $\Psi_{\mathrm{Gh+}}$ describes one of the doubly degenerate ground states. We have taken the one with total momentum π. Analogously, a proper choice can also be made in the case of $Q + M$ and N even. The excited state after the hole annihilation is given explicitly as the polynomial factors multiplying Ψ_{G}. Assuming the hole removed at the origin, $z_0 = 1$, we expand the polynomial factor as

$$\prod_{i=1}^{M+Q} (1 - z_i) = \sum_{m=0}^{M+Q} (-1)^m e_m(z), \qquad (6.184)$$

where z_i refers to either hole or magnon coordinate, and $e_m(z)$ is the elementary symmetric function of order m. It is defined by

$$e_m(z) = \sum_{i_1 < i_2 < \cdots < i_m \in I} z_{i_1} \cdots z_{i_m},$$

with the interval $I = [1, M + Q]$. Now the problem is reduced to rewriting $e_m(z)\Psi_G$ in terms of the U(1,1) Jack polynomials $J_\kappa^{(+-)}(z)$. As in the case of the dynamical structure factor, we present here the results in the thermodynamic limit, and relegate the mathematical details to Section 6.8.

The final results for $A^+(k, \omega)$ with $0 \leq k < 2\pi$ consist of the following components:

$$A^+(k, \omega) = A_{\mathrm{R}}(k, \omega) + A_{\mathrm{L}}(k, \omega) + A_{\mathrm{U}}(k, \omega), \tag{6.185}$$

where $A_{\mathrm{R}}(k, \omega)$ involves spinons and holons in the R branch, and $A_{\mathrm{L}}(k, \omega)$ involves them in the L branch. The third component $A_{\mathrm{U}}(k, \omega)$ includes an unfractionalized electron, as explained below. The contribution to $A_{\mathrm{R}}(k, \omega)$ can be understood in the following way. The added electron, or a removed hole, decomposes to a spinon and a holon, leaving an antiholon. According to the illustration in Fig. 6.6, the threshold momentum q_{th} is determined by an antiholon with momentum $\pi - 2p_{\mathrm{c}} = \pi n$, a right holon with momentum $-\pi/2 + p_{\mathrm{c}} = -\pi n/2$, and a right spinon with momentum 0, all of which have zero energy. Then we obtain

$$q_{\mathrm{th}} = \pi/2 - p_{\mathrm{c}} = \pi n/2 = k_{\mathrm{F}}, \tag{6.186}$$

which agrees with the threshold in the non-interacting Fermi gas.

The spectrum in the region R is given by

$$A_{\mathrm{R}}(k, \omega) = \int_{D_{\mathrm{R}}} dq_{\mathrm{s}} dq_{\mathrm{h}} dq_{\mathrm{a}} F_{\mathrm{R}}(q_{\mathrm{s}}, q_{\mathrm{h}}, q_{\mathrm{a}}) \delta\left(k - \sum_{i \in \mathrm{R}} q_i\right) \delta\left(\omega - \sum_{i \in \mathrm{R}} \epsilon_i(q_i)\right), \tag{6.187}$$

where D_{R} is the integral region defined by

$$D_{\mathrm{R}} = \{0 < q_{\mathrm{s}} < k_{\mathrm{F}}, \ -k_{\mathrm{F}} < q_{\mathrm{h}} < 0, \ |q_{\mathrm{a}} - \pi| < 2p_{\mathrm{c}}\}. \tag{6.188}$$

It is not allowed to shift q_{h} by π to bring it to the positive momentum region, although the periodicity of $\epsilon_{\mathrm{h}}(q)$ is π. This is because a spinon momentum and a holon one should be shifted together in this case. Note that the semionic particles appear only in pairs in physical excitations. In (6.187), the squared form factor $F_{\mathrm{R}}(q_{\mathrm{s}}, q_{\mathrm{h}}, q_{\mathrm{a}})$ is given by

$$F_{\mathrm{R}}(q_{\mathrm{s}}, q_{\mathrm{h}}, q_{\mathrm{a}}) = \frac{\epsilon_{\mathrm{s}}(q_{\mathrm{s}})^{g_{\mathrm{s}}-1} \epsilon_{\mathrm{h}}(q_{\mathrm{h}})^{g_{\mathrm{h}}-1} \epsilon_{\mathrm{a}}(q_{\mathrm{a}})^{g_{\mathrm{a}}-1}}{(q_{\mathrm{h}} + q_{\mathrm{a}}/2)^2}, \tag{6.189}$$

where $g_s = 1/2$, $g_h = 1/2$, and $g_a = 2$ correspond to statistical parameters of spinons, holons, and antiholons, respectively [72, 101]. Thus the matrix element in (6.187) can be interpreted in terms of the fractional exclusion statistics as in the case of correlation functions in related continuum systems [103]. The left-branch contribution is simply given by the relation $A_L(k, \omega) = A_R(2\pi - k, \omega)$ from (6.187).

The third component is given by

$$A_U(k, \omega) = \sqrt{\frac{\epsilon_a(k)}{\epsilon_0(k)}} \delta(\omega - \omega_{aU}(k)), \qquad (6.190)$$

which contributes only in the region $2k_F \leq k \leq 2\pi - 2k_F$. Here $\epsilon_0(k) \equiv k(\pi - k/2)$ describes the spectrum of non-interacting electrons, and $\omega_{aU}(k) \equiv \epsilon_s(k_F) + \epsilon_a(k)$. The contribution $A_U(k, \omega)$ comes from antiholons with fixed energies of spinons and holons. Such an entity may be regarded as an unfractionalized electron. A similar contribution with the delta-function peak also appears in the particle addition spectrum in the Sutherland model, as discussed in Section 2.6.2. The difference here is that $A_U(k, \omega)$ has the finite threshold for energy. This threshold is naturally understood by assuming that a spinon and a holon with fixed momenta $(q_{sR}, q_{hR}) = (0, 0)$ or $(q_{sL}, q_{hL}) = (\pi, \pi)$ are attached to the antiholon to make up an electron. In either case, the energy $\epsilon_s(k_F)$ is added in the spectrum of an antiholon, while no momentum is added. We note that the characteristic energy $\epsilon_s(k_F) = (\pi/2)^2 n(2 - n)$ corresponds to the $m = 0$ case of $T_{mix}/2$ defined by (6.117), which controls the spin–charge separation in thermodynamics.

In the dilute limit with $k_F \to 0$, both $\omega_{aU}(k)$ and $\epsilon_a(k)$ tend to $\epsilon_0(k)$. Hence the coefficient in (6.190) becomes unity, and $A_U(k, \omega)$ tends to the spectral intensity of non-interacting electrons. The other contributions $A_{R,L}(k, \omega)$ can be neglected in this limit.

Figure 6.20 shows the results of analytic calculation for $N = 60$, $Q = 29$, and $M = 15$ [8]. The spectral edges are shown by solid lines, and the spectral intensities for a finite system are shown by the area of ovals. The threshold behavior in $A_R(k, \omega)$ is summarized in Table 6.4.

6.7.6 Electron removal spectrum

The electron removal spectral function is defined by

$$A^-(k, \omega) = \sum_\nu |\langle \nu; N_e - 1|c_{k\sigma}|0; N_e \rangle|^2 \delta(\omega + E_\nu - E_0 + \mu), \qquad (6.191)$$

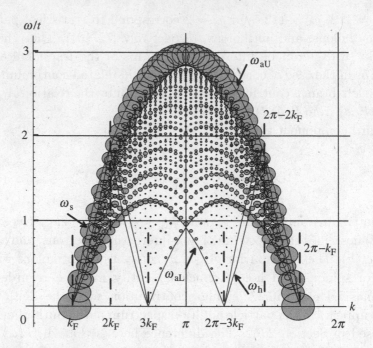

Fig. 6.20. The electron addition spectrum $A^+(k,\omega)$ in the case of $N = 60$, $Q = 29$, and $M = 15$ with Fermi momentum $k_{\rm F} = \pi/4$ [8]. The intensity is proportional to the area of each oval. The solid lines are determined by dispersion relations of the elementary excitations in the thermodynamic limit. The equations for the upper boundaries $\omega_{\rm aU}, \omega_{\rm s}$ and lower boundaries $\omega_{\rm aL}, \omega_{\rm h}$ are described in Table 6.4. Copyright (2001) by the American Physical Society. Reproduced with permission.

Table 6.4. *Threshold behaviors in $A^+(k,\omega)$ for $0 < k < \pi$. The results for*
$\pi < k < 2\pi$ can be obtained by the reflection symmetry
$A(k,\omega) = A(2\pi - k,\omega)$. Power-law behavior near the upper threshold is
indicated by $(-\Delta\omega)^\alpha$, while that near the lower threshold by $\Delta\omega^\alpha$. The step
function is represented by the exponent $\alpha = 0$.

Range of k	Threshold energy	Singularity
$[k_{\rm F}, 2k_{\rm F}]$	$\omega_{\rm s} = \epsilon_{\rm s}(k + k_{\rm F})$	$(-\Delta\omega)^{-1/2}$
$[2k_{\rm F}, \pi]$	$\omega_{\rm aU}(k)$	$(-\Delta\omega)^0$
$[k_{\rm F}, \pi]$	$\omega_{\rm aL} = \epsilon_{\rm a}(k + k_{\rm F})$	$\Delta\omega^0$
$[2k_{\rm F}, 3k_{\rm F}]$	$\omega_{\rm h} = \epsilon_{\rm h}(k - 2k_{\rm F})$	$\Delta\omega^{3/2}$

where the notation is the same as in the previous subsection. It is finite for negative ω. In the free electron gas, $A^-(k,\omega)$ is reduced to $\delta(\omega - \epsilon_0(k) + \mu)$, which probes the one-particle states below the Fermi level. In the presence of interactions, $A^-(k, \omega)$ is in general finite over a certain range of ω for each k. Experimentally the photoemission is the best way to measure $A^-(k, \omega)$.

By choosing $\sigma = \downarrow$, we make the replacement $c_{i\downarrow} \to h_i^\dagger b_i$. The electron removal spectrum is the most difficult quantity to derive analytically. The reason is that the final states do not belong to YHWS. This is seen by the commutator

$$[\Lambda^{\uparrow\downarrow}, h_i^\dagger b_i] = \sum_{j(\neq 0)} w_{ij}(h_i^\dagger + h_j^\dagger)b_j b_i, \qquad (6.192)$$

which does not vanish, nor annihilates Ψ_{G}. Moreover, in contrast with the case of $A^+(k, \omega)$, there is no region in the momentum space where the YHWS exhausts the spectrum. As a result, analytic study is limited to the YHWS contribution which constitutes a part of the spectrum.

6.7.6.1 Electron removal from spin chain

Let us first consider the spectrum from the high-density limit. It is easy to conclude that the YHWS is exhausted by one-spinon plus one-holon (1s1h) states; the charge neutrality excludes two or more holons in the absence of antiholons, and the full polarization in the YHWS constrains to a single spinon. Within this 1s1h sector, analytic solution has been obtained. The idea of derivation is as follows [104]. With the ground state Ψ_{G} with $M = N/2$ magnons and no holes, one can explicitly obtain the wave function

$$\langle z_-^{\mathrm{s}} | b_0 \Psi_{\mathrm{G}} \rangle = z_0 \prod_{i=1}^{M-1} (z_i - z_0)^2 \Psi_{\mathrm{Gs}-}(z_-^{\mathrm{s}}), \qquad (6.193)$$

where $\Psi_{\mathrm{Gs}-}$ is the ground state with $M-1$ magnons. The magnon annihilated by b_0 is located at $z = z_0$. The set z_-^{s} describes the $M-1$ magnon coordinates. Similarly, if an eigenfunction Ψ_{1h} is known for the mapped U(1,1) Sutherland model with $M-1$ magnons and one hole, one can obtain the wave function $\langle z_-^{\mathrm{s}} | h_0 \Psi_{1h} \rangle$ explicitly. Combining these known wave functions, one can derive

$$\langle \Psi_{1h} | c_{0\downarrow} | \Psi_{\mathrm{G}} \rangle = \sum_{z^{\mathrm{s}}} \langle \Psi_{1h} | h_0^\dagger | z_-^{\mathrm{s}} \rangle \langle z_-^{\mathrm{s}} | b_0 | \Psi_{\mathrm{G}} \rangle, \qquad (6.194)$$

by summation over z_-^{s}. The complete set of eigenfunctions Ψ_{1h} in the mapped Sutherland model is given in terms of Jack polynomials. We can then derive

$$\langle z_-^{\mathrm{s}} | h_0 | \Psi_{1h} \rangle = \Psi_{1h}(z^{\mathrm{h}} = 1, z_-^{\mathrm{s}}), \qquad (6.195)$$

in (6.194), and the RHS can be computed for an arbitrary size of system [25, 104]. The result in the thermodynamic limit is given by

$$
A_{1s1h}^{-}(k, \omega) = \frac{2}{\pi} \int_0^{\pi} dq_s \int_{-\pi}^{-q_s} dq_h \sqrt{\frac{\pi - q_s}{q_s}}
$$
$$
\times \left[\delta(k - q_s - q_h) + \delta(k - 2\pi - q_s - q_h) \right]
$$
$$
\times \delta(\omega + \epsilon_s(q_s) + \epsilon_h(q_h)), \tag{6.196}
$$

where $\epsilon_s(q) = q(\pi - q)$ for $q > 0$ and $\epsilon_h(q) = (q + \pi/2)^2$ for $q < 0$. The integration can be carried out to give [104]

$$
A_{1s1h}^{-}(k, \omega) = \frac{\Theta(-\omega - \epsilon_h(k - \pi)) \, \Theta(\epsilon_s(k) + \pi^2/4 + \omega)}{\pi k} \sqrt{\frac{\epsilon_h(k) + \omega}{-\omega - \epsilon_h(k - \pi)}},
$$
$$
\tag{6.197}
$$

for $0 < k < \pi$. The result for $\pi < k < 2\pi$ can be obtained by the reflection symmetry $A^{-}(k, \omega) = A^{-}(2\pi - k, \omega)$.

In addition to the contribution from the YHWS derived above, excitation contents involve three-spinon plus one-holon (3s1h) states [72]. Figure 6.21

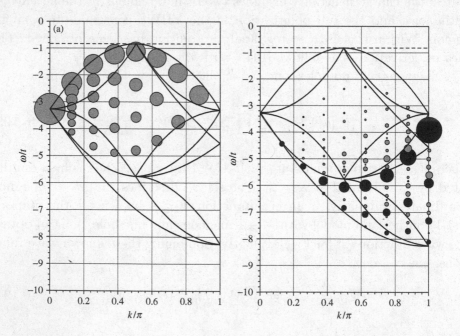

Fig. 6.21. Numerical results for the electron removal spectrum $A^{-}(k, \omega)$ from the high-density limit with $N = 16$ [155]. (a) One spinon plus one holon contribution, (b) three spinons plus one holon contribution. The intensity is proportional to the area of each circle. The solid lines are determined by dispersion relations of the elementary excitations in the thermodynamic limit.

shows the numerical result of $A^-(k,\omega)$ for 16 sites [155]. The chemical potential is given by $\mu^- = E_0(N_e) - E_0(N_e - 1)$ for the finite-size system. The region with finite intensity, i.e., support, is determined by the spectrum of spinons and holons as indicated in the thermodynamic limit.

For each momentum, the pole positions of 3s1h states with dominant intensity are indicated by the black circles in Fig. 6.21(b). Remarkably, the reversed energy $-\omega$ finds good correspondence with the two-spinon excitations in the Haldane–Shastry spin chain. Namely, the values of $-\omega - \epsilon_s(k_F)$, where $-\epsilon_s(k_F)$ gives the position of the single pole at $k = 0$, agree with the pole positions of $S(q,\omega)$ for the spin chain. These facts lead to the following interpretation: The 3s1h state with dominant intensity consists of two moving spinons together with another spinon with zero energy and a holon with maximal energy $\epsilon_s(k_F) = \epsilon_h(0)$. The 3s1h intensity is suppressed compared with the two-spinon intensity $S(q,\omega)$ in the spin chain, since addition of a further spinon and a holon reduces the matrix element. By combining the previous remark on the 1s1h contribution, we find that the dominant weight in $A^-(k,\omega)$ involves either two spinons with fixed momenta $(0,\pi)$, or a spinon–holon pair with fixed momenta $(0,\pi)$.

Electron removal with holes

In the presence of holes, $A^-(k,\omega)$ has a contribution from charge excitations. It is found by numerical study that the maximum excitation content is given by $(s_L, h_L) + \bar{h} + 2(s_R, h_R)$, plus the mirror states where R and L are interchanged [72]. However, no analytic results are available. We show numerical results for $N = 16$ sites with electron numbers $N_e = 14$ and $N_e = 6$ in Fig. 6.22(a) and (b), respectively [155]. The electron addition spectrum obtained in Section 6.7.5 is also shown in the positive energy region. The removal and addition spectra together give the spectral intensity of the single-particle Green function. The chemical potential for the finite-sized system is defined as $\mu^- = E_0(N_e) - E_0(N_e - 1)$ for the electron removal, and as $\mu^+ = E_0(N_e + 1) - E_0(N_e)$ for the electron addition. In the thermodynamic limit, μ^- and μ^+ should merge to μ, which is given by

$$\mu = -\frac{1}{12}\pi^2(3n^2 - 6n + 4). \qquad (6.198)$$

This is obtained by differentiating the ground-state energy E_0 with respect to N_e. The broken line of Fig. 6.22 indicates the value given by (6.198). The support, which was first given in [72], is shown by the solid lines.

As the hole density becomes larger, the main intensity accumulates along the upper boundary of the compact support. This boundary corresponds

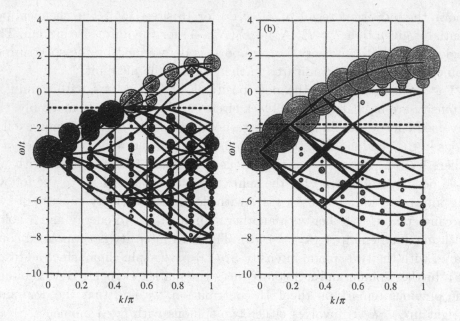

Fig. 6.22. Numerical results [155] for $A^-(k,\omega)$ for $\omega < \mu$ and $A^+(k,\omega)$ for $\omega > \mu$ with $N = 16$ and (a) $Q = 2$ ($n = 7/8$) or (b) $Q = 10$ ($n = 3/8$). The dashed line denotes the chemical potential μ, and the intensity is proportional to the area of each circle. The solid lines are determined by dispersion relations of the elementary excitations in the thermodynamic limit.

approximately to the energy band in the low-density limit $n \to 0$. The intensity of $A^+(k,\omega)$ increases with increasing ω at each k. On the other hand, the ω-dependence of $A^-(k,\omega)$ is more intricate because of the large number of elementary excitations involved.

6.7.7 Momentum distribution

By integrating the electron addition spectrum, we can derive the momentum distribution function $n_\sigma(k)$ of electrons with spin σ. Let us define the density matrix of electrons with spin σ, and that of holes in terms of X-operators by

$$\rho_\sigma(i-j) = \langle X_i^{\sigma 0} X_j^{0\sigma} \rangle, \quad \rho_{\mathrm{h}}(i-j) = \langle h_i^\dagger h_j \rangle, \tag{6.199}$$

with $h_i = X_i^{\uparrow 0}$. Then $n_\sigma(k)$ is given by the Fourier transform of $\rho_\sigma(i-j)$. The holes behave as hard-core fermions with the anticommutation rule: $\{h_i^\dagger, h_j\} = \delta_{ij}\left(X_i^{00} + X_i^{\uparrow\uparrow}\right)$. We thus obtain

$$\rho_\uparrow(i-j) + \rho_{\mathrm{h}}(i-j) = \delta_{ij}\left(1 - \langle X_i^{\downarrow\downarrow} \rangle\right). \tag{6.200}$$

In the singlet state we have $\langle X_i^{\downarrow\downarrow}\rangle = n/2$, and write $n(k) = n_\uparrow(k)$. Then we obtain

$$n(k) = 1 - n/2 - A^+(k), \tag{6.201}$$

where the Fourier transformed density matrix $A^+(k)$ of holes is given by integration of the electron addition spectral function. Namely, we have

$$A^+(k) = \int_0^\infty d\omega A^+(k,\omega)$$
$$= n_1(k) + n_1(2\pi - k) + n_2(k), \tag{6.202}$$

where the component $n_1(k)$ originates from $A_R(k,\omega)$, and is given by

$$n_1(k) = \int_{D_R} dq_s dq_h dq_a F_R(q_s, q_h, q_a)\delta\left(k - \sum_{i\in R} q_i\right), \tag{6.203}$$

where the integration range D_R has been defined in (6.188). In accordance with the support of $A_R(k,\omega)$, $n_1(k)$ is nonzero for $k_F < k < 2\pi - 2k_F$. The term $n_1(2\pi - k)$ in (6.201) originates from $A_L(k,\omega)$, and $n_2(k)$ from $A_U(k,\omega)$. We obtain from (6.190)

$$n_2(k) = \sqrt{\frac{\epsilon_a(k)}{\epsilon_0(k)}}\theta(k - 2k_F). \tag{6.204}$$

Since $n_1(k) = 0$ for $k < k_F$, $n(k)$ is a constant $1 - n/2$ below the Fermi wave number k_F. Obviously the distribution tends to that of non-interacting fermions as $n \to 0$. The distribution $n(k)$ is also given by the frequency integral of the electron removal spectrum $A^-(k,\omega)$. Hence (6.201) is regarded as the following sum rule [155]:

$$\int_{-\infty}^\infty d\omega\left[A^+(k,\omega) + A^-(k,\omega)\right] = 1 - \frac{1}{2}n, \tag{6.205}$$

which is independent of k.

Another form of the exact expression of $n(k)$ for Ψ_G was first derived by a diagrammatic perturbation theory [131]. The results presented above should correspond to the integral representation of the same results. Unfortunately, no direct proof is available that the perturbative result in terms of infinite series is equivalent to the integral representation presented above. Figure 6.23 shows the numerical results of momentum distributions [11]. The solid lines represent results obtained from the infinite series [131] by truncation, and the dots are obtained from $A^+(k)$ for a system with $N = 100$. The agreement of both results is excellent. This gives strong support for the equivalence of the integral representation presented above and the infinite series.

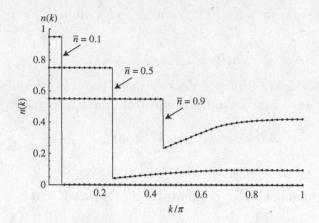

Fig. 6.23. The momentum distributions for electron densities $n = 0.1, 0.5$ and $n = 0.9$ [11]. The solid lines represent the results of perturbative theory [115,131] and the dotted ones those obtained from the integral representation. The actual numerics was performed for a finite system with $N = 100$.

The presence of a discontinuity $\sqrt{1-n}$ at $k = k_F$ reflects the fact that the SUSY t–J model has exponent $K_\rho = 1$ in terms of the Tomonaga–Luttinger theory [119]. In other words, unlike most one-dimensional fermion systems, the ground state of the SUSY t–J model can be adiabatically connected with the free-fermion ground state. However, in contrast with the Fermi liquid, the coefficient of the δ-function peak of $A_U(k, \omega)$ cannot be expressed simply in terms of the discontinuity in $n(k)$.

With a completely different method, which directly manipulates the wave function Ψ_G, a closed-form solution has been obtained in the real space [140, 141]. Although the derivation does not use any information on the excited states, the result is interpreted naturally in terms of elementary excitations.

6.8 *Derivation of dynamics for finite-sized t–J model

In this section we outline how to derive exact dynamics for the t–J model. Fortunately, the relevant states for dynamical response are mostly confined to a family called "separated states", which is specified in detail in Section 7.5.3. In this case, calculations of matrix elements and the norm of the wave function are factorized into the product of hole and spin parts. Each part can be evaluated on the basis of corresponding results for symmetric Jack polynomials. In the following, we outline the procedure to derive the electron addition spectrum and the dynamical structure factor by using the U(1,1) Jack polynomials.

6.8.1 Electron addition spectrum

In order to calculate $A^+(k,\omega)$ of (6.181) for a finite-sized (N) system, we take the singlet initial state with M magnons and Q holes, satisfying the condition $N = 2M+Q$. We consider the state (6.183) after hole annihilation. We use the binomial formula of $J_\kappa^{(+-)}(z)$ given in Section 7.5.3 for the case $(p,q) = (1,1)$ with the expansion coefficient $b_\kappa^{(+-)}(1,1)$. Since the relevant excited states belong to a special class of spin and charge excitations, called *separated states* [11], the binomial formula is factorized into the spin part and the charge part. Taking $z_0 = 1$, the wave function is expanded as

$$\prod_{i=1}^{M+Q-1} (1 - z_i)\Psi_{\mathrm{G}} = \sum_\kappa b_\kappa^{(+-)}(1,1) J_\kappa^{(+-)}(z)\Psi_{\mathrm{h}}(z), \qquad (6.206)$$

where z_i represents both hole and magnon coordinates, and

$$\Psi_{\mathrm{h}}(z) = \prod_{i<j}(z_i - z_j) \prod_i z_i^{-(2M+Q-1)/2} \qquad (6.207)$$

gives the ground state for $M+Q-1$ holes. The function $\Psi_{\mathrm{h}}(z)$ also generates the similarity transformation for the $U(1,1)$ Sutherland model. Namely, we take $\mathcal{O}(z) = |\Psi_{\mathrm{h}}(z)|$ in (3.163) with $\lambda = 1$.

In the expansion (6.206), each composition κ describes the momentum distributions of magnons and holes, which are subject to the Galilean shift. Namely, the ground state of M magnons and $Q-1$ holes corresponds to the composition

$$\kappa_i = \begin{cases} (M-1)/2, & (i=1,2,\ldots,Q-1), \\ M+Q-1-i, & (i=Q,Q+1,\ldots,M+Q-1), \end{cases} \qquad (6.208)$$

which is referred to as κ_{GS}. As seen from (6.208), the amount of the Galilean shift is $(M-1)/2$ per particle. The total momentum associated with a general excited state κ is given by $|\kappa| = |\kappa_{\mathrm{GS}}| + m$ with $|\kappa| = \sum_j \kappa_j$. The difference from the case in Section 7.5.3 is the nonzero value of κ_i for the bosonic momentum. The result $b_\kappa^{(+-)}(1,1)$, however, remains the same as that in Section 7.5.3 for separated states. Note that the binomial coefficient for symmetric Jack polynomials does not depend on the Galilean shift.

The momentum k in the electron addition spectrum $A^+(k,\omega)$ is given by

$$k = 2\pi \left[m + (M+1)/2\right]/N,$$

where the integer m describes the deviation from the Fermi momentum. The electron addition spectrum $A^+(k,\omega)$ is then expressed as

$$A^+(k,\omega) = Q\sum_{\kappa}{}' \delta(\omega - E_\kappa + E_0 - \kappa)|b_\kappa^{(+-)}|^2 \frac{\langle J_\kappa^{(+-)}\Psi_{\rm h}, J_\kappa^{(+-)}\Psi_{\rm h}\rangle_0}{\langle \Psi_{\rm G}, \Psi_{\rm G}\rangle_0}, \quad (6.209)$$

where the prime indicates restriction of the summation over κ by the momentum conservation. The norm $\langle \Psi_{\rm G}, \Psi_{\rm G}\rangle_0$ in the denominator can be obtained by a procedure described in Section 7.5.6.

In the spin part of $b_\kappa^{(+-)}$, relevant excited states are limited to those that are represented by

$$\kappa_{\rm ex}^{\rm s} = \kappa_{\rm GS}^{\rm s} + (1^{\kappa_{\rm s}}, 0^{M-\kappa_{\rm s}}). \quad (6.210)$$

Here $1^{\kappa_{\rm s}}$ means the sequence $1,\dots,1$ with the number of 1's being $\kappa_{\rm s}$. Namely, the spin part can be parameterized by the single variable $\kappa_{\rm s}$. In this case, the momenta of magnons and holes are not mixed in the expansion coefficient, hence the name of "separated states" [11]. Then $b_\kappa^{(+-)}$ is determined only by hole excitations.

The hole excitations are specified by

$$\kappa^{\rm h} = \kappa_{\rm GS}^{\rm s} + \nu = (1^{\kappa_{\rm a}-1}, 0^{Q-\kappa_{\rm a}}). \quad (6.211)$$

Correspondingly we obtain [11]

$$b_\kappa^{(+-)}(1,1) = \prod_{s\in D(\nu)} \frac{-a'(s) + 1 + \lambda'[l'(s) - 1]}{a(s) + 1 + \lambda' l(s)}, \quad (6.212)$$

where $\lambda' = \lambda/(\lambda + 1) = 1/2$ and $a(s), a'(s), l(s), l'(s)$ are the combinatorial quantities introduced in Section 7.1.6. The renormalization $\lambda \to \lambda'$ is due to the spin degrees of freedom in the background, which is discussed fully in Section 7.5.3.

The result (6.212) should be compared with (2.227) derived for the hole propagator in the Sutherland model. Namely, $b_\kappa^{(+-)}(1,1)$ takes the same form as the coefficient of binomial expansion in terms of symmetric Jack polynomials. Although $b_\kappa^{(+-)}(1,1)$ is independent of $\kappa_{\rm s}$, spin excitations enter $A^+(k,\omega)$ through the norm $\langle J_\kappa^{(+-)}\Psi_{\rm h}, J_\kappa^{(+-)}\Psi_{\rm h}\rangle_0$ of U(1,1) Jack polynomials in (6.209). Consequently, $A^+(k,\omega)$ is determined by the three parameters $\kappa_{\rm s}$, $\kappa_{\rm h}$, and $\kappa_{\rm a}$, which are related directly to the momenta of elementary excitations of spinons, holons, and antiholons, respectively.

In this way we obtain the finite-size version of (6.187):

$$A_{\rm R}(k,\omega) = \sum_{\kappa}{}' I_{\rm R}(\kappa)\delta(\omega - \Delta E_\kappa) \quad (6.213)$$

and similar results with the suffix R replaced by L and U. In (6.213) we have

$$I_R(\kappa) = \frac{1}{2\,(\Gamma(1/2))^2} \frac{\Gamma(\kappa_h - 1/2)\Gamma(\kappa_h + Q/2)}{\Gamma(\kappa_h)\Gamma(\kappa_h + (Q+1)/2)}$$

$$\times \frac{\kappa_a(Q - \kappa_a + 1)}{(2\kappa_h + \kappa_a - 1)(2\kappa_h + \kappa_a - 2)}$$

$$\times \frac{\Gamma(\kappa_s + 1/2)}{\Gamma(\kappa_s + 1)} \frac{\Gamma(N/2 - \kappa_s)}{\Gamma((N+1)/2 - \kappa_s)}, \tag{6.214}$$

with $Nk/(2\pi) = \kappa_h + \kappa_a + \kappa_s + (M-1)/2$. The Gamma functions appear in analogy with (2.233). The excitation energy is given by

$$\Delta E_\kappa = \frac{2\pi^2}{N^2}\Big[\kappa_s(N - 1 - 2\kappa_s)$$

$$+\kappa_h(2\kappa_h + Q - 2) + (\kappa_a - 1)(Q - \kappa_a)\Big]. \tag{6.215}$$

The maximum increase of the momentum is $m \leq M+Q$ by the elementary symmetric function $e_m(z)$, which appears in (6.184). In this increase, the sum of spinon and holon momenta in the positive direction is limited by $\kappa_s + \kappa_h \leq (M-1)/2$. The parameter κ_a varies from zero to Q. For the case where $\kappa_s > (M-1)/2$, we use the reflection symmetry about $k = \pi$. We then obtain $I_L(\kappa) = I_R(\kappa^-)$, where the components in $\kappa^- = (\kappa_s^-, \kappa_h^-, \kappa_a^-)$ are given by $\kappa_s^- = M - \kappa_s$, and by the relation

$$\kappa^h + (0^{Q-\kappa_a^-}, 1^{\kappa_a^- - 1}, \kappa_h^-) = \kappa_{GS}^h + (1^Q).$$

We obtain the allowed range $0 \leq \kappa_s^- + \kappa_h^- \leq (M+1)/2$.

The special case where $\kappa_s = (M-1)/2$ should be considered separately. In this case, there are no states with the same momentum that mix with κ_{ex}^s given by (6.210). This comes from the ordering and squeezing of the basis set as discussed in Section 2.1.3, and may be called the triangular structure. Hence $b_\kappa^{(+-)}$ should be unity, and $\kappa^h = \kappa_{GS}^h + (1^{\kappa_a}, 0^{Q-\kappa_a})$ with $1 \leq \kappa_a \leq Q$. We then obtain for this special case

$$I_U(\kappa) = \frac{\Gamma[(\kappa_a + 2)/2]\Gamma[(Q - \kappa_a + 2)/2]}{\Gamma[(\kappa_a + 1)/2]\Gamma[(Q - \kappa_a + 1)/2]}$$

$$\times \frac{\Gamma[(M + \kappa_a)/2]\Gamma[(M + Q - \kappa_a + 2)/2]}{\Gamma[(M + \kappa_a + 1)/2]\Gamma[(M + Q - \kappa_a + 2)/2]}, \tag{6.216}$$

where $Nk/(2\pi) = \kappa_a + M$. The excitation energy ΔE_κ is given by (6.215) in the case

$$(\kappa_s, \kappa_h, \kappa_a) = ((M-1)/2, 1, \kappa_a).$$

This means that the part of the addition spectrum is determined completely by an antiholon; the other constituents, spinon and holon, take a fixed momentum for each. As a result, (6.216) tends to the delta function in the thermodynamic limit.

In terms of I_R, I_L, and I_U, we obtain the finite-size version of $A^+(k,\omega)$. For $I_R(\kappa)$ and $I_L(\kappa)$, the thermodynamic limit is taken by essentially the same procedure as that in the Sutherland model. Details for the latter have been discussed in Section 2.7.1. Then we obtain the results described in Section 6.7.5.

6.8.2 Dynamical spin structure factor

In order to derive $S(q,\omega) = S^{-+}(q,\omega)/2$ for a finite size with $M = (N - Q)/2$ in the singlet ground state, we expand (6.161) in terms of U(1,1) Jack polynomials as

$$\Psi_{\text{flip}} = \sum_\kappa b_\kappa^{(+-)}(1,2) J_\kappa^{(+-)}(z)\Psi_{\text{h}}(z), \qquad (6.217)$$

where

$$\Psi_{\text{h}}(z) = \prod_{i<j}(z_i - z_j)\prod_i z_i^{-(2M+Q-2)/2} \qquad (6.218)$$

gives the ground state for $M + Q - 1$ holes, but with the total momentum different from (6.207). Since Ψ_{flip} contains $M - 1$ magnons and Q holes, the corresponding ground state is written as $\Psi_{\text{GS}-}$. The expansion coefficient $b_\kappa^{(+-)}(1,2)$ can be derived from the binomial formula in Section 7.5.3 for the case $(p,q) = (1,2)$.

We then obtain

$$2S(q,\omega) = M\sum_\kappa{}' \delta(\omega - \omega_\kappa)|b_\kappa^{(+-)}(1,2)|^2 \frac{\langle J_\kappa^{(+-)}\Psi_{\text{h}}, J_\kappa^{(+-)}\Psi_{\text{h}}\rangle_0}{\langle\Psi_{\text{GS}-}, \Psi_{\text{GS}-}\rangle_0}, \qquad (6.219)$$

where $\omega_\kappa = E_\kappa - E_0$ is the excitation energy. The expansion coefficient $b_\kappa^{(+-)}(1,2)$ can be derived easily if the momentum shift $|\kappa - \kappa_{\text{GS}}|$ is smaller than $Nk_F/(2\pi)$. Namely, $b_\kappa^{(+-)}$ factorizes into the hole part $b_{\kappa\text{h}}^{(+-)}$ and the magnon part $b_{\kappa\text{s}}^{(+-)}$ as $b_\kappa^{(+-)} = b_{\kappa\text{s}}^{(+-)}b_{\kappa\text{h}}^{(+-)}$. Hence we specify the excitation in terms of quantities ν_{h} and ν_{s} defined by $\kappa = \kappa_{\text{GS}} + (\nu_{\text{h}}, \nu_{\text{s}})$, and obtain the form analogous to (2.227) as

$$b_{\kappa h}^{(+-)}(1,2) = \prod_{s \in \nu_h} \frac{1 - a_h'(s) + \lambda'(l_h'(s) - 2)}{a_h(s) + 1 + \lambda' l_h(s)}, \tag{6.220}$$

$$b_{\kappa s}^{(+-)}(1,2) = \prod_{s \in \nu_s} \frac{2 - a_s'(s) + \lambda_+ l_s'(s)}{a_s(s) + 1 + \lambda_+ l_s(s)}. \tag{6.221}$$

Details of the derivation are given in Section 7.5.6.

If ν_h contains $s = (1,1)$, the coefficient $b_{\kappa h}^{(+-)}$ vanishes because we have

$$\lambda' = \lambda/(\lambda + 1) = 1/2, \quad a_s'(s) = l_s'(s) = 0.$$

Namely, the charge excitation does not enter the dynamical spin response in the small momentum region. On the other hand, if ν_s contains $s = (3,1)$, the coefficient $b_{\kappa s}^{(+-)}$ vanishes because $a_s'(s) = 2$ and $l_s'(s) = 0$. Therefore only such states contribute to $S(q,\omega)$ as are characterized by

$$\nu_h = 0, \quad \nu_s = (2^{\kappa_{s2}}, 1^{\kappa_{s1} - \kappa_{s2}}, 0^{M - \kappa_{s1} - 1}).$$

The parameters κ_{s1} and κ_{s2} ($\kappa_{s1} \geq \kappa_{s2}$) are related to the momenta of spinons. By using these parameters we can derive the expansion coefficient $b_\kappa^{(+-)}$ and norm explicitly. The analytic expression of $S(q,\omega)$ from this contribution is written in the form

$$S_{2s_R}(q,\omega) = \sum_\kappa I_{2s_R}(\kappa)\delta(\omega - \omega_\kappa), \tag{6.222}$$

where

$$I_{2s_R}(\kappa) = \frac{1}{4}\left(\kappa_{s1} - \kappa_{s2} + \frac{1}{2}\right)$$
$$\times \frac{\Gamma(\kappa_{s1} + 1)}{\Gamma(\kappa_{s1} + 3/2)} \frac{\Gamma(N/2 - \kappa_{s1} - 1/2)}{\Gamma(N/2 - \kappa_{s1})} \frac{\Gamma(\kappa_{s2} + 1/2)}{\Gamma(\kappa_{s2} + 1)} \frac{\Gamma(N/2 - \kappa_{s2})}{\Gamma(N/2 - \kappa_{s2} + 1/2)}, \tag{6.223}$$

$$\omega_\kappa = 2\left(\frac{\pi}{N}\right)^2 \left[N - 1 + (N-3)\kappa_{s1} - 2\kappa_{s1}^2 + (N-1)\kappa_{s2} - 2\kappa_{s2}^2\right]. \tag{6.224}$$

The momentum q is given by

$$q = \frac{2\pi}{N}(\kappa_{s1} + \kappa_{s2} + 1). \tag{6.225}$$

The analytic result obtained above is actually almost the same as that for $S(q, \omega)$ in the Haldane–Shastry spin chain. The only difference lies in the allowed range of momentum parameters: $(M - 3)/2 \geq \kappa_{s1} \geq \kappa_{s2} \geq 0$ instead of $N/2 - 1 \geq \kappa_{s1} \geq \kappa_{s2} \geq 0$ in the spin chain. The difference comes from the smaller k_F in the presence of holes. In the thermodynamic limit, this contribution gives $S_{2s_R}(q, \omega)$. Contributions from other regions of κ are more complicated, but can be derived analytically in the thermodynamic limit [10].

Part II
Mathematics related to $1/r^2$ systems

7

Jack polynomials

In this chapter, we discuss the mathematical properties of Jack polynomials. Since the content is a little intricate, we first explain the scope of the chapter, and the relationship between different kinds of Jack polynomials.

In deriving dynamics of the single-component Sutherland model, the symmetric Jack polynomial $J_\kappa(z)$ plays a fundamental role since each eigenfunction is a product of a Jack polynomial and a power of the Vandermonde determinant. Here $z = (z_1, \ldots, z_N)$ represent complex coordinates, and κ is a partition specifying a set of momenta for particles. The product of a Vandermonde determinant $\Delta(z)$ and a Jack polynomial is an antisymmetric polynomial, which is called an antisymmetric Jack polynomial $J_\kappa^{(-)}(z)$. Thus the fermionic eigenfunctions of the Sutherland model are constructed as the product of an antisymmetric Jack polynomial and an even power of the Vandermonde determinant. The Yangian highest-weight states (YHWS) of the Haldane–Shastry spin chain are also expressed in terms of the symmetric Jack polynomials with the particular value $\lambda = 2$ of the repulsion parameter.

In the multi-component Sutherland model, on the other hand, proper eigenfunctions must be symmetric or antisymmetric against exchange of coordinates with the same internal quantum number. Such eigenfunctions can be constructed from non-symmetric Jack polynomials $E_\eta(z)$, which do not have any symmetry against exchange of coordinates, but which are eigenfunctions of the Hamiltonian. Here η is a composition specifying a set of momenta for particles without ordering of magnitudes. For example, with the YHWS in the supersymmetric t–J model, eigenfunctions are constructed from Jack polynomials which are odd against exchange of hole coordinates, and even against exchange of magnon coordinates. These Jack polynomials are called U(1,1) Jack polynomials $J_\mu^{(+-)}(z)$, where U(1,1) refers to the supersymmetry between holes and magnons. It should be noted that the t–J

311

model itself has the SU(2,1) supersymmetry. The reduction to U(1,1) is analogous to the reduction from SU(2) to U(1) in the spin chain for describing the YHWS.

From a mathematical point of view, the non-symmetric Jack polynomial $E_\eta(z)$ is the most fundamental among a family of Jack polynomials, in the sense that every type of Jack polynomial can be constructed from $E_\eta(z)$ by antisymmetrization with respect to complex coordinates. Hence we organize the chapter beginning from the non-symmetric Jack polynomials. As a set-up, we define the composition and the Cherednik–Dunkl operators. Two kinds of inner-product, integral and combinatorial, between polynomials are introduced, and it is shown that the non-symmetric Jack polynomials form an orthogonal set of polynomials for both norms. We evaluate $E_\eta(z)$ for a special value $z = (1, \ldots, 1)$, which is useful in deriving matrix elements in dynamics. Then we introduce $J_\kappa^{(-)}(z)$ as an auxiliary object for deriving the norms of $E_\eta(z)$, and derive the norms explicitly. Following this analysis of non-symmetric Jack polynomials, we proceed to various symmetrized versions of Jack polynomials in succeeding sections. The chapter is self-contained, and is intended to be understandable without recourse to other references.

7.1 Non-symmetric Jack polynomials

7.1.1 Composition

A composition is any sequence

$$\eta = (\eta_1, \eta_2, \ldots, \eta_N) \tag{7.1}$$

of non-negative integers η_i for $i = 1, 2, \ldots, N$. The length $l(\eta)$ of a composition η is the maximum number i for nonzero η_i. The weight $|\eta|$ of a composition η is defined by $|\eta| = \sum_{i=1}^{N} \eta_i$. Let $\Lambda_N = \{\eta = (\eta_1, \eta_2, \ldots, \eta_N) \,|\, \eta_i \in \mathbf{Z}_{\geq 0}, 1 \leq i \leq N\}$ denote the set of all compositions with $l(\eta) \leq N$. A partition

$$\kappa = (\kappa, \kappa_2, \ldots, \kappa_N) \tag{7.2}$$

is a composition satisfying

$$\kappa_1 \geq \kappa_2 \geq \cdots \geq \kappa_N. \tag{7.3}$$

For example, $(2, 1, 0, 3)$ is a composition with length 3 but it is not a partition. $(3, 3, 2, 1, 1, 1, 0, 0, 0)$ is a partition (and hence a composition) with length 6. The set of all partitions with $l(\kappa) \leq N$ is written as

$$\Lambda_N^+ = \{\kappa = (\kappa_1, \kappa_2, \ldots, \kappa_N) \in \Lambda_N \,|\, \kappa_1 \geq \kappa_2 \geq \cdots \geq \kappa_N \geq 0\}.$$

(a)   (b)

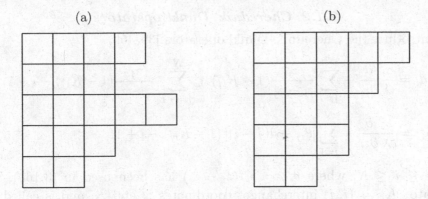

Fig. 7.1. Diagrams of (a) $\eta = (4, 3, 5, 3, 2)$ and (b) partition $\kappa = (5, 4, 3, 3, 2)$.

For a composition $\eta \in \Lambda_N$, we denote by η^+ the (unique) partition which is a rearrangement of the composition η. For example, for the composition $\eta = (2, 1, 0, 3)$, we obtain $\eta^+ = (3, 2, 1, 0)$.

It is convenient to define a diagram $D(\eta)$ of composition η, which is similar to a Young diagram. The diagram of a composition η consists of squares, the coordinates of which are (i, j), $1 \leq i \leq l(\eta)$ and $1 \leq j \leq \eta_i$. In drawing diagrams, the first coordinate i (the row index) increases as one goes downwards and the second coordinate j (the column index) increases from left to right. Figure 7.1(a) is an example of a composition derived from the partition shown in (b).

We define an order of partition and composition. The dominance order ($\leq$) on partitions is defined as follows: for κ, $\mu \in \Lambda_N^+$ satisfying $|\kappa| = |\mu|$, $\kappa \leq \mu$ if $|\kappa| = |\mu|$ and $\sum_{i=1}^{k} \kappa_i \leq \sum_{i=1}^{k} \mu_i$ for all $k = 1, \ldots, N$. $\kappa < \mu$ means that $\kappa \leq \mu$ and $\kappa \neq \mu$. For example, we obtain

$$(3, 0, 0) > (2, 1, 0) > (1, 1, 1).$$

The dominance order is a *partial order*; the order cannot necessarily be defined for any pair of partitions. For example, the dominance order between $(3,1,1,1)$ and $(2,2,2)$ cannot be defined.

We define a partial order $\prec$ on compositions as follows: for $\nu, \eta \in \Lambda_N$, $\nu \prec \eta$ if $\nu^+ < \eta^+$ with dominance ordering on partitions, or if $\nu^+ = \eta^+$ and $\sum_{i=1}^{k} \nu_i \leq \sum_{i=1}^{k} \eta_i$ for all $k = 1, \ldots, N$. For example, we obtain

$$(1, 1, 1) \prec (0, 2, 1) \prec (1, 2, 0) \prec (0, 3, 0). \tag{7.4}$$

7.1.2 Cherednik–Dunkl operators

We introduce the Cherednik–Dunkl operators [39, 45]

$$\hat{d}_i = \frac{z_i}{\lambda}\frac{\partial}{\partial z_i} + \sum_{j=1}^{i-1}\frac{z_i}{z_i - z_j}(1 - K_{ij}) + \sum_{j=i+1}^{N}\frac{z_j}{z_i - z_j}(1 - K_{ij}) - i + 1$$

$$= \frac{z_i}{\lambda}\frac{\partial}{\partial z_i} + \sum_{j=1}^{N}\left[\theta_{ij} - \theta(j - i)\right](1 - K_{ij}) - i + 1, \qquad (7.5)$$

for $1 \leq i \leq N$, where $\theta_{ij} = z_i/(z_i - z_j)$ has been used in (3.163). The operator $K_{ij} = (i,j)$ interchanges coordinates z_i and z_j, and is called the coordinate exchange operator [148]. The operator $\hat{d}_i$ is generally a mapping within homogeneous polynomials of $z = (z_1, \ldots, z_N)$. As shown below, all the operators $\{\hat{d}_i\}$ commute with each other, and hence can be diagonalized simultaneously. In terms of the Cherednik–Dunkl operators (7.5), the Hamiltonian $\mathcal{H}$ (3.16) can be written as

$$\mathcal{H} = \lambda^2 \sum_{i=1}^{N}\left(\hat{d}_i + \frac{N-1}{2}\right)^2 - \mathcal{E}_{0,N} \qquad (7.6)$$

with

$$\mathcal{E}_{0,N} = \lambda^2 \sum_{i=1}^{N}\left(\frac{N+1-2i}{2}\right)^2 = \frac{\lambda^2 N(N^2 - 1)}{12}. \qquad (7.7)$$

From this expression, we can see that joint eigenfunctions of $\hat{d}_i$ are eigenfunctions of the Hamiltonian. This is the reason why we introduce $\hat{d}_i$. The Cherednik–Dunkl operators take a slightly more compact form through a similarity transformation as given by (9.1) in Chapter 9.

Let us take monomials $z^\eta = z_1^{\eta_1}\cdots z_N^{\eta_N}$ as a basis of polynomials. For $\eta \in \Lambda_N$, the action of $\hat{d}_i$ on z^η is given by

$$\hat{d}_i(z^\eta) = \bar{\eta}_i z^\eta + \sum_{\nu \in \Lambda_N, \nu \prec \eta} v_{\eta\nu} z^\nu. \qquad (7.8)$$

Here the $v_{\eta\nu}$'s are constants, dependent on λ, and the diagonal elements $\bar{\eta}_i$ are given by

$$\bar{\eta}_i = \frac{\eta_i}{\lambda} - \left(k_i' + k_i''\right) \equiv \frac{\eta_i}{\lambda} - r_i. \qquad (7.9)$$

Here k_i' and k_i'' are numbers of η_l specified by

$$k_i' = \sharp\{l \in \{1, \ldots, i-1\}|\eta_l \geq \eta_i\}, \qquad (7.10)$$

$$k_i'' = \sharp\{l \in \{i+1, \ldots, N\}|\eta_l > \eta_i\}, \qquad (7.11)$$

where $\sharp$ denotes the number of members in the set. The sum $r_i = k_i' + k_i''$ indicates the ranking of the row in the partial order ($\prec$). Intuitively, $\lambda\bar{\eta}_i$ describes the rapidity which has the minimum separation of λ from neighboring ones. We call the property (7.8) *triangularity* because all matrix elements of $\hat{d}_i$ below (or above) the diagonal are zero in the ordered basis defined in the previous subsection.

Examples.

For $i = 1$, $N = 3$, direct calculations show that

$$\hat{d}_1(z_2^3) = -z_2^3 - z_1 z_2^2 - z_2^2 z_3,$$

$$\hat{d}_1(z_1 z_2^2) = \left(\frac{1}{\lambda} - 1\right) z_1 z_2^2 + z_2^2 z_3,$$

$$\hat{d}_1(z_2^2 z_3) = -2z_2^2 z_3 - z_1 z_2 z_3,$$

$$\hat{d}_1(z_1 z_2 z_3) = \frac{z_1 z_2 z_3}{\lambda}.$$

These results are written out in the following matrix form with ordered basis (7.4):

$$\hat{d}_1 \begin{pmatrix} z_2^3 \\ z_1 z_2^2 \\ z_2^2 z_3 \\ z_1 z_2 z_3 \end{pmatrix} = \begin{pmatrix} -1 & -1 & -1 & 0 \\ 0 & \frac{1}{\lambda} - 1 & 1 & 0 \\ 0 & 0 & -2 & -1 \\ 0 & 0 & 0 & \frac{1}{\lambda} \end{pmatrix} \begin{pmatrix} z_2^3 \\ z_1 z_2^2 \\ z_2^2 z_3 \\ z_1 z_2 z_3 \end{pmatrix}.$$

Proof of (7.8).

We introduce an operator

$$\hat{N}_{ij} = \frac{1}{z_i - z_j} (1 - K_{ij}) \tag{7.12}$$

for $i \neq j$. In terms of $\hat{N}_{ij}$, the operator $\hat{d}_i$ is rewritten as

$$\hat{d}_i = \frac{z_i}{\lambda} \frac{\partial}{\partial z_i} + \sum_{j=1}^{i-1} \hat{N}_{ij} z_j + \sum_{j=i+1}^{N} z_j \hat{N}_{ij}. \tag{7.13}$$

For positive integers (a, b), the actions of $\hat{N}_{ij}z_j$ and $z_j\hat{N}_{ij}$ on $z_i^a z_j^b$ are given, respectively, by

$$\hat{N}_{ij}z_j\left(z_i^a z_j^b\right) = \begin{cases} \sum_{l=1}^{a-b-1} z_i^{a-l}z_j^{b+l}, & \text{for} \quad a > b+1, \\ 0, & \text{for} \quad a = b+1, \\ -z_i^a z_j^a, & \text{for} \quad a = b, \\ -z_i^a z_j^b - \sum_{l=1}^{b-a} z_i^{a+l}z_j^{b-l}, & \text{for} \quad a < b \end{cases} \qquad (7.14)$$

and

$$z_j\hat{N}_{ij}\left(z_i^a z_j^b\right) = \begin{cases} \sum_{l=1}^{a-b} z_i^{a-l}z_j^{b+l}, & \text{for} \quad a > b, \\ 0, & \text{for} \quad a = b, \\ -z_i^a z_j^{a+1}, & \text{for} \quad a = b-1, \\ -z_i^a z_j^b - \sum_{l=1}^{b-a-1} z_i^{a+l}z_j^{b-l}, & \text{for} \quad a < b-1. \end{cases} \qquad (7.15)$$

First we consider the diagonal part of $\sum_{j=1}^{i-1} \hat{N}_{ij}z_j(z^\eta)$, which is given, from (7.14), by

$$(-1) \times \sharp\{l \in \{1, \ldots, i-1\}|\eta_l \geq \eta_i\} \qquad (7.16)$$

and that of $\sum_{j=i+1}^{N} z_j\hat{N}_{ij}(z^\eta)$, which is given, from (7.15), by

$$(-1) \times \sharp\{l \in \{i+1, \ldots, N\}|\eta_l > \eta_i\}. \qquad (7.17)$$

These results, combined with $z_i\partial z^\eta/\partial z_i = \eta_i z^\eta$, yield (7.9) for $\bar{\eta}_i$.

Next we consider the off-diagonal part of $\sum_{j=1}^{i-1} \hat{N}_{ij}z_j$. For composition η satisfying $\eta_i > \eta_j + 1$ $(j \in \{1, i-1\})$, we obtain

$$\hat{N}_{ij}z_j\left(z^\eta\right) = \sum_{l=1}^{\eta_i-\eta_j-1} \left(\cdots z_j^{\eta_j+l} \cdots z_i^{\eta_i-l} \cdots\right). \qquad (7.18)$$

Here the *triangularity* $\eta = (\cdots \eta_j \cdots \eta_i \cdots) > (\cdots \eta_j + l \cdots \eta_i - l \cdots)$ is obvious. Similarly, for η satisfying $\eta_i < \eta_j$ $(j \in \{1, i-1\})$, we see that

$$\hat{N}_{ij}z_j\left(z^\eta\right) = -\sum_{l=0}^{\eta_i-\eta_j-1} \left(\cdots z_j^{\eta_j-l} \cdots z_i^{\eta_i+l} \cdots\right) \qquad (7.19)$$

and $\eta = (\cdots \eta_j \cdots \eta_i \cdots) > (\cdots \eta_j - l \cdots \eta_i + l \cdots)$ are satisfied for $1 \leq l \leq \eta_i - \eta_j - 1$. The triangularity of $\sum_{j=i+1}^{N} z_j\hat{N}_{ij}$ can be shown in a similar way. From these results, we see that the action of $\hat{d}_i$ on z^η has triangularity. ∎

Now we discuss the mutual commutativity of the Cherednik–Dunkl operators. Prior to Cherednik [38], Dunkl originally introduced the following operators [45]:

$$\hat{T}_i = \frac{1}{\lambda}\frac{\partial}{\partial z_i} + \sum_{j(\neq i)} \hat{N}_{ij}, \quad i = 1, \ldots, N. \tag{7.20}$$

By a similarity transformation generated by $\mathcal{O} = \prod_{i<j} |z_i - z_j|^\lambda$, the Dunkl operator acquires another form:

$$\mathcal{O}\hat{T}_i\mathcal{O}^{-1} = \frac{1}{\lambda}\frac{\partial}{\partial z_i} - \sum_{j(\neq i)} \frac{1}{z_i - z_j}K_{ij} \equiv \pi_i. \tag{7.21}$$

A Hermitian operator similar to (7.21) has been utilized by Polychronakos [148], who derived the conserved quantities of the Calogero and Sutherland models with internal symmetry.

In order to prove the commutativity $[\hat{d}_i, \hat{d}_j] = 0$ of the Cherednik–Dunkl operators, we first prove that the Dunkl operators commute:

$$\left[\hat{T}_i, \hat{T}_j\right] = 0 = [\pi_i, \pi_j]. \tag{7.22}$$

To show this, we begin with the fundamental commutation property

$$\Delta_{ij} \equiv \sum_{l(\neq i)} \sum_{m(\neq j)} \left[\frac{1}{z_i - z_l}K_{il}, \frac{1}{z_j - z_m}K_{jm}\right] = 0, \tag{7.23}$$

which can be proved by straightforward calculation. Namely, with $i \neq j$, nonzero contributions in the summation of (7.23) come only from cases: (i) $l = m$, (ii) $l = j$, and (iii) $m = i$. In the case (i), we obtain by using the notation $q_{ij} = (z_i - z_j)^{-1}$,

$$[q_{il}K_{il}, q_{jl}K_{jl}] = q_{il}q_{ji}K_{il}K_{jl} - q_{jl}q_{ij}K_{jl}K_{il}. \tag{7.24}$$

Similarly we obtain for (ii) and (iii) the following results:

$$[q_{ij}K_{ij}, q_{jl}K_{jl}] = q_{ij}q_{il}K_{ij}K_{jl} - q_{jl}q_{il}K_{jl}K_{ij}, \tag{7.25}$$

$$[q_{il}K_{il}, q_{ji}K_{ji}] = q_{il}q_{jl}K_{il}K_{ji} - q_{ji}q_{jl}K_{ji}K_{il}, \tag{7.26}$$

where l denotes the remaining summation index which differs from i and j. Furthermore the product of K_{ij} has the following property:

$$K_{ij}K_{jl} = K_{jl}K_{il} \equiv K_{ijl} = K_{jli} = K_{lij}, \tag{7.27}$$

which means that only two kinds of K_{abc} appear in Δ_{ij} of (7.23). Using these results, we obtain

$$\Delta_{ij} = \sum_l (q_{ij}q_{il} + q_{ji}q_{jl} + q_{li}q_{lj})(K_{ijl} - K_{jil}), \tag{7.28}$$

which indeed vanishes because $q_{ij}q_{il} + q_{ji}q_{jl} + q_{li}q_{lj} = 0$ for $i \neq j \neq l \neq i$.

It is now easy to prove the commutativity of Dunkl operators. By introducing the notation $Q_i = \sum_{j\,(\neq i)} q_{ij}K_{ij}$, we obtain

$$\lambda\,[\pi_i, \pi_j] = \left[\frac{\partial}{\partial z_i}, Q_j\right] + \left[Q_i, \frac{\partial}{\partial z_j}\right] \tag{7.29}$$

$$= \left[\frac{\partial}{\partial z_i} + \frac{\partial}{\partial z_j}, q_{ji}K_{ji}\right] = 0, \tag{7.30}$$

where we have used $[Q_i, Q_j] = 0$ and $q_{ij}K_{ij} = -q_{ji}K_{ji}$. Obviously $\partial/\partial z_i + \partial/\partial z_j$ commutes with K_{ji}, and hence it also commutes with $q_{ji}K_{ji}$. As a result, we obtain (7.22).

In terms of (7.20), the Cherednik–Dunkl operators are expressed as

$$\hat{d}_i = z_i\hat{T}_i + \sum_{j=i+1}^{N} K_{ij} + 1 - N.$$

Algebraic properties of constituent operators are summarized as follows:

$$K_{ij}\hat{T}_j = \hat{T}_iK_{ij}, \quad K_{ij}\hat{T}_k = \hat{T}_kK_{ij}, \quad \text{for } k \neq \{i, j\}, \tag{7.31}$$

$$\left[\hat{T}_i, z_j\right] = \delta_{ij}\left(\frac{1}{\lambda} + \sum_{l(\neq i)} K_{il}\right) - (1 - \delta_{ij})K_{ij}. \tag{7.32}$$

From these relations, it is easy to derive the following relation:

$$\left[z_i\hat{T}_i, z_j\hat{T}_j\right] = -(z_i\hat{T}_i - z_j\hat{T}_j)K_{ij}. \tag{7.33}$$

The commutator of $\hat{d}_i$'s is then given by

$$\left[\hat{d}_i, \hat{d}_j\right] = \left[z_i\hat{T}_i, z_j\hat{T}_j\right] + \left[z_i\hat{T}_i, \sum_{k=j+1}^{N} K_{jk}\right] - \left[z_i\hat{T}_i, \sum_{l=i+1}^{N} K_{il}\right]. \tag{7.34}$$

We assume $i < j$ without loss of generality. The second term on the RHS then vanishes, and only $l = j$ in the third term on the RHS survives. Consequently, (7.34) becomes

$$\left[\hat{d}_i, \hat{d}_j\right] = \left[z_i\hat{T}_i, z_j\hat{T}_j\right] - \left[z_i\hat{T}_i, K_{ij}\right] = 0. \tag{7.35}$$

The last equality is due to (7.33).

Further, using (7.31) and (7.32), we obtain the commutation relation between $K_i (\equiv K_{ii+1})$ and $\hat{d}_j$. The algebraic relations between K_i and $\hat{d}_j$ are given by

$$\hat{d}_j K_i - K_i \hat{d}_j = 0, \qquad \text{for } j \neq i, i+1,$$
$$\hat{d}_i K_i - K_i \hat{d}_{i+1} = 1, \qquad (7.36)$$
$$\hat{d}_{i+1} K_i - K_i \hat{d}_i = -1.$$

The set of commutation relations (7.35) and (7.36) is called a degenerate affine Hecke algebra [38]. Alternatively, one may proceed in the opposite direction of logic; if one starts from the defining relations (7.36) for the degenerate affine Hecke algebra, a representation of the element $\hat{d}_j$ is explicitly constructed as the Cherednik–Dunkl operator defined by (7.5). One may then naturally ask about the explicit basis set composed of simultaneous eigenfunctions of $\hat{d}_j$ with $1 \leq j \leq N$. The basis set can be constructed in terms of non-symmetric Jack polynomials, as discussed in the following.

7.1.3 Definition of non-symmetric Jack polynomials

Now we define the non-symmetric Jack polynomial $E_\eta(z)$ as the homogeneous polynomial satisfying the following two conditions [127, 147]:

(i) $E_\eta(z)$ is a simultaneous eigenfunction of $\hat{d}_i$ for $1 \leq i \leq N$.
(ii) The polynomial $E_\eta(z)$ has the form

$$E_\eta(z) = z^\eta + \sum_{\substack{\nu \in \Lambda_N \\ \nu \prec \eta}} c_{\eta\nu} z^\nu, \qquad (7.37)$$

which is consistent with (7.8). The property (ii) fixes the normalization of non-symmetric Jack polynomials.

From this definition and (7.8), we can immediately see that

$$\hat{d}_i E_\eta = \bar{\eta}_i E_\eta \qquad (7.38)$$

for $i \in [1, N]$.

7.1.4 Orthogonality

In this subsection, we discuss the orthogonality of non-symmetric Jack polynomials with respect to two kinds of inner product.

Integral norm

For functions $f(z)$ and $g(z)$ in complex variables $z = (z_1, \ldots, z_N)$, we define the inner product $\langle \cdot, \cdot \rangle_0$ by the following formula:

$$\langle f, g \rangle_0 = \int \mathcal{D}z \overline{f(z)} g(z). \qquad (7.39)$$

Here $\int \mathcal{D}z$ denotes

$$\prod_{i=1}^{N} \oint_{|z_i|=1} \frac{\mathrm{d}z_i}{2\pi \mathrm{i} z_i} |\Delta(z)|^{2\lambda}, \tag{7.40}$$

where $\Delta(z) = \prod_{i<j}(z_i - z_j)$ is the Vandermonde determinant, and $\overline{f(z)}$ denotes the complex conjugation of $f(z)$, i.e., $\overline{f(z)} = f(z)^*$. The RHS of (7.39) is rewritten as

$$\text{C.T.} \left[f(z^{-1}) g(z) \prod_{i \neq j} \left(1 - \frac{z_i}{z_j} \right)^{\lambda} \right], \tag{7.41}$$

where the symbol C.T. denotes the constant term, and z^{-1} represents $(z_1^{-1}, \ldots, z_N^{-1})$. The non-symmetric Jack polynomials are orthogonal with respect to the inner product $\langle \cdot, \cdot \rangle_0$ [16]. Namely, we have

$$\langle E_\eta, E_\nu \rangle_0 = \delta_{\eta\nu} \langle E_\eta, E_\eta \rangle_0. \tag{7.42}$$

Combinatorial inner product

Define the polynomials $\{q_\eta(z)\}_{\eta \in \Lambda_N}$ by

$$\Omega(z|y) = \prod_{j=1}^{N} (1 - z_j y_j)^{-1} \prod_{i,j=1}^{N} (1 - z_i y_j)^{-\lambda} = \sum_{\eta \in \Lambda_N} q_\eta(z) y^\eta. \tag{7.43}$$

The combinatorial inner product $\langle \cdot, \cdot \rangle^{\mathrm{c}}$ for non-symmetric polynomials is defined by [47, 154]

$$\langle q_\nu, z^\eta \rangle^{\mathrm{c}} = \delta_{\nu\eta}. \tag{7.44}$$

Note that this inner product is different from the symmetric case written as $\langle \cdot, \cdot \rangle_{\mathrm{c}}$ in (2.176). Then we can show

$$\langle E_\eta, E_\nu \rangle^{\mathrm{c}} \propto \delta_{\eta\nu}. \tag{7.45}$$

Namely, the non-symmetric Jack polynomials are orthogonal with respect to the combinatorial inner product $\langle \cdot, \cdot \rangle^{\mathrm{c}}$ [38]. The proof of this orthogonality is a goal of the subsection. We also show that (7.44) is the non-symmetric generalization of the combinatorial norm for symmetric functions defined by (2.176).

The values for the norms $\langle E_\eta, E_\eta \rangle^{\mathrm{c}}$ and $\langle E_\eta, E_\eta \rangle_0$ will be derived, respectively, in Sections 7.2.4 and 7.2.2. The combinatorial norm is related to the matrix element of correlation functions, while the integral norm is related to the norm $\langle \Psi | \Psi \rangle$ of the state vector. In the following we show the orthogonality of E_η using the properties of Cherednik–Dunkl operators in the combinatorial and integral norms. We prove the orthogonality with respect to the combinatorial norm following a few steps [154].

For $z = (z_1, \ldots, z_N)$ and $y = (y_1, \ldots, y_N)$, $\Omega(z|y)$ in (7.43) has the following property:

$$\hat{d}_i^z \Omega(z|y) = \hat{d}_i^y \Omega(z|y), \qquad (7.46)$$

for $i = 1, \ldots, N$.

Proof of (7.46).
It suffices to show that $\hat{d}_i^z \Omega(z|y)$ is symmetric against an interchange of z_j and y_j for all j. Let $\mathcal{P}$ be such an exchange operator, giving $\mathcal{P}[f(z|y)] = f(y|z)$ for a general function $f(z|y)$. In the present case we have $\mathcal{P}[\hat{d}_i^z \Omega(z|y)] = \hat{d}_i^y \Omega(z|y)$. For the constituents of $\hat{d}_i^z$, we obtain

$$\frac{z_i \partial_i \Omega}{\lambda \Omega} = +\frac{z_i y_i}{\lambda(1 - z_i y_i)} + \sum_j \frac{z_i y_j}{1 - z_i y_j} = \frac{z_i y_i}{\lambda(1 - z_i y_i)} - N + \sum_j \frac{1}{1 - z_i y_j}, \qquad (7.47)$$

$$\frac{N_{ij}\Omega}{\Omega} = \frac{1}{z_i - z_j}\left[1 - \frac{(1 - z_i y_i)(1 - z_j y_j)}{(1 - z_i y_j)(1 - z_j y_i)}\right] = \frac{y_i - y_j}{(1 - z_i y_j)(1 - z_j y_i)}, \qquad (7.48)$$

where N_{ij} has been defined in (7.12). We call the function $f(z|y)$ $\mathcal{P}$-invariant if it satisfies $f(z|y) = f(y|z)$. Writing $f \equiv g$ to denote equivalence modulo $\mathcal{P}$-invariant terms, we combine (7.47) and (7.48) to obtain

$$\frac{\hat{d}_i^z \Omega}{\Omega} \equiv \sum_j \left[\frac{1 - \delta_{ij}}{1 - z_i y_j} + \frac{\theta(i - j)z_i(y_i - y_j)}{(1 - z_i y_j)(1 - z_j y_i)} + \frac{\theta(j - i)z_j(y_i - y_j)}{(1 - z_i y_j)(1 - z_j y_i)}\right] \qquad (7.49)$$

$$\equiv \sum_j \left[\frac{1 - \delta_{ij}}{1 - z_i y_j} + \frac{\theta(i - j)}{1 - z_j y_i} - \frac{\theta(j - i)}{1 - z_i y_j}\right] \qquad (7.50)$$

$$\equiv \sum_j \theta(i - j)\left(\frac{1}{1 - z_i y_j} + \frac{1}{1 - z_j y_i}\right). \qquad (7.51)$$

In going from (7.49) to (7.50), we have replaced the numerator using

$$z_i(y_i - y_j) \to -z_i y_j + 1 - 1 \to -z_i y_j + 1 \qquad (7.52)$$

because the terms $z_i y_i$ and -1 can be removed by the $\mathcal{P}$-invariance. The third term in (7.50) is obtained in a similar manner. The final form (7.51) is obviously $\mathcal{P}$-invariant. Hence $\hat{d}_i^z \Omega(z|y)$ is proven to be $\mathcal{P}$-invariant, and is equal to $\hat{d}_i^y \Omega(z|y)$. ∎

As a result of (7.46), we can write $\Omega(z|y)$ as

$$\Omega(z|y) = \sum_{\eta \in \Lambda_N} E_\eta(z) E_\eta(y)/g_\eta. \tag{7.53}$$

The result (7.53) is equivalent to g_η being the combinatorial norm, defined by

$$\langle E_\eta, E_\nu \rangle^c = \delta_{\eta\nu} g_\eta. \tag{7.54}$$

We note that (7.53) is an example of the Cauchy product expansion formula.

Derivation of (7.53) from (7.46).
From the form (7.37), there is a one-to-one correspondence between monomials z^η and E_η. The non-symmetric Jack polynomials E_η therefore form a basis of polynomials of z. When we regard $\Omega(z|y)$ as a polynomial of z, it can be expanded with respect to $E_\eta(z)$:

$$\Omega(z|y) = \sum_{\eta \in \Lambda_N} C_\eta(y) E_\eta(z), \tag{7.55}$$

where "coefficients" $C_\eta(y)$ are polynomials of y. From (7.46) and (7.55) we obtain

$$\sum_{\eta \in \Lambda_N} \hat{d}_j^y C_\eta(y) E_\eta(z) = \sum_{\eta \in \Lambda_N} C_\eta(y) \hat{d}_j^z E_\eta(z) = \sum_{\eta \in \Lambda_N} \bar{\eta}_j C_\eta(y) E_\eta(z). \tag{7.56}$$

From this result, we see that $C_\eta(y) \propto E_\eta(y)$. By setting $C_\eta(y) = g_\eta^{-1} E_\eta(y)$, we arrive at (7.53). ∎

Proof of equivalence between (7.53) and (7.54).
Since both z^η and $q_\eta(z)$ form a basis of polynomials of z, we expand $C_\eta(z)$ and $E_\eta(z)$ using these basis functions as

$$C_\eta(z) = \sum_{\mu \in \Lambda_N} a_{\eta\mu} z^\mu, \tag{7.57}$$

$$E_\eta(z) = \sum_{\gamma \in \Lambda_N} b_{\eta\gamma} q_\gamma(z). \tag{7.58}$$

From (7.44), we obtain

$$\langle C_\eta, E_\nu \rangle^c = \sum_{\gamma \in \Lambda_N} a_{\eta\gamma} b_{\nu\gamma}. \tag{7.59}$$

Then the relation $\langle C_\eta, E_\nu \rangle^c = \delta_{\eta\nu}$ means

$$\sum_{\gamma \in \Lambda_N} a_{\eta\gamma} b_{\nu\gamma} = \delta_{\eta\nu}. \tag{7.60}$$

Now the equivalence between (7.53) and (7.54) is translated into the equivalence between (7.60) and (7.55), which we shall show in two directions.

(i) $(7.60) \to (7.55)$

Let us introduce the (infinite-dimensional) matrices $A = \{a_{\mu\nu}\}$ and $B = \{b_{\mu\nu}\}$. Then (7.59) means $AB^{\mathrm{T}} = 1$, or the transpose of B is the inverse matrix of A. Hence we obtain $B^{\mathrm{T}}A = 1$, which gives the relation

$$\sum_{\gamma \in \Lambda_N} b_{\gamma\nu} a_{\gamma\eta} = \delta_{\eta\nu}. \tag{7.61}$$

With use of (7.61), we obtain

$$\sum_{\eta \in \Lambda_N} C_\eta(y) E_\eta(z) = \sum_{\eta,\mu,\gamma \in \Lambda_N} a_{\eta\mu} b_{\eta\gamma} q_\gamma(z) y^\mu = \sum_{\gamma \in \Lambda_N} q_\gamma(z) y^\gamma = \Omega(z|y). \tag{7.62}$$

(ii) $(7.55) \to (7.60)$

If (7.55) holds, we take the combinatorial inner product of z^μ and both sides of

$$\sum_{\eta \in \Lambda_N} C_\eta(y) E_\eta(z) = \sum_{\gamma \in \Lambda_N} q_\gamma(z) y^\gamma. \tag{7.63}$$

Then the orthogonality $\langle q_\gamma(z), z^\mu \rangle^{\mathrm{c}} = \delta_{\gamma\mu}$ leads to the relation

$$\sum_{\eta \in \Lambda_N} \langle E_\eta(z), z^\mu \rangle^{\mathrm{c}} \, C_\eta(y) = y^\mu, \tag{7.64}$$

which gives the inverse expansion of the first equality in (7.57). Hence the coefficient should satisfy

$$\langle E_\eta(z), z^\mu \rangle^{\mathrm{c}} = (A^{-1})_{\mu\eta}. \tag{7.65}$$

On the other hand, expansion (7.58) in terms of $q_\mu(z)$ leads to

$$\langle E_\eta(z), z^\mu \rangle^{\mathrm{c}} = b_{\eta\mu} = (B^{\mathrm{T}})_{\mu\eta}. \tag{7.66}$$

Since we have $(B^{\mathrm{T}})_{\mu\eta} = (A^{-1})_{\mu\eta}$ for all matrix elements, (7.60) follows. ∎

Let us turn to the integral norm. First we show that the Cherednik–Dunkl operators are self-adjoint with respect to the integral norm

$$\langle f, \hat{d}_j g \rangle_0 = \langle \hat{d}_j f, g \rangle_0. \tag{7.67}$$

Straightforward calculations show that

$$\left\langle f, z_i \frac{\partial g}{\partial z_i} \right\rangle_0 = \left\langle z_i \frac{\partial f}{\partial z_i}, g \right\rangle_0 - \lambda \sum_{j(\neq i)} \left\langle f, \frac{z_i + z_j}{z_i - z_j} g \right\rangle_0 \qquad (7.68)$$

and

$$\langle f, z_i \hat{N}_{ij} g \rangle_0 = \langle z_i \hat{N}_{ij} f, g \rangle_0 + \left\langle f, \frac{z_i + z_j}{z_i - z_j} g \right\rangle_0. \qquad (7.69)$$

In the second terms on the RHS in (7.68) and (7.69), $(z_i + z_j)g/(z_i - z_j)$ is not necessarily a polynomial. However, these two unwanted terms cancel with each other when we consider the action of the Cherednik–Dunkl operators on g. Using (7.68) and (7.69), we can readily show (7.67). The eigenfunctions of a self-adjoint operator with different eigenvalues are mutually orthogonal. Therefore, the non-symmetric Jack polynomials form an orthogonal basis with respect to the inner product given by (7.39).

7.1.5 Generating operators

To deduce mathematical properties of the non-symmetric Jack polynomials, we introduce operators that generate one non-symmetric Jack polynomial from another [114]. As a prerequisite, we introduce cyclic permutation $\hat{\tau}$ on the homogeneous polynomials of $z = (z_1, \ldots, z_N)$:

$$\hat{\tau} = K_{N-1} \cdots K_2 K_1,$$

with $K_i = K_{i,i+1}$. The operator $\hat{\tau}$ has the following properties:

$$z_i \hat{\tau} = \hat{\tau} z_{i+1}, \text{ for } i = 1, 2, \ldots, N-1, \qquad (7.70)$$

$$z_N \hat{\tau} = \hat{\tau} z_1, \qquad (7.71)$$

$$\partial_i \hat{\tau} = \hat{\tau} \partial_{i+1}, \text{ for } i = 1, 2, \ldots, N-1, \qquad (7.72)$$

$$\partial_N \hat{\tau} = \hat{\tau} \partial_1. \qquad (7.73)$$

Using $\hat{\tau}$, we introduce an operator Θ

$$\Theta = z_N \hat{\tau}. \qquad (7.74)$$

Namely, the action of Θ on an N-variable function $f(z_1, \ldots, z_N)$ is given by

$$(\Theta f)(z_1, \ldots, z_N) = z_N f(z_N, z_1, \ldots, z_{N-1}).$$

The operator Θ satisfies the following properties:

$$\hat{d}_i \Theta = \Theta \hat{d}_{i+1}, \text{ for } i = 1, 2, \ldots, N-1, \qquad (7.75)$$

$$\hat{d}_N \Theta = \Theta \left(\hat{d}_1 + 1 \right). \qquad (7.76)$$

These properties are proved as follows. First, from (7.73), it follows that

$$z_i \partial_i \Theta = \Theta z_{i+1} \partial_{i+1}, \text{ for } i = 1, 2, \ldots, N - 1, \tag{7.77}$$

$$z_N \partial_N \Theta = \Theta \left(z_1 \partial_1 + 1 \right). \tag{7.78}$$

We further obtain for $1 \le i \ne j < N$,

$$\begin{aligned} \hat{N}_{i,j} z_j \Theta &= \Theta \hat{N}_{i+1,j+1} z_{j+1}, \\ z_j \hat{N}_{i,j} \Theta &= \Theta z_{j+1} \hat{N}_{i+1,j+1}, \end{aligned} \tag{7.79}$$

while for $i = 1, 2, \ldots, N - 1$, we have

$$z_N \hat{N}_{i,N} \Theta = \Theta \hat{N}_{i+1,1} z_1, \tag{7.80}$$

$$\hat{N}_{N,i} z_i \Theta = \Theta z_{i+1} \hat{N}_{1,i+1}.$$

These results yield the algebraic relations (7.76) between $\hat{d}_i$ and Θ; the operator $\hat{d}_i$ consists of $z_i \partial / \partial z_i$, $z_i \hat{N}_{i,j}$, and $\hat{N}_{i,j} z_i$.

The operator Θ will turn out to be a generating operator of non-symmetric Jack polynomials, as shown later in this subsection. Further, we will show that the transposition K_i,

$$K_i f(z_1, \ldots, z_i, z_{i+1}, \ldots, z_N) = f(z_1, \ldots, z_{i+1}, z_i, \ldots, z_N), \tag{7.81}$$

is another generating operator of non-symmetric Jack polynomials. The algebraic relations between K_i and $\hat{d}_j$ have been given by (7.36).

So far we have defined Θ and K_i as operators on the N-variable function of $z_1, \ldots, z_N$. For convenience, we further define the actions on compositions $\eta \in \Lambda_N$ as

$$K_i \eta = (\eta_1, \ldots, \eta_{i+1}, \eta_i, \ldots, \eta_N), \tag{7.82}$$

$$\Theta \eta = (\eta_2, \ldots, \eta_N, \eta_1 + 1). \tag{7.83}$$

In the following, Θ and K_i are regarded as operators acting on both functions and compositions. An example of correspondence between η and $\Theta \eta$ is illustrated in Fig. 7.2.

Remarkably, the same form of relation holds among the eigenvalues $\bar{\eta}_i$ defined in (7.9) of Cherednik–Dunkl operators $\hat{d}_i$:

$$\overline{(\Theta \eta)}_i = \bar{\eta}_{i+1}, \tag{7.84}$$

$$\overline{(\Theta \eta)}_N = \bar{\eta}_1 + 1, \tag{7.85}$$

for $i = 1, \ldots, N - 1$.

Proof of (7.84) and (7.85).
First we show (7.84) by counting the numbers k_i' and k_i'' in (7.9). The reader is advised to consider an example shown in Fig. 7.2. Let $\sharp A$ denote the

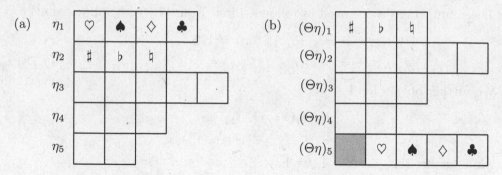

Fig. 7.2. Diagrams of (a) $\eta = (4, 3, 5, 3, 2)$ and (b) $\Theta\eta$. The cells marked by the same symbols such as $\heartsuit$, ..., $\sharp$, ... in (b), respectively, illustrate the mapping (7.120) and (7.121), to be presented later. The shaded cell in (b) represents the cell s'.

number of elements in a set A. For $i = 1, \ldots, N - 1$, we obtain

$$\sharp\{j \in \{1, \ldots, i - 1\}|\,(\Theta\eta)_j \geq (\Theta\eta)_i\} = \sharp\{j \in \{2, \ldots, i\}|\eta_j \geq \eta_{i+1}\}, \quad (7.86)$$

from the definition of $\Theta\eta$. Furthermore, we have

$$\sharp\{j \in \{i + 1, \ldots, N\}|\,(\Theta\eta)_j > (\Theta\eta)_i\}$$
$$= \sharp\{j \in \{i + 1, \ldots, N - 1\}|\,(\Theta\eta)_j > (\Theta\eta)_i\} + \sharp\{j = N|\,(\Theta\eta)_N > (\Theta\eta)_i\}$$
$$= \sharp\{j \in \{i + 2, \ldots, N\}|\eta_j > \eta_{i+1}\} + \begin{cases} 1, & (\text{for } \eta_1 + 1 > \eta_{i+1}), \\ 0, & (\text{otherwise}). \end{cases} \quad (7.87)$$

From (7.86) and (7.87), the relation (7.84) follows. Next we consider (7.85), which follows from the definition (7.83) and

$$\sharp\{j \in \{1, \ldots, N - 1\}|\,(\Theta\eta)_j \geq (\Theta\eta)_N\}$$
$$= \sharp\{j \in \{1, \ldots, N - 1\}|\eta_{j+1} \geq \eta_1 + 1\}$$
$$= \sharp\{j \in \{2, \ldots, N\}|\eta_j > \eta_1\}. \quad (7.88)$$

∎

We now show that Θ generates one non-symmetric Jack polynomial from another. Namely, the action of Θ on E_η is given by

$$\Theta E_\eta = E_{\Theta\eta}. \quad (7.89)$$

Proof of (7.89).
From (7.76), it follows that

$$\hat{d}_i\,(\Theta E_\eta) = \Theta\hat{d}_{i+1}E_\eta = \bar{\eta}_{i+1}\Theta E_\eta$$

for $i = 1, \ldots, N - 1$ and

$$\hat{d}_N (\Theta E_\eta) = \Theta \left(\hat{d}_1 + 1 \right) E_\eta = (\bar{\eta}_1 + 1) \Theta E_\eta.$$

These results show that ΘE_η is a simultaneous eigenfunction of $\hat{d}_i$ for $i = 1, \ldots, N$, i.e., $\Theta E_\eta \propto E_\nu$, where ν is the composition satisfying

$$\bar{\nu}_i = \bar{\eta}_{i+1} \text{ for } i = 1, \ldots, N - 1 \text{ and } \bar{\nu}_N = \bar{\eta}_1 + 1. \qquad (7.90)$$

Comparing (7.90) with (7.84) and (7.85), we see that $\nu = \Theta \eta$. Next we write the action of Θ on E_η in the following form:

$$\Theta E_\eta(z) = \Theta \left(z^\eta + \sum_{\substack{\nu \in \Lambda_N \\ \nu \prec \eta}} c_{\eta\nu} z^\nu \right) = z^{\Theta\eta} + \sum_{\substack{\nu \in \Lambda_N \\ \nu \prec \eta}} c_{\eta\nu} z^{\Theta\nu}.$$

The action of Θ on a monomial z^η is given by $\Theta(z^\eta) = z^{\Theta\eta}$, and obviously $z^{\Theta\nu} \neq z^{\Theta\eta}$ if $\nu \neq \eta$. Thus the coefficient of $z^{\Theta\eta}$ in ΘE_η is unity. We therefore obtain (7.89). ∎

Now we turn to the other generating operator K_i. The action of the transposition $K_i \equiv K_{ii+1} = (i, i+1)$ on E_η is given by [114, 154]

$$K_i E_\eta = \begin{cases} \xi_i E_\eta + \left(1 - \xi_i^2\right) E_{K_i\eta}, & \eta_i > \eta_{i+1}, \\ E_\eta, & \eta_i = \eta_{i+1}, \\ \xi_i E_\eta + E_{K_i\eta}, & \eta_i < \eta_{i+1}, \end{cases} \qquad (7.91)$$

where

$$\xi_i = 1/ \left(\bar{\eta}_i - \bar{\eta}_{i+1} \right). \qquad (7.92)$$

Proof of (7.91).
(A) The case of $\eta_i = \eta_{i+1}$.
Here we set $\tilde{E} = K_i E_\eta - E_\eta$. From the algebraic relation (7.36) between K_i and $\hat{d}_j$, we obtain

$$\begin{aligned} \hat{d}_j \tilde{E} &= \bar{\eta}_j \tilde{E}, \quad \text{for } j \neq i, i+1, \\ \hat{d}_i \tilde{E} &= \bar{\eta}_{i+1} \tilde{E} \\ \hat{d}_{i+1} \tilde{E} &= \bar{\eta}_i \tilde{E}. \end{aligned} \qquad (7.93)$$

From (7.93), it may seem that $\tilde{E}$ is a joint eigenfunction of $\hat{d}_j$ for $j = 1, \ldots, N$. However, there is no composition ν satisfying

$$\bar{\nu}_i = \bar{\eta}_{i+1}, \quad \bar{\nu}_{i+1} = \bar{\eta}_i, \quad \bar{\nu}_j = \bar{\eta}_j, \quad \text{for } j \neq i, i+1. \tag{7.94}$$

From this fact, we see that $\tilde{E} = 0$, i.e., $K_i E_\eta = E_\eta$. ∎

(B) The case of $\eta_i \neq \eta_{i+1}$.

In order to prove (7.91) for the case of $\eta_i \neq \eta_{i+1}$, we set

$$E' = (K_i + \xi_i) E_{K_i \eta}. \tag{7.95}$$

The function E' satisfies the following relation:

$$\hat{d}_j E' = \bar{\eta}_j E' \tag{7.96}$$

for $j = 1, \ldots, N$. We show (7.96) for (i) $j \neq i, i+1$, (ii) $j = i$, and (iii) $j = i+1$, separately.

(i) For $j \neq i, i+1$, we obtain from (7.36)

$$\begin{aligned}
\hat{d}_j E' &= \hat{d}_j \left(K_i + \xi_i \right) E_{K_i \eta} = \left(K_i + \xi_i \right) \hat{d}_j E_{K_i \eta} \\
&= \overline{(K_i \eta)}_j E' = \bar{\eta}_j E'.
\end{aligned} \tag{7.97}$$

(ii) For $j = i$, we obtain

$$\begin{aligned}
\hat{d}_i E' &= \hat{d}_i \left(K_i + \xi_i \right) E_{K_i \eta} \\
&= \left(K_i \hat{d}_{i+1} + \xi_i \hat{d}_i + 1 \right) E_{K_i \eta} \\
&= \overline{(K_i \eta)}_{i+1} K_i E_{K_i \eta} + \left(\xi_i \overline{(K_i \eta)}_i + 1 \right) E_{K_i \eta} \\
&= \bar{\eta}_i K_i E_{K_i \eta} + \left(\xi_i \bar{\eta}_{i+1} + 1 \right) E_{K_i \eta} \\
&= \bar{\eta}_i E',
\end{aligned} \tag{7.98}$$

where the last equality follows from the relation $\left(\xi_i \bar{\eta}_{i+1} + 1 \right) = \bar{\eta}_i \xi_i$.

(iii) Similarly, we obtain for $j = i + 1$,

$$
\begin{aligned}
\hat{d}_{i+1} E' &= \hat{d}_{i+1} \left(K_i + \xi_i \right) E_{K_i \eta} \\
&= \left(K_i \hat{d}_i + \xi_i \hat{d}_{i+1} - 1 \right) E_{K_i \eta} \\
&= \overline{(K_i \eta)}_i K_i E_{K_i \eta} + \left(\xi_i \overline{(K_i \eta)}_{i+1} - 1 \right) E_{K_i \eta} \\
&= \overline{\eta}_{i+1} K_i E_{K_i \eta} + \left(\xi_i \overline{\eta}_i - 1 \right) E_{K_i \eta} \\
&= \overline{\eta}_{i+1} E'.
\end{aligned}
\tag{7.99}
$$

From (7.96), we see that $E' \propto E_\eta$. Now we are ready to prove (7.91) for η with $\eta_i \neq \eta_{i+1}$, following the same classification as above.

(i) If $\eta_i > \eta_{i+1}$ is satisfied, then $K_i \eta < \eta$ and hence $E_{K_i \eta}$ does not contain z^η. The monomial z^η in $K_i E_{K_i \eta}$ comes only from the top term $z^{K_i \eta}$ in $E_{K_i \eta}$. From these facts, the coefficient of z^η in E' is unity, i.e.,

$$
\left(K_i + \xi_i \right) E_{K_i \eta} = E_\eta,
\tag{7.100}
$$

with $\eta_i > \eta_{i+1}$. Further, we apply $\left(K_i - \xi_i \right)$ on both sides of (7.100). After rearrangement, we can arrive at (7.91) for $\eta_i > \eta_{i+1}$.

(ii) For η satisfying $\eta_i < \eta_{i+1}$, we use (7.100) by noting $(K_i \eta)_i > (K_i \eta)_{i+1}$, and obtain

$$
\left(K_i + \tilde{\xi}_i \right) E_\eta = E_{K_i \eta},
\tag{7.101}
$$

where $\tilde{\xi}_i = 1 \Big/ \left(\overline{(K_i \eta)}_i - \overline{(K_i \eta)}_{i+1} \right)$. When $\eta_i \neq \eta_{i+1}$, the relation $\overline{\eta}_i = \overline{(K_i \eta)}_{i+1}$ holds. We hence obtain the relation $\tilde{\xi}_i = -\xi_i$ and

$$
\left(K_i - \xi_i \right) E_\eta = E_{K_i \eta},
\tag{7.102}
$$

which is nothing but (7.91) for η satisfying $\eta_i < \eta_{i+1}$. ∎

7.1.6 Arms and legs of compositions

In this subsection, we introduce additional notation to describe properties for non-symmetric Jack polynomials.

For a given composition $\eta = (\eta_1, \eta_2, \ldots, \eta_N) \in \Lambda_N$ and pairs of integers $s = (i, j)$ satisfying $1 \leq i \leq l(\eta)$ and $1 \leq j \leq \eta_i$, we define the following quantities:

$$a(s) = \eta_i - j, \tag{7.103}$$

$$a'(s) = j - 1, \tag{7.104}$$

$$l(s) = ul(s) + ll(s), \tag{7.105}$$

$$l'(s) = ul'(s) + ll'(s), \tag{7.106}$$

where we have defined the numbers ($\sharp$) in a set associated with s as follows:

$$ul(s) = \sharp\{k \in \{1, \ldots, i-1\} | j - 1 \le \eta_k < \eta_i\}, \tag{7.107}$$

$$ll(s) = \sharp\{k \in \{i+1, \ldots, N\} | j \le \eta_k \le \eta_i\}, \tag{7.108}$$

$$ul'(s) = \sharp\{k \in \{1, \ldots, i-1\} | \eta_k \ge \eta_i\}, \tag{7.109}$$

$$ll'(s) = \sharp\{k \in \{i+1, \ldots, N\} | \eta_k > \eta_i\}, \tag{7.110}$$

where $j - 1$ in (7.107) is not a misprint.

In the above expressions, $a(s)$, $a'(s)$, $l(s)$, and $l'(s)$ are called arm length, arm colength, leg length, and leg colength, respectively and they reduce to (2.84) when η is a partition. In terms of these quantities for each s, we construct the following quantities for a composition η:

$$d_\eta = \prod_{s \in \eta} d(s), \ d(s) \ = (a(s) + 1)/\lambda + l(s) + 1, \tag{7.111}$$

$$d'_\eta = \prod_{s \in \eta} d'(s), \ d'(s) = d(s) - 1 = (a(s) + 1)/\lambda + l(s), \tag{7.112}$$

$$e_\eta = \prod_{s \in \eta} e(s), \ e(s) \ = (a'(s) + 1)/\lambda + N - l'(s), \tag{7.113}$$

$$e'_\eta = \prod_{s \in \eta} e'(s), \ e'(s) = (a'(s) + 1)/\lambda + N - l'(s) - 1. \tag{7.114}$$

The quantities $d_\eta(s)$ and $d'_\eta(s)$ may be called generalized hook lengths since $d'(s)$ reduces to the upper hook length $h^*_\kappa(s)$ defined by (2.186) if η corresponds to a partition κ, and e_η and e'_η may be called generalized shifted factorials by the relations (7.174) and (7.180) to be discussed later.

As we shall prove, they satisfy the following recursion relations [154]:

$$\frac{d_{\Theta\eta}}{d_\eta} = \bar{\eta}_1 + \frac{1}{\lambda} + N, \quad \frac{d'_{\Theta\eta}}{d'_\eta} = \bar{\eta}_1 + \frac{1}{\lambda} + N - 1, \tag{7.115}$$

$$\frac{e_{\Theta\eta}}{e_\eta} = \bar{\eta}_1 + \frac{1}{\lambda} + N, \quad \frac{e'_{\Theta\eta}}{e'_\eta} = \bar{\eta}_1 + \frac{1}{\lambda} + N - 1, \tag{7.116}$$

$$e_{K_i\eta} = e_\eta, \quad e'_{K_i\eta} = e'_\eta, \tag{7.117}$$

for all η. In particular, the invariance (7.117) against permutations follows from the fact that $a'(s)$ and $a'(s)$ are completely determined by the corresponding partition η^+.

On the other hand, we have for $\eta_i > \eta_{i+1}$

$$\frac{d_{K_i\eta}}{d_\eta} = 1 + \xi_i, \quad \frac{d'_{K_i\eta}}{d'_\eta} = \frac{1}{1 - \xi_i}, \qquad (7.118)$$

and for $\eta_i < \eta_{i+1}$

$$\frac{d_{K_i\eta}}{d_\eta} = \frac{1}{1 - \xi_i}, \quad \frac{d'_{K_i\eta}}{d'_\eta} = 1 + \xi_i. \qquad (7.119)$$

These formulae will give a concise description of several quantities derived in the following subsections.

Proof of (7.115).
The diagram of $\Theta\eta$ is generated from that of η by (1) removing the top row, (2) moving all the other rows one unit up, (3) appending the removed row to the bottom of the diagram, and (4) adding a cell to the bottom row. From these procedures, a mapping from a square in a diagram of η to a square in a diagram of $\Theta\eta$

$$(i, j) \text{ in } \eta \rightarrow (i - 1, j) \text{ in } \Theta\eta \qquad (7.120)$$

for $i = 2, \ldots, l(\eta)$ and $j = 1, \ldots, \eta_i$ and

$$(1, j) \text{ in } \eta \rightarrow (N, j + 1) \text{ in } \Theta\eta \qquad (7.121)$$

for $j = 1, \ldots, \eta_1$ naturally follows. The implication of (7.120) and (7.121) can be understood from Fig. 7.2. It can easily be checked that the above mapping leaves the arm $a(s)$, $l(s)$ and hence $d(s)$ of a square $s = (i, j)$ unchanged. From this, we immediately see that

$$d_{\Theta\eta}/d_\eta = d(s'), \qquad (7.122)$$

where s' is the "surplus" cell $(N, 1)$ in the diagram of $\Theta\eta$ and it is shown by the shaded square in Fig. 7.2. An inspection shows that $a(s') = \eta_1$ and

$$l(s') = \sharp\{j > 1 | \eta_j \leq \eta_1\} = N - 1 - \sharp\{1 < j \leq N | \eta_j > \eta_1\}.$$

These results lead to $d(s') = \bar\eta_1 + 1/\lambda + N$. The quantity $d'_{\Theta\eta}/d'_\eta$ can be evaluated in a similar way. ∎

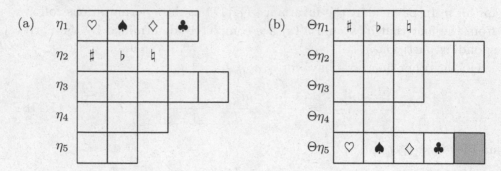

Fig. 7.3. Diagrams of (a) $\eta = (4, 3, 5, 3, 2)$ and (b) $\Theta\eta$. The cells marked by the same symbols such as $\heartsuit, \ldots, \sharp, \ldots$ in (b), respectively, illustrate the mapping (7.123) and (7.124). The shaded cell in (b) represents the cell s''.

Proof of (7.116).

Now we consider another mapping from a square in $D(\eta)$ to a square in $D(\Theta\eta)$

$$(i, j) \text{ in } D(\eta) \to (i - 1, j) \text{ in } D(\Theta\eta) \tag{7.123}$$

for $i = 2, \ldots, l(\eta)$ and $j = 1, \ldots, \eta_i$ and

$$(1, j) \text{ in } D(\eta) \to (N, j) \text{ in } D(\Theta\eta) \tag{7.124}$$

for $j = 1, \ldots, \eta_1$.

This mapping is described in Fig. 7.3. This mapping does not change the arm colength $a'(s)$ and leg colength $l'(s)$ and hence $e(s)$. Consequently, we obtain

$$\frac{e_{\Theta\eta}}{e_\eta} = e(s''),$$

where s'' is the cell (N, η_1) in the diagram of $\Theta\eta$, shown by the shaded square in Fig. 7.3. We can see that $e(s'') = \bar{\eta}_1 + 1N + N$ from the facts that $a'(s'') = \eta_1$ and $l'(s'') = \sharp\{j > 1 | \eta_j > \eta_1\}$. We thus arrive at the first equation of (7.116). The second equation of (7.116) can be shown in a similar way.   ∎

Proof of (7.118).

Now we consider a mapping from a square in $D(\eta)$ to a square in $D(K_i\eta)$

$$(i', j) \text{ in } D(\eta) \to (i', j) \text{ in } D(K_i\eta) \tag{7.125}$$

for $i' \neq i, i + 1$ and

$$(i, j) \text{ in } D(\eta) \to (i + 1, j) \text{ in } D(K_i\eta), \tag{7.126}$$

$$(i + 1, j) \text{ in } D(\eta) \to (i, j) \text{ in } D(K_i\eta). \tag{7.127}$$

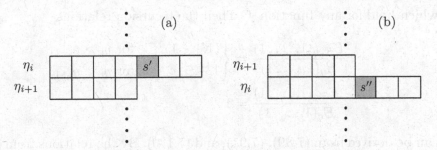

Fig. 7.4. Two rows in the diagram of (a) η and (b) $K_i\eta$. The shaded squares in (a) and (b) represent, respectively, s' and s''.

This mapping does not change the arm $a(s)$ and leg $l(s)$ except for $s' = (i, 1 + \eta_{i+1})$ in η. We thus obtain

$$\frac{d_{K_i\eta}}{d_\eta} = \frac{d(s'')}{d(s')},$$

where $s'' = (i + 1, 1 + \eta_{i+1})$ is the image of s' in the above mapping. The squares s' and s'' are shown in Fig. 7.4.

Straightforward calculations show that

$$a(s') = \eta_i - \eta_{i+1} - 1, \quad l(s') = -\left(k_i' + k_i''\right) + \left(k_{i+1}' + k_{i+1}''\right) - 1, \quad (7.128)$$

and

$$d(s') = \bar\eta_i - \bar\eta_{i+1}.$$

Further we obtain

$$d(s'') = 1 + \bar\eta_i - \bar\eta_{i+1}$$

in a similar way. From these two results, the assertion on d_η in (7.118) follows. The expression $d'_{K_i\eta}/d'_\eta$ can be evaluated in a similar way. ∎

7.1.7 Evaluation formula

Here we prove the following [154]:

$$E_\eta(1, \ldots, 1) = e_\eta/d_\eta, \tag{7.129}$$

which is called the evaluation formula.

Proof of (7.129).
We first quote the (almost trivial) relations

$$\Theta f(z_1, \ldots, z_N)|_{z_1 = \cdots = z_N = 1} = f(1, \ldots, 1),$$
$$K_i f(z_1, \ldots, z_N)|_{z_1 = \cdots = z_N = 1} = f(1, \ldots, 1), \tag{7.130}$$

which hold for any function f. Then the recursion relations

$$\frac{E_{K_i\eta}(1,\ldots,1)}{E_\eta(1,\ldots,1)} = \begin{cases} (1+\xi_i)^{-1}, & \text{for } \eta_i > \eta_{i+1}, \\ 1-\xi_i, & \text{for } \eta_i < \eta_{i+1}, \end{cases} \tag{7.131}$$

$$\frac{E_{\Theta\eta}(1,\ldots,1)}{E_\eta(1,\ldots,1)} = 1 \tag{7.132}$$

can be derived from (7.89), (7.91), and (7.130). By the relations from (7.115) to (7.119), we see that (7.131) and (7.132) have the same form as the recursion relations of e_η/d_η. Further, $E_\eta(1,\ldots,1)$ and e_η/d_η have the same value (=1) for $\eta = (0,\ldots,0)$. Thus we arrive at the expression (7.129) for any η by repeated application of Θ and K_i.   ∎

7.2 Antisymmetrization of Jack polynomials

7.2.1 *Antisymmetric Jack polynomials*

We now proceed to derive the integral and combinatorial norms of non-symmetric Jack polynomials. It turns out that the antisymmetric Jack polynomials $J_\kappa^{(-)}(z)$ are helpful for this purpose where a partition κ consists of distinct entities: $\kappa_1 > \kappa_2 > \cdots > \kappa_N$. The set of those κ is denoted by $\Lambda_N^{+>}$. Then $J_\kappa^{(-)}(z)$ is defined by the following two conditions:

(i) The polynomial $J_\kappa^{(-)}$ has the form

$$J_\kappa^{(-)}(z) = \sum_\eta a_\eta^{(-)} E_\eta(z) \tag{7.133}$$

with the normalization $a_\kappa^{(-)} = 1$. Here the sum with respect to η is taken over such composition as satisfies $\eta^+ = \kappa$.

(ii) Under the action of the transposition K_i, we obtain

$$K_i J_\kappa^{(-)}(z) = -J_\kappa^{(-)}(z), \tag{7.134}$$

for $i \in 1,\ldots,N-1$. Namely, $J_\kappa^{(-)}(z)$ is odd against the transposition.

We first derive the following recursion relation for the coefficient a_η:

$$\frac{a_{K_i\eta}^{(-)}}{a_\eta^{(-)}} = \begin{cases} -1-\xi_i, & \text{for } \eta_i > \eta_{i+1}, \\ 1/(\xi_i-1), & \text{for } \eta_i < \eta_{i+1}, \end{cases} \tag{7.135}$$

where $i = 1,\ldots,N-1$. Note that the RHS is the same as $-d_{K_i\eta}/d_\eta$.

Proof of (7.135).

Let us introduce the notation

$$\mathcal{A}_{i,\eta} = \xi_i + 1, \quad \mathcal{B}_{i,\eta} = \begin{cases} 1 - \xi_i^2, & \text{for } \eta_i > \eta_{i+1}, \\ 1, & \text{for } \eta_i < \eta_{i+1}. \end{cases} \tag{7.136}$$

Using (7.91), we then obtain

$$(K_i + 1) J_\kappa^{(-)} = \sum_\eta a_\eta^{(-)} (K_i + 1) E_\eta$$

$$= \sum_\eta a_\eta^{(-)} (\mathcal{A}_{i,\eta} E_\eta + \mathcal{B}_{i,\eta} E_{K_i\eta})$$

$$= \sum_\eta \left(a_\eta^{(-)} \mathcal{A}_{i,\eta} + a_{K_i\eta}^{(-)} \mathcal{B}_{i,K_i\eta} \right) E_\eta = 0. \tag{7.137}$$

Thus we obtain $a_{K_i\eta}^{(-)}/a_\eta^{(-)} = -\mathcal{A}_{i,\eta}/\mathcal{B}_{i,K_i\eta}$, and the recursion relation (7.135) follows. ∎

Let sign(η) be the sign of the permutation $\hat{\sigma}$ which gives $\hat{\sigma}\eta = \eta^+$ where η^+ is the corresponding partition. Then $a_\eta^{(-)}$ is expressed by

$$a_\eta^{(-)} = \text{sign}(\eta) d_\eta / d_{\eta^+}. \tag{7.138}$$

We can express $a_\eta^{(-)}$ in another way as

$$a_\eta^{(-)} = \text{sign}(\eta) \prod_{i<j;\,\text{s.t.}\,\eta_i<\eta_j} \left(1 + \frac{1}{\overline{\eta}_j - \overline{\eta}_i} \right), \tag{7.139}$$

where the notation with s.t. means that the product is taken over j and $i\ (<j)$ such that $\eta_i < \eta_j$. We call (i,j) a "reversed pair" if $1 \leq i < j \leq N$ and $\eta_i < \eta_j$. The product in (7.139) is then taken over the reversed pair.

Proof of (7.138).

Both sides of (7.138) obey the same recursion relations (7.118), (7.119), and (7.135) and satisfy the same initial condition (=1) for $\eta = \eta^+$. From this fact, the relation (7.138) follows. ∎

Proof of (7.139).

Both sides of

$$\frac{d_\eta}{d_{\eta^+}} = \prod_{i<j;\,\text{s.t.}\,\eta_i<\eta_j} \left(1 + \frac{1}{\overline{\eta}_j - \overline{\eta}_i} \right) \tag{7.140}$$

obey the same recursion relations (7.118) and (7.119), and satisfy the same initial condition (= 1) for $\eta = \eta^+$. From this fact, (7.139) follows. ∎

Example 7.1.

Recalling the relation $K_1\kappa = (\kappa_2, \kappa_1, \kappa_3, \ldots)$, we obtain

$$a^{(-)}_{K_1\kappa} = -1 - \frac{1}{\bar{\kappa}_1 - \bar{\kappa}_2}. \tag{7.141}$$

Example 7.2.

By repeated application of transpositions, we obtain

$$a^{(-)}_{K_2K_1\kappa}/a^{(-)}_{K_1\kappa} = -1 - \frac{1}{(K_1\kappa)_2 - (K_1\kappa)_3} = -1 - \frac{1}{\bar{\kappa}_1 - \bar{\kappa}_3}, \tag{7.142}$$

from which

$$a^{(-)}_{K_2K_1\kappa} = \left(-1 - \frac{1}{\bar{\kappa}_1 - \bar{\kappa}_2}\right)\left(-1 - \frac{1}{\bar{\kappa}_1 - \bar{\kappa}_3}\right) \tag{7.143}$$

immediately follows.

Example 7.3.

Let κ^{R} be the composition given by

$$\kappa^{\mathrm{R}} = (\kappa_N, \ldots, \kappa_2, \kappa_1). \tag{7.144}$$

We then obtain

$$a^{(-)}_{\kappa^{\mathrm{R}}} = \prod_{i<j}\left(-1 - \frac{1}{\bar{\kappa}_i - \bar{\kappa}_j}\right). \tag{7.145}$$

This expression can be obtained immediately from (7.139).

Now we define the operator $\hat{\sigma} \in S_N$ on a function f as

$$\hat{\sigma}f(z_1, \ldots, z_N) = f(z_{\sigma(1)}, \ldots, z_{\sigma(N)}). \tag{7.146}$$

Then the antisymmetrization operator on a function f is defined as

$$\mathrm{Asym}\, f(z) = \sum_{\hat{\sigma}\in S_N} \mathrm{sign}(\hat{\sigma})\hat{\sigma}f(z). \tag{7.147}$$

We can then write $J^{(-)}_\kappa$ as

$$\rho^{(-)}_\eta J^{(-)}_\kappa(z) = \mathrm{Asym}\, E_\eta(z), \tag{7.148}$$

where η is a composition such that $\eta^+ = \kappa$, and $\rho^{(-)}_\eta$ is given by

$$\rho^{(-)}_\eta = \mathrm{sign}(\eta)d'_\eta/d'_{\kappa^{\mathrm{R}}}. \tag{7.149}$$

Proof of (7.149).

Applying K_i with $i \in [1, N-1]$ on both sides of (7.148) for η such that $\eta_i > \eta_{i+1}$, we obtain

$$\rho_\eta^{(-)} K_i J_\kappa^{(-)}(z) = -\rho_\eta^{(-)} J_\kappa^{(-)}(z) \tag{7.150}$$

on the LHS and

$$\begin{aligned} K_i \mathrm{Asym}\, E_\eta &= \mathrm{Asym}\, K_i E_\eta \\ &= \mathrm{Asym}\, \left(\xi_i E_\eta + \left(1 - \xi_i^2 \right) E_{K_i \eta} \right) \\ &= \left(\xi_i \rho_\eta^{(-)} + \left(1 - \xi_i^2 \right) \rho_{K_i \eta}^{(-)} \right) J_\kappa^{(-)} \end{aligned} \tag{7.151}$$

on the RHS. From (7.150) and (7.151), the recursion relation

$$\frac{\rho_{K_i \eta}^{(-)}}{\rho_\eta^{(-)}} = \frac{1}{\xi_i - 1}, \quad \text{for } \eta_i > \eta_{i+1}, \tag{7.152}$$

follows. Performing a similar calculation for $\eta_i < \eta_{i+1}$, and comparing both results with (7.118) and (7.119), we find that

$$\rho_\eta^{(-)} \propto \mathrm{sign}(\eta) d'_\eta. \tag{7.153}$$

The coefficient of proportionality in (7.153) can be obtained by considering $\rho_{\kappa^R}^{(-)}$. On the RHS of (7.148) for $\eta = \kappa^R$, E_κ comes only from the term

$$\mathrm{sign}\left(\hat{\sigma}^R \right) \hat{\sigma}^R E_{\kappa^R}, \tag{7.154}$$

where

$$\hat{\sigma}^R = (N, N-1, \ldots, 2, 1).$$

With successive use of (7.91) for $\eta_i < \eta_{i+1}$, the coefficient of E_κ in (7.154) is derived as $(-1)^{N(N-1)/2}$. The coefficient of E_κ on the LHS of (7.148) for $\eta = \kappa^R$ is given by $\rho_{\kappa^R}^{(-)}$. Thus we obtain

$$\rho_{\kappa^R}^{(-)} = (-1)^{N(N-1)/2}. \tag{7.155}$$

From (7.153), (7.155), and using $\mathrm{sign}(\kappa^R) = (-1)^{N(N-1)/2}$, we obtain (7.149). Alternatively, $\rho_\kappa^{(-)}$ can be expressed as

$$\rho_\kappa^{(-)} = \prod_{i<j} \left(\frac{\bar{\kappa}_i - \bar{\kappa}_j - 1}{\bar{\kappa}_i - \bar{\kappa}_j} \right), \tag{7.156}$$

Using (7.152) and (7.155). ∎

7.2.2 Integral norm

Now we are ready to calculate the integral norm $\langle E_\eta, E_\eta \rangle_0$. First we note that Θ introduced in (7.74) and K_i ($\equiv K_{i,i+1}$) are isometric:

$$\langle \Theta f, \Theta g \rangle_0 = \langle f, g \rangle_0, \quad \langle K_i f, K_i g \rangle_0 = \langle f, g \rangle_0, \tag{7.157}$$

i.e., Θ and K_i preserve the inner product with respect to $\langle \cdot, \cdot \rangle_0$. Using this property and (7.89), we obtain

$$\langle E_{\Theta\eta}, E_{\Theta\eta} \rangle_0 = \langle E_\eta, E_\eta \rangle_0. \tag{7.158}$$

Using (7.91), we have for η satisfying $\eta_i > \eta_{i+1}$,

$$\begin{aligned}
\langle E_{K_i\eta}, E_{K_i\eta} \rangle_0 &= \left(1 - \xi_i^2\right)^{-2} \langle K_i E_\eta - \xi_i E_\eta, K_i E_\eta - \xi_i E_\eta \rangle_0 \\
&= \frac{(1 + \xi_i^2) \langle E_\eta, E_\eta \rangle_0}{(1 - \xi_i^2)^2} - \frac{2\xi_i \langle E_\eta, K_i E_\eta \rangle_0}{(1 - \xi_i^2)^2} \\
&= \frac{\langle E_\eta, E_\eta \rangle_0}{(1 - \xi_i^2)}.
\end{aligned} \tag{7.159}$$

On the other hand, for η satisfying $\eta_i < \eta_{i+1}$, we have

$$\begin{aligned}
\langle E_{K_i\eta}, E_{K_i\eta} \rangle_0 &= \langle K_i E_\eta - \xi_i E_\eta, K_i E_\eta - \xi_i E_\eta \rangle_0 \\
&= \left(1 + \xi_i^2\right) \langle E_\eta, E_\eta \rangle_0 - 2\xi_i \langle E_\eta, K_i E_\eta \rangle_0 \\
&= \left(1 - \xi_i^2\right) \langle E_\eta, E_\eta \rangle_0.
\end{aligned} \tag{7.160}$$

Recall that Θ and K_i are sufficient to generate the arbitrary composition η from the trivial composition $(0, \ldots, 0)$. Using these recursion relations, we obtain

$$\frac{\langle E_\eta, E_\eta \rangle_0}{\langle 1, 1 \rangle_0} = \frac{d'_\eta e_\eta}{d_\eta e'_\eta} \tag{7.161}$$

for arbitrary η. Here the denominator $\langle 1, 1 \rangle_0$ denotes the norm of the zeroth-order non-symmetric Jack polynomial for the composition $(0, \ldots, 0)$.

Proof of (7.161).
Here we set the RHS of (7.161) to be A_η. The recursion formula for A_η,

$$\frac{A_{\Theta\eta}}{A_\eta} = 1, \quad \frac{A_{K_i\eta}}{A_\eta} = \begin{cases} 1/\left(1 - \xi_i^2\right), & \text{for } \eta_i > \eta_{i+1}, \\ 1, & \text{for } \eta_i = \eta_{i+1}, \\ 1 - \xi_i^2, & \text{for } \eta_i < \eta_{i+1}, \end{cases} \tag{7.162}$$

follows from (7.118) and (7.119). Both sides of (7.161) are equal because these two satisfy the same recursion formula and have the same value ($=1$) at the trivial composition $(0, \ldots, 0)$. ∎

In the following, the denominator of the LHS of (7.161) is shown to be

$$\langle 1, 1 \rangle_0 = \frac{\Gamma(N\lambda + 1)}{[\Gamma(1 + \lambda)]^N} \qquad (7.163)$$

for positive integer λ. This relation generalizes Dyson's conjecture given by (4.76) for the particular case of $\lambda = 2$.

Before going into details of the proof, we explain the outline. First we show that

$$J_\delta^{(-)}(z) = \prod_{1 \le i < j \le N} (z_i - z_j) \equiv \Delta(z), \qquad (7.164)$$

where $\delta = (N - 1, N - 2, \ldots, 1, 0)$. From this fact, the relation

$$\frac{\langle J_\delta^{(-)}, J_\delta^{(-)} \rangle_0}{\langle 1, 1 \rangle_0} = \frac{I(\lambda + 1, N)}{I(\lambda, N)} \qquad (7.165)$$

follows. Here we have introduced the notation

$$I(\lambda, N) \equiv \langle 1, 1 \rangle_0$$

to make explicit the λ-dependence of $\langle 1, 1 \rangle_0$. Next we evaluate the LHS of (7.165) and obtain

$$\frac{\langle J_\delta^{(-)}, J_\delta^{(-)} \rangle_0}{\langle 1, 1 \rangle_0} = \frac{\Gamma[N(\lambda + 1) + 1]}{\Gamma[N\lambda + 1](\lambda + 1)^N}, \qquad (7.166)$$

as shown below. Using (7.165), (7.166), and the result $I(\lambda = 1, N) = N!$ for free fermions, we can prove (7.163) by mathematical induction.

Proof of (7.164).
Any antisymmetric homogeneous polynomial of z can be written as a product of the difference-product $\Delta(z)$ and a symmetric homogeneous polynomial. Both $J_\delta^{(-)}(z)$ and $\Delta(z)$ have the same degree ($= N(N-1)/2$) and hence we can put $J_\delta^{(-)}(z) = c\Delta(z)$ with a constant to be determined. We expand this expression with respect to monomials and consider the coefficient z^δ of both sides. On the RHS, obviously the coefficient of z^δ in $\Delta(z)$ is unity. When we expand $J_\delta^{(-)}(z)$ in terms of non-symmetric Jack polynomials, only E_δ contains the monomial z^δ and the coefficient of E_δ in $J_\delta^{(-)}(z)$ is unity. Thus the coefficient of z^δ in $J_\delta^{(-)}(z)$ is also unity. From these facts, we obtain $c = 1$. ∎

Proof of (7.166).

We note the following relation:

$$\frac{\langle J_\kappa^{(-)}, J_\kappa^{(-)} \rangle_0}{\langle E_\kappa, E_\kappa \rangle_0} = \frac{N!}{\rho_\kappa^{(-)}}, \tag{7.167}$$

which results from

$$(\rho_\kappa^{(-)})^2 \langle J_\kappa^{(-)}, J_\kappa^{(-)} \rangle_0 = \langle \mathrm{Asym}\, E_\kappa, \mathrm{Asym}\, E_\kappa \rangle_0 = \langle E_\kappa, \mathrm{Asym}(\mathrm{Asym}\, E_\kappa) \rangle_0$$
$$= N! \langle E_\kappa, (\mathrm{Asym}\, E_\kappa) \rangle_0 = N! \rho_\kappa^{(-)} \langle E_\kappa, J_\kappa^{(-)} \rangle_0$$
$$= N! \rho_\kappa^{(-)} \langle E_\kappa, E_\kappa \rangle_0. \tag{7.168}$$

Using (7.161) and (7.167), we rewrite the LHS of (7.166) as follows:

$$\frac{\langle J_\delta^{(-)}, J_\delta^{(-)} \rangle_0}{\langle 1, 1 \rangle_0} = \frac{\langle J_\delta^{(-)}, J_\delta^{(-)} \rangle_0}{\langle E_\delta, E_\delta \rangle_0} \frac{\langle E_\delta, E_\delta \rangle_0}{\langle 1, 1 \rangle_0} = \frac{N!}{\rho_\delta^{(-)}} \frac{d_\delta' e_\delta}{d_\delta e_\delta'} = N! \frac{d_{\delta^{\mathrm{R}}}' e_\delta}{d_\delta e_\delta'}. \tag{7.169}$$

Figure 7.5 shows (a) the diagram $D(\delta)$ of the partition $\delta = (N-1, \ldots, 0)$ and (b) the diagram $D(\delta^{\mathrm{R}})$ of δ^{R} for $N = 5$. In the one-to-one correspondence between $s' = (i, j)$ in $D(\delta)$ and $s'' = (N + i + 1, j)$ in $D(\delta^{\mathrm{R}})$, we find that $a(s') = a(s'')$ and $l(s') + 1 = l(s'')$. We then obtain $d(s') = d'(s'')$ and hence $d_\delta = d'_{\delta^{\mathrm{R}}}$. For the RHS of (7.169), we next derive

$$\frac{e_\delta}{e_\delta'} = \frac{\Gamma[N(\lambda + 1) + 1]}{N! \Gamma(N\lambda + 1)(\lambda + 1)^N}. \tag{7.170}$$

Then (7.166) follows. ∎

Proof of (7.170).

For the partition δ, arm colength and leg colength are respectively given by

$$a'(s) = j - 1, \quad l'(s) = i - 1. \tag{7.171}$$

From (7.171), we obtain

$$\frac{e_\delta}{e_\delta'} = \prod_{i=1}^{N-1} \prod_{j=1}^{N-i} \frac{(N - i + 1)\lambda + j}{(N - i)\lambda + j}$$

$$= \prod_{i=1}^{N-1} \frac{\Gamma[\lambda(N - i) + 1]\Gamma[(1 + \lambda)(N - i + 1)]}{\Gamma[\lambda(N - i + 1) + 1]\Gamma[(1 + \lambda)(N - i) + 1]}. \tag{7.172}$$

In the second equality, we have taken the product over j. We can confirm that the expression on the rightmost is equal to that of (7.170) by noting

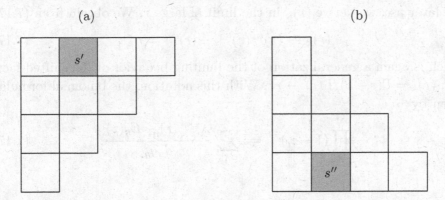

Fig. 7.5. (a) Diagram of the partition $\delta = (N, N-1, \ldots, 0)$ and (b) diagram of δ^{R} for $N = 5$. For the two shaded squares, the relation $d(s') = d'(s'')$ holds.

the following relations:

$$\prod_{i=1}^{N-1} \frac{\Gamma[\lambda(N-i)+1]}{\Gamma[\lambda(N-i+1)+1]} = \frac{\Gamma[\lambda+1]}{\Gamma[N\lambda+1]},$$

$$\prod_{i=1}^{N-1} \frac{\Gamma[(1+\lambda)(N-i+1)]}{\Gamma[(1+\lambda)(N-i)+1]} = \frac{\Gamma[(1+\lambda)N+1]}{N!\,(1+\lambda)^N\,\Gamma[\lambda+1]}. \qquad (7.173)$$

∎

7.2.3 Binomial formula

To describe the binomial formula, we first introduce the generalized shifted factorial $(r)_\kappa$ for an indeterminate r and a partition $\kappa = \eta^+$ as

$$(r)_\kappa = \prod_{s \in D(\kappa)} \left[r + a'(s) - \lambda l'(s) \right]. \qquad (7.174)$$

In the special case of a single row with $\kappa_1 = k$, we have $l'(s) = 0$ and obtain

$$(r)_k = r(r+1)\cdots(r+k-1) = \Gamma(r+k)/\Gamma(r), \qquad (7.175)$$

which justifies its name. In the general case, (7.174) can be written as

$$(r)_\kappa = \prod_{i=1}^{N} \frac{\Gamma[r-\lambda(i-1)+\kappa_i]}{\Gamma[r-\lambda(i-1)]} = (-1)^{|\kappa|} \prod_{i=1}^{N} \frac{\Gamma[-r+\lambda(i-1)+1]}{\Gamma[-r+\lambda(i-1)+1-\kappa_i]}. \qquad (7.176)$$

For later use, we derive $(r)_\kappa$ in the limit of large r. We obtain from (7.176)

$$(r)_\kappa \to (r)^{\kappa_1} (r)^{\kappa_2} \cdots (r)^{\kappa_N} = r^{|\kappa|}, \qquad (7.177)$$

which is again a generalization of the limiting behavior of the shifted factorial: $(r)_k = \Gamma(r+k)/\Gamma(r) \to r^k$. With this notation, the binomial formula is given by

$$\prod_{i=1}^{N} (1 - z_i)^{-r} = \sum_{\eta \in \Lambda_N} \frac{\lambda^{-|\eta|} (r)_{\eta^+} E_\eta(z)}{d_\eta g_\eta}. \qquad (7.178)$$

Proof of (7.178).

We start with the formula (7.53) for $z = (z_1, \dots, z_M)$ and $y = (y_1, \dots, y_M)$, where M is an integer larger than N. Setting $y_1 = \cdots = y_M = 1$ and $z_{N+1} = \cdots = z_M = 0$, the expression (7.53) reduces to

$$\prod_{i=1}^{N} (1 - z_i)^{-M\lambda-1} = \sum_{\eta} g_\eta^{-1} E_\eta(\overbrace{1, \dots, 1}^{M}) E_\eta(z_1, \dots, z_N). \qquad (7.179)$$

On the RHS, we have the evaluation formula $E_\eta(1^M) = e_\eta/d_\eta$ with reference to (7.129). By comparing (7.113) and (7.174), we obtain

$$e_\eta = (M\lambda + 1)_{\eta^+} \lambda^{-|\eta|}. \qquad (7.180)$$

From this expression, we see that the RHS of (7.179) depends on $M\lambda + 1$ only through $(M\lambda + 1)_{\eta^+}$, and hence it is a polynomial of $M\lambda + 1$. Since the LHS is an analytic function of $M\lambda + 1$, the relation (7.179) holds for arbitrary complex values of $M\lambda + 1$. Replacing $M\lambda + 1$ by r, we obtain (7.178). ∎

7.2.4 Combinatorial norm

We are in a position to derive explicitly the combinatorial norm g_η of non-symmetric Jack polynomials. We want to show [16] that

$$\langle E_\eta, E_\eta \rangle^{\mathrm{c}} = g_\eta = d_\eta'/d_\eta. \qquad (7.181)$$

The key quantity for the derivation is the expression

$$\frac{\int \mathcal{D}z \prod_{i=1}^{N} (1 - z_i)^a \left(1 - z_i^{-1}\right)^a E_\eta(z)}{\int \mathcal{D}z \prod_{i=1}^{N} (1 - z_i)^a \left(1 - z_i^{-1}\right)^a}, \qquad (7.182)$$

where the weighted integral defined by (7.40) has been used. In the limit $a \to \infty$, the integrand in the numerator is dominated by the contribution around the neighborhood of $z_1 = z_2 = \cdots = z_N = -1$. Namely, (7.182) reduces to

$$E_\eta(-1, -1, \ldots, -1) = (-1)^{|\eta|} e_\eta / d_\eta, \qquad (7.183)$$

where we have used (7.129). Furthermore, (7.182) can be expressed in terms of the integral norm $\langle E_\eta, E_\eta \rangle_0$, which is known, and the combinatorial norm g_η. Therefore, by comparing this with (7.183), we can extract g_η as shown below.

To represent the integrals in (7.182) concisely, we introduce the notation

$$F(a, b, \eta) = \text{C.T.} \left[\prod_{i=1}^N (1 - z_i)^a \left(1 - z_i^{-1} \right)^b E_\eta(z) w(z) \right], \qquad (7.184)$$

where $w(z) \equiv \prod_{i \neq j} (1 - z_i / z_j)^\lambda$ is the weight function and C.T. $[X(z)]$ means the constant term in the Laurent expansion of $X(z)$. We further introduce the notation

$$\eta^{\text{R}} = (\eta_N, \eta_{N-1}, \ldots, \eta_1), \quad z^{\text{R}} = (z_N, z_{N-1}, \ldots, z_1). \qquad (7.185)$$

Then the relation

$$F(a, b, \eta) = F(b, a, -\eta^{\text{R}}) \qquad (7.186)$$

can be derived using

$$\text{C.T.} \left[f(z) E_\eta(z) w(z) \right] = \text{C.T.} \left[f(z) E_\eta(z^{\text{R}}) w(z) \right], \qquad (7.187)$$

$$\text{C.T.} \left[f(z) g(z^{-1}) w(z) \right] = \text{C.T.} \left[f(z^{-1}) g(z) w(z) \right], \qquad (7.188)$$

$$E_\eta(z^{-1}) = E_{-\eta^{\text{R}}}(z^{\text{R}}), \qquad (7.189)$$

where $f(z)$ and $g(z)$ are symmetric functions. The relations (7.187) and (7.188) are easy to prove. The relation (7.189) is derived below.

Proof of (7.189).
Let $y \equiv 1/z = (1/z_1, \ldots, 1/z_N)$. Direct calculation shows that

$$\hat{d}_i^{z^{\text{R}}} = -\hat{d}_{N+1-i}^y + 1 - N. \qquad (7.190)$$

Using this relation, we obtain

$$\hat{d}_i^{z^{\text{R}}} E_\eta(y) = \left\{ -\overline{\eta}_{N+1-i} + 1 - N \right\} E_\eta(y) = \overline{(-\eta^{\text{R}})}_i E_\eta(y), \qquad (7.191)$$

which means that $E_\eta(y)$ is proportional to $E_{-\eta^{\text{R}}}(z^{\text{R}})$. The coefficients of $y^\eta = (z^{\text{R}})^{-\eta^{\text{R}}}$ in $E_\eta(y)$ and $E_{-\eta^{\text{R}}}(z^{\text{R}})$ are both unity. Hence we obtain (7.189). ∎

From the binomial formula (7.178), we obtain

$$\prod_{i=1}^{N} \left(1 - z_i^{-1}\right)^a = \sum_{\eta \in \Lambda_N} \lambda^{-|\eta|} (-a)_{\eta^+} E_\eta(z^{-1})/(d_\eta g_\eta), \qquad (7.192)$$

where we have set $r = -a$ and replaced z by z^{-1}. Multiplying this expression by $E_\eta(z)w(z)$, and extracting the constant term, we obtain

$$F(0, a, \eta) = \lambda^{-|\eta|} \frac{(-a)_\kappa}{d_\eta g_\eta} \langle E_\eta, E_\eta \rangle_0, \qquad (7.193)$$

where $\kappa = \eta^+$. According to (7.186), the same quantity can also be expressed in terms of η^{R}. We use the following relation corresponding to the Galilean boost:

$$z^a E_\eta(z) = E_{\eta+a}(z), \qquad (7.194)$$

where $\eta + a$ represents $(\eta_1 + a, \ldots, \eta_N + a)$. In order to prove (7.194), we first note that $z^a E_\eta$ is the eigenfunction of $\hat{d}_i$ with eigenvalue $\overline{\eta}_i + a_\lambda$. Furthermore, the coefficient of $z^{\eta+a}$ in $z^a E_\eta(z)$ is unity. Then (7.194) follows. Furthermore, from (7.189), we have

$$\langle E_\eta, E_\eta \rangle_0 = \langle E_{-\eta^{\mathrm{R}}+a}, E_{-\eta^{\mathrm{R}}+a} \rangle_0. \qquad (7.195)$$

Using the relation

$$\prod_{i=1}^{N} \left(1 - z_i^{-1}\right)^a = \prod_{i=1}^{N} (-z_i)^{-a} \prod_{i=1}^{N} (1 - z_i)^a, \qquad (7.196)$$

we obtain

$$F(a, 0, -\eta^{\mathrm{R}}) = \mathrm{C.T.} \left[\prod_{i=1}^{N} (1 - z_i)^a E_{-\eta^{\mathrm{R}}}(x) w(z) \right]$$

$$= (-1)^{Na} \mathrm{C.T.} \left[\prod_{i=1}^{N} \left(1 - z_i^{-1}\right)^a E_{-\eta^{\mathrm{R}}+a}(z) w(z) \right]$$

$$= (-1)^{Na} \lambda^{|\eta|-Na} \frac{(-a)_{-\kappa^{\mathrm{R}}+a} \langle E_\eta, E_\eta \rangle_0}{d_{-\eta^{\mathrm{R}}+a} g_{-\eta^{\mathrm{R}}+a}}, \qquad (7.197)$$

where we have used (7.195) in the last step. Since $F(a, 0, -\eta^{\mathrm{R}}) = F(0, a, \eta)$ from (7.186), we now have the following identity for any composition and number a:

$$(-1)^{Na} \lambda^{|\eta|-Na} \frac{(-a)_{-\kappa^{\mathrm{R}}+a}}{d_{-\eta^{\mathrm{R}}+a} g_{-\eta^{\mathrm{R}}+a}} = \lambda^{-|\eta|} \frac{(-a)_\kappa}{d_\eta g_\eta}. \qquad (7.198)$$

In (7.184), on the other hand, the factor $(1 - z_i)^{2a}$ emerges inside C.T.$[\cdots]$, by setting $a = b$ and using (7.196). Namely, we obtain

$$F(a, a, \eta) = (-1)^{Na} \lambda^{|\eta| - Na} \frac{(-2a)_{-\kappa^{R} + a}}{d_{-\eta^{R} + a} g_{-\eta^{R} + a}} \langle E_{\eta}, E_{\eta} \rangle_0$$

$$= \lambda^{-|\eta|} \frac{(-a)_{\kappa} (-2a)_{-\kappa^{R} + a}}{(-a)_{-\kappa^{R} + a} d_{\eta} g_{\eta}} \langle E_{\eta}, E_{\eta} \rangle_0, \qquad (7.199)$$

where (7.198) has been used in the second equality. In the special case of $\eta = (0, \ldots, 0)$, (7.199) reduces to

$$F(a, a, 0) = \frac{(-2a)_{a^+} \langle 1, 1 \rangle_0}{(-a)_{a^+}}, \qquad (7.200)$$

using $d_{\eta} = g_{\eta} = 1$ for $\eta = (0, \ldots, 0)$. Here we have written $a^+ = (a, a, \ldots, a)$ to avoid confusion between the partition and the number. In the explicit expression using (7.176):

$$(-a)_{-\kappa^{R} + a} = (-1)^{|\kappa|} \prod_{i=1}^{N} \frac{\Gamma[a + \lambda(i - 1) + 1]}{\Gamma[a + \lambda(i - 1) + 1 - a + \kappa_i^{R}]}, \qquad (7.201)$$

we observe that the a's in the denominator cancel each other. Using (7.201), we divide (7.199) by (7.200), and find a ratio

$$\frac{(-a)_{a^+}}{(-a)_{-\kappa^{R} + a}} = (-1)^{|\kappa|} \prod_{i=1}^{N} \frac{\Gamma[\lambda(i - 1) + 1 + \kappa_i^{R}]}{\Gamma[\lambda(i - 1) + 1]}$$

$$= (-\lambda)^{|\kappa|} \prod_{j=1}^{N} \frac{\Gamma[N - j + (1 + \kappa_j)/\lambda]}{\Gamma[N - j + 1/\lambda]}$$

$$= (-\lambda)^{|\kappa|} e'_{\kappa} = (-\lambda)^{|\eta|} e'_{\eta}, \qquad (7.202)$$

where we have put $i = N - j + 1$ in the second equality, and used the invariance property (7.117). Using (7.176), we obtain in the limit of $a \to \infty$,

$$\frac{(-2a)_{-\kappa^{R} + a}}{(-2a)_{a^+}} \to (-a)^{-|\kappa|}, \qquad (7.203)$$

which cancels $(-a)_{\kappa} \to (-a)^{|\kappa|}$ in $F(a, a, \eta)$. Then we find the limiting value

$$\lim_{a \to \infty} \frac{F(a, a, \eta)}{F(a, a, 0)} = (-1)^{|\eta|} \frac{e'_{\eta}}{d_{\eta} g_{\eta}} \frac{\langle E_{\eta}, E_{\eta} \rangle_0}{\langle 1, 1 \rangle_0} = (-1)^{|\eta|} \frac{d'_{\eta} e_{\eta}}{d_{\eta}^2 g_{\eta}}. \qquad (7.204)$$

Since the LHS of (7.204) is given by (7.183), we finally obtain the combinatorial norm

$$g_{\eta} = d'_{\eta}/d_{\eta}. \qquad (7.205)$$

7.3 Symmetric Jack polynomials

7.3.1 Relation to non-symmetric Jack polynomials

We discuss symmetric Jack polynomials $J_\kappa(z)$ for a partition κ, which has been defined in Section 2.5 as the unique eigenfunctions of (3.16), or equivalently of the similarity-transformed Hamiltonian

$$\mathcal{H} = \lambda^2 \sum_{i=1}^{N} \left(\hat{d}_i + \frac{N-1}{2} \right)^2 - \mathcal{E}_{0,N} \tag{7.206}$$

satisfying the triangularity condition (2.172). Alternatively, on the basis of non-symmetric Jack polynomials, the symmetric Jack polynomials can be defined as the homogeneous symmetric polynomials satisfying the following two conditions:

(i) The polynomial J_κ has the form

$$J_\kappa(z) = \sum_\eta a_\eta^{(+)} E_\eta(z) \tag{7.207}$$

with the normalization $a_\kappa^{(+)} = 1$. Here the sum with respect to η is taken over compositions such that $\eta^+ = \kappa$.

(ii) Under the action of the transposition K_i, the polynomial $J_\kappa(z)$ is invariant: $K_i J_\kappa(z) = J_\kappa(z)$ for $i \in 1, \dots, N-1$.

Every E_η in (7.207) is an eigenfunction of (7.206) which has the common eigenvalue

$$\lambda^2 \sum_{i=1}^{N} \left(\bar{\kappa}_i + \frac{N-1}{2} \right)^2 - \mathcal{E}_{0,N},$$

and hence J_κ in (7.207) is the symmetric eigenfunction of (7.206). Further, from the normalization and triangularity of non-symmetric Jack polynomials, J_κ in (7.207) satisfies the normalization and triangularity (2.172). Thus the definition of J_κ in the present subsection is equivalent to that in Section 2.5.

The relation between E_η and J_κ is useful to derive the mathematical formulae for J_κ in the following subsections. Further, the symmetric Jack polynomials can be related to the antisymmetric Jack polynomials as

$$J_\kappa^{(-)}(z; \lambda) = \Delta(z) J_\mu(z; \lambda + 1), \tag{7.208}$$

where $\kappa = \mu + \delta$ and $\delta = (N-1, N-2, \dots, 0) \in \Lambda_N^+$. We have made explicit here the λ-dependence of $J_\kappa^{(-)}$ and J_μ.

Now we derive the recursion relation for the coefficient $a_\eta^{(+)}$:

$$\frac{a_{K_i\eta}^{(+)}}{a_\eta^{(+)}} = \frac{d_\eta'}{d_{K_i\eta}'} = \begin{cases} 1 - \xi_i, & \text{for } \eta_i > \eta_{i+1}, \\ 1/(1 + \xi_i), & \text{for } \eta_i < \eta_{i+1}, \end{cases} \qquad (7.209)$$

with $i = 1, \ldots, N - 1$.

Proof of (7.209).

From the symmetric property of $J_\kappa(z)$ and (7.91), we obtain

$$(K_i - 1) J_\kappa = \sum_\eta a_\eta^{(+)} (K_i - 1) E_\eta$$

$$= \sum_\eta a_\eta^{(+)} (\mathcal{A}_{i,\eta} E_\eta + \mathcal{B}_{i,\eta} E_{K_i\eta})$$

$$= \sum_\eta \left(a_\eta^{(+)} \mathcal{A}_{i,\eta} + a_{K_i\eta}^{(+)} \mathcal{B}_{i,(K_i\eta)} \right) E_\eta = 0, \qquad (7.210)$$

where

$$\mathcal{A}_{i,\eta} = \begin{cases} \xi_i - 1, & \text{for } \eta_i \neq \eta_{i+1}, \\ 0, & \text{for } \eta_i = \eta_{i+1} \end{cases} \quad \text{and} \quad \mathcal{B}_{i,(K_i\eta)} = \begin{cases} 1 - \xi_i^2, & \text{for } \eta_i < \eta_{i+1}, \\ 0, & \text{for } \eta_i = \eta_{i+1}, \\ 1, & \text{for } \eta_i > \eta_{i+1}. \end{cases} \qquad (7.211)$$

The only difference between $\mathcal{B}_{i,\eta}$ and $\mathcal{B}_{i,(K_i\eta)}$ is the interchange of the cases $\eta_i < \eta_{i+1}$ and $\eta_i > \eta_{i+1}$. From this, and using (7.118) and (7.119), the recursion relation (7.209) follows. ∎

It is instructive to compare (7.209) with (7.135) for antisymmetric Jack polynomials. The relation (7.209) suffices to generate a_η for every η satisfying $\eta^+ = \kappa$ for a partition κ. We obtain from (7.118) and (7.119) [154]

$$a_\eta^{(+)} = \frac{d_{\eta^+}'}{d_\eta'} = \prod_{i<j;\,\text{s.t.}\,\eta_i<\eta_j} \left(1 - \frac{1}{\bar{\eta}_j - \bar{\eta}_i} \right) \qquad (7.212)$$

for η satisfying $\eta^+ = \kappa$. The product in (7.212) is taken over the reversed pairs as in (7.139).

For later convenience, we give an example of $a_\eta^{(+)}$. For a partition $\kappa \in \Lambda_N^+$, let κ^{R} be the composition given by

$$\kappa^{\mathrm{R}} = (\kappa_N, \kappa_{N-1}, \ldots, \kappa_1). \qquad (7.213)$$

We then obtain

$$a_{\kappa^{\mathrm{R}}}^{(+)} = \prod_{i<j;\,\text{s.t.}\,\kappa_i>\kappa_j} \left(1 - \frac{1}{\bar{\kappa}_i - \bar{\kappa}_j} \right), \qquad (7.214)$$

which excludes the pairs with $\kappa_i = \kappa_j$, and can be obtained immediately from (7.212).

We introduce the symmetrization operator on a function $f(z_1, \ldots, z_N)$ as

$$\mathrm{Sym}\, f(z) = \sum_{\hat{\sigma} \in S_N} \hat{\sigma} f(z), \tag{7.215}$$

where $\hat{\sigma} f$ has been defined in (7.146). We can then write $J_\kappa(z)$ in terms of $E_\eta(z)$ with $\kappa = \eta^+$ as

$$\rho_\eta^{(+)} J_\kappa(z) = \mathrm{Sym}\, E_\eta(z), \tag{7.216}$$

with a constant $\rho_\eta^{(+)}$. Putting $\kappa = \eta^+$, we shall derive

$$\rho_\kappa^{(+)} = \prod_{1 \le i < j \le N} \frac{\bar{\kappa}_i - \bar{\kappa}_j + 1}{\bar{\kappa}_i - \bar{\kappa}_j} = \left(\prod_r p_r! \right) \prod_{\substack{1 \le i < j \le N; \\ \text{s.t.} \kappa_i > \kappa_j}} \frac{\bar{\kappa}_i - \bar{\kappa}_j + 1}{\bar{\kappa}_i - \bar{\kappa}_j}. \tag{7.217}$$

Here the second expression excludes the pairs with $\kappa_i = \kappa_j$, and p_r is the number of r in κ; the factor $(\prod_r p_r!)$ is the number of elements in S_N which leave κ invariant. We can express $\rho_\kappa^{(+)}$ in another way as

$$\rho_\kappa^{(+)} = \left(\prod_r p_r! \right) \frac{d_{\kappa^R}}{d_\kappa} = N! \frac{d_\kappa'' e_\kappa}{d_\kappa e_\kappa''}. \tag{7.218}$$

Proof of (7.217).
We compare the coefficient of E_{κ^R} on both sides of (7.216). On the RHS, E_{κ^R} comes only from

$$\sum_{\hat{\sigma}; \text{s.t.}\ \hat{\sigma}\kappa = \kappa^R} \hat{\sigma} E_\kappa. \tag{7.219}$$

We set $\hat{\sigma}^R$ as

$$\hat{\sigma}^R = \left(\ldots, \overbrace{1 + p_1 + p_2, \ldots, p_1 + p_2 + p_3}^{p_3}, \overbrace{1 + p_1, \ldots, p_1 + p_2}^{p_2}, \overbrace{1, \ldots, p_1}^{p_1} \right), \tag{7.220}$$

for a given partition κ. The expression (7.219) then turns into

$$\left(\prod_r p_r! \right) \hat{\sigma}^R E_\kappa. \tag{7.221}$$

With successive use of (7.91) for $\eta_i > \eta_{i+1}$, the coefficient of E_{κ^R} in $\hat{\sigma}^R E_\kappa$ is derived as

$$\prod_{i<j;\,\text{s.t.}\,\kappa_i>\kappa_j} \left\{ 1 - \frac{1}{(\overline{\kappa}_i - \overline{\kappa}_j)^2} \right\}, \tag{7.222}$$

which is to be compared with (7.214). The coefficient of E_{κ^R} on the LHS of (7.216) is given by $\rho_\kappa^{(+)} a_{\kappa^R}^{(+)}$. From these results and (7.222), we obtain (7.217). ∎

Proof of (7.218).
For composition η, we have the relation

$$\frac{d_\eta}{d_{\eta^+}} = \prod_{i<j;\,\text{s.t.}\,\eta_i<\eta_j} \left(1 + \frac{1}{\overline{\eta}_j - \overline{\eta}_i} \right). \tag{7.223}$$

This is proved by observing that both sides obey the same recursion relation and satisfy the same initial condition ($= 1$) for $\eta = \eta^+$. Setting $\eta = \kappa^R$, we obtain the first equality of (7.218). Next we show the following equality:

$$\left(\prod_r p_r! \right) \frac{d_{\kappa^R}}{d_\kappa''} = N! \frac{e_\kappa}{e_\kappa''}, \tag{7.224}$$

where we have introduced

$$d_\eta'' = \prod_{s\in D(\eta)} d''(s), \qquad d''(s) = \left(\frac{a(s)}{\lambda} + l(s) + 1 \right), \tag{7.225}$$

$$e_\eta'' = e_\kappa'' = \prod_{s\in D(\kappa)} e''(s), \qquad e''(s) = \frac{a'(s)}{\lambda} + N - l'(s). \tag{7.226}$$

Note that $d''(s)$ reduces to the lower hook length $h_*^\kappa(s)$ in the case of η being a partition κ.

Figure 7.6 shows diagrams of partitions κ (a) and κ^R (b). For the two squares s' and s'' in the two diagrams, the relations $a(s') = a(s'') + 1$ and $l(s') = l(s'')$ are satisfied, from which $d''(s') = d(s'')$ follows and further

$$\prod_{s:\,\text{white squares in }D(\kappa^R)} d(s) \Bigg/ \prod_{s:\,\text{white squares in }D(\kappa)} d''(s) = 1 \tag{7.227}$$

follows. We thus obtain

$$\frac{d_{\kappa^R}}{d_\kappa''} = \prod_{s:\,\text{shaded squares in }D(\kappa^R)} d(s) \Bigg/ \prod_{s:\,\text{shaded squares in }D(\kappa)} d''(s). \tag{7.228}$$

(a)   (b)

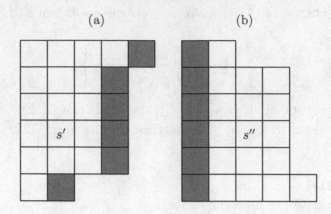

Fig. 7.6. (a) Diagram of a partition κ and (b) that of κ^{R}. In the two squares s' in $D(\kappa)$ and s'' in $D(\kappa^{\mathrm{R}})$, we note that $a(s') = a(s'') + 1$ and $l(s') = l(s'')$ and hence $d''(s') = d(s'')$.

For the shaded squares in $D(\kappa)$, i.e., the rightmost squares in each row, the arm $a(s)$ is zero and $l(s)$ measures the number of squares located below s. The denominator (7.228) hence becomes

$$\prod_{s:\ \text{shaded squares in } D(\kappa)} d''(s) = \prod_{s:\ \text{shaded squares in } D(\kappa)} (l(s) + 1) = \prod_r p_r!.$$
(7.229)

For the shaded squares in $D(\kappa^{\mathrm{R}})$, i.e., the leftmost squares in each row, we have $a(s) = \kappa_i^{\mathrm{R}} - 1$ and $l(s) = i - 1$ as determined by the ranking r_i. The numerator (7.228) then becomes

$$\prod_{s:\ \text{shaded squares in } D(\kappa^{\mathrm{R}})} d(s) = \prod_{i=1}^{N} \left(\frac{\kappa_i}{\lambda} + N - i + 1 \right),$$
(7.230)

to which the LHS of (7.224) reduces. By a similar argument, on the RHS of (7.224) we obtain

$$\frac{e_\kappa}{e_\kappa''} = \prod_{s:\ \text{lightly shaded squares in } D(\kappa)} e(s) \Bigg/ \prod_{s:\ \text{heavily shaded squares in } D(\kappa)} e''(s),$$
(7.231)

where lightly and heavily shaded squares are shown in Fig. 7.7. In the numerator of (7.231), $a'(s) = \kappa_i - 1$ and $l'(s) = i - 1$ from which

$$\prod_{s:\ \text{lightly shaded squares in } D(\kappa)} e(s) = \prod_{i=1}^{\kappa_1'} \left(\frac{\kappa_i}{\lambda} + N - i + 1 \right)$$
(7.232)

(a)

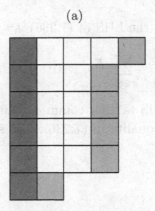

Fig. 7.7. Diagram of a partition κ. The leftmost squares in each row are heavily shaded and the rightmost squares in each row are lightly shaded. These are used in (7.231).

follows. In the denominator of (7.231), $a'(s) = 0$ and $l'(s) = i - 1$ lead to

$$\prod_{s:\text{ heavily shaded squares in } D(\kappa)} e(s) = \prod_{i=1}^{\kappa'_1} (N - i + 1). \qquad (7.233)$$

Using (7.232) and (7.233), the RHS of (7.224) is given by

$$\frac{N! \prod_{i=1}^{\kappa'_1} (\kappa_i/\lambda + N - i + 1)}{\prod_{i=1}^{\kappa'_1} (N - i + 1)} = \prod_{i=1}^{N} \left(\frac{\kappa_i}{\lambda} + N - i + 1 \right). \qquad (7.234)$$

From (7.230) and (7.234), we obtain (7.224). Multiplying both sides of (7.224) by d''_κ/d_κ, we arrive at

$$\left(\prod_r p_r! \right) d_{\kappa^R}/d_\kappa = N! d''_\kappa e_\kappa/(d_\kappa e''_\kappa), \qquad (7.235)$$

which is nothing but the second equality of (7.218). ∎

7.3.2 Evaluation formula

The evaluation formula for symmetric Jack polynomials is given by

$$J_\kappa(1, \ldots, 1) = \frac{N! e_\kappa}{\rho_\kappa^{(+)} d_\kappa} = \frac{e''_\kappa}{d''_\kappa}, \qquad (7.236)$$

with $\kappa \in \Lambda_N^+$.

Proof of (7.236).
From (7.215) and (7.216), the LHS of (7.236) is written as

$$J_\kappa(1,\ldots,1) = \frac{1}{\rho_\kappa^{(+)}} \sum_{\sigma \in S_N} E_\kappa(z_{\sigma(1)},\ldots,z_{\sigma(N)})\Big|_{z \to 1^N}. \qquad (7.237)$$

Each summand on the RHS takes a common value e_κ/d_κ owing to (7.129). Thus we obtain the first equality in (7.236). The second equality follows by (7.218). ∎

7.3.3 Symmetry-changing operator

We introduce an operator $\hat{O}^{(-)}$ which transforms $J_{\kappa_-}^{(-)}$ into J_κ, namely from the fermionic function to a bosonic one. An analogous operator for the U(2) Jack polynomials will be used in order to derive the evaluation formula, as explained later. Since the nature of the symmetry change becomes clearest in the single-component case, we discuss here its simplest version. The operator $\hat{O}^{(-)}$ is defined by [63]

$$\hat{O}^{(-)} = \prod_{i<j} \left(\hat{d}_i - \hat{d}_j + 1 \right) \qquad (7.238)$$

in terms of the Cherednik–Dunkl operators. We shall prove the following relation:

$$\hat{O}^{(-)} J_\kappa^{(-)}(z) = \pi_\kappa J_\kappa(z), \qquad (7.239)$$

with

$$\pi_\kappa = \prod_{i<j} \left(\bar{\kappa}_i - \bar{\kappa}_j + 1 \right). \qquad (7.240)$$

Proof of (7.239).
A non-symmetric Jack polynomial E_η is an eigenfunction of $\hat{O}^{(-)}$ with the eigenvalue

$$\prod_{i<j} \left(\bar{\eta}_i - \bar{\eta}_j + 1 \right). \qquad (7.241)$$

In the case of $\eta^+ = \kappa$, this eigenvalue is rewritten as

$$\pi_\kappa \prod_{i<j;\, \text{s.t.}\, \eta_i < \eta_j} \frac{(\bar{\eta}_i - \bar{\eta}_j + 1)}{(\bar{\eta}_j - \bar{\eta}_i + 1)} = \pi_\kappa \frac{a_\eta^{(+)}}{a_\eta^{(-)}}, \qquad (7.242)$$

in terms of (7.139), (7.212), and (7.240). With use of this, we obtain

$$\hat{O}^{(-)} J_\kappa^{(-)} = \sum_\eta a_\eta^{(-)} \hat{O}^{(-)} E_\eta$$

$$= \pi_\kappa \sum_\eta a_\eta^{(+)} E_\eta = \pi_\kappa J_\kappa. \tag{7.243}$$

∎

Let us introduce a symmetric polynomial $\tilde{J}_\kappa^{(-)}(z)$ by

$$J_\kappa^{(-)}(z) = \tilde{J}_\kappa^{(-)}(z) \Delta(z), \tag{7.244}$$

where $\Delta(z) = \prod_{i<j} (z_i - z_j)$. Consideration of the integral norm and triangularity shows that $\tilde{J}_\kappa^{(-)}(z)$ is actually equal to the symmetric Jack polynomial with shifted λ and κ:

$$\tilde{J}_\kappa^{(-)}(z; \lambda) = J_{\kappa-\delta}(z; \lambda+1), \tag{7.245}$$

where $\delta = (N-1, N-2, \ldots, 1, 0)$, and we have made explicit the dependence on the repulsion parameter λ.

In the particular case of $\kappa = \delta$, we obtain $\tilde{J}_\delta^{(-)}(z) = 1$ from (7.245) and $J_\delta^{(-)}(z) = \Delta(z)$ from (7.164). We shall prove the following evaluation formula for $\tilde{J}_\kappa^{(-)}(z, \lambda)$ with $\kappa \in \Lambda_N^{+>}$:

$$\tilde{J}_\kappa^{(-)}(1^N) = \frac{\tilde{\pi}_\kappa e_\kappa}{d_\kappa e_\delta}, \tag{7.246}$$

where

$$\tilde{\pi}_\kappa = \prod_{i<j} (\bar{\kappa}_i - \bar{\kappa}_j). \tag{7.247}$$

Since $\tilde{J}_\delta^{(-)}(1^N) = 1$, (7.246) requires the relation

$$\tilde{\pi}_\delta = d_\delta, \tag{7.248}$$

which can easily be checked. Namely, by putting explicit values for $\bar{\kappa}_i = (N-i)/\lambda - (i-1)$ for the case of $\kappa = \delta$, we obtain $\bar{\kappa}_i - \bar{\kappa}_j = (1+1/\lambda)(j-i)$ and

$$\tilde{\pi}_\delta = \left(\frac{\lambda+1}{\lambda}\right)^{N(N-1)/2} (N-1)!(N-2)! \cdots 2!. \tag{7.249}$$

On the other hand, we have $a(s) = l(s)$ for $\kappa = \delta$, and the product of $d(s) = [a(s)+1]/\lambda + l(s) + 1$ over s leads to the same expression as (7.249). Hence we obtain (7.248).

In order to prove (7.246), we begin with the evaluation

$$\rho_\kappa^{(-)} \hat{d}_i J_\kappa^{(-)}(z) = \hat{d}_i \sum_{\sigma \in S_N} \text{sgn}(\sigma) E_\kappa(z_{\sigma(1)} \ldots z_{\sigma(N)})$$

$$= \sum_{\sigma \in S_N} \text{sgn}(\sigma) \bar{\kappa}_{\sigma^{-1}(i)} E_\kappa(z_{\sigma(1)} \ldots z_{\sigma(N)})$$

$$\underset{z=1^N}{\to} \frac{e_\kappa}{d_\kappa} \sum_{\sigma \in S_N} \text{sgn}(\sigma) \bar{\kappa}_{\sigma(i)} = 0, \qquad (7.250)$$

where we have used the relation $\text{sgn}(\sigma) = \text{sgn}(\sigma^{-1})$. The RHS vanishes for $N > 2$ after summation over permutation. More generally, for any set of non-negative exponents $n(i)$, we obtain

$$\rho_\kappa^{(-)} \hat{d}_1^{n(1)} \hat{d}_2^{n(2)} \cdots \hat{d}_N^{n(N)} J_\kappa^{(-)}(z) \underset{z=1^N}{\to} \frac{e_\kappa}{d_\kappa} \sum_{\sigma \in S_N} \text{sgn}(\sigma) \bar{\kappa}_{\sigma(1)}^{n(1)} \cdots \bar{\kappa}_{\sigma(N)}^{n(N)}. \quad (7.251)$$

Note that the RHS vanishes unless the $n(i)$ are distinct from each other. The set with the minimum sum of exponents is given by $n(i) = N - i$ and its permutations. In this case we obtain

$$\sum_{\sigma \in S_N} \text{sgn}(\sigma) \bar{\kappa}_{\sigma(1)}^{N-1} \bar{\kappa}_{\sigma(2)}^{N-2} \cdots \bar{\kappa}_{\sigma(N)}^0 = \prod_{i<j} (\bar{\kappa}_i - \bar{\kappa}_j) = \tilde{\pi}_\kappa. \qquad (7.252)$$

In the next stage, we shall show

$$\hat{O}^{(-)} f(z) \Delta(z) \underset{z=1^N}{\to} f(1^N) \hat{O}^{(-)} \Delta(z) \Big|_{z \to 1^N}. \qquad (7.253)$$

The derivation uses the identity

$$(K_{ij} - 1)[\Delta(z)f(z)] = f(z)(K_{ij} - 1)\Delta(z) + [K_{ij}\Delta(z)](K_{ij} - 1)f(z), \qquad (7.254)$$

which is analogous to the chain rule in differential calculus. Because of the odd property $K_{ij}\Delta = -\Delta$, the RHS can also be written as $-\Delta(K_{ij} + 1)f$. Using (7.254), we obtain for any polynomial $f(z)$

$$\hat{d}_i [f(z)\Delta(z)] = f(z)\hat{d}_i\Delta(z) + g(z)\Delta(z), \qquad (7.255)$$

where

$$g(z) = \left[\frac{2z_i}{\lambda} \frac{\partial}{\partial z_i} - \hat{d}_i \right] f(z) \qquad (7.256)$$

is also a polynomial of z. We can show that the second term in the RHS of (7.255) can be neglected in the limit of $z^n \to 1$ after operation of $\hat{O}^{(-)}$. Namely, in the expansion of $\hat{O}^{(-)}$ in terms of $\hat{d}_i$, any product $\hat{d}_1^{n(1)} \hat{d}_2^{n(2)} \cdots$ $\hat{d}_i^{n(i)-1} \cdots \hat{d}_N^{n(N)}$ operating on $g(z)\Delta(z)$ has a sum of exponents less than

$N(N-1)/2$. Then the resultant terms vanish in the limit $z = 1^N$, owing to the antisymmetric property of $\Delta(z) = J_\delta^{(-)}(z)$. The only surviving term contains all operators $\hat{d}_i$ acting on $\Delta(z)$. Namely, any term containing $\hat{d}_i$ acting on $f(z)$ need not be kept in the evaluation $z = 1^N$. This leads to (7.253).

In the final stage, we put $f(z) = \tilde{J}_\kappa^{(-)}(z)$ and obtain

$$\hat{O}^{(-)} J_\kappa^{(-)}(z) \underset{z=1^N}{\to} \tilde{J}_\kappa^{(-)}(1^N) \hat{O}^{(-)} \Delta(z) \Big|_{z \to 1^N}. \tag{7.257}$$

On the other hand, using (7.239) gives another evaluation:

$$\hat{O}^{(-)} J_\kappa^{(-)}(z) \underset{z=1^N}{\to} \frac{N! \pi_\kappa e_\kappa}{\rho_\kappa^{(+)} d_\kappa} = \frac{N! \tilde{\pi}_\kappa e_\kappa}{d_\kappa}, \tag{7.258}$$

where we have used $\rho_\kappa^{(+)} = \pi_\kappa / \tilde{\pi}_\kappa$. In particular, the case $\kappa = \delta$ gives

$$\hat{O}^{(-)} J_\delta^{(-)}(z) \underset{z=1^N}{\to} \hat{O}^{(-)} \Delta(z) \Big|_{z \to 1^N} = \frac{N! \tilde{\pi}_\delta e_\delta}{d_\delta} = N! e_\delta. \tag{7.259}$$

Comparison between (7.257), (7.258), and (7.259) finally leads to the evaluation formula (7.246).

7.3.4 Bosonic description of partitions

The evaluation formula of (7.246) is rewritten as

$$\tilde{J}_\kappa(1^N) = \left(e_\mu'' / d_\mu'' \right) \Big|_{\lambda \to \lambda + 1} \tag{7.260}$$

in terms of the quantities defined on the Young diagram for the partition $\mu = \kappa - \delta$. Here e_μ'' and d_μ'' have been defined in (7.226) and (7.225), respectively. The expressions (7.260) and (7.245) reproduce the evaluation formula (7.236) for symmetric Jack polynomials. More generally, we obtain

$$d_\kappa / \tilde{\pi}_\kappa = \left(\frac{\lambda + 1}{\lambda} \right)^{|\mu|} d_\mu'' \Big|_{\lambda \to \lambda + 1} \tag{7.261}$$

and

$$e_\kappa / e_\delta = \left(\frac{\lambda + 1}{\lambda} \right)^{|\mu|} e_\mu'' \Big|_{\lambda \to \lambda + 1}. \tag{7.262}$$

The relations (7.261) and (7.262) involve the transition from a fermionic partition $\kappa \in \Lambda_N^{+>}$ to a subpartition $\mu \in \Lambda_N^+$ [17]. This procedure amounts to concentrating only on the excitations from the ground state, and is analogous to bosonization in field theory. The increase $\lambda \to \lambda + 1$ of the coupling constant follows the transition.

We shall first derive the relation between $\tilde{\pi}_\kappa$ and d_κ. In the special case of $\kappa = \delta$, we have already shown their equality in (7.248). Let us take a

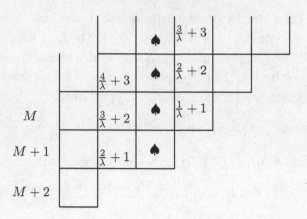

Fig. 7.8. Part of a fermionic partition $\kappa \in \Lambda^{+>}$ with indication of $d(s)$ for two columns flanking the column (♠) representing an excitation. The row index is written as M, etc.

fermionic partition $\kappa \in \Lambda_N^{+>}$ which has a part as shown in Fig. 7.8, and consider a cell $s = (M, \kappa_M)$ in the figure. Then we obtain

$$d(s) = \frac{a(s) + 1}{\lambda} + l(s) + 1 = \frac{1}{\lambda} + 1. \tag{7.263}$$

Figure 7.8 shows the value of $d(s)$ for each cell with column $j = \kappa_M = \kappa_{M+1} + 1$. Let us factorize $\tilde{\pi}_\kappa$ as $\tilde{\pi}_\kappa = \prod_{M=1}^{N-1} P_M$, with

$$P_M = \prod_{i=1}^{M} (\bar{\kappa}_i - \bar{\kappa}_{M+1}). \tag{7.264}$$

Then we obtain for the case in Fig. 7.8,

$$P_M = \left(\frac{1}{\lambda} + 1\right)\left(\frac{2}{\lambda} + 2\right)\cdots\left(\frac{M}{\lambda} + M\right). \tag{7.265}$$

The result is precisely the same as the product of $d(s)$ taken over the column with $j = \kappa_{M+1} + 1$.

Figure 7.8 also shows some values of $d(s)$ for another column with $j = \kappa_{M+2} + 1$. The $d(s)$-product taken over the column agrees now with

$$P_{M+1} = \prod_{i}^{M+1} (\bar{\kappa}_i - \bar{\kappa}_{M+2}) = \left(\frac{2}{\lambda} + 1\right)\left(\frac{3}{\lambda} + 2\right)\cdots\left(\frac{M+2}{\lambda} + M + 1\right). \tag{7.266}$$

As indicated by ♠ in Fig. 7.8, d_κ can have a remaining column that does not correspond to any of the P_i.

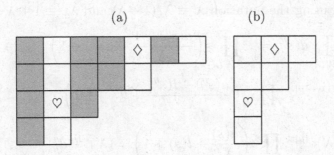

Fig. 7.9. Diagrams of (a) a partition $\kappa = (7, 5, 3, 2, 0)$ and (b) the partition $\mu = (3, 2, 1, 1)$ corresponding to $W(\kappa)$, with the relation $\kappa = \mu + \delta$. In (a), white squares form the set W, while shaded squares form the set S.

Generalizing this result for $1 \le i < j \le N$, we find the relation

$$d(s = (i, \kappa_j + 1)) = \frac{\kappa_i - \kappa_j}{\lambda} - j + i = \bar{\kappa}_i - \bar{\kappa}_j. \qquad (7.267)$$

For $\kappa \in \Lambda_N^{+>}$, we define a subset $S(\kappa)$ of $D(\kappa)$ as

$$S(\kappa) = \{s = (i, \kappa_j + 1) | 1 \le i < j \le N\} \qquad (7.268)$$

and $W(\kappa)$ as the complementary set of $S(\kappa)$ in $D(\kappa)$. Figure 7.9(a) shows an example for a partition $\kappa = (7, 5, 3, 2, 0)$, where shaded squares belong to $S(\kappa)$ and unshaded squares $W(\kappa)$. From (7.267), we obtain $\tilde{\pi}_\kappa = \prod_{s \in S(\kappa)} d(s)$ and

$$\frac{d_\kappa}{\tilde{\pi}_\kappa} = \prod_{s \in W(\kappa)} d(s). \qquad (7.269)$$

By comparing Fig. 7.9(a) and (b), we see that there is a one-to-one correspondence between $s'' \in W(\kappa)$ and $s \in D(\mu)$. Namely, we obtain Fig. 7.9(b) by removing shaded columns in Fig. 7.9(a), and putting the remaining ones together by a horizontal shift. We parameterize $s'' \in W(\kappa)$ as

$$s'' = (i, N - l(s) - i + j) \in W(\kappa), \quad \text{where } s = (i, j) \in D(\mu), \qquad (7.270)$$

as illustrated in Fig. 7.9(a) and (b). The squares marked by $\heartsuit$ in the two diagrams represent s and s'' related by (7.270). The same applies to squares marked by $\diamondsuit$. This parameterization does not change the leg lengths, but does change the arm lengths by the number of shaded columns to the right of the square s'' in $D(\kappa)$. We thus obtain

$$a(s'')|_{D(\kappa^{\mathrm{R}})} = (a(s) + l(s))|_{D(\mu)}, \quad l(s'')|_{D(\kappa^{\mathrm{R}})} = l(s)|_{D(\mu)}.$$

We obtain, using the notation $\lambda' \equiv \lambda/(1+\lambda)$ and $\lambda_+ \equiv 1 + \lambda$,

$$
\prod_{s'' \in W(\kappa)} d(s'') = \prod_{s'' \in W(\kappa)} \left(\frac{a(s'') + 1}{\lambda} + l(s'') + 1 \right)
$$

$$
= (\lambda')^{-|\mu|} \prod_{s'' \in W(\kappa)} \left(\frac{a(s'') - l(s'')}{\lambda_+} + l(s'') + 1 \right)
$$

$$
= (\lambda')^{-|\mu|} \prod_{s \in D(\mu)} \left(\frac{a(s)}{\lambda_+} + l(s) + 1 \right) = (\lambda')^{-|\mu|} d''_\mu \Big|_{\lambda \to \lambda_+}. \tag{7.271}
$$

A similar bosonic description is applicable to e_κ/e_δ. The LHS of (7.262) is written as

$$
\frac{e_\kappa}{e_\delta} = \prod_{s \in D(\kappa) \backslash D(\delta)} \left(\frac{a'(s) + 1}{\lambda} + N - l'(s) \right). \tag{7.272}
$$

We consider the one-to-one correspondence between squares in $D(\kappa) \backslash D(\delta)$ and $D(\mu)$ as

$$
s = (i, j + N - i) \text{ in } D(\kappa) \backslash D(\delta) \leftrightarrow s' = (i, j) \text{ in } D(\mu) \tag{7.273}
$$

so that the relations

$$
a'(s) \Big|_{D(\kappa) \backslash D(\delta)} = (a'(s') + N - l'(s') - 1) \Big|_{D(\mu)},
$$

$$
l'(s) \Big|_{D(\kappa) \backslash D(\delta)} = l'(s') \Big|_{D(\mu)} \tag{7.274}
$$

follow. From (7.274), we obtain

$$
\frac{e_\kappa}{e_\delta} = \prod_{s' \in D(\mu)} \left(\frac{a'(s') + N - l'(s') - 1 + 1}{\lambda} + N - l'(s') \right)
$$

$$
= (\lambda')^{-|\mu|} \prod_{s' \in D(\mu)} \left(\frac{a'(s')}{\lambda_+} + N - l'(s') \right) = (\lambda')^{-|\mu|} e''_\mu \Big|_{\lambda \to \lambda_+}
$$

$$
= (\lambda')^{-|\mu|} \dot{e}_\mu(\lambda_+; \lambda_+ N), \tag{7.275}
$$

where we have introduced the notation

$$
\dot{e}_\mu(\lambda; \lambda M) \equiv \prod_{s \in D(\mu)} (a'(s)/\lambda + M - l's)). \tag{7.276}
$$

Note the relation

$$
\lambda^{|\mu|} \dot{e}_\mu(\lambda; \lambda M) = (\lambda M)_\mu, \tag{7.277}
$$

where the generalized factorial on the RHS has been introduced by (7.174).

For later use in (7.306) and (7.5.4) we shall also derive the relation

$$d'_\kappa/\pi^{\mathrm{F}}_\kappa = \prod_{s\in W(\kappa)} d'(s) = (\lambda')^{-|\mu|} \prod_{s'\in W(\mu)} \left(\frac{a(s')+1}{\lambda_+} + l(s')\right), \qquad (7.278)$$

where

$$\pi^{\mathrm{F}}_\kappa = \prod_{i<j} (\bar\kappa_i - \bar\kappa_j - 1). \qquad (7.279)$$

As a special case, we obtain $d'_\delta/\pi^{\mathrm{F}}_\delta = 1$. To derive (7.278), we carry out the factorization $\pi^{\mathrm{F}}_\kappa = \prod_{M=1}^{N-1} Q_M$ such that

$$Q_M = \prod_{i=1}^{M} (\bar\kappa_i - \bar\kappa_{M+1} - 1), \qquad (7.280)$$

and make a correspondence with d'_κ. For example, we obtain for κ in Fig. 7.8

$$Q_M = \frac{1}{\lambda} \left(\frac{2}{\lambda} + 1\right) \cdots \left(\frac{M}{\lambda} + M - 1\right), \qquad (7.281)$$

which agrees precisely with the product of

$$d'(s) = \frac{a(s)+1}{\lambda} + l(s), \qquad (7.282)$$

taken over the column with $j = \kappa_{M+1} + 1$. The product of $d'(s)$ taken over another column with $j = \kappa_{M+2} + 1$ agrees with

$$Q_{M+1} = \frac{2}{\lambda} \left(\frac{3}{\lambda} + 1\right) \cdots \left(\frac{M+2}{\lambda} + M\right). \qquad (7.283)$$

Then, following the same argument as that for $d_\kappa/\tilde\pi$, we obtain the first equality in (7.278).

The second equality can be shown by following the same step as in (7.271).

7.3.5 Integral norm

In this subsection, we discuss the integral norm of J_κ, which satisfies

$$\frac{\langle J_\kappa, J_\kappa\rangle_0}{\langle E_\kappa, E_\kappa\rangle_0} = \frac{N!}{\rho^{(+)}_\kappa}. \qquad (7.284)$$

In the derivation of (7.284), the following two relations are used:

$$\mathrm{Sym}\,(\mathrm{Sym}\,f) = N!\mathrm{Sym}\,f, \qquad (7.285)$$

$$\langle f, \mathrm{Sym}\,g\rangle_0 = \langle \mathrm{Sym}\,f, g\rangle_0. \qquad (7.286)$$

The relation (7.285) is obvious, and (7.286) is derived as follows. Replacing f, g by $K_i f$ and $K_i g$ in the second equation of (7.157), we obtain

$$\langle f, K_i g \rangle_0 = \langle K_i f, g \rangle_0, \tag{7.287}$$

since $K_i^2 = 1$. A permutation $\hat{\sigma}$ in S_N can be expressed in the form of a product of $\{K_i\}$, and hence

$$\langle f, \hat{\sigma} g \rangle_0 = \langle \hat{\sigma}^{-1} f, g \rangle_0 \tag{7.288}$$

follows from the above relation. Summing both sides of (7.288) with respect to $\sigma \in S_N$, we obtain

$$\left\langle f, \sum_{\sigma \in S_N} \hat{\sigma} g \right\rangle = \left\langle \sum_{\sigma \in S_N} \hat{\sigma}^{-1} f, g \right\rangle, \tag{7.289}$$

from which (7.286) follows. The definitions (7.215) and (7.288) lead to (7.286). Using (7.216), (7.285), and (7.286), we obtain

$$\left(\rho_\kappa^{(+)} \right)^2 \langle J_\kappa, J_\kappa \rangle_0 = \langle \operatorname{Sym} E_\kappa, \operatorname{Sym} E_\kappa \rangle_0 = \langle E_\kappa, \operatorname{Sym} (\operatorname{Sym} E_\kappa) \rangle_0.$$

$$= N! \langle E_\kappa, (\operatorname{Sym} E_\kappa) \rangle_0 = N! \rho_\kappa^{(+)} \langle E_\kappa, J_\kappa \rangle_0$$

$$= N! \rho_\kappa^{(+)} \langle E_\kappa, E_\kappa \rangle_0, \tag{7.290}$$

where the last equality follows from the monic property (7.207). Then we obtain (7.284). Since we know $\langle E_\kappa, E_\kappa \rangle_0$ by (7.161) and (7.163), and $\rho_\kappa^{(+)}$ by (7.218), the integral norm (2.191) of symmetric Jack polynomials is derived as

$$\langle J_\kappa, J_\kappa \rangle_0 = \frac{\Gamma(N\lambda + 1)}{\Gamma(\lambda + 1)^N} \cdot \frac{d_\kappa' e_\kappa''}{d_\kappa'' e_\kappa'}. \tag{7.291}$$

7.3.6 Combinatorial norm

We shall prove (2.185) written as

$$j_\kappa \equiv \langle J_\kappa, J_\kappa \rangle_c = \prod_{s \in D(\kappa)} \frac{a(s) + \lambda l(s) + 1}{a(s) + \lambda l(s) + \lambda} = \prod_{s \in D(\kappa)} \frac{h_\kappa^*(s)}{h_*^\kappa(s)} \tag{7.292}$$

for $\kappa \in \Lambda_N^+$. We first show that the combinatorial inner product for symmetric Jack polynomials can be formulated in a manner analogous to the nonsymmetric version given by (7.44). For two bases of symmetric polynomials

$\{u_\kappa(z)\}$ and $\{v_\kappa(z)\}$ indexed by partitions κ, the following relations can be shown to be equivalent:

$$\langle u_\kappa, v_\mu \rangle_c = \delta_{\kappa,\mu}; \text{ for all } \kappa, \mu \in \Lambda_N^+$$

$$\Longleftrightarrow \sum_{\kappa \in \Lambda_N^+} u_\kappa(z) v_\kappa(y) = \prod_{i=1}^N \prod_{j=1}^N (1 - z_i y_j)^{-\lambda}, \qquad (7.293)$$

where the combinatorial inner product is defined by (2.176). We set $u_\kappa = J_\kappa$, $v_\kappa = (h_*^\kappa / h_\kappa^*) J_\kappa$ where the upper and lower hook lengths, h_κ^* and h_*^κ, are given by (2.186). Then we shall show

$$\sum_\kappa \left(\frac{h_*^\kappa}{h_\kappa^*} \right) J_\kappa(z) J_\kappa(y) = \prod_{i=1}^N \prod_{j=1}^N (1 - z_i y_j)^{-\lambda}, \qquad (7.294)$$

which amounts to proving (7.292).

Proof of (7.293).
We extend the number of variables to infinity by setting $z_j = y_j = 0$ for $j > N$ in the end. Such a device is necessary in dealing with power-sum symmetric polynomials. The RHS of the second relation in (7.293) is rewritten as

$$\prod_{i=1}^\infty \prod_{j=1}^\infty (1 - z_i y_j)^{-\lambda} = \exp\left(-\lambda \sum_{i,j=1}^\infty \ln(1 - z_i y_j) \right)$$

$$= \exp\left(\lambda \sum_{n=1}^\infty \sum_{i=1}^\infty \sum_{j=1}^\infty \frac{z_i^n y_j^n}{n} \right) = \exp\left(\lambda \sum_{n=1}^\infty \frac{p_n(z) p_n(y)}{n} \right)$$

$$= \left(\sum_{l_1=0}^\infty \frac{\lambda^{l_1}}{1^{l_1} l_1!} p_1(z)^{l_1} p_1(y)^{l_1} \right) \left(\sum_{l_2=0}^\infty \frac{\lambda^{l_2}}{2^{l_2} l_2!} p_2(z)^{l_2} p_2(y)^{l_2} \right) \cdots$$

$$= \sum_{\kappa \in \Lambda_\infty^+} \frac{\lambda^{l(\kappa)}}{\zeta_\kappa} p_\kappa(z) p_\kappa(y), \qquad (7.295)$$

where Λ_∞^+ is the $N \to \infty$ version of Λ_N^+, and ζ_κ in the last equality is defined in (2.175). The second equation in (7.293) is then equivalent to

$$\sum_{\kappa \in \Lambda_\infty^+} u_\kappa(z) v_\kappa(y) = \sum_{\kappa \in \Lambda_\infty^+} \frac{\lambda^{l(\kappa)}}{\zeta_\kappa} p_\kappa(z) p_\kappa(y). \qquad (7.296)$$

The power-sum polynomials $\{p_\kappa\}$ form a basis of symmetric polynomials.

Hence we can expand $u_\kappa(z)$, $v_\kappa(y)$ as

$$u_\kappa(z) = \sum_{\mu \in \Lambda_\infty^+} a_{\kappa\mu} p_\mu(z), \quad v_\kappa(y) = \sum_{\mu \in \Lambda_\infty^+} \frac{\lambda^{l(\kappa)}}{\zeta_\kappa} b_{\kappa\mu} p_\mu(z). \tag{7.297}$$

Using (7.297) and (2.176), the first equation of (7.293) can be rewritten as

$$\sum_{\kappa \in \Lambda_\infty^+} a_{\kappa\mu} b_{\kappa\nu} = \delta_{\mu\nu}. \tag{7.298}$$

The equivalence between (7.298) and the second relation of (7.293) is now obvious by imposing $z_j = y_j = 0$ for $j > N$. ∎

Proof of (7.294).
We need a few steps for the proof and begin with the relation

$$\prod_{j=1}^N \prod_{k=1}^N (1 - z_j y_k) \ \mathrm{Asym}_{(z)} \prod_{i=1}^N (1 - z_i y_i)^{-1}$$

$$= \prod_{j,k=1}^N (z_j y_k - 1) \ \mathrm{Asym}_{(z)} \prod_{i=1}^N (z_i y_j - 1)^{-1} = \Delta(z)\Delta(y), \tag{7.299}$$

where $\mathrm{Asym}_{(z)}$ is the antisymmetrization operator of the variables $\{z_i\}$. The relation (7.299) is a variant of Cauchy's double alternant formula, where the alternant refers to the Vandermonde determinant on the RHS. To confirm (7.299), we note that the LHS is an antisymmetric polynomial of $\{z_j\}$, and also of $\{y_j\}$. Hence the LHS is divisible by $\Delta(z)\Delta(y)$, and the quotient must be a symmetric polynomial of $\{z_j\}$ and $\{y_j\}$. Since the maximum power of z_j is $N - 1$ on both sides, the symmetric polynomial is actually of zeroth order, namely a constant. The constant is found to be unity by inspecting the coefficient of $z_1^{N-1} y_1^{N-1}$, which is unity on both sides.

Secondly we quote the result

$$\prod_{i=1}^N (1 - z_i y_i)^{-1} \prod_{i,j=1}^N (1 - z_i y_j)^{-\lambda} = \sum_\eta \frac{d_\eta}{d'_\eta} E_\eta(z) E_\eta(y), \tag{7.300}$$

which follows from (7.43), (7.53), and (7.181). By antisymmetrization of (7.300) with respect to z and using (7.299), we obtain

$$\Delta(z)\Delta(y)\prod_{j=1}^{N}\prod_{k=1}^{N}(1-z_j y_k)^{-1-\lambda} = \sum_{\eta\in\Lambda_N}\frac{d_\eta}{d'_\eta}\left[\text{Asym}\, E_\eta(z)\right]E_\eta(y)$$

$$= \sum_{\kappa\in\Lambda_N^>} J_\kappa^{(-)}(z)\sum_{\eta;\,\text{s.t.}\,\eta^+=\kappa}\frac{\rho_\eta^{(-)}d_\eta}{d'_\eta}E_\eta(y). \tag{7.301}$$

The leftmost side of (7.301) is antisymmetric with respect to y. Hence the expression

$$\sum_{\eta;\,\text{s.t.}\,\eta^+=\kappa}\frac{\rho_\eta^{(-)}d_\eta}{d'_\eta}E_\eta(y) \tag{7.302}$$

in the rightmost side of (7.301) is also antisymmetric, and is proportional to $J_\kappa^{(-)}(y)$. Recalling $a_\kappa^{(-)} = 1$ in (7.133), (7.302) is rewritten as

$$\sum_{\eta;\,\text{s.t.}\,\eta^+=\kappa}\frac{\rho_\eta^{(-)}d_\eta}{d'_\eta}E_\eta(y) = \frac{\rho_\kappa^{(-)}d_\kappa}{d'_\kappa}J_\kappa^{(-)}(y). \tag{7.303}$$

Substituting (7.303) into (7.301), we obtain

$$\Delta(z)\Delta(y)\prod_{j=1}^{N}\prod_{k=1}^{N}(1-z_j y_k)^{-1-\lambda} = \sum_{\kappa\in\Lambda_N^{+>}}\frac{\rho^{(-)}d_\kappa}{d'_\kappa}J_\kappa^{(-)}(z;\lambda)J_\kappa^{(-)}(y;\lambda). \tag{7.304}$$

Using the relation (7.208) to factor out $\Delta(z)\Delta(y)$ in (7.304), and putting $\lambda\to\lambda-1$, we obtain

$$\prod_{j=1}^{N}\prod_{k=1}^{N}(1-z_j y_k)^{-\lambda} = \sum_{\mu\in\Lambda_N^+}\left.\frac{\rho^{(-)}d_\kappa}{d'_\kappa}\right|_{\lambda\to\lambda-1}J_\mu(z;\lambda)J_\mu(y;\lambda), \tag{7.305}$$

where $\mu = \kappa - \delta$. Here we have again made the λ-dependence of $J_\kappa(z)$ explicit. Referring to (7.293), we have here proven that the symmetric Jack polynomials are orthogonal with respect to the combinatorial inner product defined by (2.176).

We now obtain

$$\left.\frac{\rho_\kappa^{(-)}d_\kappa}{d'_\kappa}\right|_{\lambda\to\lambda-1} = \left.\frac{\pi_\kappa^{\text{F}}d_\kappa}{d'_\kappa\,\tilde{\pi}_\kappa}\right|_{\lambda\to\lambda-1} = \frac{d''_\mu}{d'_\mu}, \tag{7.306}$$

where (7.156) has been used with π_κ^{F} defined by (7.278). The last equality is derived by use of the bosonic interpretation in Section 7.3.4:

$$\frac{d_\kappa}{\tilde{\pi}_\kappa}\Big|_{\lambda\to\lambda-1} = \left(1 - \frac{1}{\lambda}\right)^{-|\mu|} d_\mu'', \qquad \frac{d_\kappa'}{\pi_\kappa^{\mathrm{F}}}\Big|_{\lambda\to\lambda-1} = \left(1 - \frac{1}{\lambda}\right)^{-|\mu|} d_\mu', \qquad (7.307)$$

with $\mu = \kappa - \delta$. By using the relation $d_\mu'' = h_*^\mu$ and $d_\mu' = h_\mu^*$ for a case of μ being a partition, we arrive at (7.294). ∎

7.3.7 Binomial formula

Proof of the binomial formula (2.227).
We can now prove the binomial formula by using the corresponding results for non-symmetric Jack polynomials. From (7.181) and (7.178), we obtain the relation [154]

$$\prod_{i=1}^{N} (1 - z_i)^r = \sum_{\eta \in \Lambda_N} \frac{\lambda^{-|\eta|} (-r)_{\eta^+} E_\eta(z)}{d_\eta'}$$

$$= \sum_{\kappa \in \Lambda_N^+} \frac{\lambda^{-|\kappa|} (-r)_\kappa}{d_\kappa'} \sum_{\eta \in \Lambda_N^+; \text{ s.t. } \eta^+ = \kappa} \frac{d_{\eta^+}'}{d_\eta'} E_\eta(z)$$

$$= \sum_{\kappa \in \Lambda_N^+} (-1)^{|\kappa|} b_\kappa(r) J_\kappa(z), \qquad (7.308)$$

with

$$b_\kappa(r) = \frac{1}{d_\kappa'} \prod_{s \in D(\kappa)} \left(\frac{r - a'(s)}{\lambda} + l'(s)\right). \qquad (7.309)$$

The relation (7.308) reduces to (2.227) when $r = \lambda$. ∎

7.3.8 Power-sum decomposition

We shall derive the power-sum decomposition given by

$$p_m(z) = \frac{m}{\lambda} \sum_{\kappa \in \Lambda_\infty^+; \text{s.t.}|\kappa|=m} \vartheta_\kappa J_\kappa(z), \qquad (7.310)$$

where

$$\vartheta_\kappa = \frac{1}{d_\kappa'} \prod_{s'(\neq(1,1))\in D(\kappa)} \left(\frac{a'(s')}{\lambda} - l'(s')\right). \qquad (7.311)$$

Note that d_κ' for a partition κ is the same as the product of the upper hook length $h_\kappa^*(s)$ over $s \in D(\kappa)$. This formula has appeared in (2.308)

Proof of (7.310) [12, 65].

The power-sum decomposition used in Chapter 2 can be derived from the preceding binomial formula. We start with (7.308). By extending the number of variables to infinity, we rewrite the LHS of (7.308) as

$$\prod_{i=1}^{\infty} (1 - z_i)^r = \exp\left(-r \sum_{i=1}^{\infty} \ln(1 - z_i)\right) = \exp\left(-r \sum_{m=1}^{\infty} \frac{p_m(z)}{m}\right) \quad (7.312)$$

in terms of the power-sum symmetric functions $p_m(z)$. Then we find the following relation:

$$\frac{\partial}{\partial r} \exp\left(-r \sum_{m=1}^{\infty} \frac{p_m(z)}{m}\right)\bigg|_{r=0} = -\sum_{m=1}^{\infty} \frac{p_m(z)}{m}. \quad (7.313)$$

From (7.308) and (7.313), we obtain

$$p_m(z) = m \sum_{\kappa \in \lambda_\infty^+; \text{s.t.}|\kappa|=m} (-1)^{|\kappa|+1} \frac{\partial b_\kappa(r)}{\partial r}\bigg|_{r=0} J_\kappa(z). \quad (7.314)$$

It remains to find the expression of $\partial b_\kappa(r)/\partial r|_{r=0}$. Noting that d'_κ is independent of r, we get

$$(-1)^{|\kappa|+1} \frac{\partial b_\kappa(r)}{\partial r}\bigg|_{r=0} = \frac{1}{\lambda d'_\kappa} \sum_{s \in D(\kappa)} \prod_{s'(\neq s) \in D(\kappa)} \left(\frac{a'(s')}{\lambda} - l'(s')\right). \quad (7.315)$$

The factor $1/\lambda$ on the RHS of (7.315) comes from the derivative of $[a'(s) - r]/\lambda - l'(s)$ with respect to r. Because $a'(s') = l'(s') = 0$ for $s' = (1,1)$, all the terms but $s = (1,1)$ vanish in the summation with respect to s. Thus (7.315) becomes

$$\frac{1}{\lambda d'_\kappa} \prod_{s(\neq(1,1)) \in D(\kappa)} \left[\frac{a'(s)}{\lambda} - l'(s)\right] \equiv \frac{\vartheta_\kappa}{\lambda}. \quad (7.316)$$

From (7.314), (7.315), and (7.316), we obtain (7.310). ∎

7.3.9 Duality

In the particle–hole excitations in the Sutherland model, the particles have exclusion statistics characterized by λ. On the other hand, the statistical parameter for hole excitations is $1/\lambda$, and the holes are represented by conjugate partitions. Thus we have encountered the correspondence

$$\text{particles: } (\lambda, \kappa) \Longleftrightarrow \text{holes: } \left(\frac{1}{\lambda}, \kappa'\right),$$

where κ' is the conjugate partition of κ. In this subsection, we describe the above correspondence at the level of wave functions. As in the previous subsection, the variables z may include the infinite sequence. Hence we prefer the name of Jack functions rather than polynomials. By restricting the number of variables to a finite value, we recover the Jack polynomials.

Writing the λ-dependence explicitly, we define the duality transformation in the following manner:

$$\omega_\lambda J_\kappa(z;\lambda) = j_\kappa(\lambda) J_{\kappa'}(z;1/\lambda), \tag{7.317}$$

where j_κ is the combinatorial norm given by

$$j_\kappa(\lambda) = \frac{d'_\kappa(\lambda)}{d''_\kappa(\lambda)} = \prod_{s\in D(\kappa)} \frac{a(s) + \lambda l(s) + 1}{a(s) + \lambda l(s) + \lambda} = \prod_{s\in D(\kappa)} \frac{h^*_\kappa(s;\lambda)}{h_*^\kappa(s;\lambda)}. \tag{7.318}$$

Noting that the arm and leg lengths are interchanged in $D(\kappa')$, we obtain

$$j_{\kappa'}\left(\frac{1}{\lambda}\right) = \prod_{s\in D(\kappa')} \frac{h^*_{\kappa'}(s;1/\lambda)}{h_*^{\kappa'}(s;1/\lambda)} = \prod_{s\in D(\kappa)} \frac{a(s) + \lambda l(s) + \lambda}{a(s) + \lambda l(s) + 1} = \frac{1}{j_\kappa(\lambda)}. \tag{7.319}$$

Namely, we have

$$j_\kappa(\lambda) j_{\kappa'}(1/\lambda) = 1, \tag{7.320}$$

which guarantees the property $\omega_\lambda \omega_{1/\lambda} = 1$, or

$$\omega_\lambda^{-1} = \omega_{1/\lambda}. \tag{7.321}$$

In the special case of $\lambda = 1$, the property is referred to as involution, where the inverse of the duality transformation ω_1 is itself. The Schur function $s_\kappa(z)$ is relevant to the case, and satisfies the duality: $\omega_1 s_\kappa(z) = s_{\kappa'}(z)$.

This duality transformation has the remarkable property

$$\omega_\lambda p_\kappa(z) = (-\lambda)^{-l(\kappa)}(-1)^{|\kappa|} p_\kappa(z), \tag{7.322}$$

which means that the power-sum symmetric functions are eigenfunctions of the duality operator ω_λ. In order to prove (7.322), we operate ω_λ on both sides of (7.310). Then we obtain

$$\omega_\lambda p_m(z) = \frac{m}{\lambda} \sum_{\kappa\in\Lambda_\infty^+;\,\mathrm{s.t.}\,|\kappa|=m} \frac{1}{d'_\kappa(\lambda)} \cdot \frac{d'_\kappa(\lambda)}{d''_\kappa(\lambda)} J_{\kappa'}\left(z;\frac{1}{\lambda}\right)$$

$$\times \prod_{s(\neq(1,1))\in D(\kappa)} \left[\frac{a'(s)}{\lambda} - l'(s)\right]$$

$$= (-1)^{-m+1}\lambda^{-1} p_m(z). \tag{7.323}$$

Note that the arm and leg colengths are interchanged in the conjugate partition, and

$$d_\kappa''(\lambda) = d_{\kappa'}'(\lambda) = \lambda^{-|\kappa|} d_{\kappa'}'(1/\lambda). \tag{7.324}$$

Then the second equality in (7.323) follows by summing over κ'. For a partition $\kappa = (\kappa_1, \kappa_2, \ldots)$ with length $l(\kappa)$, we obtain p_κ as the product of p_m with $m = \kappa_i \neq 0$. Then we apply ω_λ on the product to obtain (7.322).

We next show

$$\omega_\lambda(z) \prod_{ij}(1 - z_i y_j)^{-\lambda} = \omega_\lambda(y) \prod_{ij}(1 - z_i y_j)^{-\lambda} = \prod_{ij}(1 + z_i y_j), \tag{7.325}$$

where $\omega_\lambda(z)$ means the duality operation on functions of z. It can be confirmed by the expansion

$$\prod_{ij}(1 + z_i y_j) = \exp\left(\sum_{i,j} \ln(1 + z_i y_j)\right) = \exp\left(-\sum_{n=1}^{\infty} \sum_{i,j} \frac{(-z_i y_j)^n}{n}\right)$$

$$= \exp\left(\sum_n (-1)^{n+1} \frac{p_n(z) p_n(y)}{n}\right)$$

$$= \left(\sum_{l_1=0}^{\infty} \frac{(-1)^{(1+1)l_1}}{1^{l_1} l_1!} p_1(z)^{l_1} p_1(y)^{l_1}\right) \left(\sum_{l_2=0}^{\infty} \frac{(-1)^{(2+1)l_2}}{2^{l_2} l_2!} p_2(z)^{l_2} p_2(y)^{l_2}\right) \cdots$$

$$= \sum_{\kappa \in \Lambda_\infty^+} \frac{(-1)^{|\kappa|+l(k)}}{\zeta_\kappa} p_\kappa(z) p_\kappa(y). \tag{7.326}$$

The final form is precisely the result we obtain by the duality transformation (7.322) for each term on the rightmost side of (7.295). Conversely, we obtain

$$\omega_{1/\lambda}(z) \prod_{ij}(1 + z_i y_j) = \prod_{ij}(1 - z_i y_j)^{-\lambda}. \tag{7.327}$$

If we use the Jack polynomials instead of power-sum polynomials for the expansion, we obtain the relation

$$\sum_\kappa J_\kappa(y; \lambda) J_{\kappa'}(z; 1/\lambda) = \prod_{i,j}(1 + y_j z_k), \tag{7.328}$$

where the RHS does not depend on λ.

7.3.10 Skew Jack functions and Pieri formula

For partitions $\mu \subseteq \kappa$, we define the skew Jack function by the following combinatorial inner product:

$$\langle J_{\kappa/\mu}, J_\nu \rangle_c \equiv \langle J_\kappa, J_\mu J_\nu \rangle_c \equiv g^\kappa_{\mu\nu}, \tag{7.329}$$

which also defines the quantity $g^\kappa_{\mu\nu}$. Equivalent relations can be written in terms of the combinatorial norm $j_\kappa = \langle J_\kappa, J_\kappa \rangle_c$ as

$$J_{\kappa/\mu}(z) = \sum_\nu g^\kappa_{\mu\nu} J_\nu(z)/j_\nu, \tag{7.330}$$

$$J_\mu(z) J_\nu(z) = \sum_\kappa g^\kappa_{\mu\nu} J_\kappa(z)/j_\kappa, \tag{7.331}$$

where z is the infinite sequence of variables. We have suppressed the repulsion parameter λ in the Jack functions $J_\kappa(z)$ for notational simplicity. We shall first derive the relation [169]

$$J_\kappa(x, y) = \sum_\mu J_{\kappa/\mu}(x) J_\mu(y) j_\mu^{-1}. \tag{7.332}$$

Proof of (7.332).
By a variant of the Cauchy product expansion formula (7.305) for variables $x = (x_1, \ldots, x_N)$, $y = (y_1, \ldots, y_M)$, and $z = (z_1, \ldots, z_{N+M})$, we obtain

$$\sum_\mu J_\mu(x) J_\mu(z) j_\mu^{-1} \sum_\nu J_\nu(y) J_\nu(z) j_\nu^{-1}$$

$$= \prod_{j=1}^N \prod_{k=1}^{N+M} (1 - x_j z_k)^{-\lambda} \prod_{l=1}^M \prod_{m=1}^{N+M} (1 - y_l z_m)^{-\lambda}$$

$$= \sum_\kappa J_\kappa(x, y) J_\kappa(z) j_\kappa^{-1}, \tag{7.333}$$

where $J_\kappa(x, y)$ has a combined set (x, y) of variables. Note that the Cauchy product expansion formula for Jack polynomials is valid even though the number of variables in x and z is different. This can be verified from (7.295) for power-sum symmetric functions by linear transformation to Jack polynomials.

For fixed μ, we have the relation

$$\sum_\kappa J_{\kappa/\mu}(x) J_\kappa(z) j_\kappa^{-1} = \sum_\nu J_\mu(z) J_\nu(x) J_\nu(z) j_\nu^{-1}, \tag{7.334}$$

which can be confirmed by taking the combinatorial inner product of both sides with $J_\nu(x) J_\kappa(z)$. Here, for two sets of variables, we define the inner product as

$$\langle J_\kappa(x)J_\mu(z), J_\sigma(x)J_\tau(z)\rangle_c = \langle J_\kappa(x), J_\sigma(x)\rangle_c \langle J_\mu(z), J_\tau(z)\rangle_c. \qquad (7.335)$$

Let us multiply both sides of (7.334) by $J_\mu(y)/j_\mu$, and sum over μ. The RHS then becomes equal to the first side of (7.333). Comparing the coefficients of $J_\kappa(z)$ in the final side of (7.333) and the LHS of (7.334) after multiplication by $J_\mu(y)/j_\mu$, we obtain (7.332). ∎

By taking $x = 1$ in (7.332), we obtain

$$J_\kappa(1, y) = \sum_\mu f_{\kappa\mu} J_\mu(y), \qquad (7.336)$$

where the coefficient is given in (2.260) by [169]

$$f_{\kappa\mu} = g_{\mu n}^\kappa j_n^{-1} j_\mu^{-1} = \prod_{s \in C_{\kappa/\mu} \backslash R_{\kappa/\mu}} \left(\frac{h_\kappa^*}{h_*^\kappa}\right)\left(\frac{h_*^\mu}{h_\mu^*}\right), \qquad (7.337)$$

with $n = |\kappa| - |\mu|$. The first equality uses (7.330) and $J_n(1) = 1$, and the second equality is derived below.

Proof of (7.337).
In order to derive $g_{\mu n}^\kappa$, we consider a partition μ with length $l(\mu) = N$. Then we may restrict x to N variables: $x_1, x_2, \ldots, x_N$. In the special case where ν in (7.331) consists of a single row with length n, we expand the product as

$$J_\mu(x)J_n(x) = \sum_\kappa g_{\mu n}^\kappa J_\kappa(x) j_\kappa^{-1}. \qquad (7.338)$$

The result is a generalization of the corresponding result

$$s_\mu(x)s_n(x) = \sum_\kappa s_\kappa(x) \qquad (7.339)$$

for Schur functions s_κ introduced in (2.5). The nonzero contribution κ on the RHS of (7.338) or (7.339) comes only from such κ as has at most one more cell for each column of μ, and $|\kappa| = |\mu| + n$. The latter restriction can be interpreted as the momentum conservation, while the first one is due to the single row of n. The Young diagram κ/μ in this case is called the horizontal n-strip. A similar simplification occurs if ν consists of a single column. In this case, the nonzero contribution of κ in κ/μ comes only from a vertical m-strip where m is the length of the single column ν. The results (7.338), (7.339), and their vertical strip versions are called the Pieri formula.

We derive $g_{\mu n}^{\kappa}$ by induction, noting that the only relevant cases are

$$\text{(a) } l(\kappa) = l(\mu) = N, \quad \text{(b) } l(\kappa) = l(\mu) + 1 = N + 1.$$

Let us first consider case (a). By a Galilean boost, we obtain the relation

$$J_{\mu}(z) = z_1 z_2 \cdots z_N J_{\mu-I}(z) = e_N(z) J_{\mu-I}(z), \qquad (7.340)$$

with the notation $I = 1^N$. By performing the same boost on the RHS of (7.331) with $\nu = n$, we can factor out $e_N(z)$ and obtain

$$J_{\mu-I}(z) J_n(z) = \sum_{\kappa} g_{\mu n}^{\kappa} J_{\kappa-I}(z)/j_{\kappa} = \sum_{\kappa} g_{\mu-I,n}^{\kappa-I} J_{\kappa-I}(z)/j_{\kappa-I}, \qquad (7.341)$$

where the last equality results from (7.338) by replacing μ by $\mu - I$. Comparing the coefficients of $J_{\kappa-I}(z,y)$, we obtain

$$g_{\mu n}^{\kappa}/j_{\kappa} = g_{\mu-I,n}^{\kappa-I}/j_{\kappa-I}. \qquad (7.342)$$

Since we know the combinatorial norm j_{κ} for any partition, we can iterate to partitions of shorter rows.

Let us now assume another case with $l(\kappa) = N + m > N$. We apply the duality operator ω_{λ} on both sides of (7.338). Writing the Jack functions with the repulsion parameter $1/\lambda$ as $\bar{J}_{\mu'}$ for notational simplicity, we obtain

$$\bar{J}_{\mu'}(z) \bar{J}_{1^n}(z) = \sum_{\kappa} \frac{g_{\mu n}^{\kappa}}{j_{\mu} j_n} \bar{J}_{\kappa'}(z), \qquad (7.343)$$

where we have used (7.317). The maximum power of z_1 on the LHS is $N+1$, where N now corresponds to the length μ'_1 of the first row. Then such κ' as have $\kappa'_1 > N + 1$ do not enter the summation on the RHS. This confirms that we need to consider only the case with $l(\kappa) = N + 1$.

The Jack polynomials have the property

$$J_{\kappa}(z_1, z_2, \ldots) = z_1^{\kappa_1} J_{\kappa_-}(z_2, \ldots) + \cdots, \qquad (7.344)$$

where $\kappa_- = (\kappa_2, \kappa_3, \ldots)$ and the omitted terms do not have the factor $z_1^{\kappa_1}$. Then by comparing the terms with the factor z_1^{N+1} on both sides of (7.343), we obtain

$$\bar{J}_{\mu'_-}(z) \bar{J}_{1^{n-1}}(z) = \sum_{\kappa} \frac{g_{\mu n}^{\kappa}}{j_{\mu} j_n} \bar{J}_{\kappa'_-}(z), \qquad (7.345)$$

where we have relabeled the variables z to begin at z_1. Now we apply the duality operator $\omega_{1/\lambda}$ on both sides of (7.345). We obtain

$$J_{\mu-I}(z)J_{n-1}(z) = \sum_\kappa \frac{g^\kappa_{\mu n} j_{\mu-I} j_{n-1}}{j_\mu j_n j_{\kappa-I}} J_{\kappa-I}(z) = \sum_\kappa \frac{g^{\kappa-I}_{\mu-I,n-1}}{j_{\kappa-I}} J_{\kappa-I}(z),$$

(7.346)

where we interpret $I = 1^{N+1}$ if I appears together with κ. Then we have

$$g^\kappa_{\mu n} = \frac{j_\mu j_n}{j_{\mu-I} j_{n-1}} g^{\kappa-I}_{\mu-I,n-1}.$$

(7.347)

Using either (7.342) or (7.347), we can continue the iteration until $g^\gamma_{\alpha\beta}$ is reduced to zero or unity with $\alpha = 0$ or $\beta = 0$. The product of factors j_n/j_{n-1}, which goes like $(j_n/j_{n-1})(j_{n-1}/j_{n-2})\ldots$, becomes equal to j_n because of the successive cancellation. Then the resultant j_n cancels with j_n^{-1} in (7.337). Furthermore, the factor $j_\mu/j_{\mu-I}$ cancels with the contribution from j_μ^{-1} for such a column as corresponds to κ/μ. Then we finally obtain the second equality in (7.337). ∎

7.4 U(2) Jack polynomials

7.4.1 Relation to non-symmetric Jack polynomials

The coefficient $a_\eta^{(--)}$ appearing in (3.177) can be obtained in a way similar to the case of antisymmetric and symmetric Jack polynomials. From the definition (3.177), (3.178), and the recursion relation (7.91) of non-symmetric Jack polynomials, we derive the recursion relation for the coefficient $a_\eta^{(--)}$ in (3.177) as [15]

$$a_{K_i\eta}^{(--)} = -(1 + \xi_i)a_\eta^{(--)}, \quad \eta \in (\Lambda^>_{N_\uparrow}, \Lambda^>_{N_\downarrow}),$$

(7.348)

for $i \in [1, N_\uparrow - 1]U[N_\uparrow + 1, N - 1]$ and $\eta_i > \eta_{i+1}$. An analogous calculation can be performed for the case of $\eta_i < \eta_{i+1}$. From (7.348) and the initial condition $a_{\kappa^\uparrow,\kappa^\downarrow}^{(--)} = 1$, we obtain for $\eta = (\eta^\uparrow, \eta^\downarrow) \in (\Lambda_{N_\uparrow}, \Lambda_{N_\downarrow})$

$$a_\eta^{(--)} = \text{sgn}(\eta^\uparrow)\text{sgn}(\eta^\downarrow)\frac{d_\eta}{d_\kappa}$$

$$= \text{sgn}(\eta^\uparrow)\text{sgn}(\eta^\downarrow) \prod_{\sigma=\uparrow,\downarrow} \prod_{(i<j)\in I_\sigma; \text{s.t.} \eta_i<\eta_j} \left(1 + \frac{1}{\overline{\eta}_j - \overline{\eta}_i}\right)$$

$$= \prod_{\sigma=\uparrow,\downarrow} \prod_{(i<j)\in I_\sigma; \text{s.t.} \eta_i<\eta_j} \left(\frac{\overline{\eta}_j - \overline{\eta}_i + 1}{\overline{\eta}_i - \overline{\eta}_j}\right),$$

(7.349)

where $\kappa = ((\eta^\uparrow)^+, (\eta^\downarrow)^+) \in (\Lambda^{+>}_{N_\uparrow}, \Lambda^{+>}_{N_\downarrow}) I_\uparrow = [1, N_\uparrow]$ and $I_\downarrow = [N_\uparrow + 1, N]$.

The alternating property (3.178) of $J_\kappa^{(--)}$ leads to the relation

$$\rho_\eta^{(--)} J_\kappa^{(--)}(z) = \sum_{\hat{\sigma} \in S_{N_\uparrow} \times S_{N_\downarrow}} \text{sgn}(\hat{\sigma})\hat{\sigma} E_\eta(z), \qquad (7.350)$$

where we have introduced $\kappa = (\kappa^\uparrow, \kappa^\downarrow) \in (\Lambda_{N_\uparrow}^{+>}, \Lambda_{N_\downarrow}^{+>})$ and $\eta = (\eta^\uparrow, \eta^\downarrow) \in (\Lambda_{N_\uparrow}, \Lambda_{N_\downarrow})$ such that $(\eta^\sigma)^+ = \kappa^\sigma$ for $\sigma = \uparrow, \downarrow$. The recursion relation for $\rho_\eta^{(--)}$ is given by

$$\rho_{K_i\eta}^{(--)} / \rho_\eta^{(--)} \begin{cases} \dfrac{1}{-1+\xi_i} & \text{for} \quad \eta_i > \eta_{i+1} \\ 1 + \xi_i & \text{for} \quad \eta_i < \eta_{i+1} \end{cases} \qquad (7.351)$$

Furthermore, we have

$$\rho_{(\kappa^{\uparrow R}, \kappa^{\downarrow R})}^{(--)} = (-1)^{N_\uparrow(N_\uparrow - 1)/2}(-1)^{N_\downarrow(N_\downarrow - 1)/2}. \qquad (7.352)$$

We now follow a route similar to that which led to (7.138) for antisymmetric Jack polynomials. Then we obtain

$$\rho_\eta^{(--)} = \text{sign}(\eta^\uparrow)\text{sign}(\eta^\downarrow)\frac{d'_\eta}{d'_{(\kappa^{\uparrow R}, \kappa^{\downarrow R})}}, \qquad (7.353)$$

where

$$\kappa^{\uparrow R} = (\kappa_{N_\uparrow}^\uparrow, \ldots, \kappa_1^\uparrow), \quad \kappa^{\downarrow R} = (\kappa_{N_\downarrow}^\downarrow, \ldots, \kappa_1^\downarrow).$$

Particularly, for a partition $\kappa \in \left(\Lambda_{N_\uparrow}^{+>}, \Lambda_{N_\downarrow}^{+>}\right)$, we obtain

$$\rho_\kappa^{(--)} = \prod_{s=\uparrow,\downarrow} \prod_{i<j \in I_s} \left(\frac{\bar{\kappa}_i - \bar{\kappa}_j - 1}{\bar{\kappa}_i - \bar{\kappa}_j}\right). \qquad (7.354)$$

Note that $J_\kappa^{(--)}$ reduces to the antisymmetric Jack polynomials when $N_\uparrow = 0$ or $N_\downarrow = 0$.

7.4.2 Integral norm

The polynomials are orthogonal to each other with respect to the integral norm. For $\kappa \in (\Lambda_{N\uparrow}^{+>}, \Lambda_{N\downarrow}^{+>})$, the relation

$$\langle J_\kappa^{(--)}, J_\kappa^{(--)} \rangle_0 / \langle E_\kappa, E_\kappa \rangle_0 = \frac{N_\uparrow! N_\downarrow!}{\rho_\kappa^{(--)}} \qquad (7.355)$$

can be derived in the same way as the derivation of (7.167). Using (7.161), (7.163), (7.353), and (7.355), we obtain for $\kappa \in (\Lambda_{N_\uparrow}^{+>}, \Lambda_{N_\downarrow}^{+>})$

$$\langle J_\kappa^{(--)}, J_\kappa^{(--)} \rangle_0 = N_\uparrow! N_\downarrow! \frac{\Gamma(N\lambda + 1)}{\Gamma(\lambda + 1)^N} \frac{e_\kappa d_\kappa'}{\rho_\kappa^{(--)} e_\kappa' d_\kappa}$$

$$= N_\uparrow! N_\downarrow! \frac{\Gamma(N\lambda + 1)}{\Gamma(\lambda + 1)^N} \frac{e_{\bar\kappa} d'_{(\kappa\uparrow R, \kappa\downarrow R)}}{e_\kappa' d_\kappa}. \tag{7.356}$$

7.4.3 Cauchy product expansion formula

For the coordinate $z = (z_1, \ldots, z_N)$ and a fixed integer $N_\uparrow \in I = [1, N]$, we define $z^\uparrow = (z_1, \ldots, z_{N_\uparrow})$ and $z^\downarrow = (z_{N_\uparrow+1}, \ldots, z_N)$. The Cauchy product expansion formula for U(2) Jack polynomials is given by [15]

$$\prod_{s=\uparrow,\downarrow} \prod_{i,j \in I_s} (1 - z_i y_j)^{-1} \prod_{i,j \in I} (1 - z_i y_j)^{-\lambda}$$

$$= \sum_{\mu \in (\Lambda_{N_\uparrow}^{+>}, \Lambda_{N_\downarrow}^{+>})} \frac{\rho_\mu^{(--)} d_\mu}{d_\mu'} \tilde{J}_\mu^{(--)}(z) \tilde{J}_\mu^{(--)}(y), \tag{7.357}$$

where

$$\tilde{J}_\mu^{(--)}(z) \equiv J_\mu^{(--)}(z) / [\Delta(z^\uparrow) \Delta(z^\downarrow)] \tag{7.358}$$

with

$$\Delta(z^\sigma) = \prod_{\substack{i,j \in I_\sigma \\ i<j}} (z_j - z_j). \tag{7.359}$$

The proof of the Cauchy product expansion formula (7.357) is based on the corresponding formula [154] for the non-symmetric polynomials

$$\Omega(z|y) = \sum_{\eta \in \Lambda_N} \frac{d_\eta}{d_\eta'} E_\eta(z) E_\eta(y)$$

and Cauchy's double alternant formula. The proof also requires the transformation properties (7.91) and (7.209), together with those for d_η and d_η' [154].

7.4.4 $U_B(2)$ Jack polynomials

In order to derive the evaluation formula (7.369) to follow, we make a detour and introduce auxiliary polynomials $J_\kappa^{(++)}(z)$, which may be termed the $U_B(2)$ Jack polynomials, since it constitutes the eigenfunctions of the U(2) bosonic Sutherland model. Namely, $J_\kappa^{(++)}(z_1, \ldots, z_N)$ for $\kappa = (\kappa^\uparrow, \kappa^\downarrow) \in$

$(\Lambda_{N_\uparrow}^+, \Lambda_{N_\downarrow}^+)$ constitute the eigenfunctions of the U(2) bosonic Sutherland model, and satisfy the following conditions [15]:

(i) The polynomial $J_\kappa^{(++)}$ has the form

$$J_\kappa^{(++)}(z) = \prod_{\sigma=\uparrow,\downarrow} \sum_{(\eta^\sigma)^+=\kappa^\sigma} a_{\eta^\uparrow,\eta^\downarrow}^{(++)} E_{\eta^\uparrow,\eta^\downarrow}(z) \qquad (7.360)$$

with the normalization

$$a_\kappa^{(++)} = 1.$$

(ii) Under the action of the transposition K_i, the polynomial $J_\kappa^{(++)}(z)$ for $\kappa = (\kappa^\uparrow, \kappa^\downarrow) \in (\Lambda_{N_\uparrow}^+, \Lambda_{N_\downarrow}^+)$ is transformed as

$$K_i J_\kappa^{(++)}(z) = J_\kappa^{(++)}(z) \qquad (7.361)$$

for $i \in [1, N_\uparrow - 1]$ or $i \in [N_\uparrow + 1, N - 1]$.

The coefficient in (7.360) is obtained for $\eta \in (\Lambda_{N_\uparrow}, \Lambda_{N_\downarrow})$ as

$$a_\eta^{(++)} = \prod_{\sigma=\uparrow,\downarrow} \prod_{(i<j)\in I_\sigma;\text{s.t.}\eta_i<\eta_j} \left(1 - \frac{1}{\bar{\eta}_j - \bar{\eta}_i}\right), \qquad (7.362)$$

by following a route similar to that used in Section 7.3.7 to derive (7.212).

From the symmetric property (7.361) of $J_\kappa^{(++)}$, we can also write

$$\rho_\eta^{(++)} J_\kappa^{(++)}(z) = \sum_{\sigma\in S_{N_\uparrow} \times S_{N_\downarrow}} \hat{\sigma} E_\eta(z) \qquad (7.363)$$

for $\kappa = (\kappa^\uparrow, \kappa^\downarrow) \in (\Lambda_{N_\uparrow}^+, \Lambda_{N_\downarrow}^+)$ with

$$\rho_\kappa^{(++)} = \prod_{\sigma=\uparrow\downarrow} \prod_{(i<j)\in I_\sigma} \left(\frac{\bar{\kappa}_i - \bar{\kappa}_j + 1}{\bar{\kappa}_i - \bar{\kappa}_j}\right). \qquad (7.364)$$

The expression for $\rho_\kappa^{(++)}$ can be derived in a way similar to the derivation of (7.217) for symmetric Jack polynomials. The evaluation formula

$$J_\kappa^{(++)}(1,\dots,1) = \frac{N_\uparrow! N_\downarrow!}{\rho_\kappa^{(++)}} E_\kappa(1,\dots,1) = \frac{N_\uparrow! N_\downarrow! e_\kappa}{\rho_\kappa^{(++)} d_\kappa} \qquad (7.365)$$

can also be derived in analogy with (7.236).

7.4.5 Evaluation formula

As a generalization of $\hat{O}^{(-)}$ in (7.238), we introduce a symmetry-changing operator $\hat{O}^{(--)}$, which acts on $J_\kappa^{(--)}$ and gives $J_\kappa^{(++)}$. The operator $\hat{O}^{(--)}$ is a useful device to derive the evaluation formula for $\tilde{J}_\kappa^{(--)}(1,\ldots,1)$, as explained in the next subsection. The operator $\hat{O}^{(--)}$ is defined by [63]

$$\hat{O}^{(--)} = \prod_{\sigma=\uparrow,\downarrow} \prod_{(i<j)\in I_\sigma} \left(\hat{d}_i - \hat{d}_j + 1 \right) \tag{7.366}$$

in terms of the Cherednik–Dunkl operators. In analogy with (7.239), the following relation can be derived:

$$\hat{O}^{(--)} J_\kappa^{(--)} = \pi_\kappa^{(--)} J_\kappa^{(++)}, \tag{7.367}$$

with

$$\pi_\kappa^{(--)} = \prod_{\sigma=\uparrow,\downarrow} \prod_{(i<j)\in I_\sigma} (\bar{\kappa}_i - \bar{\kappa}_j + 1). \tag{7.368}$$

The evaluation formula for U(2) Jack polynomials is given, apart from the Vandermonde part, by

$$\tilde{J}_\kappa^{(--)}(\underbrace{1,\ldots,1}_{N}) = \frac{\tilde{\pi}_\kappa^{(--)} e_\kappa}{d_\kappa e_{\delta_{\uparrow\downarrow}}}, \tag{7.369}$$

where the symbol $\delta_{\uparrow\downarrow}$ is an abbreviation for $(\delta(N_\uparrow), \delta(N_\downarrow))$ and

$$\tilde{\pi}_\kappa^{(--)} = \prod_{\sigma=\uparrow,\downarrow} \prod_{\substack{i,j\in I_\sigma \\ i<j}} (\bar{\kappa}_i - \bar{\kappa}_j) \tag{7.370}$$

for $\kappa \in (\Lambda_{N_\uparrow}^{+>}, \Lambda_{N_\downarrow}^{+>})$.

The original derivation by Dunkl [46] utilized the evaluation formula [154] for the non-symmetric Jack polynomials and a certain skew symmetric operator [46]. We derive (7.369) in another way by modifying the method given in [63]. The argument is a slight generalization of that in Section 7.3.3. Namely, the following relation is derived:

$$\hat{O}^{(--)} J_\kappa^{(--)}(z_1,\ldots,z_N) \Big|_{z_1,\ldots,z_N \to 1^N}$$
$$= \tilde{J}_\kappa^{(--)}(1,\ldots,1) \hat{O}^{(--)} \Delta(z^\uparrow) \Delta(z^\downarrow) \Big|_{z_1,\ldots,z_N \to 1^N}. \tag{7.371}$$

Combining (7.367) and (7.371) and using the evaluation formula (7.365) for $J_\kappa^{(++)}(1,\ldots,1)$, we obtain

$$\tilde{J}_\kappa^{(--)}(1,\ldots,1)\hat{O}^{(--)}\Delta(z^\uparrow)\Delta(z^\downarrow)\Big|_{z_1,\ldots,z_N\to 1^N}$$

$$= \pi_\kappa^{(--)}J_\kappa^{(++)}(1,\ldots,1) = \frac{N_\uparrow!N_\downarrow!\tilde{\pi}_\kappa^{(--)}e_\kappa}{d_\kappa}. \qquad (7.372)$$

On the other hand, $\tilde{J}_{\delta_{\uparrow\downarrow}}^{(--)}$ reduces to unity, and thus (7.372) for $\kappa = \delta_{\uparrow\downarrow}$ becomes

$$\hat{O}^{(--)}\Delta(z^\uparrow)\Delta(z^\downarrow)\Big|_{z_1,\ldots,z_N\to 1^N} = \frac{N_\uparrow!N_\downarrow!\tilde{\pi}_{\delta_{\uparrow\downarrow}}^{(--)}e_{\delta_{\uparrow\downarrow}}}{d_{\delta_{\uparrow\downarrow}}}. \qquad (7.373)$$

Provided we have

$$d_{\delta_{\uparrow\downarrow}}/\tilde{\pi}_{\delta_{\uparrow\downarrow}}^{(--)} = 1, \qquad (7.374)$$

we arrive at the evaluation formula (7.369) by dividing (7.372) by (7.373).

To derive (7.374), let us first decompose $\tilde{\pi}_{\delta_{\uparrow\downarrow}}^{(--)} = \tilde{\pi}_{\delta\uparrow}\tilde{\pi}_{\delta\downarrow}$ assuming $\kappa_\sigma = \delta_\sigma$ for both spins. For an up-spin cell $s = (M,\kappa_M)$, we obtain

$$\bar{\kappa}_M - \bar{\kappa}_{M+1} = \frac{1}{\lambda} - (r_M - r_{M+1}) = \frac{1}{\lambda} + 2, \qquad (7.375)$$

since the presence of the down-spin part gives $r_{M+1} - r_M = 2$ for the ranking of the rows defined by (7.9). In general, $r_{M+1} - r_i$ becomes twice the single-component value. By evaluating $\bar{\kappa}_i - \bar{\kappa}_{M+1}$, we obtain

$$P_{M\uparrow} = \left(\frac{1}{\lambda} + 2\right)\left(\frac{2}{\lambda} + 4\right)\cdots\left(\frac{M}{\lambda} + 2M\right). \qquad (7.376)$$

On the other hand, $d(s)$ contains the leg length $l(s) = ul(s) + ll(s)$ defined by (7.105), where the upper leg length is zero: $ul(i,\kappa_M) = 0$. We obtain easily the lower leg lengths $ll(M,\kappa_M) = 1$ and $ll(1,\kappa_M) = 2M - 1$. For a general row i, we obtain

$$l(i,\kappa_M) + 1 = ll(i,\kappa_M) + 1 = r_{M+1} - r_i. \qquad (7.377)$$

Since the arm length is given by $a(i,\kappa_M) = \kappa_i - \kappa_M$, we find that $P_{M\uparrow}$ is the same as the product of $d(s)$ taken over the up-spin column with $j = \kappa_M$. A similar argument is applicable to $\tilde{\pi}_{\delta\downarrow}$. In this case the up-spin part contributes to the upper leg length of a down-spin cell $s = (i,\kappa_M)$. Namely, we obtain

$$ul(i,\kappa_M) = -k_i' + k_{M+1}', \quad ll(i,\kappa_M) + 1 = -k_i'' + k_{M+1}''. \qquad (7.378)$$

Hence $\tilde{\pi}_{\delta\downarrow}^{(--)}$ is the same as the product of $d(s)$ taken over the down-spin column with $j = \kappa_M$, and we obtain (7.374).

To derive the ratio $d_\kappa / \tilde{\pi}_\kappa^{(--)}$ for general $\kappa = (\kappa^\uparrow, \kappa^\downarrow) \in (\Lambda_{N_\uparrow}^{+>}, \Lambda_{N_\downarrow}^{+>})$, we define a subset $S(\kappa)$ of $D(\kappa)$ as

$$S(\kappa) = \{s = (i, \kappa_j + 1) | 1 \leq i < j \leq N_\uparrow\}$$
$$\cup \{s = (i, \kappa_j + 1) | N_\uparrow + 1 \leq i < j \leq N\}. \qquad (7.379)$$

As in Section 7.3.4, we also define $W(\kappa)$ as the complementary set $W(\kappa) = D(\kappa) \setminus S(\kappa)$. Then we have the relation

$$d_\kappa / \tilde{\pi}_\kappa^{(--)} = \prod_{s \in W(\kappa)} d(s). \qquad (7.380)$$

Proof of (7.380).
Let us consider an up-spin cell $s = (i, \kappa_{M+1} + 1)$ that belongs to $S(\kappa)$. From the foregoing argument, it is easy to see the following relation:

$$l(i, \kappa_{M+1} + 1) + 1 = r_{M+1} - r_i. \qquad (7.381)$$

The LHS contributes to $d(s)$, while the RHS contributes to $\bar{\kappa}_i - \bar{\kappa}_{M+1}$. Since the arm length is given by $a(i, \kappa_M) = \kappa_i - \kappa_{M+1} - 1$, we find

$$d(i, \kappa_M) = \bar{\kappa}_i - \bar{\kappa}_{M+1} \qquad (7.382)$$

and the product of $d(s)$ taken over the up-spin column with $j = \kappa_{M+1} + 1$ gives the same value as $\prod_{i=1}^{M} (\bar{\kappa}_i - \bar{\kappa}_{M+1})$. Similarly, for a down-spin cell $s = (i, \kappa_{M+1} + 1)$ that belongs to $S(\kappa)$, we obtain the result corresponding to (7.378), provided $s = (i, \kappa_M)$ is replaced by a more general expression $(i, \kappa_{M+1} + 1)$ in the presence of $W(\kappa)$. Hence the product of $d(s)$ taken over the down-spin column with $j = \kappa_{M+1} + 1$ agrees with $\prod_{i=N_\uparrow + 1}^{M} (\bar{\kappa}_i - \bar{\kappa}_{M+1})$. By combining the up- and down-spin contributions, we find that $\tilde{\pi}_\kappa^{(--)}$ cancels the product $d(s)$ taken over the set $S(\kappa)$, while the product over those $d(s)$ that belong to $W(\kappa)$ remains. ∎

7.4.6 Binomial formula

The binomial formula is derived using the Cauchy product expansion formula (7.357) and evaluation formula (7.369). For $r \in \mathbf{C}$, the binomial formula for U(2) Jack polynomials is given by [105]

$$\prod_{\sigma=\uparrow,\downarrow} \prod_{i \in I_\sigma} (1 - z_i)^{r - N_\sigma} = \sum_{\kappa \in (\Lambda_{N_\uparrow}^{+>}, \Lambda_{N_\downarrow}^{+>})} b_\kappa^{(--)}(r) \tilde{J}_\kappa^{(--)}(z), \qquad (7.383)$$

where

$$b_\kappa^{(--)}(r) = \lambda^{|\delta|-|\kappa|} \frac{(1-r)_{\kappa^+}}{(1-r)_{\delta^+}} \frac{\pi_\kappa^{\mathrm{F}}}{d'_\kappa}. \tag{7.384}$$

Here $\pi_\kappa^{\mathrm{F}(--)}$ is defined by $\pi_\kappa^{F(--)} = \prod_{\sigma=\uparrow\downarrow} \prod_{(i<j)\in I_\sigma} (\bar\kappa_i - \bar\kappa_j - 1)$ and the generalized shifted factorial $(t)_\kappa$ defined in (7.174) is used. We note that the bosonic description of (7.384) does not lead to simple renormalization of the coupling constant, in contrast with the single-component case described in Section 7.3.4. The difference comes from the entanglement of up- and down-spin components in leg lengths.

In order to derive (7.384), let N' be $N'_\uparrow + N'_\downarrow (\geq N)$ for that pair $(N'_\uparrow, N'_\downarrow)$ of positive integers $N'_\uparrow$ and $N'_\downarrow$ which satisfies $N'_\uparrow - N'_\downarrow = N_\uparrow - N_\downarrow$. We consider the Cauchy product expansion formula

$$\prod_{s=\uparrow,\downarrow} \prod_{i,j\in I'_s} (1 - z_i y_j)^{-1} \prod_{i,j\in I'} (1 - z_i y_j)^{-\lambda}$$

$$= \sum_{\nu\in(\Lambda_{N'_\uparrow}^{+>},\Lambda_{N'_\downarrow}^{+>})} \frac{d_\nu \rho_\nu^{(--)}}{d'_\nu} \tilde{J}_\nu^{(--)}(z) \tilde{J}_\nu^{(--)}(y)$$

$$= \sum_{\nu\in(\Lambda_{N'_\uparrow}^{+>},\Lambda_{N'_\downarrow}^{+>})} \frac{d_\nu \pi_\nu^{\mathrm{F}(--)}}{d'_\nu \tilde\pi_\nu^{(--)}} \tilde{J}_\nu^{(--)}(z) \tilde{J}_\nu^{(--)}(y) \tag{7.385}$$

in N' variables. Here $I'_\uparrow = [1, N'_\uparrow]$, $I'_\downarrow = [N'_\uparrow + 1, N']$, and $I' = [1, N']$. On the LHS of (7.385), we set $z_{N_\uparrow+1} = \cdots = z_{N'_\uparrow} = 0$, $z_{N'_\uparrow+N_\downarrow+1} = \cdots = z_{N'_\uparrow+N'_\downarrow} = 0$, and $y_1 = \cdots = y_{N'_\uparrow+N'_\downarrow} = 1$. In the second equality of (7.385), we have used $\rho_\nu^{(--)} = \pi_\kappa^{\mathrm{F}(--)} / \tilde\pi_\nu^{(--)}$.

Further, we replace $(z_{N'_\uparrow+1}, \ldots, z_{N'_\uparrow+N_\downarrow})$ by $(z_{N_\uparrow+1}, \ldots, z_{N_\uparrow+N_\downarrow})$. The LHS of (7.385) then turns into

$$\prod_{i\in I_\uparrow=[1,N_\uparrow]} (1 - z_i)^{-N'\lambda-N'_\uparrow} \prod_{j\in I_\downarrow=[N_\uparrow+1,N]} (1 - z_j)^{-N'\lambda-N'_\uparrow+N_\uparrow-N_\downarrow}. \tag{7.386}$$

On the RHS of (7.385), we set $y_j = 1$ for $1 \leq i \leq N'$. With use of the evaluation formula (7.369), we immediately see that the RHS of the Cauchy product expansion formula becomes

$$\sum_\nu \frac{e_\nu \pi_\nu^{\mathrm{F}(--)}}{e_{\delta'} d'_\nu} \tilde{J}_\nu^{(--)}(z). \tag{7.387}$$

Here the sum is taken over $\nu \in (\Lambda_{N'_\uparrow}^{+>}, \Lambda_{N'_\downarrow}^{+>})$. The symbol $\delta' = (\delta'^\uparrow, \delta'^\downarrow)$ denotes the composition $\delta(N'_\uparrow, N'_\downarrow) \in (\Lambda_{N'_\uparrow}^{+>}, \Lambda_{N'_\downarrow}^{+>})$. Now we set $z_{N_\uparrow+1}$

$= \cdots = z_{N'_\uparrow} = 0$ and $z_{N'_\uparrow + N_\downarrow + 1} = \cdots = z_{N'_\uparrow + N'_\downarrow} = 0$ in (7.387). Non-vanishing contributions in the sum (7.387) then come only from the compositions $\nu = (\nu^\uparrow, \nu^\downarrow) \in (\Lambda^{+>}_{N'_\uparrow}, \Lambda^{+>}_{N'_\downarrow})$ satisfying $l(\nu^\uparrow - \delta'^\uparrow) \leq N_\uparrow$ and $l(\nu^\downarrow - \delta'^\downarrow) \leq N_\downarrow$, where

$$\nu^s - \delta'^s = (\nu^s_1 - \delta'^s_1, \ldots, \nu^s_{N_s} - \delta'^s_{N_s}) \in \Lambda^+_{N'_s}$$

with $s = (\uparrow, \downarrow)$. The reason is as follows: If either $l(\nu^\uparrow - \delta'^\uparrow) > N_\uparrow$ or $l(\nu^\downarrow - \delta'^\downarrow) > N_\downarrow$ holds for the composition $\nu = (\nu^\uparrow, \nu^\downarrow) \in (\Lambda^{+>}_{N'_\uparrow}, \Lambda^{+>}_{N'_\downarrow})$, then, in each monomial of the polynomials $\tilde{J}_\nu$, the minimum power of $z_{N_\uparrow + 1}$ is 1 or that of $z_{N'_\uparrow + N_\downarrow + 1}$ is 1. Therefore, those ν do not contribute to the sum in (7.387) when both $z_{N_\uparrow + 1}$ and $z_{N'_\uparrow + N_\downarrow + 1}$ are set to zero. For a composition $\nu = (\nu^\uparrow, \nu^\downarrow) \in (\Lambda^{+>}_{N'_\uparrow}, \Lambda^{+>}_{N'_\downarrow})$ satisfying $l(\nu^\uparrow - \delta'^\uparrow) \leq N_\uparrow$ and $l(\nu^\downarrow - \delta'^\downarrow) \leq N_\downarrow$, the composition $\kappa = (\kappa^\uparrow, \kappa^\downarrow) \in (\Lambda^{+>}_{N_\uparrow}, \Lambda^{+>}_{N_\downarrow})$ can be defined as the composition satisfying the relation

$$\kappa^\uparrow - \delta^\uparrow = \nu^\uparrow - \delta'^\uparrow, \tag{7.388}$$

$$\kappa^\downarrow - \delta^\downarrow = \nu^\downarrow - \delta'^\downarrow. \tag{7.389}$$

When the relations (7.388) and (7.389) hold for compositions $\kappa \in (\Lambda^{+>}_{N_\uparrow}, \Lambda^{+>}_{N_\downarrow})$ and $\nu \in (\Lambda^{+>}_{N'_\uparrow}, \Lambda^{+>}_{N'_\downarrow})$, we can then find the following consequences. First, the two polynomials

$$\tilde{J}^{(--)}_\nu (z_1, \ldots, z_{N_\uparrow}, \underbrace{0, \ldots, 0}_{N'_\uparrow - N_\uparrow}, z_{N'_\uparrow + 1}, \ldots, z_{N'_\uparrow + N_\downarrow}, \underbrace{0, \ldots, 0}_{N'_\downarrow - N_\downarrow}) \tag{7.390}$$

and

$$\tilde{J}^{(--)}_\kappa (z_1, \ldots, z_{N_\uparrow}, z_{N'_\uparrow + 1}, \ldots, z_{N'_\uparrow + N_\downarrow}) \tag{7.391}$$

are equal. From now on, we replace $z_{N'_\uparrow + 1}, \ldots, z_{N'_\uparrow + N_\downarrow}$ by $z_{N_\uparrow + 1}, \ldots, z_N$. Second, the relation

$$d'_\nu / \pi^{\mathrm{F}(--)}_\nu = d'_\kappa / \pi^{\mathrm{F}(--)}_\kappa \tag{7.392}$$

holds. The LHS of (7.392) is written as

$$\prod_{s \in W(\nu)} d'(s). \tag{7.393}$$

There is a one-to-one correspondence between squares in $s' \in W(\nu)$ and $s \in W(\kappa)$:

$$s' = (i, N'_\uparrow - N_\uparrow + j) \in W(\kappa), \quad \text{where } s = (i, j) \in D(\mu) \tag{7.394}$$

for $i \in [1, N_\uparrow]$ and

$$s' = (i + N'_\uparrow - N_\uparrow, N'_\uparrow - N_\uparrow + j) \in W(\kappa), \quad \text{where } s = (i, j) \in D(\kappa) \quad (7.395)$$

for $i \in [N_\uparrow + 1, N]$. In this parameterization, the arm length and leg lengths do not change and hence $d'(s')$ in $W(\nu)$ and $d'(s)$ in $W(\kappa)$ are the same. As a result, we obtain

$$\prod_{s \in W(\nu)} d'(s) = \prod_{s \in W(\kappa)} d'(s), \quad (7.396)$$

from which (7.392) follows.

Third, the expression $e_\nu/e_{\delta'}$ can be rewritten as

$$\frac{\prod_{s \in D(\nu)}((a'(s) + 1)/\lambda + N' - l'(s))}{\prod_{s \in D(\delta')}((a'(s) + 1)/\lambda + N' - l'(s))}$$
$$= \frac{\prod_{s \in D(\kappa)}((a'(s) + N'_\uparrow - N_\uparrow + 1)/\lambda + N' - l'(s))}{\prod_{s \in D(\delta)}((a'(s) + N'_\uparrow - N_\uparrow + 1)/\lambda + N' - l'(s))}$$
$$= \lambda^{|\delta| - |\kappa|} \frac{(1 + N'\lambda + N'_\uparrow - N_\uparrow)_{\kappa+}}{(1 + N'\lambda + N'_\uparrow - N_\uparrow)_{\delta+}}. \quad (7.397)$$

Notice the relation using the generalized shifted factorial defined in (7.174):

$$\prod_{s \in D(\eta)} [(a'(s) + k)/\lambda + k' - l'(s)] = \lambda^{-|\eta|}(k'\lambda + k)_{\eta+}$$

for $\eta \in \Lambda_N^+$ and integers k, k'. As a result of these relations, the expression (7.387) can be rewritten as

$$\sum_\kappa \lambda^{|\delta| - |\kappa|} \frac{(1 + N'\lambda + N'_\uparrow - N_\uparrow)_{\kappa+}}{(1 + N'\lambda + N'_\uparrow - N_\uparrow)_{\delta+}} \frac{\pi_\kappa^{F(--)}}{d'_\kappa} \tilde{J}_\kappa^{(--)}(z), \quad (7.398)$$

where the sum is taken over $\kappa \in (\Lambda_{N_\uparrow}^{+>}, \Lambda_{N_\downarrow}^{+>})$. Now we obtain the relation

$$\prod_{i \in I_\uparrow} (1 - z_j)^{-N'\lambda - N'_\uparrow} \prod_{j \in I_\downarrow} (1 - z_j)^{-N'\lambda - N'_\uparrow + N_\uparrow - N_\downarrow}$$
$$= \sum_\kappa \lambda^{|\delta| - |\kappa|} \frac{(1 + N'\lambda + N'_\uparrow - N_\uparrow)_{\kappa+}}{(1 + N'\lambda + N'_\uparrow - N_\uparrow)_{\delta+}} \frac{\pi_\kappa^{F(--)}}{d'_\kappa} \tilde{J}_\kappa^{(--)}(z), \quad (7.399)$$

from the expressions (7.386) and (7.398). We notice that the RHS of (7.399) is a polynomial of $N'\lambda + N'_\uparrow$ and hence the relation (7.399) also holds for arbitrary complex values of $N'\lambda + N'_\uparrow$. We replace $-N'\lambda - N'_\uparrow + N_\uparrow$ by a complex variable r. Consequently, we obtain the binomial formula (7.383).

7.4.7 Power-sum decomposition

The power-sum symmetric polynomials can be decomposed into a linear combination of $\tilde{J}_\kappa^{(--)}$. The power-sum decomposition formula is used in the calculation of the dynamical density correlation function of the Sutherland model with SU(2) internal symmetry.

The derivation is similar to that in Section 7.3.8. Our starting point is the binomial formula (7.384). Setting $N_\uparrow = N_\downarrow = N/2$ and $r \to N/2 + r$ in (7.383) and (7.384), we obtain

$$\prod_{i=1}^{N} (1 - z_i)^r = \sum_{\kappa \in \left(\Lambda_{N/2}^{+>}, \Lambda_{N/2}^{+>}\right)} b_\kappa^{(--)} \left(r + \frac{N}{2}\right) \tilde{J}_\kappa^{(--)}(z_1, \ldots, z_N), \quad (7.400)$$

with

$$b_\kappa^{(--)} \left(r + \frac{N}{2}\right) = \lambda^{|\delta| - |\kappa|} \frac{\pi_k^{F(--)}}{d'_\kappa} \frac{(1 - N/2 - r)_{\kappa+}}{(1 - N/2 - r)_{\delta+}}$$

$$= \frac{\pi^{F(--)}}{d'_\kappa} \prod_{s \in D(\kappa^+) \cap \overline{D}(\delta^+)} \left(\frac{a'(s) + 1 - r - N/2}{\lambda} - l'(s)\right), \quad (7.401)$$

where δ denotes $(\delta(N/2), \delta(N/2))$. We have used the property of the generalized shifted factorial defined in (7.174) for the second equality of (7.401).

Following the same argument as that in Section 7.3.8, we obtain

$$p_m(z) = -m \sum_\kappa{}' \frac{\partial b_\kappa^{(--)}(r + N/2)}{\partial r}\bigg|_{r=0} \tilde{J}_\kappa^{(--)}(z) \quad \text{for } m > 0, \quad (7.402)$$

where the summation runs over $\kappa \in \left(\Lambda_{N/2}^{+>}, \Lambda_{N/2}^{+>}\right)$ satisfying $|\kappa| - |\delta| = m$. Since π_κ^F and d'_κ do not contain r, we obtain

$$\frac{\partial}{\partial r} b_\kappa^{(--)}(r + N/2)\bigg|_{r=0}$$

$$= -\frac{1}{\lambda} \frac{\pi_\kappa^{F(--)}}{d'_\kappa} \sum_{s \in D(\kappa^+) \cap \overline{D(\delta^+)}} \prod_{s'(\neq s) \in D(\kappa^+) \cap \overline{D(\delta^+)}} \left\{\frac{a'(s) + 1 - N/2}{\lambda} - l'(s)\right\}.$$

$$(7.403)$$

A typical Young diagram of κ^+ with $\kappa \in (\Lambda_{N/2}^{+>}, \Lambda_{N/2}^{+>})$ is shown in Fig. 7.10. The shaded square is $s = (i, j) = (1, N/2)$. $D(\kappa^+)$ for κ relevant to (7.402) contains the cell $s = (1, N/2)$. Since

$$\frac{a'(s) + 1 - N/2}{\lambda} - l'(s) = 0$$

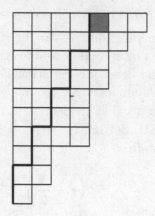

Fig. 7.10. Diagram of a partition κ^+ with $\kappa \in \left(\Lambda_{N/2}^{+>}, \Lambda_{N/2}^{+>} \right)$ for $N/2 = 5$. The boundary of $D(\delta^+)$ with $\delta = (\delta(N/2), \delta(N/2))$ is drawn with a bold line. The shaded square is $(i, j) = (1, N/2)$.

for $s = (1, N/2)$, all terms but $s = (1, N/2)$ in the summation with respect to s in (7.403) vanish. As a result, we obtain

$$p_m(z) = \frac{m}{\lambda} \sum_{\kappa}' c_\kappa \tilde{J}_\kappa^{(--)}(z) \quad \text{for } m > 0, \tag{7.404}$$

with $c_\kappa = \pi_\kappa^{\mathrm{F}(--)} c_\kappa' / d_\kappa'$ and

$$c_\kappa' = \prod_{s(\neq(1,N/2))\in D(\kappa^+)\cap\overline{D(\delta^+)}} \left\{ \frac{a'(s) + 1 - N/2}{\lambda} - l'(s) \right\}. \tag{7.405}$$

7.5 U(1,1) Jack polynomials

The U(1,1) Jack polynomials are relevant to the U(1,1) Sutherland model in Chapter 3, and the dynamics of the supersymmetric t–J model in Chapter 6. This section summarizes the fundamental properties of the U(1,1) Jack polynomials [15].

7.5.1 Relation to non-symmetric Jack polynomials

For non-negative integers N_B and N_F, we define $I_\mathrm{B} = [1, N_\mathrm{B}]$ and $I_\mathrm{F} = [N_\mathrm{B} + 1, N]$, where $N = N_\mathrm{B} + N_\mathrm{F}$ is the total number of particles. Let $S_{N_\mathrm{B}} \times S_{N_\mathrm{F}}$ be a subgroup of the symmetric group S_N that does not mix I_B and I_F. For $\kappa = (\kappa^\mathrm{B}, \kappa^\mathrm{F}) \in (\Lambda_{N_\mathrm{B}}^+, \Lambda_{N_\mathrm{F}}^{+>})$, we define the U(1,1) Jack polynomials $J_\kappa^{(+-)}(z_1, \ldots, z_N)$ by the following two conditions:

(i) The polynomial $J_\kappa^{(+-)}$ for $\kappa = (\kappa^B, \kappa^F) \in (\Lambda_{N_B}^+, \Lambda_{N_F}^{+>})$ has the form

$$J_\kappa^{(+-)}(z) = \sum_\eta a_\eta^{(+-)} E_\eta(z), \qquad (7.406)$$

where summation is over such η that satisfies $(\eta^B)^+ = \kappa^B$ together with $(\eta^F)^+ = \kappa^F$, and we set the normalization

$$a_\kappa^{(+-)} = 1.$$

(ii) Under the action of the transposition K_i, the polynomial $J_\kappa^{(+-)}(z)$ for $\kappa = (\kappa^B, \kappa^F) \in (\Lambda_{N_B}^+, \Lambda_{N_F}^{+>})$ transforms as

$$\begin{aligned} K_i J_\kappa^{(+-)}(z) &= J_\kappa^{(+-)}(z), & \text{for } i \in [1, N_B - 1], \\ K_i J_\kappa^{(+-)}(z) &= -J_\kappa^{(+-)}(z), & \text{for } i \in [N_B + 1, N - 1]. \end{aligned} \qquad (7.407)$$

The U(1,1) Jack polynomials are eigenfunctions of (3.16) and constitute eigenfunctions of the U(1,1) Sutherland model. If $N_B = 0$, the U(1,1) Jack polynomials reduce to the antisymmetric Jack polynomials $J_\kappa^{(-)}$ discussed in Section 7.2. If, on the other hand, $N_F = 0$, the U(1,1) Jack polynomials reduce to the symmetric Jack polynomials J_κ in Section 7.3.

We define Sym_B as the symmetrization operator with respect to the variables z_i for $i \in I_B$ and Asym_F as the antisymmetrization operator with respect to the variables z_i for $i \in I_F$. Owing to the property (7.407), the U(1,1) Jack polynomials can be written in the form

$$\rho_\kappa^{(+-)} J_\kappa^{(+-)}(z_1, \ldots, z_N) = \text{Sym}_B \text{Asym}_F E_\kappa(z_1, \ldots, z_N), \qquad (7.408)$$

for $\kappa \in (\Lambda_{N_B}^+, \Lambda_{N_F}^{+>})$ with the factorized coefficient $\rho_\kappa^{(+-)} = \rho_\kappa^B \rho_\kappa^F$ as

$$\rho_\kappa^B = \prod_{(i<j)\in I_B} \frac{\bar\kappa_i - \bar\kappa_j + 1}{\bar\kappa_i - \bar\kappa_j}, \qquad \rho_\kappa^F = \prod_{(i<j)\in I_F} \frac{\bar\kappa_i - \bar\kappa_j - 1}{\bar\kappa_i - \bar\kappa_j}. \qquad (7.409)$$

The expression (7.409) can be derived in a way similar to that used in the case of antisymmetric and symmetric Jack polynomials. It is convenient to express (7.409) as

$$\rho_\kappa^B = \left(\prod_r p_r!\right) \frac{d_{(\kappa^B)^R, \kappa^F}}{d_{\kappa^B, \kappa^F}}, \qquad \rho_\kappa^F = \frac{d'_{(\kappa^B)^R, \kappa^F}}{d'_{(\kappa^B)^R, (\kappa^F)^R}}, \qquad (7.410)$$

which are generalizations of (7.149) and (7.218). The U(1,1) Jack polynomials are given in the form

$$J_\kappa^{(+-)} = \Delta(z^F) \tilde{J}_\kappa^{(+-)}, \qquad (7.411)$$

where $z^{\mathrm{F}} = (z_{N_{\mathrm{B}}+1}, \ldots, z_N)$, $\Delta(z^{\mathrm{F}}) = \prod_{(i<j)\in I_{\mathrm{F}}}(z_i - z_j)$, and $\tilde{J}_\kappa^{(+-)}$ is a polynomial satisfying

$$K_i \tilde{J}_\kappa^{(+-)}(z) = \tilde{J}_\kappa^{(+-)}(z), \tag{7.412}$$

for $i \neq N_{\mathrm{B}}, N$.

7.5.2 Evaluation formula

We derive the evaluation formulae

$$\tilde{J}_\kappa^{(+-)}(1, \ldots, 1) = \frac{N_{\mathrm{B}}! \pi_\kappa^{(+-)} e_\kappa}{\rho_\kappa^{(++)} d_\kappa e_{\kappa(0)}}, \tag{7.413}$$

$$\pi_\kappa^{(+-)} = \prod_{(i<j)\in I_{\mathrm{F}}} (\bar{\kappa}_i - \bar{\kappa}_j + 1), \tag{7.414}$$

$$\kappa(0) = (0^{N_{\mathrm{B}}}, \delta(N_{\mathrm{F}})), \tag{7.415}$$

where $\kappa(0)$ is the composition corresponding to the ground state of the U(1,1) Sutherland model. Precisely speaking, a Galilean boost has been made in $\kappa(0)$ only to the fermionic component to make κ_i nonzero.

The formula (7.413) can be derived in a way similar to the corresponding formula (7.369). In terms of the symmetry-changing operator

$$\hat{O}^{(+-)} = \prod_{(i<j)\in I_{\mathrm{F}}} \left(\hat{d}_i - \hat{d}_j + 1 \right) \tag{7.416}$$

for the fermion part, the relation

$$\hat{O}^{(+-)} J_\kappa^{(+-)} = \pi_\kappa^{(+-)} J_\kappa^{(++)} \tag{7.417}$$

follows. Further, the relation

$$\hat{O}^{(+-)} J_\kappa^{(+-)} \Big|_{(z_1, \ldots, z_N) \to 1^N} = \tilde{J}_\kappa^{(+-)}(1, \ldots, 1) \hat{O}^{(+-)} \Delta(z^{\mathrm{F}}) \Big|_{z^{\mathrm{F}} \to 1^{N_{\mathrm{F}}}} \tag{7.418}$$

holds. On the other hand, we obtain from (7.417) and (7.418)

$$\pi_\kappa^{(+-)} J_\kappa^{(++)}(1, \ldots, 1) = \tilde{J}_\kappa^{(+-)}(1, \ldots, 1) \hat{O}^{(+-)} \Delta(z^{\mathrm{F}}) \Big|_{z^{\mathrm{F}} \to 1^{N_{\mathrm{F}}}}. \tag{7.419}$$

In the case of $\kappa = \kappa(0)$, $\tilde{J}^{(+-)}$ reduces to unity, and (7.419) becomes

$$\pi_{\kappa(0)}^{(+-)} J_{\kappa(0)}^{(++)}(1, \ldots, 1) = \hat{O}^{(+-)} \Delta(z^{\mathrm{F}}) \Big|_{z^{\mathrm{F}} \to 1^{N_{\mathrm{F}}}}. \tag{7.420}$$

From (7.419) and (7.420), we obtain

$$\tilde{J}_\kappa^{(+-)}(1,\ldots,1) = \frac{\pi_\kappa^{(+-)} J_\kappa^{(++)}(1,\ldots,1)}{\pi_{\kappa(0)}^{(+-)} J_{\kappa(0)}^{(++)}(1,\ldots,1)}. \tag{7.421}$$

On the RHS of (7.421), $J_\kappa^{(++)}(1,\ldots,1)$ and $J_{\kappa(0)}^{(++)}(1,\ldots,1)$ can be evaluated from (7.365).

We note the relation

$$\rho_\kappa^{(++)} = \rho_\kappa^{\mathrm{B}} \pi_\kappa^{(+-)} / \tilde{\pi}_\kappa^{(+-)}, \tag{7.422}$$

where

$$\tilde{\pi}_\kappa^{(+-)} \equiv \prod_{(i<j)\in I_{\mathrm{F}}} (\bar{\kappa}_i - \bar{\kappa}_j). \tag{7.423}$$

In the case of $\kappa = \kappa(0)$, we obtain $\rho_{\kappa(0)}^{\mathrm{B}} = N_{\mathrm{B}}!$ and

$$\tilde{\pi}_{\kappa(0)}^{(+-)} = d_{\kappa(0)}, \tag{7.424}$$

in analogy with the single-component case (7.248). Then we obtain the relation

$$\rho_{\kappa(0)}^{(++)} d_{\kappa(0)} / \pi_{\kappa(0)}^{(+-)} = N_{\mathrm{B}}!, \tag{7.425}$$

and finally arrive at the evaluation formula (7.413).

7.5.3 Bosonization for separated states

We consider the following expansion:

$$\prod_{i\in I_{\mathrm{B}}} (1-z_i)^p \prod_{j\in I_{\mathrm{F}}} (1-z_j)^q = \sum_{\kappa\in(\Lambda_{N_{\mathrm{B}}}^+, \Lambda_{N_{\mathrm{F}}}^{+>})} (-1)^{|k|-|k(0)|} b_\kappa^{(+-)}(p,q)\, \tilde{J}_\kappa^{(+-)}(z),$$

$$\tag{7.426}$$

where p, q are numbers. The coefficient $b_\kappa^{(+-)}$ in the general case has not yet been derived. However, provided κ belongs to *separated states*, which will be specified below, we can obtain explicit results. The final result for $b_\kappa^{(+-)}(p,q)$ for separated states is given by (7.448).

We characterize the composition κ for the maximum length $N = N_{\mathrm{B}} + N_{\mathrm{F}}$ by using the excitations $(\mu^{\mathrm{B}}, \mu^{\mathrm{F}})$ as

$$\kappa = \kappa(0) + (\mu^{\mathrm{B}}, \mu^{\mathrm{F}}) \in (\Lambda_{N_{\mathrm{B}}}^+, \Lambda_{N_{\mathrm{F}}}^{+>}), \tag{7.427}$$

where $\kappa(0)$ has been defined in (7.415).

We call the composition $\kappa = (\kappa^{\mathrm{B}}, \kappa^{\mathrm{F}})$ a "separated state" if it satisfies the following condition:

$$0 \leq \sharp(\text{nonzero columns in } \mu^{\mathrm{B}}) + \sharp(\text{nonzero rows in } \mu^{\mathrm{F}}) < N_{\mathrm{F}}$$

$$\Leftrightarrow 0 \leq l((\mu^{\mathrm{B}})') + l(\mu^{\mathrm{F}}) \leq N_{\mathrm{F}} - 1, \tag{7.428}$$

where $(\mu^{\mathrm{B}})'$ means the conjugate partition. The condition prevents mixing of fermionic and bosonic excitation momenta in the expansion (7.426). The condition (7.428) is visualized in the Young diagram. Namely, the bosonic excitations from the ground state $\kappa(0) = (0^{N_{\mathrm{B}}}, \delta(N_{\mathrm{F}}))$ begin from the left, and the fermionic excitations from the right. In the low-energy excitations, there is enough room in the central region of the diagram where neither bosonic nor fermionic excitations are present.

7.5.4 Factorization for separated states

In the composition κ defined by (7.427), we define the subset W^{B} as such part of the Young diagram $D(\kappa)$ that corresponds to μ^{B}, and another subset W^{F} as such part that corresponds to μ^{F}. Namely, we have $D(k) = D(k_0) + W^B + W^F$. Let us first consider $s \in W^{\mathrm{B}}$ in the presence of the fermionic component. The leg length l also counts the contribution from $\Lambda_{N_{\mathrm{F}}}^{+>}$, while the arm length a is not affected by $\Lambda_{N_{\mathrm{F}}}^{+>}$. Namely, when we consider a and l within the partition μ^{B}, we have the correspondence

$$a_\kappa(s) = a_{\mu^{\mathrm{B}}}(s'), \quad l_\kappa(s) = l_{\mu^{\mathrm{B}}}(s') + a_{\mu^{\mathrm{B}}}(s') + 1, \tag{7.429}$$

where s' denotes the cell index in the composition $\mu_{\mathrm{B}} + \mu_{\mathrm{F}}$. Then we obtain

$$d_\kappa(\lambda, s) = \frac{1}{\lambda}\left(a_{\mu^{\mathrm{B}}} + 1\right) + l_{\mu^{\mathrm{B}}} + a_{\mu^{\mathrm{B}}} + 2$$

$$= \frac{\lambda + 1}{\lambda}\left(a_{\mu^{\mathrm{B}}} + 1\right) + l_{\mu^{\mathrm{B}}} + 1 = d_{\mu^{\mathrm{B}}}(\lambda', s'), \tag{7.430}$$

$$d'_\kappa(\lambda, s) = \frac{1}{\lambda}\left(a_{\mu^{\mathrm{B}}} + 1\right) + l_{\mu^{\mathrm{B}}} + a_{\mu^{\mathrm{B}}} + 1$$

$$= \frac{\lambda + 1}{\lambda}\left(a_{\mu^{\mathrm{B}}} + 1\right) + l_{\mu^{\mathrm{B}}} + 1 = d'_{\mu^{\mathrm{B}}}(\lambda', s'). \tag{7.431}$$

We emphasize that the generalized hook lengths experience only the change $\lambda \to \lambda' \equiv \lambda/(1 + \lambda)$ through the transition from κ to μ. For the fermionic cell s, $s \in \Lambda_{N_{\mathrm{F}}}^{+>}$, $d_\kappa(\lambda, s)$ and $d'_\kappa(\lambda, s)$ are not affected by μ^{B} for separated states.

On the other hand, the quantity e_κ also has an interesting renormalization effect for fermions. Since e_κ is invariant against any change of rows, as noted in (7.117), we consider a partition κ^+ which is obtained from κ by appropriate exchange of rows. Accordingly, we define the subset T^{B} in κ^+ which originates from cells in W^{B}, and another subset T^{F} which originates from cells in W^{F}. The relevant cells $s \in T^{\mathrm{B}} + T^{\mathrm{F}}$ are at the right end of the Young diagram of κ^+. Let us first consider the case $s \in T^{\mathrm{B}}$. We have

$$a'_\kappa(s) = a'_{\mu^{\mathrm{B}}}(s'), \quad l'_\kappa(s) = l'_{\mu^{\mathrm{B}}}(s') + N_{\mathrm{F}} - a'_{\mu^{\mathrm{B}}}(s') - 1. \tag{7.432}$$

Then we obtain

$$e_\kappa(\lambda, s) = \frac{1}{\lambda}\left(a'_\kappa + 1\right) + N - l'_\kappa = \frac{1}{\lambda}\left(a'_{\mu^{\mathrm{B}}} + 1\right) + N - \left(l'_{\mu^{\mathrm{B}}} + N_{\mathrm{F}} - a'_{\mu^{\mathrm{B}}} - 1\right)$$

$$= \frac{1}{\lambda'}\left(a'_{\mu^{\mathrm{B}}} + 1\right) + N_{\mathrm{B}} - l'_{\mu^{\mathrm{B}}} = e_{\mu^{\mathrm{B}}}(\lambda', s'). \tag{7.433}$$

In the case $s \in T^{\mathrm{F}}$, we have

$$a'_\kappa(s) = a'_{\mu^{\mathrm{F}}}(s') + N_{\mathrm{F}} - l'_{\mu^{\mathrm{F}}}(s') - 1, \quad l'_\kappa(s) = l'_{\mu^{\mathrm{F}}}(s'). \tag{7.434}$$

Then we obtain

$$e_\kappa(\lambda, s) = \frac{1}{\lambda}\left(a'_{\mu^{\mathrm{F}}} + N_{\mathrm{F}} - l_{\mu^{\mathrm{F}}}\right) + N - l'_{\mu^{\mathrm{F}}}$$

$$= \frac{\lambda + 1}{\lambda}\left(\frac{a'_{\mu^{\mathrm{F}}}}{\lambda + 1} + N_{\mathrm{F}} + \lambda' N_{\mathrm{B}} - l'_{\mu^{\mathrm{F}}}\right) \tag{7.435}$$

which includes not only the change $\lambda \to \lambda_+ \equiv \lambda + 1$ as in the single component system, but also $N_{\mathrm{F}} \to N_{\mathrm{F}} + \lambda' N_{\mathrm{B}}$.

Synthesizing these steps for separated states, we obtain the following results:

$$d'_\kappa(\lambda)/\pi_\kappa^{\mathrm{F}}(\lambda) = \prod_{s \in S^{\mathrm{B}}(\kappa)} d'_\kappa(\lambda, s) \prod_{s \in S^{\mathrm{F}}(\kappa)} d'_\kappa(\lambda, s)$$

$$= (\lambda')^{-|\mu^{\mathrm{F}}|} d'_{\mu^{\mathrm{B}}}(\lambda') d'_{\mu^{\mathrm{F}}}(\lambda_+), \tag{7.436}$$

$$e_\kappa(\lambda)/e_{\kappa(0)}(\lambda) = \prod_{s \in T^{\mathrm{B}}(\kappa^+)} e_{\kappa^+}(\lambda, s) \prod_{s \in T^{\mathrm{F}}(\kappa^+)} e_{\kappa^+}(\lambda, s)$$

$$= (\lambda')^{-|\mu^{\mathrm{F}}|} \dot{e}_{\mu^{\mathrm{B}}}(\lambda'; 1 + \lambda' N_{\mathrm{B}}) \dot{e}_{\mu^{\mathrm{F}}}(\lambda_+; \lambda_+ N_{\mathrm{F}} + \lambda N_{\mathrm{B}}), \tag{7.437}$$

where $\dot{e}_\mu$ has been defined by (7.276), and is related to $\cdot b_\kappa(r)$ defined in (7.309) by

$$\dot{e}_\mu(\lambda; \lambda M) = (-1)^{|\mu|} b_\mu(-\lambda M) d'_\mu. \tag{7.438}$$

These results are useful for the binomial formula to be presented below.

7.5.5 Binomial formula for separated states

To derive the binomial formula, we follow the same strategy as that for $\tilde{J}_\kappa^{(--)}(z)$. We introduce positive integers $N_{\mathrm{B}}', N_{\mathrm{F}}'$, and N' such that $N' = N_{\mathrm{B}}' + N_{\mathrm{F}}'$, $N_{\mathrm{B}}' > N_{\mathrm{B}}$, and $N_{\mathrm{F}}' > N_{\mathrm{F}}$. Then we define I', I_{B}', and I_{F}' by

$$I' = [1, N'], \quad I_{\mathrm{B}}' = [1, N_{\mathrm{B}}'], \quad I_{\mathrm{F}}' = [N_{\mathrm{B}}' + 1, N_{\mathrm{B}}' + N_{\mathrm{F}}'].$$

Applying both $\mathrm{Sym_B}$ and $\mathrm{Asym_F}$ to (7.300), we obtain the Cauchy product expansion formula for U(1,1) Jack polynomials:

$$\sum_{\sigma \in S_{N_{\mathrm{B}}'}} \prod_{i \in I_{\mathrm{B}}'} \frac{1}{1 - z_{\sigma(i)} y_i} \prod_{i,j \in I_{\mathrm{F}}'} \frac{1}{1 - z_i y_j} \prod_{i,j \in I'} \frac{1}{(1 - z_i y_j)^\lambda}$$

$$= \sum_{\nu \in \Lambda_{N_{\mathrm{B}}'}^+ \times \Lambda_{N_{\mathrm{F}}'}^{+>}} \frac{d_\nu}{d_\nu'} \rho_\nu^{(+-)} \tilde{J}_\nu^{(+-)}(z) \tilde{J}_\nu^{(+-)}(y), \tag{7.439}$$

where $\tilde{J}_\nu^{(+-)}(z) = J_\nu^{(+-)}(z) / \Delta(z^{\mathrm{F}})$. We use the relation

$$\rho_\nu^{(++)} / \tilde{\pi}_\nu^{(+-)} = \rho_\nu^{(+-)} / \pi_{\nu^{\mathrm{F}}}^{(+-)}, \tag{7.440}$$

with $\pi_\nu^{\mathrm{F}(+-)} = \prod_{(i<j) \in I_{\mathrm{F}}'} (\bar{\nu}_i - \bar{\nu}_j - 1)$. Then the evaluation formula (7.413) of the U(1,1) Jack polynomials is rewritten as

$$\tilde{J}_\nu^{(+-)}(1^{N'}) = \frac{N_{\mathrm{B}}'! \pi_\nu^{\mathrm{F}(+-)} e_\nu}{\rho_\nu^{(+-)} d_\nu e_{\nu(0)}}. \tag{7.441}$$

Substituting $y = (1^{N'})$ in (7.439), we obtain

$$\prod_{i \in I_{\mathrm{B}}'} (1 - z_i)^{-1-\lambda N'} \prod_{j \in I_{\mathrm{F}}'} (1 - z_j)^{-\lambda N' - N_{\mathrm{F}}'} = \sum_{\nu \in \Lambda_{N_{\mathrm{B}}'}^+ + \Lambda_{N_{\mathrm{F}}'}^{+>}} \frac{\pi_\nu^{\mathrm{F}} e_\nu}{d_\nu' e_{\nu(0)}} \tilde{J}_\nu^{(+-)}(z).$$

$$\tag{7.442}$$

If one could represent the RHS of (7.442) as an analytic function of N_{B}' and N_{F}', one would establish the binomial formula. This task, however, has not yet been achieved to its full generality.

The composition $\nu \in \Lambda_{N_{\mathrm{B}}'}^+ + \Lambda_{N_{\mathrm{F}}'}^{+>}$ is characterized by the excitations $(\mu^{\mathrm{B}}, \mu^{\mathrm{F}})$ as

$$\nu = (0^{N_{\mathrm{B}}'}, \delta(N_{\mathrm{F}}')) + (\mu^{\mathrm{B}}, 0^{N_{\mathrm{B}}' - N_{\mathrm{B}}}, \mu^{\mathrm{F}}, 0^{N_{\mathrm{F}}' - N_{\mathrm{F}}}). \tag{7.443}$$

We have the following relation:

$$\tilde{J}_\mu^{(+-)}(z_1^{\mathrm{B}},\ldots,z_{N_{\mathrm{B}}}^{\mathrm{B}},\underbrace{0,\ldots,0}_{N_{\mathrm{B}}'-N_{\mathrm{B}}},z_1^{\mathrm{F}},\ldots,z_{N_{\mathrm{F}}}^{\mathrm{F}},\underbrace{0,\ldots,0}_{N_{\mathrm{F}}'-N_{\mathrm{F}}})$$

$$= \tilde{J}_\kappa^{(+-)}(z_1^{\mathrm{B}},\ldots,z_{N_{\mathrm{B}}}^{\mathrm{B}},z_1^{\mathrm{F}},\ldots,z_{N_{\mathrm{F}}}^{\mathrm{F}}). \tag{7.444}$$

For a separated state $\kappa \in \Lambda_{N_{\mathrm{B}}}^+ + \Lambda_{N_{\mathrm{F}}}^{+>}$, the expansion coefficient for the composition $\nu \in \Lambda_{N_{\mathrm{B}}'}^+ + \Lambda_{N_{\mathrm{F}}'}^{+>}$ can be reduced to

$$\frac{\pi_\nu^{\mathrm{F}}(\lambda)e_\nu}{d_\nu'(\lambda)e_{\nu(0)}} = \frac{\dot{e}_{\mu^{\mathrm{B}}}(\lambda';1+\lambda' N_{\mathrm{B}}')}{d_{\mu^{\mathrm{B}}}'(\lambda')}\frac{\dot{e}_{\mu^{\mathrm{F}}}(\lambda_+;\lambda N_{\mathrm{B}}'+\lambda_+ N_{\mathrm{F}}')}{d_{\mu^{\mathrm{F}}}'(\lambda_+)}, \tag{7.445}$$

by reference to (7.436) and (7.437). We note that the RHS of (7.445) is an analytic (polynomial) function of N_{B}' and N_{F}'. Then we identify

$$-p = 1 + \lambda(N_{\mathrm{B}}' + N_{\mathrm{F}}'), \tag{7.446}$$

$$-q = \lambda N' + N_{\mathrm{F}}' = \lambda N_{\mathrm{B}}' + \lambda_+ N_{\mathrm{F}}'. \tag{7.447}$$

In this way we finally obtain the binomial expansion coefficient in (7.426) for separated states:

$$b_\kappa^{(+-)}(p,q) = (-1)^{|\mu^{\mathrm{B}}|+|\mu^{\mathrm{F}}|}\frac{\dot{e}_{\mu^{\mathrm{B}}}(\lambda';-p+\lambda' q)}{d_{\mu^{\mathrm{B}}}'(\lambda')}\frac{\dot{e}_{\mu^{\mathrm{F}}}(\lambda_+;-q)}{d_{\mu^{\mathrm{F}}}'(\lambda_+)}, \tag{7.448}$$

which takes a form factorized into bosonic and fermionic contributions. According to (7.438), both contributions reduce to the binomial coefficient (7.308) for the symmetric Jack polynomials.

7.5.6 Integral norm

The U(1,1) Jack polynomials are orthogonal with each other with respect to the integral norm. For $\kappa \in (\Lambda_{N_{\mathrm{B}}}^+, \Lambda_{N_{\mathrm{F}}}^{+>})$, we have the relation

$$\langle J_\kappa^{(+-)}, J_\kappa^{(+-)}\rangle_0/\langle E_\kappa, E_\kappa\rangle_0 = \frac{N_{\mathrm{B}}!N_{\mathrm{F}}!}{\rho_\kappa^{(+-)}}, \tag{7.449}$$

which can be derived in the same way as (7.167). By using explicit results for $\langle E_\kappa, E_\kappa\rangle_0$, as presented in Section 7.2.2, we obtain

$$\langle J_\kappa^{(+-)}, J_\kappa^{(+-)}\rangle_0 = N_{\mathrm{B}}!N_{\mathrm{F}}!\frac{\Gamma(N\lambda+1)}{\Gamma(\lambda+1)^N}\frac{d_\kappa' e_\kappa}{\rho_\kappa^{(+-)}d_\kappa e_\kappa'}. \tag{7.450}$$

It is instructive to compare the result with the norm

$$\langle J_\kappa, J_\kappa\rangle_0 = \frac{\Gamma(N\lambda+1)}{\Gamma(\lambda+1)^N}\cdot\frac{d_\kappa' e_\kappa''}{d_\kappa'' e_\kappa'} \tag{7.451}$$

for the symmetric Jack polynomial.

In the special case of $\kappa = \kappa(0)$, we can derive the following result:

$$\langle J_{\kappa(0)}^{(+-)}), J_{\kappa(0)}^{(+-)}\rangle_0 = \frac{N_\mathrm{F}!}{(\lambda_+)^{N_\mathrm{F}-1}} \cdot \frac{\Gamma(N\lambda + N_\mathrm{F})\Gamma(1 + \lambda' N_\mathrm{B})}{\Gamma(\lambda_+)^N \Gamma(N_\mathrm{F} + \lambda' N_\mathrm{B})}, \qquad (7.452)$$

which has been obtained in [10]. Note that the norm reproduces (7.451) with $N_\mathrm{F} = 0$ and $\kappa(0) = 0^{N_\mathrm{B}}$. It also reproduces (7.166) with $N_\mathrm{B} = 0$ and $\kappa(0) = \delta$. To derive (7.452), we first note

$$\rho_{\kappa(0)}^{(+-)} = N_\mathrm{B}! d'_{\kappa(0)}/d_{\kappa(0)}, \qquad (7.453)$$

which is obtained from (7.408) and (7.409). Then we only need to evaluate $e_{\kappa(0)}/e'_{\kappa(0)}$. This ratio can be obtained by following a procedure similar to that used in deriving (7.169).

The integral norm for a separated state can be written in a form that is convenient for taking the thermodynamic limit. Using (7.445) and putting the relations of combinatorial factors discussed in Section 7.1.6, we obtain the ratio of norms as

$$\frac{\langle J_\kappa^{(+-)}, J_\kappa^{(+-)}\rangle_0}{\langle J_{\kappa(0)}^{(+-)}, J_{\kappa(0)}^{(+-)}\rangle_0} = \mathcal{N}_\mathrm{B}\mathcal{N}_\mathrm{F}, \qquad (7.454)$$

which takes a factorized form of bosonic and fermionic parts. The bosonic part is given by

$$\mathcal{N}_\mathrm{B} = \frac{d'_{\mu^\mathrm{B}}(\lambda')\dot{e}_{\mu^\mathrm{B}}(\lambda'; 1 + \lambda' N_\mathrm{B})}{d''_{\mu^\mathrm{B}}(\lambda')\dot{e}_{\mu^\mathrm{B}}(\lambda'; 1 + \lambda'(N_\mathrm{B} - 1))}. \qquad (7.455)$$

Here we note the relation

$$\dot{e}_\mu(\lambda; \lambda N) = e''_\mu(N), \quad \dot{e}_\mu(\lambda; \lambda N + 1) = e'_\mu(N + 1), \qquad (7.456)$$

which results from the definition (7.276). Then (7.455) corresponds to the ratio of norms for symmetric Jack polynomials given by (7.451), provided the coupling constant is taken as λ'.

The fermionic part is given by

$$\mathcal{N}_\mathrm{F} = \frac{d'_{\mu^\mathrm{F}}(\lambda_+)\dot{e}_{\mu^\mathrm{F}}(\lambda_+; \lambda_+(N_\mathrm{F} + \lambda' N_\mathrm{B}))}{d''_{\mu^\mathrm{F}}(\lambda_+)\dot{e}_{\mu^\mathrm{F}}(\lambda_+; 1 + \lambda_+(N_\mathrm{F} + \lambda' N_\mathrm{B} - 1))}, \qquad (7.457)$$

which again takes the form (7.451), provided we take not only

  (i) the renormalized coupling λ_+, but also
  (ii) the effective number of particles as $N_\mathrm{F} + \lambda' N_\mathrm{B}$.

8

Yang–Baxter relations and orthogonal eigenbasis

The spectrum of the Sutherland model for fermions with spin has a degeneracy. The orthogonal set of the basis of the Fock space of degenerate states is usually related to a Lie algebra such as su_2 and su_3. In our case, however, the degeneracy extends to states with different values of total spin, and the ordinary Lie algebra is insufficient. The relevant algebra is called Yangian [43,44], and is the algebra controlling those operators satisfying the Yang–Baxter relation [118,178]. In this chapter we describe how the Yangian naturally emerges in the many-particle states described by Jack polynomials, and constructs an orthogonal basis of degenerate states. We concentrate here on the simplest case of the gl_2 Yangian, corresponding to spin 1/2 particles. More general cases will be discussed in Chapter 9.

We introduce a variant of the R-matrix in Section 8.1, and its standard form in Section 8.2. We then introduce a monodromy matrix and discuss its algebraic properties, such as the Yang–Baxter relation and the intertwining property. In Section 8.4, we consider the action of the monodromy matrix on the set of states constituting the degenerate eigenstates. We will see that an element of the monodromy matrix plays the role of a creation or annihilation operator of orthogonal eigenfunctions. This means that all eigenfunctions can be generated algebraically from the highest-weight state, for example, by successive action of an element of the monodromy matrix. Thus, the degeneracy beyond the Lie algebra is accounted for by the symmetry of the monodromy matrix, which is nothing but the Yangian.

A demonstration of this construction will be given in Section 8.5 for the two-particle and three-particle systems. The general expression for the orthogonal set of eigenfunctions is also given, which corresponds to the so-called Yangian Gelfand–Zetlin basis. In Section 8.6, we give the formula for the norm of the Yangian Gelfand–Zetlin basis. The contents of this chapter relies heavily on the paper by Takemura and Uglov [180].

8.1 Fock condition and R-matrix

In the present section, we will see that an R-matrix appears naturally as a consequence of the Fock condition imposed on the wave function whose spatial part consists of non-symmetric Jack polynomials. This R-matrix is an operator acting on the spin space of two particles, and obeys the Yang–Baxter relation. This finding gives us a hint on the procedure of construction of the orthogonal set of eigenfunctions of the U(2) fermionic Sutherland model. The resultant eigenfunctions are called the Yangian Gelfand–Zetlin basis [143, 144], as will be explained later.

We use the notation $z = (z_1, \ldots, z_N)$ and $\sigma = (\sigma_1, \ldots, \sigma_N)$ for the set of complex spatial and spin coordinates. The real coordinate x_i is related to z_i through $z_i = \exp(2\pi i x_i / L)$. The eigenfunctions of (3.5) for the U(2) fermion model are written in the form

$$\Psi(z, \sigma) = \Phi(z, \sigma)\Psi_0^{\mathrm{B}}(z), \tag{8.1}$$

$$\Phi(z, \sigma) = \sum_{\eta;\ \mathrm{s.t.}\eta^+ = \kappa} E_\eta(z)\varphi_\eta(\sigma), \tag{8.2}$$

where $\Psi_0^{\mathrm{B}}(z)$ is the symmetric function defined by (2.5). Let $\mathcal{F}_{N,\kappa}$ be the set of (8.2) satisfying the fermionic Fock condition

$$\hat{K}_{ij}\Phi = -\hat{P}_{ij}\Phi. \tag{8.3}$$

For a given partition $\kappa \in \Lambda_N^+$, the energy level in general has a degeneracy. We define $\mathcal{F}_{N,\kappa}$ to specify the degenerate wave functions with the same κ.

Let $\tilde{\Psi} \in \mathcal{F}_{N,\kappa}$ be a wave function of the form

$$\tilde{\Psi}(z, \sigma) = \tilde{\Phi}(z, \sigma)\Psi_0^{\mathrm{B}}(z), \tag{8.4}$$

$$\tilde{\Phi}(z, \sigma) = \sum_{\eta;\ \mathrm{s.t.}\eta^+ = \kappa} E_\eta(z)\tilde{\varphi}_\eta(\sigma), \tag{8.5}$$

where the spin part is specified by $\tilde{\varphi}_\eta(\sigma)$. The inner product $\langle \tilde{\Psi}, \Psi \rangle$ is given by

$$\langle \tilde{\Psi}, \Psi \rangle = \sum_{\sigma_1 = 1,2} \cdots \sum_{\sigma_N = 1,2} \int_0^L \mathrm{d}x_1 \cdots \int_0^L \mathrm{d}x_N \tilde{\Psi}^*(z, \sigma)\Psi(z, \sigma)$$

$$= L^N \sum_{\eta;\ \mathrm{s.t.}\eta^+ = \kappa} \langle E_\eta, E_\eta \rangle_0 \, (\tilde{\varphi}_\eta, \varphi_\eta), \tag{8.6}$$

where the spin part is given by

$$(\tilde{\varphi}, \varphi) \equiv \sum_{\sigma_1 = 1}^2 \cdots \sum_{\sigma_N = 1}^2 \tilde{\varphi}(\sigma)\varphi(\sigma). \tag{8.7}$$

For later convenience, we introduce the following notation:

$$\langle \tilde{\Phi}, \Phi \rangle' = \sum_{\eta;\ \text{s.t.}\eta^+ = \kappa} \langle E_\eta, E_\eta \rangle_0 (\tilde{\varphi}_\eta, \varphi_\eta), \tag{8.8}$$

for Φ (8.2) and $\tilde{\Phi}$ (8.5). We shall find an orthogonal basis of eigenfunctions in $\mathcal{F}_{N,\kappa}$. In the present subsection, we first derive a non-orthogonal basis of eigenfunctions in $\mathcal{F}_{N,\kappa}$. By imposing the Fock condition (8.3), we can derive the relation between φ_η in (8.2) and φ_κ with $\kappa = \eta^+$. Let us begin with some examples.

Example 8.1. $N = 2$, $\kappa = (\kappa_1, \kappa_2)$ with $\kappa_1 = \kappa_2$.
(8.2) in this case reduces to

$$E_{\kappa_1\kappa_1}(z_1, z_2)\varphi_{\kappa_1\kappa_1}(\sigma_1, \sigma_2), \tag{8.9}$$

where the spatial part is a symmetric function of z_1 and z_2. Therefore the Fock condition (8.3) requires

$$\hat{P}_{12}\varphi_{\kappa_1\kappa_1} = -\varphi_{\kappa_1\kappa_1}, \tag{8.10}$$

which means the singlet spin pair.

Example 8.2. $N = 2$, $\kappa = (\kappa_1, \kappa_2)$ with $\kappa_1 > \kappa_2$.
The role of the recursion relation (3.169) of non-symmetric Jack polynomials becomes clear in this case. Namely, the action of $\hat{K}_{12}$ on the LHS of the Fock condition (8.3) yields

$$\hat{K}_{12} \left(E_{\kappa_1\kappa_2}\varphi_{\kappa_1\kappa_2} + E_{\kappa_2\kappa_1}\varphi_{\kappa_2\kappa_1} \right)$$

$$= E_{\kappa_1\kappa_2} \left(\frac{\varphi_{\kappa_1\kappa_2}}{\overline{\kappa}_1 - \overline{\kappa}_2} + \varphi_{\kappa_2\kappa_1} \right) + E_{\kappa_2\kappa_1} \left[\left(1 - \frac{1}{(\overline{\kappa}_1 - \overline{\kappa}_2)^2} \right) \varphi_{\kappa_1\kappa_2} + \frac{\varphi_{\kappa_2\kappa_1}}{\overline{\kappa}_2 - \overline{\kappa}_1} \right].$$

The result should be equal to

$$-\hat{P}_{12}\Phi = - \left(E_{\kappa_1\kappa_2}\hat{P}_{12}\varphi_{\kappa_1\kappa_2} + E_{\kappa_2\kappa_1}\hat{P}_{12}\varphi_{\kappa_2\kappa_1} \right). \tag{8.11}$$

Since $E_{\kappa_1\kappa_2}$ and $E_{\kappa_2\kappa_1}$ are linearly independent, we obtain

$$\check{R}_{12}(\overline{\kappa}_1 - \overline{\kappa}_2)\varphi_{\kappa_1\kappa_2} = -\varphi_{\kappa_2\kappa_1} \tag{8.12}$$

$$\check{R}_{12}(\overline{\kappa}_2 - \overline{\kappa}_1)\varphi_{\kappa_2\kappa_1} = - \left[1 - (\overline{\kappa}_1 - \overline{\kappa}_2)^{-2} \right] \varphi_{\kappa_1\kappa_2}, \tag{8.13}$$

where we have introduced an operator

$$\check{R}_{ii+1}(u) = \frac{1}{u} + \hat{P}_{ii+1}. \tag{8.14}$$

Because of the relation

$$\check{R}_{12}(\overline{\kappa}_2 - \overline{\kappa}_1)\check{R}_{12}(\overline{\kappa}_1 - \overline{\kappa}_2) = 1 - (\overline{\kappa}_1 - \overline{\kappa}_2)^{-2}, \tag{8.15}$$

(8.12) and (8.13) are compatible. Consequently, we obtain

$$\Phi = \left(E_{\kappa_1\kappa_2} - E_{\kappa_2\kappa_1}\check{R}_{12}(\overline{\kappa}_1 - \overline{\kappa}_2)\right)\varphi_{\kappa_1\kappa_2}. \tag{8.16}$$

Example 8.3. $N = 3$, $\kappa = (\kappa_1, \kappa_2, \kappa_3)$ with $\kappa_1 > \kappa_2 > \kappa_3$.
The Fock condition (8.3) takes the same form as (8.12) from the recursion relation (3.169) of non-symmetric Jack polynomials, as long as the spectral parameter involved in $\check{R}_{ij}$ is positive. Hence we obtain

$$\varphi_{\kappa_2\kappa_1\kappa_3} = -\check{R}_{12}\left(\overline{\kappa}_1 - \overline{\kappa}_2\right)\varphi_{\kappa_1\kappa_2\kappa_3},$$
$$\varphi_{\kappa_3\kappa_2\kappa_1} = -\check{R}_{12}\left(\overline{\kappa}_2 - \overline{\kappa}_3\right)\varphi_{\kappa_2\kappa_3\kappa_1},$$
$$\varphi_{\kappa_3\kappa_1\kappa_2} = -\check{R}_{12}\left(\overline{\kappa}_1 - \overline{\kappa}_3\right)\varphi_{\kappa_1\kappa_3\kappa_2},$$

and analogous relations with $\check{R}_{12}$ replaced by $\check{R}_{23}$ or $\check{R}_{13}$. By combining these relations, all φ_η for $\eta^+ = \kappa$ are generated from φ_κ. We take $\varphi_{\kappa_3\kappa_2\kappa_1}$ as an example. There are two routes from $(\kappa_1\kappa_2\kappa_3)$ to $(\kappa_3\kappa_2\kappa_1)$, such as

route 1: $(\kappa_1\kappa_2\kappa_3) \rightarrow (\kappa_1\kappa_3\kappa_2) \rightarrow (\kappa_3\kappa_1\kappa_2) \rightarrow (\kappa_3\kappa_2\kappa_1)$,

route 2: $(\kappa_1\kappa_2\kappa_3) \rightarrow (\kappa_2\kappa_1\kappa_3) \rightarrow (\kappa_3\kappa_1\kappa_2) \rightarrow (\kappa_3\kappa_2\kappa_1)$.

Namely we obtain

$$\varphi_{\kappa_3\kappa_2\kappa_1} = -\check{R}_{23}\left(\overline{\kappa}_1 - \overline{\kappa}_2\right)\check{R}_{12}\left(\overline{\kappa}_1 - \overline{\kappa}_3\right)\check{R}_{23}\left(\overline{\kappa}_2 - \overline{\kappa}_3\right)\varphi_{\kappa_1\kappa_2\kappa_3} \tag{8.17}$$

via route 1 and

$$\varphi_{\kappa_3\kappa_2\kappa_1} = -\check{R}_{12}\left(\overline{\kappa}_2 - \overline{\kappa}_3\right)\check{R}_{23}\left(\overline{\kappa}_1 - \overline{\kappa}_3\right)\check{R}_{12}\left(\overline{\kappa}_1 - \overline{\kappa}_2\right)\varphi_{\kappa_1\kappa_2\kappa_3} \tag{8.18}$$

via route 2. The compatibility of (8.17) and (8.18) is guaranteed by the relation

$$\check{R}_{23}\left(u_1 - u_2\right)\check{R}_{12}\left(u_1 - u_3\right)\check{R}_{23}\left(u_2 - u_3\right)$$
$$= \check{R}_{12}\left(u_2 - u_3\right)\check{R}_{23}\left(u_1 - u_3\right)\check{R}_{12}\left(u_1 - u_2\right). \tag{8.19}$$

This relation (8.19) is a variant of the Yang–Baxter relation [118], as discussed in detail in Section 8.2.

Now we introduce a related operator $R_\eta^{(\lambda)}$ by the recursive relations [180]

$$R_\kappa^{(\lambda)} = 1, \tag{8.20}$$

$$R_{K_{ii+1}\eta}^{(\lambda)} = -\check{R}_{ii+1}(\overline{\eta}_i - \overline{\eta}_{i+1})R_\eta^{(\lambda)} \quad \text{for} \quad \eta_i > \eta_{i+1}. \tag{8.21}$$

We further introduce an operator $U(\kappa; \lambda)$ which creates Φ with spatial dependence from a spin-only wave function. The operator $U(\kappa; \lambda)$ is defined in terms of $R_\eta^{(\lambda)}$ by the relation [180]

$$\Phi = U(\kappa; \lambda)\varphi_\kappa \equiv \sum_{\eta; \text{ s.t. } \eta^+ = \kappa} E_\eta R_\eta^{(\lambda)} \varphi_\kappa, \tag{8.22}$$

where the sum is taken over all distinct rearrangements of $\kappa \in \Lambda_N^+$.

Example 8.4. $N = 3$, $\kappa = (\kappa_1, \kappa_2, \kappa_3)$ with $\kappa_1 > \kappa_2 = \kappa_3$.
In the present case, (8.2) becomes

$$\Phi = \sum_{\eta = (\kappa_1\kappa_2\kappa_2), (\kappa_2\kappa_1\kappa_2), (\kappa_2\kappa_2\kappa_1)} E_\eta R_\eta^{(\lambda)} \varphi_{\kappa_1\kappa_2\kappa_2}, \tag{8.23}$$

where $\varphi_{\kappa_1\kappa_2\kappa_2}$ satisfies

$$\hat{P}_{23}\varphi_{\kappa_1\kappa_2\kappa_2} = -\varphi_{\kappa_1\kappa_2\kappa_2}. \tag{8.24}$$

Example 8.5. $N = 3$, $\kappa = (\kappa_1, \kappa_2, \kappa_3)$ with $\kappa_1 = \kappa_2 = \kappa_3$.
In the present case, (8.2) becomes

$$E_{\kappa_1\kappa_1\kappa_1} \varphi_{\kappa_1\kappa_1\kappa_1}. \tag{8.25}$$

From the Fock condition, the $\varphi_{\kappa_1\kappa_1\kappa_1}$ have to satisfy

$$\hat{P}_{12}\varphi_{\kappa_1\kappa_1\kappa_1} = -\varphi_{\kappa_1\kappa_1\kappa_1} \quad \hat{P}_{23}\varphi_{\kappa_1\kappa_1\kappa_1} = -\varphi_{\kappa_1\kappa_1\kappa_1}, \tag{8.26}$$

which requires $\varphi_{\kappa_1\kappa_1\kappa_1} = 0$. Namely, in the present case, non-vanishing eigenfunctions do not exist.

We can generalize the above results as follows. The set of one-particle spin wave functions is given by C^2, where a basis $\{v_1(\sigma), v_2(\sigma)\}$ satisfies the orthonormal condition

$$(v_\alpha, v_\beta) = \sum_{\sigma=1}^{2} v_\alpha^*(\sigma)v_\beta(\sigma) = \delta_{\alpha\beta} \quad \text{for } \alpha, \beta \in [1, 2]. \tag{8.27}$$

$v_1(\sigma)$ and $v_2(\sigma)$ can be regarded as spin wave functions corresponding to $|\uparrow\rangle$ and $|\downarrow\rangle$. The set of N-particle spin wave functions is given by the tensor product of C^2:

$$\overbrace{C^2 \otimes \cdots \otimes C^2}^{N} = \otimes^N C^2.$$

We introduce a subset V_κ in $\otimes^N C^2$ such that two spins with the same momentum form a singlet. In other words, we define

$$V_\kappa = \left\{ \varphi \subset \otimes^N C^2 | \forall i \text{ satisfying } \kappa_i = \kappa_{i+1} \to \hat{P}_{i,i+1}\varphi = -\varphi \right\}. \tag{8.28}$$

For example, in the case of $N = 3$ and $\kappa_1 > \kappa_2 = \kappa_3$, $V_{\kappa_1\kappa_2\kappa_2}$ consists of two functions

$$v_1 \otimes (v_1 \otimes v_2 - v_2 \otimes v_1), \quad v_2 \otimes (v_1 \otimes v_2 - v_2 \otimes v_1). \tag{8.29}$$

Let us further introduce the set $\Lambda^+_{N,2} \subset \Lambda^+_N$ as those partitions where at most two κ_i can take the same value. Namely, we define

$$\Lambda^+_{N,2} = \left\{ \kappa = (\kappa_1, \kappa_2, \ldots, \kappa_N) \in \Lambda^+_N | \sharp \{\kappa_i \mid \kappa_i = \forall s \in \mathbb{Z}\} \leq 2 \right\}. \tag{8.30}$$

For $\kappa \in \Lambda^+_{N,2}$, the set of states

$$\Phi = U(\kappa; \lambda)\varphi \in \mathcal{F}_{N,\kappa}, \quad \varphi \in V_\kappa \tag{8.31}$$

satisfies the Fock condition (8.3). The operator $U(\kappa; \lambda)$ makes the wave function antisymmetric with respect to particle exchange.

We can obtain all the eigenfunctions of the form (8.2) by taking φ to be a basis of V_κ. For $N = 2$ and $\kappa_1 > \kappa_2$, for example, the basis of spin space V_κ is given by

$$v_{\alpha\beta} \equiv v_\alpha \otimes v_\beta, \quad \alpha, \beta = 1, 2 \tag{8.32}$$

and the wave functions

$$\phi_{\kappa,\alpha\beta} \equiv U(\kappa; \lambda)v_\alpha \otimes v_\beta, \quad \alpha, \beta = 1, 2 \tag{8.33}$$

are eigenfunctions of the Hamiltonian (3.16).

For $N = 3$ and $\kappa_1 > \kappa_2 = \kappa_3$, two wave functions

$$U(\kappa; \lambda)v_\alpha \otimes (v_{12} - v_{21}), \quad \alpha = 1, 2 \tag{8.34}$$

give eigenfunctions of the Hamiltonian (3.16).

Owing to the orthogonality of E_η, the functions $\Phi \in \mathcal{F}_{N,\kappa}$ and $\Phi' \in \mathcal{F}_{N,\kappa'}$ are orthogonal when $\kappa \neq \kappa'$. Within $\mathcal{F}_{N,\kappa}$, however, different basis functions are not necessarily orthogonal. Indeed, the two eigenfunctions (8.34) are mutually orthogonal but the four functions (8.33) are not. We shall show that the orthogonalization is naturally provided by the Yangian algebra in the following.

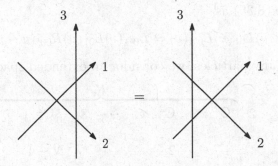

Fig. 8.1. Yang–Baxter relation for the R-matrix.

8.2 R-matrix and monodromy matrix

In order to obtain an orthogonal basis, we look for a set of self-adjoint operators $\hat{A}_1, \hat{A}_2, \ldots$ satisfying the following conditions:

- For $\kappa \in \Lambda_{N,2}^+$, $\Phi \in \mathcal{F}_{N,\kappa}$, $\hat{A}_i \Phi \in \mathcal{F}_{N,\kappa}$.
- $\left[\hat{A}_i, \hat{A}_j \right] = 0$.
- The set of eigenvalues of $\hat{A}_1, \hat{A}_2, \ldots$ specifies uniquely each eigenstate; the joint spectrum of those operators is non-degenerate.

Those operators are given by the monodromy matrix, which will be introduced in the next section. As a set up, we now define the R-matrix and the monodromy matrix relevant to the Sutherland model.

Let us start with (8.19). By multiplying both sides of (8.19) by $\hat{P}_{12}$ from the left and by $\hat{P}_{23}\hat{P}_{13}$ from the right and setting $u = u_1 - u_3$ and $v = u_2 - u_3$, we obtain

$$R_{13}(u - v)R_{23}(u)R_{12}(v) = R_{12}(v)R_{23}(u)R_{13}(u - v), \qquad (8.35)$$

where

$$R_{ij}(u) = 1 + \frac{\hat{P}_{ij}}{u} \qquad (8.36)$$

is called the R-matrix. The relation (8.35) is called the Yang–Baxter relation for the R-matrix, and is described graphically in Fig. 8.1.

The relation (8.35) is the functional relation of the operators acting on $\otimes^3 \mathbb{C}^2$. Now we rename each factor of $\otimes^3 \mathbb{C}^2$: we call the first factor $0'$, the second 0, and the third 1. The first two are regarded as auxiliary spaces and the third factor as a physical (spin) space. We introduce the operators $L_{01}(u)$ and $L_{0'1}(u)$ by

$$L_{01}(u) = 1 + \hat{P}_{01}/u, \quad L_{0'1}(u) = 1 + \hat{P}_{0'1}/u. \qquad (8.37)$$

We then rewrite (8.35) as

$$R_{00'}(u-v)L_{01}(u)L_{0'1}(v) = L_{0'1}(v)L_{01}(u)R_{00'}(u-v). \qquad (8.38)$$

In the case of many particles, we consider an extended space $\otimes^N \mathbf{C}^2$:

$$\overbrace{\mathbf{C}^2 \ \otimes \ \mathbf{C}^2 \ \otimes \ \mathbf{C}^2 \ \otimes \ \mathbf{C}^2 \ \otimes \ \cdots \ \otimes \ \mathbf{C}^2 \ \otimes \ \mathbf{C}^2}^{N} \qquad (8.39)$$

$$0 \qquad 0' \qquad 1 \qquad 2 \qquad \cdots \qquad N-1 \qquad N.$$

As an operator acting on the space (8.39), we introduce the monodromy operator

$$T_0(u; f) \equiv L_{01}(u+f_1)L_{02}(u+f_2)\cdots L_{0N}(u+f_N), \qquad (8.40)$$

where $f = (f_1, f_2, \ldots, f_N)$ are called the spectral parameters and

$$L_{0i}(u) = 1 + \frac{\hat{P}_{0i}}{u}, \quad \text{for } i = 1, \ldots, N. \qquad (8.41)$$

We shall derive the Yang–Baxter relation for the monodromy operators

$$R_{00'}(u-v)T_0(u; f)T_{0'}(v; f) = T_{0'}(v; f)T_0(u; f)R_{00'}(u-v). \qquad (8.42)$$

This relation follows from repeated use of (8.38) with replacement of $L_{01}, L_{0'1}$ by $L_{0i}, L_{0'i}$. When $N = 1$, (8.42) reduces to (8.38). When $N = 2$, we see that

$$R_{00'}(u-v)T_0(u; f)T_{0'}(v; f)$$
$$= R_{00'}(u-v)L_{01}(u+f_1)\underbrace{L_{02}(u+f_2)L_{0'1}(v+f_1)}_{\text{commute}}L_{0'2}(v+f_2)$$
$$= \underbrace{R_{00'}(u-v)L_{01}(u+f_1)L_{0'1}(v+f_1)}_{L_{0'1}(v+f_1)L_{01}(u+f_1)R_{00'}(u-v)}L_{02}(u+f_2)L_{0'2}(v+f_2)$$
$$= L_{0'1}(v+f_1)L_{01}(u+f_1)\underbrace{R_{00'}(u-v)L_{02}(u+f_2)L_{0'2}(v+f_2)}_{L_{0'2}(v+f_2)L_{02}(u+f_2)R_{00'}(u-v)}$$
$$= L_{0'1}(v+f_1)\underbrace{L_{01}(u+f_1)L_{0'2}(v+f_2)}_{\text{commute}}L_{02}(u+f_2)R_{00'}(u-v)$$
$$= T_{0'}(v; f)T_0(u; f)R_{00'}(u-v).$$

The relation (8.42) for a general value of N can be proved similarly. Figure 8.2 shows a graphical description of (8.42).

Now we define the monodromy matrix that has operators as its elements. The monodromy operator (8.40) acts in an auxiliary space $0'$ as the identity

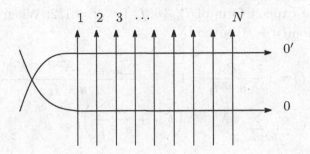

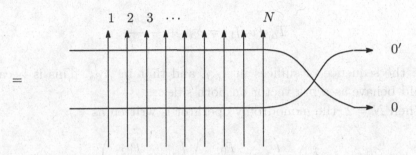

Fig. 8.2. Yang–Baxter relation $RTT = TTR$ (8.42) for the monodromy operators.

operator and hence it is natural to treat $T_0(u; f)$ as the operator on the tensor product of an auxiliary space 0 and the physical space:

$$\mathrm{C}^2 \;\otimes\; \left(\otimes^N \mathrm{C}^2\right)$$

$$0 \qquad 1 \sim N.$$

(8.43)

Regarding operators $T_{\alpha\beta}(u; f)$ as matrix elements, the monodromy matrix is defined by

$$T_0(u; f) = \sum_{\alpha,\beta=1,2} X^{\alpha\beta} \otimes T_{\alpha\beta}(u; f),$$

(8.44)

where $X^{\alpha\beta}$ is an operator acting on spin states in the auxiliary space 0:

$$X^{\alpha\beta} v_\gamma = \delta_{\beta\gamma} v_\alpha, \quad \alpha, \beta, \gamma = 1, 2.$$

(8.45)

In terms of spin operators, the X-operators are expressed as

$$X^{11} = \left(\frac{1}{2} + \hat{S}_z\right), \quad X^{22} = \left(\frac{1}{2} - \hat{S}^z\right), \quad X^{12} = \hat{S}^+, \quad X^{21} = \hat{S}^-.$$

(8.46)

Let us find the explicit form of $T_{\alpha\beta}(u;f)$ for $N = 1, 2$. When $N = 1$, we have $T_0(u;f) = L_{01}(u + f_1)$ given by

$$T_0(u;f) = 1 + \frac{\hat{P}_{01}}{u + f_1} = 1 \otimes 1 + \frac{\sum_{\alpha,\beta=1,2} X^{\alpha\beta} \otimes X^{\beta\alpha}}{u + f_1}$$

$$= \sum_{\alpha,\beta=1,2} X^{\alpha\beta} \otimes \left(\delta_{\alpha\beta} + \frac{X^{\beta\alpha}}{u + f_1} \right),$$

which means

$$T_{\alpha\beta}(u; f_1) = \delta_{\alpha\beta} + \frac{X^{\beta\alpha}}{u + f_1}. \tag{8.47}$$

Note the sequence of suffices in $T_{\alpha\beta}$, and that in $X_{\beta\alpha}$. This is because α should behave as a bra vector on both sides.

When $N = 2$, the monodromy operator is written as

$$T_0(u; f_1, f_2) = \left(1 + \frac{\hat{P}_{01}}{u + f_1} \right) \left(1 + \frac{\hat{P}_{02}}{u + f_2} \right)$$

$$= 1 + \frac{\hat{P}_{01}}{u + f_1} + \frac{\hat{P}_{02}}{u + f_2} + \frac{\hat{P}_{01}\hat{P}_{02}}{(u + f_1)(u + f_2)}. \tag{8.48}$$

We use the relations

$$\hat{P}_{01} = \sum_{\alpha,\beta=1,2} X^{\alpha\beta} \otimes X^{\beta\alpha} \otimes 1, \quad \hat{P}_{02} = \sum_{\alpha,\beta=1,2} X^{\alpha\beta} \otimes 1 \otimes X^{\beta\alpha}, \tag{8.49}$$

$$\hat{P}_{01}\hat{P}_{02} = \sum_{\alpha,\beta=1,2} X^{\alpha\beta} \otimes \sum_{\alpha'=1,2} X^{\alpha'\alpha} \otimes X^{\beta\alpha'}. \tag{8.50}$$

Then we rewrite (8.48) as

$$T_0(u; f_1, f_2) = \sum_{\alpha,\beta=1,2} X^{\alpha\beta} \otimes \sum_{\alpha'=1,2} \left(\delta_{\alpha\alpha'} + \frac{X^{\alpha'\alpha}}{u + f_1} \right) \otimes \left(\delta_{\alpha'\beta} + \frac{X^{\beta\alpha'}}{u + f_2} \right),$$

$$\tag{8.51}$$

from which we find

$$T_{\alpha\beta}(u; f_1, f_2) = \sum_{\alpha'=1,2} \left(\delta_{\alpha\alpha'} + \frac{X^{\alpha'\alpha}}{u + f_1} \right) \otimes \left(\delta_{\alpha'\beta} + \frac{X^{\beta\alpha'}}{u + f_2} \right). \tag{8.52}$$

Note again the sequence of α and β on both sides of (8.52).

For a general integer N, the monodromy matrix is given by [22]

$$T_{\alpha\beta}(u; f_1, f_2, \ldots, f_N)$$

$$= \sum_{\alpha_1, \alpha_2, \cdots, \alpha_{N-1}} \left(\delta_{\alpha\alpha_1} + \frac{X^{\alpha_1\alpha}}{u + f_1} \right) \otimes \left(\delta_{\alpha_1\alpha_2} + \frac{X^{\alpha_2\alpha_1}}{u + f_2} \right) \otimes$$

$$\cdots \otimes \left(\delta_{\alpha_{N-1}\beta} + \frac{X^{\beta\alpha_{N-1}}}{u + f_N} \right), \tag{8.53}$$

which can be proved by mathematical induction. The procedure (8.53) to construct $T_{\alpha\beta}(u)$ acting on $V \otimes \cdots \otimes V$ from that acting on V is called *co-product.*

8.3 Yangian gl_2

We will examine detailed algebraic properties of the monodromy matrix. When the monodromy matrix acts on $V(= \otimes^N \mathbf{C}^2)$, the R-matrix is rewritten as

$$R_{00'}(u - v) = 1 + \sum_{\gamma, \delta} \frac{X^{\gamma\delta} \otimes X^{\delta\gamma} \otimes 1}{u - v}. \tag{8.54}$$

Using this and (8.44), the LHS of (8.42) is written as

$$R_{00'}(u - v)T_0(u; f)T_{0'}(v; f)$$

$$= \sum_{\alpha\beta} \sum_{\gamma\delta} X^{\alpha\beta} \otimes X^{\gamma\delta} \otimes \left(T_{\alpha\beta}(u; f)T_{\gamma\delta}(v; f) + \frac{T_{\gamma\beta}(u; f)T_{\alpha\delta}(v; f)}{u - v} \right). \tag{8.55}$$

Similarly, the RHS of (8.42) is written as

$$T_{0'}(v; f)T_0(u; f)R_{00'}(u - v)$$

$$= \sum_{\alpha\beta} \sum_{\gamma\delta} X^{\alpha\beta} \otimes X^{\gamma\delta} \otimes \left(T_{\gamma\delta}(u; f)T_{\alpha\beta}(v; f) + \frac{T_{\gamma\beta}(v; f)T_{\alpha\delta}(u; f)}{u - v} \right). \tag{8.56}$$

Since (8.55) and (8.56) are equal, we obtain for $\alpha, \beta, \gamma, \delta = 1, 2$

$$(u - v)\left[T_{\alpha\beta}(u), T_{\gamma\delta}(v) \right] = T_{\gamma\beta}(v)T_{\alpha\delta}(u) - T_{\gamma\beta}(u)T_{\alpha\delta}(v), \tag{8.57}$$

which is equivalent to

$$(u - v)\left[T_{\gamma\delta}(u), T_{\alpha\beta}(v) \right] = T_{\gamma\beta}(u)T_{\alpha\delta}(v) - T_{\gamma\beta}(v)T_{\alpha\delta}(u). \tag{8.58}$$

The algebra for operator-valued matrices satisfying (8.58) is called the Yangian gl_2, and written $Y(gl_2)$. The form given by (8.53) constitutes a solution of the functional equation (8.58). The quantum determinant is defined by

$$\text{qdet}\{T_{ij}(u)\} \equiv T_{11}(u)T_{22}(u-1) - T_{12}(u)T_{21}(u-1), \tag{8.59}$$

which commutes with all elements of $T_{ij}(u)$. We shall generalize the quantum determinant beyond gl_2 in Chapter 9.

The elements of the monodromy matrix are defined by

$$T_{11}(u) \equiv A_1(u), \; T_{12}(u) \equiv B(u), \; T_{21}(u) \equiv C(u), \; T_{22}(u) \equiv D(u). \tag{8.60}$$

Then (8.58) is rewritten in terms of components. For example, we obtain

$$(u-v)\,[A_1(u), B(v)] = B(u)A_1(v) - B(v)A_1(u), \tag{8.61}$$

$$(u-v)\,[C(u), B(v)] = D(u)A_1(v) - D(v)A_1(u), \tag{8.62}$$

where (8.61) corresponds to the case $(\alpha, \beta, \gamma, \delta) = (1, 2, 1, 1)$ in (8.58), and (8.62) to the case $(1, 2, 2, 1)$. It is convenient to draw an analogy with the angular momentum to understand the properties of the Yangian. The algebra for angular momentum is characterized by the following:

(i) $\hat{\boldsymbol{L}}^2 = \hat{L}_x^2 + \hat{L}_y^2 + \hat{L}_z^2$ commutes with all elements L_x, L_y, and L_z.

(ii) Simultaneous eigenstates can be constructed for $\hat{\boldsymbol{L}}^2$ and $\hat{L}_z$ as an orthogonal basis.

(iii) As shown by the property

$$\left[\hat{L}^{\pm}, \hat{\boldsymbol{L}}^2\right] = 0, \quad \left[\hat{L}^{\pm}, \hat{L}_z\right] = \mp L^{\pm}, \tag{8.63}$$

the eigenvalue of $\hat{L}_z$ is raised by $\hat{L}^+$, and lowered by $\hat{L}^-$.

The properties of the Yangian gl_2 are summarized in an analogous way:

(i) The quantum determinant, $\text{qdet}\{T_{ij}(u)\} \equiv A_2(u)$, commutes with all elements $A_1(u), B(u), C(u)$, and $D(u)$.

(ii) Simultaneous eigenfunctions can be constructed for $A_2(u)$ and $A_1(u)$.

(iii) Let $|\chi\rangle$ be an eigenstate of $A_1(u)$ with eigenvalue $\omega(u)$. By choosing v so that $\omega(v) = 0$, we obtain from (8.61):

$$(u-v)\,[A_1(u), B(v)]\,|\chi\rangle = -\omega(u)B(v)|\chi\rangle. \tag{8.64}$$

By comparing (8.64) and (8.63), we see that $B(v)$ for a particular value of v acts as a raising or lowering operator of the eigenvalue of $A_1(u)$. A similar statement applies to $C(v)$ as a raising or lowering operator of the eigenvalue of $A_1(u)$. It should be remembered that $B(v)$ decreases the spin S_z of the system by one, while $C(v)$ increases S_z by one.

The correspondence between the angular momentum and $Y(gl_2)$ is summarized as follows:

$$\begin{array}{ccc} A_2(u) & & \hat{\boldsymbol{L}}^2 \\ A_1(u) & & \hat{L}^z \\ B(u) & \Longleftrightarrow & \hat{L}^- \\ C(u) & & \hat{L}^+ \end{array} \tag{8.65}$$

for a particular value of u.

8.4 Relation to U(2) Sutherland model

The monodromy matrix has a set of parameters $f_1, \ldots, f_N$ called the spectral parameters. The U(2) Sutherland model fits nicely in the Yangian symmetry by identifying the spectral parameters as the rapidities of particles. The close relation between the Yangian and the Sutherland model is also anticipated from our construction of the Yang–Baxter relation using the non-symmetric Jack polynomials.

Let us begin with the proof of the following statement [180]:

$$T_{\alpha\beta}(u; f_1, \ldots, f_{i+1}, \overset{i}{f_i}, \overset{i+1}{\ldots}, f_N)\check{R}_{ii+1}(f_i - f_{i+1})$$

$$= \check{R}_{ii+1}(f_i - f_{i+1})T_{\alpha\beta}(u; f_1, \ldots, \overset{i}{f_i}, \overset{i+1}{f_{i+1}}, \ldots, f_N), \tag{8.66}$$

which will turn out to be useful in the next section. It shows that $\check{R}_{ii+1}(f_i - f_{i+1})$ *intertwines* the parameters f_i, f_{i+1} of the monodromy matrix.

Proof of (8.66).
First we regard $\check{R}_{ii+1}(f_i - f_{i+1})$ as an operator acting on $\otimes^N \mathbb{C}^2$ and construct $1 \otimes \check{R}_{ii+1}(f_i - f_{i+1})$ acting on the tensor product of the auxiliary space 0 and $\otimes^N \mathbb{C}^2$. The relation

$$L_{0,i}(u + f_{i+1})L_{0,i+1}(u + f_i)\left(1 \otimes \check{R}_{ii+1}(f_i - f_{i+1})\right)$$

$$= \left(1 \otimes \check{R}_{ii+1}(f_i - f_{i+1})\right)L_{0,i}(u + f_i)L_{0,i+1}(u + f_{i+1}) \tag{8.67}$$

is nothing but (8.38) in a slightly different form. The relation

$$T_0(u; f_1, \ldots, f_{i+1}, \overset{i}{f_i}, \overset{i+1}{\ldots}, f_N)\left(1 \otimes \check{R}_{ii+1}(f_i - f_{i+1})\right)$$

$$= \left(1 \otimes \check{R}_{ii+1}(f_i - f_{i+1})\right)T_0(u; f_1, \ldots, \overset{i}{f_i}, \overset{i+1}{f_{i+1}}, \ldots, f_N) \tag{8.68}$$

follows from (8.67). Further, we rewrite both sides of (8.68) in terms of $T_{\alpha\beta}(u; f_1, \dots)$:

$$T_0(u; f_1, \dots, f_{i+1}, f_i, \dots, f_N) \left[1 \otimes \check{R}_{ii+1}(f_i - f_{i+1})\right]$$
$$= \sum_{\alpha,\beta} X^{\alpha\beta} \otimes T_{\alpha\beta}(u; f_1, \dots, f_{i+1}, f_i, \dots, f_N) \left[1 \otimes \check{R}_{ii+1}(f_i - f_{i+1})\right]$$
$$= \sum_{\alpha,\beta} X^{\alpha\beta} \otimes T_{\alpha\beta}(u; f_1, \dots, f_{i+1}, f_i, \dots, f_N) \check{R}_{ii+1}(f_i - f_{i+1}) \qquad (8.69)$$

and

$$\left[1 \otimes \check{R}_{ii+1}(f_i - f_{i+1})\right] T_0(u; f) = \sum_{\alpha,\beta} X^{\alpha\beta} \otimes \check{R}_{ii+1}(f_i - f_{i+1}) T_{\alpha\beta}(u; f).$$
$$(8.70)$$

The relation (8.66) follows from (8.69) and (8.70). ∎

We want to find a commuting set of self-adjoint operators acting on $\mathcal{F}_{N,\kappa}$. As a trial, let us see the action of the monodromy matrix $T_{\alpha\beta}(u; \hat{d}_1, \dots, \hat{d}_N)$ on $\mathcal{F}_{N,\kappa}$. Note that the spectral parameters $f_1, \dots, f_N$ are replaced by the Cherednik–Dunkl operators $\hat{d}_1, \dots, \hat{d}_N$. This is a crucial trick in the following argument. Recalling that non-symmetric Jack polynomials are eigenfunctions of $\hat{d}_i$, we easily see that

$$T_{\alpha\beta}(u; \hat{d}_1, \dots, \hat{d}_N)\Phi \in \mathcal{F}_{N,\kappa}, \qquad (8.71)$$

provided that $\kappa \in \Lambda_{N,2}^+$ and $\Phi \in \mathcal{F}_{N,\kappa}$. Further, we find the conjugate property [180]

$$T_{\alpha\beta}^\dagger(u; \hat{d}_1, \dots, \hat{d}_N) = T_{\beta\alpha}(u; \hat{d}_1, \dots, \hat{d}_N), \qquad (8.72)$$

with respect to the inner product (2.189). This property is derived by recalling that $T_{\alpha\beta}(u; \hat{d}_1, \dots, \hat{d}_N)$ consists of $X^{\alpha\beta}$, and that $\hat{d}_1, \dots, \hat{d}_N$ are self-adjoint.

From (8.71) and (8.72), we construct a commuting set of operators in $\mathcal{F}_{N,\kappa}$ from $T_{\alpha\beta}(u; \hat{d}_1, \dots, \hat{d}_N)$. The action of $T_{\alpha\beta}(u; \hat{d}_1, \dots, \hat{d}_N)$ on $\Phi \in \mathcal{F}_{N,\kappa}$ is processed as

$$T_{\alpha\beta}(u; \hat{d}_1, \ldots, \hat{d}_N)U(\kappa; \lambda)\varphi$$

$$= \sum_{\eta;\ \text{s.t.}\ \eta^+ = \kappa} T_{\alpha\beta}(u; \hat{d}_1, \ldots, \hat{d}_N)E_\eta R_\eta^{(\lambda)}\varphi$$

$$= \sum_{\eta;\ \text{s.t.}\ \eta^+ = \kappa} E_\eta T_{\alpha\beta}(u; \overline{\eta}_1, \ldots, \overline{\eta}_N)R_\eta^{(\lambda)}\varphi \tag{8.73}$$

$$= \sum_{\eta;\ \text{s.t.}\ \eta^+ = \kappa} E_\eta R_\eta^{(\lambda)}T_{\alpha\beta}(u; \overline{\kappa}_1, \ldots, \overline{\kappa}_N)\varphi \tag{8.74}$$

$$= U(\kappa; \lambda)T_{\alpha\beta}(u; \overline{\kappa}_1, \ldots, \overline{\kappa}_N)\varphi. \tag{8.75}$$

Equality of (8.73) and (8.74) can be understood by examining the following example. For a composition η given by

$$\eta = K_{ii+1}\kappa = (\kappa_1, \ldots, \kappa_{j-1}, \kappa_{j+1}, \kappa_j, \ldots, \kappa_N),$$

$R_\eta^{(\lambda)}$ reduces to $R_{ii+1}(\overline{\kappa}_i - \overline{\kappa}_{i+1})$. When we set $(\overline{\kappa}_1, \overline{\kappa}_2, \ldots) = (f_1, f_2, \ldots)$, the LHS and RHS of (8.66) turn into (8.73) and (8.74), respectively.

Let us take another example. For $N = 3$, $\eta = (\kappa_3, \kappa_2, \kappa_1)$ satisfying $\kappa_1 > \kappa_2 > \kappa_3$, we obtain

$$(8.73) = -T_{\alpha\beta}(u; \overline{\kappa}_3, \overline{\kappa}_2, \overline{\kappa}_1)\check{R}_{23}(\overline{\kappa}_1 - \overline{\kappa}_2)\check{R}_{12}(\overline{\kappa}_1 - \overline{\kappa}_3)\check{R}_{23}(\overline{\kappa}_2 - \overline{\kappa}_3)$$

$$= -\check{R}_{23}(\overline{\kappa}_1 - \overline{\kappa}_2)T_{\alpha\beta}(u; \overline{\kappa}_3, \overline{\kappa}_1, \overline{\kappa}_2)\check{R}_{12}(\overline{\kappa}_1 - \overline{\kappa}_3)\check{R}_{23}(\overline{\kappa}_2 - \overline{\kappa}_3)$$

$$= -\check{R}_{23}(\overline{\kappa}_1 - \overline{\kappa}_2)\check{R}_{12}(\overline{\kappa}_1 - \overline{\kappa}_3)T_{\alpha\beta}(u; \overline{\kappa}_1, \overline{\kappa}_3, \overline{\kappa}_2)\check{R}_{23}(\overline{\kappa}_2 - \overline{\kappa}_3)$$

$$= -\check{R}_{23}(\overline{\kappa}_1 - \overline{\kappa}_2)\check{R}_{12}(\overline{\kappa}_1 - \overline{\kappa}_3)\check{R}_{23}(\overline{\kappa}_2 - \overline{\kappa}_3)T_{\alpha\beta}(u; \overline{\kappa}_1, \overline{\kappa}_2, \overline{\kappa}_3)$$

$$= (8.74)$$

with repeated use of the intertwining property (8.66) of $\check{R}_{ii+1}(u)$.

From (8.75), we can act $T_{\alpha\beta}(u; \overline{\kappa}_1, \ldots, \overline{\kappa}_N)$ on $\varphi \in V_\kappa$, instead of acting $T_{\alpha\beta}(u; \hat{d}_1, \ldots, \hat{d}_N)$ on $\Phi \in \mathcal{F}_{N,\kappa}$. If the eigenvalues of $T_{11}(u; \overline{\kappa}_1, \ldots, \overline{\kappa}_N)$ are non-degenerate for certain $\chi \in V_\kappa$, then those $\{\chi\}$ constitute an orthogonal set of basis in V_κ.

Now we show the following theorem [180]:

Theorem 8.6.
If the spin function φ belongs to V_κ, then the action of $T_{\alpha\beta}$ is stable: $T_{\alpha\beta}(u; \overline{\kappa}_1, \overline{\kappa}_2, \ldots)\varphi \in V_\kappa$.

Proof.
When $f_i = f_{i+1} - 1$ in (8.14), $\check{R}_{ii+1}(f_i - f_{i+1})$ becomes $-1 + \hat{P}_{ii+1}$, which antisymmetrizes the state vector. Then for $\varphi \in \otimes^N \mathbb{C}^2$, the intertwining relation (8.66) becomes

$$
T_{\alpha\beta}(u; f_1, \ldots, \overset{i}{f_i + 1}, \overset{i+1}{f_i}, \ldots, f_N)(1 - \hat{P}_{ii+1})\varphi
$$
$$
= 2T_{\alpha\beta}(u; f_1, \ldots, \overset{i}{f_i + 1}, \overset{i+1}{f_i}, \ldots, f_N)\varphi
$$
$$
= (1 - \hat{P}_{ii+1})T_{\alpha\beta}(u; f_1, \ldots, \overset{i}{f_i}, \overset{i+1}{f_i + 1}, \ldots, f_N)\varphi. \tag{8.76}
$$

From this relation, we see that

$$
\hat{P}_{ii+1}\varphi = -\varphi \rightarrow \hat{P}_{ii+1}\varphi' = -\varphi'
$$

with

$$
\varphi' \equiv T_{\alpha\beta}(u; f_1, \ldots, f_i + 1, f_i, \ldots, f_N)\varphi.
$$

Any vectors $\varphi \in V_\kappa$ satisfy $\hat{P}_{ii+1}\varphi = -\varphi$ for i such that $\overline{\kappa}_i = \overline{\kappa}_{i+1} + 1$, which is equivalent to $\kappa_i = \kappa_{i+1}$. Thus

$$
\varphi' = T_{\alpha\beta}(u; \overline{\kappa}_1, \ldots, \overline{\kappa}_N)\varphi \tag{8.77}
$$

satisfies $\hat{P}_{ii+1}\varphi' = -\varphi'$ for such i as has $\kappa_i = \kappa_{i+1}$. Therefore, (8.77) belongs to V_κ. ∎

8.5 Construction of orthogonal set of eigenbasis

8.5.1 *Examples for small systems*

First we consider some examples for small systems ($N = 1, 2, 3$), and later generalize the results. From now on, we use the following notation:

$$
\rho_1(u) = \prod_{i=1}^{N}(u + \overline{\kappa}_i), \quad \rho_2(u) = \prod_{i=1}^{N}(u + \overline{\kappa}_i)(u - 1 + \overline{\kappa}_i), \tag{8.78}
$$

$$
a_1(u) = \rho_1(u)T_{11}(u), \quad b(u) = \rho_1(u)T_{12}(u),
$$
$$
c(u) = \rho_1(u)T_{21}(u), \quad d(u) = \rho_1(u)T_{22}(u). \tag{8.79}
$$

Furthermore we define

$$
a_2(u) = \rho_2(u)\left[T_{11}(u)T_{22}(u - 1) - T_{12}(u)T_{21}(u - 1)\right]
$$
$$
= a_1(u)d(u - 1) - b(u)c(u - 1), \tag{8.80}
$$

which is proportional to the quantum determinant introduced in (8.59), but without singularity because of the factor $\rho_2(u)$. Hence, $a_2(u)$ is referred to as the "regular" quantum determinant in this book. Other quantities are also related to $A_1, B, C,$ and D by the factor $\rho_1(u)$. We denote a basis of $\otimes^N \mathbb{C}^2$ by

$$v_{\alpha_1 \alpha_2 \cdots \alpha_N} \equiv v_{\alpha_1} \otimes v_{\alpha_2} \otimes \cdots \otimes v_{\alpha_N}. \tag{8.81}$$

In the rest of this subsection, we give Examples 8.7 to 8.10.

Example 8.7. $N = 1$.
From (8.47), we obtain

$$a_1(u; \overline{\kappa}_1) = u + \overline{\kappa}_1 + X^{11}, \quad b(u; \overline{\kappa}_1) = X^{21},$$
$$c(u; \overline{\kappa}_1) = X^{12}, \quad d(u; \overline{\kappa}_1) = u + \overline{\kappa}_1 + X^{22}. \tag{8.82}$$

From these expressions, we obtain

$$a_2(u; \overline{\kappa}_1) = \rho_2(u) \left(1 + \frac{X^{11} + X^{22}}{u + \overline{\kappa}_1} \right) = (u + \overline{\kappa}_1)^2 - 1, \tag{8.83}$$

which is a c-number and commutes with $a_1(u), b(u), c(u), d(u)$. For $N = 1$, V_κ reduces to $\mathbb{C}^2$ where v_1 and v_2 form a basis. Obviously, both v_1 and v_2 are eigenvectors of $a_1(u)$:

$$a_1(u; \overline{\kappa}_1)v_1 = (u + 1 + \overline{\kappa}_1) v_1, \quad a_1(u; \overline{\kappa}_1)v_2 = (u + \overline{\kappa}_1) v_2. \tag{8.84}$$

The highest-weight state v_1 is annihilated by $c_1(u)$; $c_1(u)v_1 = 0$. The other state v_2 is generated by action of $b(u)$ on v_1.

Example 8.8. $N = 2$.
This is the simplest nontrivial case. We obtain

$$a_1(u; \overline{\kappa}_1, \overline{\kappa}_2) = \rho_1(u) \left(T_{11}(u) \otimes T_{11}(u) + T_{12}(u) \otimes T_{21}(u) \right)$$
$$= \left(u + \overline{\kappa}_1 + X^{11} \right) \otimes \left(u + \overline{\kappa}_2 + X^{11} \right) + X^{21} \otimes X^{12}, \tag{8.85}$$
$$b(u; \overline{\kappa}_1, \overline{\kappa}_2) = \left(u + \overline{\kappa}_1 + X^{11} \right) \otimes X^{21} + X^{21} \otimes \left(u + \overline{\kappa}_2 + X^{22} \right), \tag{8.86}$$
$$c(u; \overline{\kappa}_1, \overline{\kappa}_2) = X^{12} \otimes \left(u + \overline{\kappa}_2 + X^{11} \right) + \left(u + \overline{\kappa}_1 + X^{22} \right) \otimes X^{12}, \tag{8.87}$$
$$d(u; \overline{\kappa}_1, \overline{\kappa}_2) = X^{12} \otimes X^{21} + \left(u + \overline{\kappa}_1 + X^{22} \right) \otimes \left(u + \overline{\kappa}_2 + X^{22} \right). \tag{8.88}$$

The regular quantum determinant in the present case is obtained as a scalar

$$a_2(u; \overline{\kappa}_1, \overline{\kappa}_2) = \prod_{i=1}^{2} \left((u + \overline{\kappa}_i)^2 - 1 \right). \tag{8.89}$$

For $N = 2$, $\kappa_1 > \kappa_2$, V_κ is $\mathbb{C}^2 \otimes \mathbb{C}^2$, where the basis is given by

$$v_{\alpha\beta} \equiv v_\alpha \otimes v_\beta, \quad \alpha, \beta = 1, 2. \tag{8.90}$$

The action of $a_1(u; \overline{\kappa}_1, \overline{\kappa}_2)$ is, for the basis (8.90), given by

$$a_1(u; \overline{\kappa}_1, \overline{\kappa}_2)v_{11} = (u + 1 + \overline{\kappa}_1)(u + 1 + \overline{\kappa}_2)v_{11}, \tag{8.91}$$

$$a_1(u; \overline{\kappa}_1, \overline{\kappa}_2)v_{22} = (u + \overline{\kappa}_1)(u + \overline{\kappa}_2)v_{22}, \tag{8.92}$$

$$a_1(u; \overline{\kappa}_1, \overline{\kappa}_2) \begin{pmatrix} v_{12} \\ v_{21} \end{pmatrix} \tag{8.93}$$

$$= \begin{pmatrix} (u + 1 + \overline{\kappa}_1)(u + \overline{\kappa}_2) & 1 \\ 0 & (u + \overline{\kappa}_1)(u + 1 + \overline{\kappa}_2) \end{pmatrix} \begin{pmatrix} v_{12} \\ v_{21} \end{pmatrix}. \tag{8.94}$$

Since the matrix is triangular, the eigenvectors $\chi_{\alpha\beta}$ are easily obtained as

$$\chi_{11} = v_{11}, \quad \chi_{12} = v_{12} - \frac{v_{21}}{\overline{\kappa}_1 - \overline{\kappa}_2}, \quad \chi_{21} = v_{21}, \quad \chi_{22} = v_{22}. \tag{8.95}$$

The basis formed by (8.95) is an example of a Yangian Gelfand–Zetlin basis [143, 144], and these are *not* classified by the total spin of two particles. The eigenvalue of each eigenstate is obtained from the diagonal element of the matrix in (8.94). Owing to

$$\overline{\kappa}_1 - \overline{\kappa}_2 = (\kappa_1 - \kappa_2)/\lambda + 1 > 1,$$

all the eigenvalues are distinct.

The four wave functions

$$\Phi_{\kappa;\alpha\beta} \equiv U(\kappa; \lambda)\chi_{\alpha\beta}, \quad \alpha, \beta = 1, 2 \tag{8.96}$$

with $\kappa = (\kappa_1, \kappa_2)$ are eigenfunctions of the self-adjoint operator $T_{11}(u; \hat{d}_1, \hat{d}_2)$ and the eigenvalues are non-degenerate. Thus the wave functions (8.96) constitute an orthogonal basis of eigenfunctions in $\mathcal{F}_{N,\kappa}$.

Now we rederive the eigenbasis (8.95) algebraically starting from the highest-weight state $\chi_{11} = v_{11}$ [143, 144]. The Yangian relation (8.61) is rewritten as

$$(u - v)[a_1(u), b(v)] = b(u)a_1(v) - b(v)a_1(u). \tag{8.97}$$

Applying both sides of (8.97) to χ_{11}, we obtain

$$(u - v)[a_1(u), b(v)]\chi_{11} = \varpi_{\text{hws}}(v)b(u)\chi_{11} - \varpi_{\text{hws}}(u)b(v)\chi_{11}, \tag{8.98}$$

where we have introduced the notation

$$\varpi_{\text{hws}}(u) = (u + \overline{\kappa}_1 + 1)(u + \overline{\kappa}_2 + 1).$$

In (8.98), we take $v = -1 - \bar{\kappa}_i$ so that $\varpi_{\text{hws}}(v) = 0$ for $i = 1, 2$. Then we obtain

$$(u + 1 + \bar{\kappa}_i) [a_1(u), b(-1 - \bar{\kappa}_i)] \chi_{11} = -\varpi_{\text{hws}}(u) b(-1 - \bar{\kappa}_i) \chi_{11}, \qquad (8.99)$$

which is rewritten as

$$a_1(u) b(-1 - \bar{\kappa}_i) \chi_{11} = \frac{(u + \bar{\kappa}_i) \varpi_{\text{hws}}(u)}{(u + 1 + \bar{\kappa}_i)} b(-1 - \bar{\kappa}_i) \chi_{11}. \qquad (8.100)$$

Thus $b(-1 - \bar{\kappa}_i) \chi_{11}$ turns out to be an eigenfunction of $a_1(u)$. The eigenvalue is given by

$$\frac{(u + \bar{\kappa}_i) \varpi_{\text{hws}}(u)}{(u + 1 + \bar{\kappa}_i)} = \begin{cases} (u + \bar{\kappa}_1)(u + 1 + \bar{\kappa}_2), & (i = 1), \\ (u + 1 + \bar{\kappa}_1)(u + \bar{\kappa}_2), & (i = 2). \end{cases} \qquad (8.101)$$

We introduce $\tilde{\chi}_{ij}$ and, using (8.86), confirm the following relations:

$$\tilde{\chi}_{21} \equiv b(-1 - \bar{\kappa}_1) \chi_{11} = -(\bar{\kappa}_1 - \bar{\kappa}_2 + 1) \chi_{21}, \qquad (8.102)$$

$$\tilde{\chi}_{12} \equiv b(-1 - \bar{\kappa}_2) \chi_{11} = (\bar{\kappa}_1 - \bar{\kappa}_2) \chi_{12}. \qquad (8.103)$$

The basis function χ_{22} can be generated similarly. Namely, we apply both sides of (8.97) to $\tilde{\chi}_{21}$, setting $v = -1 - \bar{\kappa}_2$. Then we obtain

$$(u + 1 + \bar{\kappa}_2) [a_1(u), b(-1 - \bar{\kappa}_2)] \tilde{\chi}_{21}$$
$$= -(u + \bar{\kappa}_1)(u + 1 + \bar{\kappa}_2) b(-1 - \bar{\kappa}_2) \tilde{\chi}_{21}, \qquad (8.104)$$

which is rewritten as

$$a_1(u) b(-1 - \bar{\kappa}_2) \tilde{\chi}_{21} = (u + \bar{\kappa}_1)(u + \bar{\kappa}_2) b(-1 - \bar{\kappa}_2) \tilde{\chi}_{21}. \qquad (8.105)$$

Thus we identify the lowest-weight eigenfunction as

$$\tilde{\chi}_{22} \equiv b(-1 - \bar{\kappa}_2) \tilde{\chi}_{21} \propto \chi_{22}. \qquad (8.106)$$

The basis function $\tilde{\chi}_{22}$ can be rewritten in another way. Namely, by setting $(\alpha, \beta, \gamma, \delta) = (1, 2, 1, 2)$ in (8.58), we find the commuting property

$$[b(u), b(v)] = 0, \quad \text{for } |u - v| \neq 1. \qquad (8.107)$$

We then obtain

$$\begin{aligned} \tilde{\chi}_{22} &= b(-1 - \bar{\kappa}_2) \tilde{\chi}_{21} \\ &= b(-1 - \bar{\kappa}_2) b(-1 - \bar{\kappa}_1) \chi_{11} \\ &= b(-1 - \bar{\kappa}_1) b(-1 - \bar{\kappa}_2) \chi_{11} \\ &= b(-1 - \bar{\kappa}_1) \tilde{\chi}_{12}. \end{aligned} \qquad (8.108)$$

Hence the basis function $\tilde{\chi}_{22}$ can be generated from $\tilde{\chi}_{12}$ as well as from $\tilde{\chi}_{21}$.

We can confirm (8.105) and (8.108) directly using (8.86). Namely, we derive

$$b(-1 - \bar{\kappa}_2)b(-1 - \bar{\kappa}_1)v_{11} = b(-1 - \bar{\kappa}_1)b(-1 - \bar{\kappa}_2)v_{11}$$
$$= \left(1 - (\bar{\kappa}_1 - \bar{\kappa}_2)^2\right)\chi_{22}. \tag{8.109}$$

Thus the eigenbasis (8.95) is equivalent to

$$\chi_{11}, \quad b(-1 - \bar{\kappa}_1)\chi_{11}, \quad b(-1 - \bar{\kappa}_2)\chi_{11}, \quad b(-1 - \bar{\kappa}_1)b(-1 - \bar{\kappa}_2)\chi_{11}, \tag{8.110}$$

up to a constant factor. By starting with the highest-weight state $\chi_{11} = v_{11}$, all the eigenstates $\tilde{\chi}_{\alpha\beta}$ have been generated algebraically by the lowering operators. Further lowering gives $b(u)\chi_{22} = b(u)v_{22} = 0$.

We note that χ_{12} and χ_{21} are also annihilated as

$$b(-\bar{\kappa}_2)\chi_{12} = 0, \quad b(-\bar{\kappa}_1)\chi_{21} = 0. \tag{8.111}$$

To derive (8.111), we apply both sides of (8.97) to $\tilde{\chi}_{21}$. Setting $v = -\bar{\kappa}_1$, we obtain

$$[a_1(u), b(-\bar{\kappa}_1)]\tilde{\chi}_{21} = -(u + 1 + \bar{\kappa}_2)b(-\bar{\kappa}_1)\tilde{\chi}_{21}, \tag{8.112}$$

from which

$$a_1(u)b(-\bar{\kappa}_1)\tilde{\chi}_{21} = (u + \bar{\kappa}_1 - 1)(u + \bar{\kappa}_2 + 1)b(-\bar{\kappa}_1)\tilde{\chi}_{21} \tag{8.113}$$

follows. However, the eigenvalues of $a_1(u)$ are exhausted by

$$(u + \bar{\kappa}_1 + 1)(u + \bar{\kappa}_2 + 1), \qquad (u + \bar{\kappa}_1 + 1)(u + \bar{\kappa}_2),$$
$$(u + \bar{\kappa}_1)(u + \bar{\kappa}_2 + 1), \qquad (u + \bar{\kappa}_1)(u + \bar{\kappa}_2).$$

Thus, (8.113) requires $b(-\bar{\kappa}_1)\tilde{\chi}_{21} = 0$. The first equality of (8.111) can be derived similarly.

Figure 8.3 illustrates the relations (8.103), (8.109), and (8.111). As we shall derive, the following actions:

$$c(-\bar{\kappa}_1)\tilde{\chi}_{22} = \left[(\bar{\kappa}_1 - \bar{\kappa}_2)^2 - 1\right]\tilde{\chi}_{12}, \tag{8.114}$$

$$c(-\bar{\kappa}_2)\tilde{\chi}_{22} = \left[(\bar{\kappa}_1 - \bar{\kappa}_2)^2 - 1\right]\tilde{\chi}_{21}, \tag{8.115}$$

$$c(-\bar{\kappa}_1)\tilde{\chi}_{21} = \left[(\bar{\kappa}_1 - \bar{\kappa}_2)^2 - 1\right]\chi_{11}, \tag{8.116}$$

$$c(-\bar{\kappa}_2)\tilde{\chi}_{12} = \left[(\bar{\kappa}_1 - \bar{\kappa}_2)^2 - 1\right]\chi_{11}. \tag{8.117}$$

are also included in the figure. Of these relations, (8.114) can be derived in

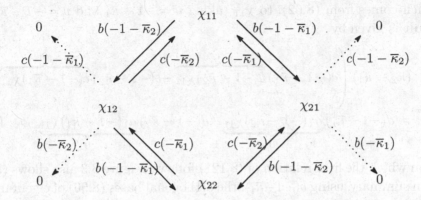

Fig. 8.3. Algebraic structure of Yangian Gelfand–Zetlin basis for $N=2$ and $\kappa_1 > \kappa_2$.

the following way. Setting $(\alpha, \beta, \gamma, \delta) = (2, 2, 1, 1)$, and $v = u - 1$ in (8.58), we obtain

$$a_1(u)d(u-1) - b(u)c(u-1) = d(u-1)a_1(u) - b(u-1)c(u). \qquad (8.118)$$

Setting $v = u - 1$ in (8.62), we obtain

$$c(u)b(u-1) - b(u)c(u-1) = d(u)a_1(u-1) - d(u-1)a_1(u).$$

From this, (8.62) and (8.80), we obtain

$$c(u)b(u-1) = d(u)a_1(u-1) - a_2(u). \qquad (8.119)$$

Using (8.108) and (8.119), the LHS of (8.114) becomes

$$
\begin{aligned}
&c(-\bar{\kappa}_1)b(-1 - \bar{\kappa}_1)\tilde{\chi}_{12} \\
&= d(-\bar{\kappa}_1)\underbrace{a_1(-1 - \bar{\kappa}_1)\tilde{\chi}_{12}}_{=0} - \underbrace{a_2(-\bar{\kappa}_1)}_{1-(\bar{\kappa}_1 - \bar{\kappa}_2)^2}\tilde{\chi}_{12}
\end{aligned}
$$

$$= \text{RHS of (8.114)}. \qquad (8.120)$$

Other relations in (8.114)–(8.117) can be derived similarly.

Besides the trivial relation $c(u)\chi_{11} = 0$, the operators $c(-1 - \bar{\kappa}_i)$ for $i = 1, 2$ annihilate χ_{12} and χ_{21}:

$$c(-1 - \bar{\kappa}_1)\chi_{12} = 0, \quad c(-1 - \bar{\kappa}_2)\chi_{21} = 0. \qquad (8.121)$$

The first relation of (8.121) can be obtained in the following way. We apply

$$(u - v)\,[c(u), b(v)] = d(u)a_1(v) - d(v)a_1(u), \qquad (8.122)$$

which comes from (8.62), to χ_{11} taking $u = -1 - \bar{\kappa}_1$ and $v = -1 - \bar{\kappa}_2$. The result is given by

$$
(\bar{\kappa}_2 - \bar{\kappa}_1) \left(c(-1 - \bar{\kappa}_1) \overbrace{b(-1 - \bar{\kappa}_2)\chi_{11}}^{\tilde{\chi}_{12}} - b(-1 - \bar{\kappa}_2) \overbrace{c(-1 - \bar{\kappa}_1)\chi_{11}}^{0} \right)
$$

$$
= d(-1 - \bar{\kappa}_1) \underbrace{a_1(-1 - \bar{\kappa}_2)\chi_{11}}_{0} - d(-1 - \bar{\kappa}_2) \underbrace{a_1(-1 - \bar{\kappa}_1)\chi_{11}}_{0}, \qquad (8.123)
$$

from which the first relation of (8.121) follows. Figure 8.3 also shows (8.121).

In summary, using $b(-1 - \bar{\kappa}_i)$, the orthogonal bases (8.96) of eigenfunctions are given by

$$
U_\kappa v_{11}, \qquad\qquad\qquad U_\kappa b(-1 - \bar{\kappa}_1) v_{11},
$$
$$
U_\kappa b(-1 - \bar{\kappa}_2) v_{11}, \qquad U_\kappa b(-1 - \bar{\kappa}_1) b(-1 - \bar{\kappa}_2) v_{11}. \qquad (8.124)
$$

Example 8.9. $N = 3$, $\kappa_1 > \kappa_2 > \kappa_3$.

In the present case, the functions

$$
v_{\alpha\beta\gamma} = v_\alpha \otimes v_\beta \otimes v_\gamma, \quad \alpha, \beta, \gamma = 1, 2 \qquad (8.125)
$$

constitute a basis of $V_\kappa = \otimes^3 \mathbb{C}^2$. Explicit expressions for monodromy matrices are given by the co-product as

$$
\begin{aligned}
a_1(u; \bar{\kappa}_1, \bar{\kappa}_2, \bar{\kappa}_3) &= \left(u + \bar{\kappa}_1 + X^{11} \right) \otimes \left(u + \bar{\kappa}_2 + X^{11} \right) \otimes \left(u + \bar{\kappa}_3 + X^{11} \right) \\
&\quad + \left(u + \bar{\kappa}_1 + X^{11} \right) \otimes X^{21} \otimes X^{12} + X^{21} \otimes X^{12} \otimes \left(u + \bar{\kappa}_3 + X^{11} \right) \\
&\quad + X^{21} \otimes \left(u + \bar{\kappa}_2 + X^{22} \right) \otimes X^{12},
\end{aligned}
$$

$$
\begin{aligned}
b(u; \bar{\kappa}_1, \bar{\kappa}_2, \bar{\kappa}_3) &= \left(u + \bar{\kappa}_1 + X^{11} \right) \otimes \left(u + \bar{\kappa}_2 + X^{11} \right) \otimes X^{21} \\
&\quad + \left(u + \bar{\kappa}_1 + X^{11} \right) \otimes X^{21} \otimes \left(u + \bar{\kappa}_3 + X^{22} \right) + X^{21} \otimes X^{12} \otimes X^{21} \\
&\quad + X^{21} \otimes \left(u + \bar{\kappa}_2 + X^{22} \right) \otimes \left(u + \bar{\kappa}_3 + X^{22} \right),
\end{aligned}
$$

$$
\begin{aligned}
c(u; \bar{\kappa}_1, \bar{\kappa}_2, \bar{\kappa}_3) &= X^{12} \otimes \left(u + \bar{\kappa}_2 + X^{11} \right) \otimes \left(u + \bar{\kappa}_3 + X^{11} \right) \\
&\quad + X^{12} \otimes X^{21} \otimes X^{12} + \left(u + \bar{\kappa}_1 + X^{22} \right) \otimes X^{12} \otimes \left(u + \bar{\kappa}_3 + X^{11} \right) \\
&\quad + \left(u + \bar{\kappa}_1 + X^{22} \right) \otimes \left(u + \bar{\kappa}_2 + X^{22} \right) \otimes X^{12}.
\end{aligned}
$$

The regular quantum determinant $a_2(u; \bar{\kappa}_1, \bar{\kappa}_2, \bar{\kappa}_3)$ is given by the tensor product of the quantum determinant for $N = 1$. The action of $a_1(u)$ on the basis is block diagonal:

$$
a_1(u) v_{111} = (u + 1 + \bar{\kappa}_1)(u + 1 + \bar{\kappa}_2)(u + 1 + \bar{\kappa}_3) v_{111}, \qquad (8.126)
$$
$$
a_1(u) v_{222} = (u + \bar{\kappa}_1)(u + \bar{\kappa}_2)(u + \bar{\kappa}_3) v_{222}. \qquad (8.127)
$$

For a more complicated block with the basis vector $\mathbf{v}$, we obtain

$$a_1(u)\mathbf{v} = \rho_1(u)U\mathbf{v}, \tag{8.128}$$

where U is a matrix. In the case of $\mathbf{v} = \mathbf{v}_1 \equiv (v_{112}, v_{121}, v_{211})^{\mathrm{T}}$, we obtain $U = U_1$ in (8.128) with

$$U_1 = \begin{pmatrix} \frac{(u+1+\overline{\kappa}_1)(u+1+\overline{\kappa}_2)}{(u+\overline{\kappa}_1)(u+\overline{\kappa}_2)} & \frac{(u+1+\overline{\kappa}_1)}{(u+\overline{\kappa}_1)(u+\overline{\kappa}_2)(u+\overline{\kappa}_3)} & \frac{1}{(u+\overline{\kappa}_1)(u+\overline{\kappa}_3)} \\ 0 & \frac{(u+1+\overline{\kappa}_1)(u+1+\overline{\kappa}_3)}{(u+\overline{\kappa}_1)(u+\overline{\kappa}_3)} & \frac{(u+1+\overline{\kappa}_3)}{(u+\overline{\kappa}_1)(u+\overline{\kappa}_2)(u+\overline{\kappa}_3)} \\ 0 & 0 & \frac{(u+1+\overline{\kappa}_2)(u+1+\overline{\kappa}_3)}{(u+\overline{\kappa}_2)(u+\overline{\kappa}_3)} \end{pmatrix}. \tag{8.129}$$

For another case of $\mathbf{v}_2 \equiv (v_{122}, v_{212}, v_{221})^{\mathrm{T}}$, we obtain $U = U_2$ in (8.128) with

$$U_2 = \begin{pmatrix} \frac{(u+1+\overline{\kappa}_1)}{(u+\overline{\kappa}_1)} & \frac{1}{(u+\overline{\kappa}_1)(u+\overline{\kappa}_2)} & \frac{(u+1+\overline{\kappa}_2)}{(u+\overline{\kappa}_1)(u+\overline{\kappa}_2)(u+\overline{\kappa}_3)} \\ 0 & \frac{(u+1+\overline{\kappa}_2)}{(u+\overline{\kappa}_2)} & \frac{1}{(u+\overline{\kappa}_2)(u+\overline{\kappa}_3)} \\ 0 & 0 & \frac{(u+1+\overline{\kappa}_3)}{(u+\overline{\kappa}_3)} \end{pmatrix}. \tag{8.130}$$

Both matrices U_1 and U_2 are triangular, and hence they can be diagonalized. The eigenvectors are given by

$$\chi_{112} = v_{112} - \frac{v_{121}}{\overline{\kappa}_2 - \overline{\kappa}_3} - \frac{(\overline{\kappa}_2 - \overline{\kappa}_3 - 1)\, v_{211}}{(\overline{\kappa}_2 - \overline{\kappa}_3)(\overline{\kappa}_1 - \overline{\kappa}_3)}, \tag{8.131}$$

$$\chi_{121} = v_{121} - \frac{v_{211}}{\overline{\kappa}_1 - \overline{\kappa}_2}, \tag{8.132}$$

$$\chi_{211} = v_{211}, \tag{8.133}$$

$$\chi_{122} = v_{122} - \frac{v_{212}}{\overline{\kappa}_1 - \overline{\kappa}_2} - \frac{(\overline{\kappa}_1 - \overline{\kappa}_2 - 1)\, v_{221}}{(\overline{\kappa}_1 - \overline{\kappa}_3)(\overline{\kappa}_1 - \overline{\kappa}_2)}, \tag{8.134}$$

$$\chi_{212} = v_{212} - \frac{v_{221}}{\overline{\kappa}_2 - \overline{\kappa}_3}, \tag{8.135}$$

$$\chi_{221} = v_{221}. \tag{8.136}$$

Each eigenvalue is obtained from the diagonal element of the matrix in (8.129) or (8.130). Owing to the conditions $\overline{\kappa}_1 > \overline{\kappa}_2 + 1$ and $\overline{\kappa}_2 > \overline{\kappa}_3 + 1$, all eigenvalues are distinct. All eigenvectors are generated by the successive action of lowering operators $b(-1 - \overline{\kappa}_i)$ $(i = 1, 2, 3)$ on the highest-weight state $\chi_{111} = v_{111}$. Namely, we obtain

$$b(-1 - \overline{\kappa}_1)v_{111} = (-1 - \overline{\kappa}_1 + \overline{\kappa}_2)(-1 - \overline{\kappa}_1 + \overline{\kappa}_3)\chi_{211}, \qquad (8.137)$$

$$b(-1 - \overline{\kappa}_2)v_{111} = (\overline{\kappa}_1 - \overline{\kappa}_2)(-1 - \overline{\kappa}_2 + \overline{\kappa}_3)\chi_{121}, \qquad (8.138)$$

$$b(-1 - \overline{\kappa}_3)v_{111} = (\overline{\kappa}_1 - \overline{\kappa}_3)(\overline{\kappa}_2 - \overline{\kappa}_3)\chi_{112}. \qquad (8.139)$$

Then the double action of $b(-1 - \overline{\kappa}_i)$ gives

$$b(-1 - \overline{\kappa}_1)b(-1 - \overline{\kappa}_2)v_{111}$$
$$= \left(1 - (\overline{\kappa}_1 - \overline{\kappa}_2)^2\right)(-1 - \overline{\kappa}_1 + \overline{\kappa}_3)(-1 - \overline{\kappa}_2 + \overline{\kappa}_3)\chi_{221}, \qquad (8.140)$$

$$b(-1 - \overline{\kappa}_1)b(-1 - \overline{\kappa}_3)v_{111}$$
$$= (-1 - \overline{\kappa}_1 + \overline{\kappa}_2)\left(1 - (\overline{\kappa}_1 - \overline{\kappa}_3)^2\right)(\overline{\kappa}_2 - \overline{\kappa}_3)\chi_{212}, \qquad (8.141)$$

$$b(-1 - \overline{\kappa}_2)b(-1 - \overline{\kappa}_3)v_{111}$$
$$= (\overline{\kappa}_1 - \overline{\kappa}_2)(\overline{\kappa}_1 - \overline{\kappa}_3)\left(1 - (\overline{\kappa}_2 - \overline{\kappa}_3)^2\right)\chi_{122}. \qquad (8.142)$$

Further application of $b(-1 - \overline{\kappa}_i)$ gives

$$b(-\overline{\kappa}_1 - 1)b(-\overline{\kappa}_2 - 1)b(-\overline{\kappa}_3 - 1)v_{111} = \prod_{1 \leq i < j \leq 3}\left(1 - (\overline{\kappa}_i - \overline{\kappa}_j)^2\right)\chi_{222}. \qquad (8.143)$$

Conversely, all eigenvectors are generated by the successive action of $c(-\overline{\kappa}_i)$ $(i = 1, 2, 3)$ on the lowest-weight state $\chi_{222} = v_{222}$. Namely, we obtain

$$c(-\overline{\kappa}_1)v_{222} = (-\overline{\kappa}_1 + \overline{\kappa}_2)(-\overline{\kappa}_1 + \overline{\kappa}_3)\chi_{122}, \qquad (8.144)$$

$$c(-\overline{\kappa}_2)v_{222} = (\overline{\kappa}_1 - \overline{\kappa}_2 + 1)(-\overline{\kappa}_2 + \overline{\kappa}_3)\chi_{212}, \qquad (8.145)$$

$$c(-\overline{\kappa}_3)v_{222} = (\overline{\kappa}_1 - \overline{\kappa}_3 + 1)(\overline{\kappa}_2 - \overline{\kappa}_3 + 1)\chi_{221}. \qquad (8.146)$$

Then the double action of $c(-\overline{\kappa}_i)$ gives

$$c(-\overline{\kappa}_1)c(-\overline{\kappa}_2)v_{222}, \qquad (8.147)$$

$$= \left(1 - (\overline{\kappa}_i - \overline{\kappa}_j)^2\right)(-\overline{\kappa}_1 + \overline{\kappa}_3)(-\overline{\kappa}_2 + \overline{\kappa}_3)\chi_{112}, \qquad (8.148)$$

$$c(-\overline{\kappa}_1)c(-\overline{\kappa}_3)v_{222}, \qquad (8.149)$$

$$= (-\overline{\kappa}_1 + \overline{\kappa}_2)\left(1 - (\overline{\kappa}_1 - \overline{\kappa}_3)^2\right)(-\overline{\kappa}_2 + \overline{\kappa}_3 + 1)\chi_{121},$$

$$c(-\overline{\kappa}_2)c(-\overline{\kappa}_3)v_{222}, \qquad (8.150)$$

$$= (\overline{\kappa}_1 - \overline{\kappa}_2 + 1)(\overline{\kappa}_1 - \overline{\kappa}_3 + 1)\left(1 - (\overline{\kappa}_2 - \overline{\kappa}_3)^2\right)\chi_{211}.$$

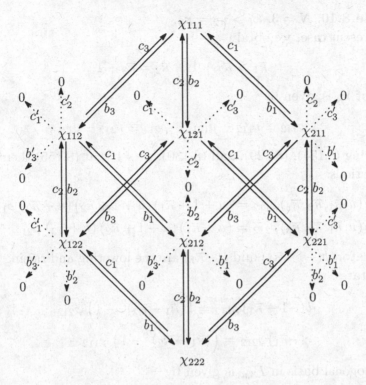

Fig. 8.4. Algebraic structure of Yangian Gelfand–Zetlin basis for $N = 3$ and $\kappa_1 > \kappa_2 > \kappa_3$. The symbols b_i and c_i stand for the lowering operator $b(-1 - \overline{\kappa}_i)$ and the raising one $c(-\overline{\kappa}_i)$, respectively. The symbols b_i' and c_i' stand for the annihilation operators $b(-\overline{\kappa}_i)$ and $c(-1 - \overline{\kappa}_i)$, respectively.

Further application of $c(-\overline{\kappa}_i)$ gives

$$c(-\overline{\kappa}_1)c(-\overline{\kappa}_2)c(-\overline{\kappa}_3)v_{222} = \prod_{1 \le i < j \le 3} \left(1 - (\overline{\kappa}_i - \overline{\kappa}_j)^2\right) \chi_{111}. \qquad (8.151)$$

In addition to the raising and lowering operators described above, the operators $b(-\overline{\kappa}_i)$ annihilate eigenstates whose ith subscript is 2. Namely we obtain

$$b(-\overline{\kappa}_1)\chi_{2\alpha\beta} = 0, \quad b(-\overline{\kappa}_2)\chi_{\alpha 2\beta} = 0, \quad b(-\overline{\kappa}_3)\chi_{\alpha\beta 2} = 0, \qquad (8.152)$$

for $\alpha, \beta = 1, 2$. On the other hand, the $c(-1 - \overline{\kappa}_i)$ annihilate eigenstates whose ith subscript is 1:

$$c(-1 - \overline{\kappa}_1)\chi_{1\alpha\beta} = 0, \quad c(-1 - \overline{\kappa}_2)\chi_{\alpha 1\beta} = 0, \quad c(-1 - \overline{\kappa}_3)\chi_{\alpha\beta 1} = 0. \qquad (8.153)$$

Figure 8.4 summarizes the algebraic structure.

Example 8.10. $N = 3$, $\kappa_1 > \kappa_2 = \kappa_3$.

In the present case, we obtain

$$\overline{\kappa}_1 > \overline{\kappa}_2 + 1, \quad \overline{\kappa}_2 = \overline{\kappa}_3 + 1. \tag{8.154}$$

A basis of V_κ is given by

$$\varphi_{112} \equiv v_{112} - v_{121}, \quad \varphi_{212} \equiv v_{212} - v_{221}. \tag{8.155}$$

Considering (8.154), (8.129), and (8.130), we see that (8.155) themselves are eigenfunctions:

$$a_1(u; \overline{\kappa}_1, \overline{\kappa}_2, \overline{\kappa}_3)\varphi_{112} = (u + 1 + \overline{\kappa}_1)(u + 1 + \overline{\kappa}_2)(u + \overline{\kappa}_3)\varphi_{112},$$
$$a_1(u; \overline{\kappa}_1, \overline{\kappa}_2, \overline{\kappa}_3)\varphi_{212} = (u + \overline{\kappa}_1)(u + 1 + \overline{\kappa}_2)(u + \overline{\kappa}_3)\varphi_{212}.$$

The operators $b(-1 - \overline{\kappa}_1)$ and $c(-\overline{\kappa}_1)$ are the lowering and raising operators of eigenstates:

$$b(-1 - \overline{\kappa}_1)\varphi_{112} = \left((\overline{\kappa}_1 - \overline{\kappa}_2)^2 - 1\right)\varphi_{212}, \tag{8.156}$$

$$c(-\overline{\kappa}_1)\varphi_{212} = \left((\overline{\kappa}_1 - \overline{\kappa}_2)^2 - 1\right)\varphi_{112}. \tag{8.157}$$

The orthogonal basis in $F_{3,\kappa}$ is given by

$$\{U_\kappa\varphi_{112}, U_\kappa\varphi_{212}\}, \tag{8.158}$$

or equivalently by

$$\{U_\kappa\varphi_{112}, U_\kappa b(-1 - \overline{\kappa}_1)\varphi_{112}\}. \tag{8.159}$$

8.5.2 *Orthogonal eigenbasis for N-particle systems*

We are now ready to construct an orthogonal set of eigenbasis of $\mathcal{F}_{N,\kappa}$ with arbitrary N. For $\kappa \in \Lambda_{N,2}^+$, we consider a subset I_κ of $I = \{1, 2, \ldots, N\}$ by removing all adjacent pairs $\{i, i+1\}$ with $\kappa_i = \kappa_{i+1}$. Namely, I_κ is a subset that picks up unpaired spins from I. On the other hand, let $\mathcal{I}_\kappa \subset [1, \ldots, N]$ be the set of i such that $[i, i+1]$ forms a singlet. For example, with $N = 3$ and $\kappa_1 > \kappa_2 > \kappa_3$, we have $I_\kappa = \{1, 2, 3\}$. For $N = 3$ and $\kappa_1 > \kappa_2 = \kappa_3$, we have $I_\kappa = \{1\}$ and $\mathcal{I}_\kappa = \{2\}$.

More generally, for a given $I_\kappa = (i(1), i(2), \ldots)$, a basis function φ_α of V_κ is defined as

$$\varphi_\alpha \equiv (v_{12} - v_{21}) \otimes \cdots \otimes (v_{12} - v_{21}) \otimes \overset{i(1)}{v_{\alpha_{i(1)}}} \otimes$$
$$(v_{12} - v_{21}) \otimes \cdots \otimes (v_{12} - v_{21}) \otimes \overset{i(2)}{v_{\alpha_{i(2)}}} \otimes \cdots. \tag{8.160}$$

These functions are uniquely specified by the index $\alpha \in W_\kappa$, where

$$W_\kappa \equiv \left\{ \alpha = (\alpha_1, \dots, \alpha_N) \in [1, 2]^N \,\middle|\, (\alpha_i, \alpha_{i+1}) = (1, 2) \text{ if } \kappa_i = \kappa_{i+1} \right\}.$$
$$(8.161)$$

For example, we put $\alpha_1 = 1, \alpha_2 = 2$ if $\kappa_1 = \kappa_2$. The corresponding state φ_α with $N = 2$ is the spin singlet.

The order of $\alpha, \tilde{\alpha} \in W_\kappa$ with $\sum_i \alpha_i = \sum_i \tilde{\alpha}_i$ is defined as

$$\tilde{\alpha} < \alpha, \text{ if the first non-vanishing } \tilde{\alpha}_i - \alpha_i \text{ is positive.} \quad (8.162)$$

For $N = 3$ and $\kappa_1 > \kappa_2 > \kappa_3$, the order of $\alpha \in W_\kappa$ with $\sum_i \alpha_i = 4$ is given by

$$112 > 121 > 211.$$

The action of $a_1(u)$ on this basis is given in a triangular form:

$$a_1(u)\varphi_\alpha = \varpi_\alpha(u)\varphi_\alpha + \sum_{\tilde{\alpha} < \alpha} c_{\alpha\tilde{\alpha}}\varphi_{\tilde{\alpha}}, \quad (8.163)$$

because $a_1(u)$ is the sum of the terms

$$(u + \bar{\kappa}_1 + X^{11}) \otimes \cdots \otimes (u + \bar{\kappa}_{i-1} + X^{11}) \otimes X^{21} \otimes \cdots. \quad (8.164)$$

The diagonal element $\varpi_\alpha(u)$ can be obtained by comparing the coefficient of φ_α on both sides of (8.163). On the LHS, the coefficient of φ_α comes from

$$(u + \bar{\kappa}_1 + X^{11}) \otimes \cdots \otimes (u + \bar{\kappa}_N + X^{11})\varphi_\alpha. \quad (8.165)$$

Then the diagonal element $\varpi_\alpha(u)$ is given by

$$\varpi_\alpha(u) = \varpi_\alpha(u; \bar{\kappa}_1, \dots, \bar{\kappa}_N) = \prod_{i \in \mathcal{I}_\kappa \cup (I_\kappa / J(\alpha))} (u + \bar{\kappa}_i + 1) \prod_{i \in J(\alpha)} (u + \bar{\kappa}_i), \quad (8.166)$$

where a subset $J(\alpha)$ of I_κ is occupied by down spins. Namely, it is defined as

$$J(\alpha) \equiv \{i \in I_\kappa | \alpha_i = 2\}. \quad (8.167)$$

The eigenstates of $a_1(u)$ are written in the form

$$\chi_\alpha = \varphi_\alpha + \sum_{\tilde{\alpha} < \alpha} \tilde{c}_{\alpha\tilde{\alpha}}\varphi_{\tilde{\alpha}} \quad (8.168)$$

with coefficient $\tilde{c}_{\alpha\tilde{\alpha}}$. The eigenvalue of $a_1(u)$ in (8.168) is $\varpi_\alpha(u)$. In terms of (8.168), we construct a set of orthogonal bases of $\mathcal{F}_{N,\kappa}$ as

$$\Phi_{\kappa,\alpha} = U_\kappa \chi_\alpha, \quad (\alpha \in W_\kappa). \quad (8.169)$$

The eigenbases can be derived in another way. The highest-weight state φ_{hws} in V_κ with the maximum eigenvalue of S^z_{tot} is described as

$$
\varphi_{\mathrm{hws}} = (v_{12} - v_{21}) \otimes \cdots \otimes (v_{12} - v_{21}) \otimes \overset{i(1)}{v_1} \otimes
$$
$$
(v_{12} - v_{21}) \otimes \cdots \otimes (v_{12} - v_{21}) \otimes \overset{i(2)}{v_1} \otimes \cdots . \tag{8.170}
$$

For φ_{hws}, all the coefficients $c_{\alpha\bar\alpha}$ in (8.163) are zero because there are no states with the same S^z_{tot} for a given κ. Hence we obtain

$$
a_1(u)\varphi_{\mathrm{hws}} = \varpi_\alpha(u)\varphi_{\mathrm{hws}}, \tag{8.171}
$$

with $\varpi_\alpha(u)$ given by (8.166). Then the state

$$
\tilde\chi_\alpha = \prod_{i \in J(\alpha)} b(-1 - \bar\kappa_i)\varphi_{\mathrm{hws}} \tag{8.172}
$$

is also an eigenfunction of $a_1(u)$ which satisfies [143, 144]

$$
a_1(u)\tilde\chi_\alpha = \varpi_\alpha(u)\tilde\chi_\alpha, \tag{8.173}
$$

as will be proved below. The orthogonal basis in $\mathcal{F}_{N,\kappa}$ for $\kappa \in \Lambda^+_{N,2}$ is given by the set [180]

$$
\tilde\Phi_{\kappa,\alpha} \equiv U_\kappa\tilde\chi_\alpha, \quad (\alpha \in W_\kappa). \tag{8.174}
$$

The basis functions (8.169) and (8.174) are equivalent up to a constant factor: $\Phi_{\kappa,\alpha} \propto \tilde\Phi_{\kappa,\alpha}$.

Proof of (8.173).
We shall use mathematical induction for the proof. The statement is true for the highest-weight state φ_{hws} as given by (8.171). Suppose that the above statement holds for a function $\tilde\chi_\alpha$ if the number $\sharp J(\alpha)$ of elements in the set $J(\alpha)$ is less than or equal to an integer $r \in [0, \sharp I_\kappa - 1]$. Any function $\tilde\chi_{\alpha'}$ for α' satisfying $\sharp J(\alpha') = r + 1$ is written as

$$
\tilde\chi_{\alpha'} = b(-1 - \bar\kappa_j)\tilde\chi_\alpha, \tag{8.175}
$$

with α such that $J(\alpha') = J(\alpha) \oplus \{j\}$. Let us apply (8.97) to both sides of (8.175). Setting $v = -1 - \bar\kappa_j$, we obtain

$$
(u + 1 + \bar\kappa_j) \left[a_1(u)\tilde\chi_{\alpha'} - \varpi_\alpha(u)\tilde\chi_{\alpha'}\right] = -\varpi_\alpha(u)\tilde\chi_{\alpha'}. \tag{8.176}
$$

Here we have used the relation

$$
a_1(-1 - \bar\kappa_i)\tilde\chi_\alpha = 0 \quad \text{for} \quad i \in I_\kappa/J(\alpha) \oplus \mathcal{I}_\kappa.
$$

From (8.176), we obtain

$$a_1(u)\tilde{\chi}_{\alpha'} = \varpi_\alpha(u)\frac{(u+\overline{\kappa}_j)}{(u+\overline{\kappa}_j+1)}\tilde{\chi}_{\alpha'} = \varpi_{\alpha'}(u)\tilde{\chi}_{\alpha'}, \qquad (8.177)$$

which shows that $\tilde{\chi}_{\alpha'}$ is also an eigenfunction. By induction, (8.173) is hence proved. ∎

We note that the eigenvalues $\varpi_\alpha(u)$ for $\alpha \in W_\kappa$ are non-degenerate. From (8.173), it follows that $\tilde{\chi}_\alpha$ is the eigenfunction of $a_1(u)$ with an eigenvalue distinct from the others. Recalling the results in Section 8.4, we see that (8.174), which belongs to $\mathcal{F}_{N,\kappa}$, is an eigenfunction of the self-adjoint operator $a_1(u, \hat{d}_1, \ldots, \hat{d}_N)$ with a distinct eigenvalue. This means that (8.174) forms an orthogonal set of eigenbases in $\mathcal{F}_{N,\kappa}$. Let $\mathcal{F}_N$ be the collection of $\mathcal{F}_{N,\kappa}$:

$$\mathcal{F}_N \equiv \oplus_{\kappa \in \mathcal{L}_{N,2}^+} \mathcal{F}_{N,\kappa}. \qquad (8.178)$$

The orthogonal basis of $\mathcal{F}_N$ is then given by [180]

$$\left\{\Phi_{\kappa,\alpha} | \kappa \in \mathcal{L}_{N,2}^+, \ \alpha \in W_\kappa\right\}. \qquad (8.179)$$

8.6 Norm of Yangian Gelfand–Zetlin basis

The norm of the set of orthogonal bases of $\mathcal{F}_{N,\kappa}$ can be obtained algebraically [180] starting from the simpler case of the highest-weight state. In the case of $\alpha \in W_\kappa$, $i \in I_\kappa$, and $\alpha_i = 1$, $\Phi_{\kappa,\alpha}$ has a descendant state

$$b(-1 - \hat{d}_i)\Phi_{\kappa,\alpha}. \qquad (8.180)$$

Suppose we know the norm of the highest-weight state $\hat{U}_\kappa \varphi_{\text{hws}}$ in $\mathcal{F}_{N,\kappa}$. We then obtain the norm

$$\|\Phi_{\kappa,\beta}\|^2 \equiv \langle \Phi_{\kappa,\beta}, \Phi_{\kappa,\beta}\rangle' \qquad (8.181)$$

for any basis function in $\mathcal{F}_{N,\kappa}$ from the recursion formula [180]

$$\left\|b(-1-\hat{d}_i)\Phi_{\kappa,\alpha}\right\|^2 / \|\Phi_{\kappa,\alpha}\|^2 = \frac{\tilde{\varpi}(-\overline{\kappa}_i)}{\varpi_\alpha(-\overline{\kappa}_i)} \lim_{k \to \overline{\kappa}_i} \frac{\varpi_\alpha(-1-k)}{k - \overline{\kappa}_i}, \qquad (8.182)$$

which will be verified below. Here $\tilde{\varpi}(-\overline{\kappa}_i)$ represents the eigenvalue of $a_2(u)$ for the eigenstate χ_α:

$$a_2(u)\chi_\alpha = \tilde{\varpi}(u)\chi_\alpha, \qquad (8.183)$$

where $\tilde{\varpi}(u)$ actually does not depend on α since $a_2(u)$ is a scalar, namely the regular quantum determinant. As an example of formula (8.182), the case of two particles with $\kappa = (\kappa_1, \kappa_2)$ satisfying $\kappa_1 > \kappa_2$ is explicitly written as

$$\left\| \Phi_{\kappa_1 \kappa_2, 12} \right\|^2 / \left\| \Phi_{\kappa_1 \kappa_2, 11} \right\|^2 = \frac{\bar{\kappa}_1 - \bar{\kappa}_2 - 1}{\bar{\kappa}_1 - \bar{\kappa}_2 - 1}, \tag{8.184}$$

$$\left\| \Phi_{\kappa_1 \kappa_2, 21} \right\|^2 / \left\| \Phi_{\kappa_1 \kappa_2, 11} \right\|^2 = \frac{\bar{\kappa}_1 - \bar{\kappa}_2}{\kappa_1 - \kappa_2 + 1}, \tag{8.185}$$

$$\left\| \Phi_{\kappa_1 \kappa_2, 22} \right\|^2 / \left\| \Phi_{\kappa_1 \kappa_2, 11} \right\|^2 = 1. \tag{8.186}$$

In the derivation of (8.184) and (8.185), we have used (8.102) and (8.103).

Proof of (8.182).
Let us introduce an inner product $\langle\langle \varphi, \tilde{\varphi} \rangle\rangle$ for $\varphi, \tilde{\varphi} \in V_\kappa$, via the inner product $\langle \cdot, \cdot \rangle'$ in $\mathcal{F}_{N,\kappa}$. Namely, we define

$$\langle\langle \varphi, \tilde{\varphi} \rangle\rangle \equiv \langle \hat{U}_\kappa \varphi, \hat{U}_\kappa \tilde{\varphi} \rangle. \tag{8.187}$$

It then follows that

$$\langle\langle \chi_\alpha, \chi_\beta \rangle\rangle = \langle \Phi_{\kappa,\alpha}, \Phi_{\kappa,\beta} \rangle', \tag{8.188}$$

for $\alpha, \beta \in W_\kappa$. The adjoint property

$$b^\dagger(u; \bar{\kappa}_1, \ldots, \bar{\kappa}_N) = c(u; \bar{\kappa}_1, \ldots, \bar{\kappa}_N) \tag{8.189}$$

holds with respect to $\langle\langle \cdot, \cdot \rangle\rangle$. This can be confirmed as follows:

$$\begin{aligned}
\langle\langle \varphi, b(u; \bar{\kappa}_1, \ldots, \bar{\kappa}_N) \tilde{\varphi} \rangle\rangle &= \langle \hat{U}_\kappa \varphi, b(u; \hat{d}_1, \ldots, \hat{d}_N) \hat{U}_\kappa \tilde{\varphi} \rangle' \\
&= \langle c(u; \hat{d}_1, \ldots, \hat{d}_N) \hat{U}_\kappa \varphi, \hat{U}_\kappa \tilde{\varphi} \rangle' \\
&= \langle \hat{U}_\kappa c(u; \bar{\kappa}_1, \ldots, \bar{\kappa}_N) \varphi, \hat{U}_\kappa \tilde{\varphi} \rangle' \\
&= \langle\langle c(u; \bar{\kappa}_1, \ldots, \bar{\kappa}_N) \varphi, \tilde{\varphi} \rangle\rangle, \tag{8.190}
\end{aligned}$$

where we have used (8.72) in the second equality. We then use the relation

$$\langle\langle b(-1 - \bar{\kappa}_i)\chi_\alpha, b(-1 - \bar{\kappa}_i)\chi_\alpha \rangle\rangle = \langle\langle \chi_\alpha, c(-1 - \bar{\kappa}_i)b(-1 - \bar{\kappa}_i)\chi_\alpha \rangle\rangle, \tag{8.191}$$

for the LHS of (8.182). Calculation of the RHS requires some steps. First, (8.62) is transformed as

$$c(u)b(v) = b(v)c(u) + (u - v)^{-1}[d(u)a_1(v) - d(v)a_1(u)], \tag{8.192}$$

and set $u = -1 - \bar{\kappa}_i$ and $v = -1 - k$. Secondly, provided $\alpha_i = 1$ in $\alpha \in W_\kappa$, we have the relations

$$a_1(-1 - \kappa_i)\chi_\alpha = \varpi_\alpha(-1 - \kappa_i)\chi_\alpha = 0, \tag{8.193}$$

$$c(-1 - \kappa_i)\chi_\alpha = 0. \tag{8.194}$$

Taking the limit $k \to \bar{\kappa}_i$ with $\alpha_i = 1$, we obtain from (8.191), (8.192), (8.193) and (8.194),

$$\frac{\langle\langle \chi_\alpha, c(-1-\bar{\kappa}_i)b(-1-\bar{\kappa}_i)\chi_\alpha\rangle\rangle}{\langle\langle \chi_\alpha, d(-1-\bar{\kappa}_i)\chi_\alpha\rangle\rangle} = \lim_{k \to \bar{\kappa}_i} \frac{\varpi_\alpha(-1-k)}{k - \bar{\kappa}_i}. \tag{8.195}$$

The denominator on the LHS of (8.195) can be evaluated with use of (8.80) and (8.183). Namely, we take the expectation value of (8.80) with respect to χ_α with $u = -\bar{\kappa}_i$. The result is given by

$$\tilde{\varpi}(-\bar{\kappa}_i)\langle\langle \chi_\alpha, \chi_\alpha\rangle\rangle = \varpi_\alpha(-\bar{\kappa}_i)\langle\langle \chi_\alpha, d(-1-\bar{\kappa}_i)\chi_\alpha\rangle\rangle, \tag{8.196}$$

with use of (8.194). From (8.195) and (8.196), we finally obtain

$$\frac{\langle\langle \chi_\alpha, c(-1-\bar{\kappa}_i)b(-1-\bar{\kappa}_i)\chi_\alpha\rangle\rangle}{\langle\langle \chi_\alpha, \chi_\alpha\rangle\rangle} = \frac{\tilde{\varpi}(-\bar{\kappa}_i)}{\varpi_\alpha(-\bar{\kappa}_i)} \lim_{k \to \bar{\kappa}_i} \frac{\varpi_\alpha(-1-k)}{k - \bar{\kappa}_i}, \tag{8.197}$$

which proves (8.182). ∎

9

SU(K) and supersymmetric Yangians

We now describe the properties of the SU(K) chain in more algebraic terms. The relevant symmetry is the SU(K) Yangian, which is usually referred to as $Y(sl_K)$. On the other hand, the Yangian symmetry of the Sutherland model with spin 1/2 is given by $Y(gl_2)$. The difference between $Y(sl_K)$ and $Y(gl_K)$ is that the charge degrees of freedom U(1) is absent in $Y(sl_K)$, while the spin degrees of freedom is included in both $Y(sl_K)$ and $Y(gl_K)$. In Section 4.8.3, we have removed the U(1) component by means of the freezing trick, i.e., by taking the infinitely large repulsion parameter λ. In this chapter we take a more algebraic approach to study $Y(sl_K)$, which makes the relation between $Y(gl_K)$ and $Y(sl_K)$ clearer.

In Section 9.1, we begin with the construction of the SU(K) monodromy matrix as the product form. The commuting property of the Cherednik–Dunkl operators plays a crucial role here. The monodromy matrix is the basic building block of the Yangian algbebra, as discussed in Section 8.2. Because the Hamiltonian of the SU(K) spin chain and the monodromy matrix commute, each eigenstate of the spin chain constitutes a basis for representations of the SU(K) Yangian.

In Section 9.2 and 9.3, we discuss the quantum determinant by comparing it with the ordinary determinant. It is demonstrated in Section 9.4 that the quantum determinant has a scalar nature that leads to a simple evaluation. Because of the scalar nature, however, the quantum determinant cannot distinguish different SU(K) Yangian states. We need to come back to the monodromy matrix which gives different eigenvalues for such states within the irreducible representation. For explicit representation of the SU(K) monodromy matrix, we introduce the Lax pair method in Section 9.5. It is demonstrated that the Lax method reproduces the same monodromy matrix as that given in the product form, and that all conserved quantities are constructed as the power of the Lax operator.

A ribbon diagram without inscription is in one-to-one correspondence with a motif, and both specify an SU(K) representation. In Section 9.6, we provide another way of representing the Yangian supermultiplet in terms of the Drinfeld polynomial. Finally, in Section 9.7, we sketch an idea to extend the Yangian to arbitrary supersymmetry, including the SUSY t–J model.

9.1 Construction of monodromy matrix

We start with the following form of the Cherednik–Dunkl operators:

$$\underline{d}_i = \frac{z_i}{\lambda}\frac{\partial}{\partial z_i} + \sum_j{}' \left[\theta(j-i) - \theta_{ij}\right] K_{ij}, \tag{9.1}$$

where $\theta_{ij} = z_i/(z_i - z_j)$ and $\theta(i-j)$ is the step function. The prime in the summation means exclusion of $j = i$. The second term can also be written as

$$\sum_j{}' \left[\theta(j-i) - \theta_{ij}\right] K_{ij} = -\frac{1}{2}\sum_j{}' \left[\operatorname{sgn}(i-j) + w_{ij}\right] K_{ij}, \tag{9.2}$$

with $w_{ij} = (z_i + z_j)/(z_i - z_j)$. The operator $\underline{d}_i$ is related to $\hat{d}_i$ used in Chapters 3 and 7 by the similarity transformation, which has been displayed in (3.163). Namely, we obtain with $\mathcal{O} = \prod_{i<j} |z_i - z_j|^\lambda$,

$$\underline{d}_i = \mathcal{O}\hat{d}_i\mathcal{O}^{-1} + N - 1. \tag{9.3}$$

The form $\hat{d}_i$ is useful for discussion of non-symmetric Jack polynomials. However, we prefer here the simpler form $\underline{d}_i$ which enables us to derive the explicit form of the Yangian generators as given by (9.15). To study the SU(K) system without the spatial motion of particles, we introduce an operator $\hat{D}_i$ by

$$\hat{D}_i = \underline{d}_i - \frac{z_i}{\lambda}\frac{\partial}{\partial z_i} = \sum_j{}' \left[\theta(j-i) - \theta_{ij}\right] K_{ij}. \tag{9.4}$$

The commuting property $[\hat{D}_i, \hat{D}_j] = 0$ also holds without the differential operators, as can be deduced in analogy with the fundamental commutation property (7.23).

We can then construct the monodromy matrix as

$$\hat{T}_0(u) = \left(1 + \frac{P_{01}}{u - \hat{D}_1}\right)\cdots\left(1 + \frac{P_{0M}}{u - \hat{D}_N}\right) = \sum_{ab} X_0^{ab}\hat{T}_{ab}(u), \tag{9.5}$$

$$\hat{T}_{ab}(u) = \sum_{c\cdots e}\left(1 + \frac{X_1^{bc}}{u - \hat{D}_1}\right)\cdots\left(1 + \frac{X_M^{ea}}{u - \hat{D}_M}\right), \tag{9.6}$$

where the order of indices in $X_0^{ab}\hat{T}_{ab}(u)$ comes from the ket property of a in $X_0^{ab} = |a\rangle\langle b|$, and the bra property of a in $\hat{T}_{ab}(u) = \langle a|\hat{T}(u)|b\rangle$. Since the $\hat{D}_i$'s commute among themselves and with P_{0j}, they can be treated as if they are c-numbers. By following the same logic as that for the Sutherland model, we can confirm that $\hat{T}_0(u)$ satisfies the Yang–Baxter relation, and is qualified as the monodromy matrix. In fact, the Yang–Baxter relation holds for any values of the set $\{z_i\}$. However, an interesting conservation law appears only if the $\{z_i\}$ form the regular lattice with the same spacing on the ring. Since all coordinates are fixed, we no longer have the U(1) degrees of freedom, i.e., the phonons, and the relevant Yangian becomes $Y(sl_K)$ instead of $Y(gl_K)$.

The Hamiltonian of the SU(K) chain is obtained from the Sutherland model written in the form

$$H_S = \lambda^2 \sum_{i=1}^{N} \left(\underline{d}_i - \frac{N-1}{2}\right)^2 - \frac{\lambda^2}{12}N(N^2 - 1), \qquad (9.7)$$

where we have taken the length $L = 2\pi$.

Let us analyze the λ-dependence [183] of the commuting property:

$$[\underline{d}_i, \lambda^{-2}H_S] = 0, \qquad (9.8)$$

which holds for all i. The LHS has terms of orders ranging from $O(\lambda^0)$ to $O(\lambda^{-3})$, whose coefficients should all vanish. The coefficient of the $O(1/\lambda)$ terms in (9.8) is given by

$$z_i\frac{\partial}{\partial z_i}\sum_{j\neq l} h_{jl} + [\tilde{H}_{SU(K)}, \hat{D}_i] = 0, \qquad (9.9)$$

where $h_{ij} = -z_iz_j/(z_i - z_j)^2 = \theta_{ij}\theta_{ji}$ and

$$\tilde{H}_{SU(K)} = \sum_{i\neq j} h_{ij}K_{ij} \qquad (9.10)$$

appears as the $O(\lambda)$ term in H_S. Since the first term of (9.9) is zero for the periodic lattice, the second term should also be zero. Namely, $\hat{D}_i$ commutes with $\tilde{H}_{SU(K)}$, which is written in terms of K_{ij}. Of course the same result follows by computing $[\tilde{H}_{SU(K)}, \hat{D}_i]$ directly.

In the bosonic Fock space with $K_{ij}P_{ij}\Psi_B = \Psi_B$, $\tilde{H}_{SU(K)}$ is equivalent to $H_{SU(K)} = \sum_{i\neq j} h_{ij}P_{ij}$. In the fermionic Fock space with $K_{ij}P_{ij}\Psi_F = -\Psi_F$, on the other hand, the relevant form must be $-\sum_{i\neq j} h_{ij}K_{ij}$ to have the same Hamiltonian $H_{SU(K)} = \sum_{i\neq j} h_{ij}P_{ij}$. The fermionic form can be obtained when we replace $\lambda \rightarrow -\lambda$ in (9.7). In this section, we choose the bosonic

Fock space. We discuss the case of a combination of fermions and bosons in Section 9.7.

Since $\tilde{H}_{\mathrm{SU}(K)}$ does not involve the spin operators, it commutes with P_{0i} for any i. Hence we obtain the important property

$$[\tilde{H}_{\mathrm{SU}(K)}, \hat{T}_0(u)] = 0 = [\tilde{H}_{\mathrm{SU}(K)}, \hat{T}_{ab}(u)], \tag{9.11}$$

for any combination of a and b. Furthermore, since $\hat{T}_0(u)$ keeps the bosonic symmetry of the wave function, $H_{\mathrm{SU}(K)}$ with P_{ij} instead of K_{ij} also commutes with the monodromy matrix. Namely, we obtain

$$[H_{\mathrm{SU}(K)}, \hat{T}_{ab}(u)] = 0. \tag{9.12}$$

The monodromy matrix is expanded in terms of $1/u$ as

$$\hat{T}_{ab}(u) = \delta_{ab} + \frac{1}{u}\hat{T}_{ab}^{(1)} + \frac{1}{u^2}\hat{T}_{ab}^{(2)} + \cdots + \frac{1}{u^n}\hat{T}_{ab}^{(n)} + \cdots, \tag{9.13}$$

where any term $\hat{T}_{ab}^{(n)}$ commutes with the Hamiltonian. The first two terms are given by

$$\hat{T}_{ab}^{(1)} = \sum_i X_i^{ba} \equiv J^{ba}, \tag{9.14}$$

$$\hat{T}_{ab}^{(2)} = \sum_{i \neq j} \sum_c \theta_{ij} X_i^{bc} X_j^{ca} = \Lambda^{ba} + \frac{1}{2} \sum_c (J^{bc} J^{ca} - J^{ba}), \tag{9.15}$$

where Λ^{ba} is given by

$$\Lambda^{ba} = \frac{1}{2} \sum_{i \neq j} \sum_c w_{ij} X_i^{bc} X_j^{ca} = \sum_c f_{bac} \Lambda^c, \tag{9.16}$$

using $\theta_{ij} = (w_{ij} + 1)/2$. Here Λ^c is the Yangian generator in a form defined by (5.12). Equation (9.15) can most easily be derived by taking $N = 2$. Then we obtain

$$\hat{T}_{ab}^{(2)} = \sum_c \left(X_1^{bc} X_2^{ca} + \hat{D}_1 X_1^{ba} + \hat{D}_2 X_2^{ba} \right), \tag{9.17}$$

$$\hat{D}_1 X_1^{ba} = (1 - \theta_{12}) K_{12} X_1^{ba} = \theta_{21} \sum_c X_2^{bc} X_1^{ca}, \tag{9.18}$$

$$\hat{D}_2 X_2^{ba} = -\theta_{21} K_{21} X_2^{ba} = -\theta_{21} \sum_c X_1^{bc} X_2^{ca}, \tag{9.19}$$

where we have used the property $K_{12} = P_{12}$ when acting on bosonic states. The RHS of (9.17) indeed gives (9.15). It is clear that the same results follow for arbitrary N.

The set Λ^{ab} with K^2 components minus the U(1) component $\sum_a \Lambda^{aa}$ is equivalent to the set Λ^α with $K^2 - 1$ components, as given by (5.12). They are manifestly related by the structure constant f_{abc} of the Lie algebra SU(K). Explicit representation of $\hat{T}_{ab}^{(n)}$ of higher order will be given in Section 9.5 using the Lax pair formalism.

9.2 Quantum determinant vs. ordinary determinant

We proceed to an algebraic characterization of the supermultiplet, which has already been discussed in terms of ribbon diagrams and motifs in Section 4.9.2 of Chapter 4. The key quantity for the characterization is called the Drinfeld polynomial [43, 44]. Namely, the set of Drinfeld polynomials gives equivalent information to the motif or the ribbon diagram. The quantum determinant, which has been discussed in Chapter 8 for SU(2), becomes indispensable in the course of the discussion. Since the quantum determinant is a fundamental quantity in the Yangian theory, we explain the basic property generalizing to the $Y(sl_K)$ and $Y(gl_K)$ cases.

We begin with an interpretation of the ordinary determinant as an antisymmetrizer. Consider a set of unit vectors e_i with $1 \leq i \leq K$. Any vector in the K-dimensional space can be formed by a linear combination of e_i as

$$A_j = \sum_i a_{ij} e_i. \tag{9.20}$$

Then we can represent the determinant in terms of the antisymmetrizer as

$$\text{Asym}\, A_1 \otimes A_2 \otimes \cdots \otimes A_K = \det\{a_{ij}\}\text{Asym}\, e_1 \otimes e_2 \otimes \cdots \otimes e_K. \tag{9.21}$$

The determinant is regarded as a projection from a matrix $\{a_{ij}\}$ to a c-number. The antisymmetrized vectors on the RHS represent the SU(K) singlet. In the simplest case of $K = 2$, for example, we obtain

$$\text{Asym}\, e_1 \otimes e_2 = e_1 \otimes e_2 - e_2 \otimes e_1 = |\uparrow\downarrow\rangle - |\downarrow\uparrow\rangle, \tag{9.22}$$

with e_1 identified as spin up. We now consider a situation where a set $a_{ij}(u)$ with a spectral parameter u consists of operators that do not commute with each other, but satisfy the Yangian commutation relation

$$(u - v)\,[a_{ij}(u), a_{kl}(v)] = a_{kj}(v)a_{il}(u) - a_{kj}(u)a_{il}(v). \tag{9.23}$$

This commutation relation has appeared in (8.57) and (8.58). The elements $a_{ij}(u)$ are now written as $T_{ij}(u)$, implying that they are qualified as elements of the monodromy matrix. Then we generalize the ordinary determinant to the quantum determinant by the relation

$$\text{Asym}\,\boldsymbol{T}_1(u) \otimes \boldsymbol{T}_2(u-1) \otimes \cdots \otimes \boldsymbol{T}_K(u-K+1)$$
$$= \text{qdet}\,\{T_{ij}(u)\}\text{Asym}\,\boldsymbol{e}_1 \otimes \boldsymbol{e}_2 \otimes \cdots \otimes \boldsymbol{e}_K, \tag{9.24}$$

where $\boldsymbol{T}_j(v) = \sum_i T_{ij}(v)\boldsymbol{e}_i$ and

$$\text{qdet}\,\{T_{ij}(u)\} = \sum_P \text{sgn}(P)T_{P(1)1}(u)T_{P(2)2}(u-1)\cdots T_{P(K)K}(u-K+1). \tag{9.25}$$

Note that the antisymmetrization acts only on vectors $\boldsymbol{e}_i$ and keeps the sequence of non-commuting operators. Unlike the classical determinant, the quantum determinant is still an operator in the K^N-dimensional vector space, where N is the number of sites in the system. An alternative expression is obtained as

$$\text{qdet}\,\{T_{ij}(u)\} = \sum_P \text{sgn}(P)T_{1P(1)}(u-K+1)T_{2P(2)}(u-K+2)\cdots T_{KP(K)}(u), \tag{9.26}$$

which is equivalent to (9.25) as shown below. The spectral parameters $u - K + j$ in the quantum determinant are sequential. The reason for this choice comes from the Yangian commutation relation of (9.23), which simplifies with $u - v = \pm 1$.

9.3 Capelli determinant

In order to understand the nature of quantum determinants, we take the simplest version where only a single site is involved. This version is called the Capelli determinant in the mathematical literature [67]. Let us consider a $K \times K$ operator-valued matrix whose elements $C_{ji}(u)$ are given by

$$C_{ji}(u) = \delta_{ij} + X^{ij}/u, \tag{9.27}$$

where the operator X^{ij} acts on the basis $|k\rangle$ with $k = 1, \ldots, K$ as $X^{ij}|k\rangle = \delta_{jk}|i\rangle$. This quantity has appeared as the basic element in the monodromy matrix, as explained in Section 8.2. Note that the order ij of indices is different between C_{ji} and X^{ij}. With this definition, the i's in C_{ji} and X^{ij} both represent an index of a ket vector $|i\rangle$, while j in both C_{ji} and X^{ij} represent a bra vector $\langle j|$.

In the case of $K = 2$, the quantum (Capelli) determinant is derived according to (9.25) and (9.26) as

$$\text{qdet } \{C_{ij}(u)\} = \left(1 + \frac{X^{11}}{u}\right)\left(1 + \frac{X^{22}}{u-1}\right) - \frac{X^{21}X^{12}}{u(u-1)} \tag{9.28}$$

$$= \left(1 + \frac{X^{11}}{u-1}\right)\left(1 + \frac{X^{22}}{u}\right) - \frac{X^{12}X^{21}}{(u-1)u} = 1 + \frac{1}{u}, \tag{9.29}$$

where we have used the completeness relation $X^{11} + X^{22} = 1$.

In the next simplest case of $K = 3$, the Capelli determinant is given by

$$\text{qdet } \{C_{ij}(u)\} = \left(1 + \frac{X^{11}}{u}\right)\left(1 + \frac{X^{22}}{u-1}\right)\left(1 + \frac{X^{33}}{u-2}\right)$$

$$+ \frac{X^{31}}{u}\frac{X^{12}}{(u-1)}\frac{X^{23}}{(u-2)} + \cdots = 1 + \frac{1}{u}, \tag{9.30}$$

using the completeness relation $X^{11} + X^{22} + X^{33} = 1$. Thus *the Capelli determinant is a c-number* $1 + 1/u$ *which is independent of* K. Note that an operator-valued determinant can in general be an operator acting on K-dimensional vectors.

The mysterious scalar property of the Capelli determinant can be understood naturally in terms of the Yangian algebra. We introduce an auxiliary space, which consists of K copies of K-dimensional vectors, and interpret X^{ij} as the product $X^{ij} \otimes 1 \cdots \otimes 1$ which acts as a scalar in the auxiliary space. On the other hand, the antisymmetrizer acts on the auxiliary space, and does not touch the physical space, as is apparent in (9.25). In the simplest case of $K = 2$, the Yang–Baxter equation can be written in the form

$$\text{Asym}_2\, \boldsymbol{C}_1(u-1) \otimes \boldsymbol{C}_2(u) = \boldsymbol{C}_2(u) \otimes \boldsymbol{C}_1(u-1)\text{Asym}_2, \tag{9.31}$$

where $\boldsymbol{C}_j = C_{1j}\boldsymbol{e}_1 + C_{2j}\boldsymbol{e}_2$ as in (9.20). Note that $\text{Asym}_2 = 1 - P_{12}$ is the special case $v - u = -1$ of the R-matrix $R_{12}(v - u)$. We recognize that Asym_2 works as the antisymmetrizer for two vectors in the auxiliary space. The antisymmetrizer extracts the singlet when acting on a state in the auxiliary space. This means that $\boldsymbol{C}_2(u-1) \otimes \boldsymbol{C}_1(u)$ does not produce a triplet if acting on the singlet. The diagonal element is given by (9.29).

For general K, we obtain the antisymmetrizer Asym_K as follows [134]:

$$\text{Asym}_K = R_{K-1,K}(R_{K-2,K}R_{K-2,K-1})\cdots(R_{2K}\cdots R_{23})(R_{1K}\cdots R_{12}), \tag{9.32}$$

where $R_{ij} = R_{ij}(u_i - u_j)$ with $u_i - u_j = -1$ acts on the ith and jth vectors in the auxiliary space. Namely, action of Asym_K on the ordered set of vectors

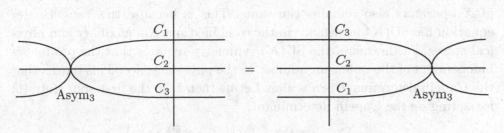

Fig. 9.1. Graphical representation of (9.35) with $K = 3$. The quantum number along the vertical line is represented by X-operators.

e_i $(1 \leq i \leq K)$ leads to the SU(K) singlet $|0_K\rangle$:

$$\text{Asym}_K(e_1 \otimes e_2 \cdots \otimes e_K) = \sum_P \text{sgn}(P)e_{P(1)} \otimes e_{P(2)} \cdots \otimes e_{P(K)} = |0_K\rangle,$$

(9.33)

which can be verified by mathematical induction, starting with $K = 2$. Note that Asym_K does not preserve the norm of a state; double action of Asym_K gives

$$\text{Asym}_K{}^2 = K!\,\text{Asym}_K. \tag{9.34}$$

We can interpret Asym_K as an explicit realization of Asym used in (9.21).

The crucial point is that Asym_K is a product of R-matrices. Hence, repeated use of the Yang–Baxter equation reverses the order of C_i, and finally leads to

$$\text{Asym}_K \boldsymbol{C}_1(u - K + 1) \otimes \cdots \boldsymbol{C}_K(u)$$
$$= \boldsymbol{C}_K(u) \otimes \cdots \boldsymbol{C}_1(u - K + 1)\text{Asym}_K, \tag{9.35}$$

which is illustrated in Fig. 9.1 for the case of $K = 3$.

Equation (9.35) shows that action of $\boldsymbol{C}_K(u) \otimes \cdots \boldsymbol{C}_1(u - K + 1)$ on the SU(K) singlet again gives the singlet, as assured by the LHS. Both sides of (9.35) define the Capelli determinant $\text{qdet}\,C(u)$. Namely, we obtain

$$\text{qdet}\,\{C_{ij}(u)\}\text{Asym}_K = \boldsymbol{C}_K(u) \otimes \cdots \boldsymbol{C}_1(u - K + 1)\text{Asym}_K \tag{9.36}$$
$$= \text{Asym}_K\boldsymbol{C}_1(u - K + 1) \otimes \cdots \boldsymbol{C}_K(u). \tag{9.37}$$

By applying (9.36) to $e_1 \otimes e_2 \cdots \otimes e_K$, we obtain (9.25). By using (9.37), on the other hand, we obtain (9.26). Hence, both definitions given by (9.25) and (9.26) are equivalent for the Capelli determinant. The equivalence applies to quantum determinants in general.

Since the singlet remains by this action, the Yang–Baxter equation requires that the quantum number in the physical space after successive application of X-operators also remains the same. This is because the Yang–Baxter equation has $SU(K)$ invariance in the combined space of auxiliary and physical states. Furthermore, the $SU(K)$ symmetry requires that the product is independent of the quantum number in the physical space. This means that the Capelli determinant is a scalar. Let us then take the first spin state $|1\rangle$ for acting on the Capelli determinant:

$$\text{qdet}\,\{C_{ij}(u)\}|1\rangle = \left(1 + \frac{X^{11}}{u}\right)\cdots\left(1 + \frac{X^{KK}}{u - K + 1}\right)|1\rangle + \cdots. \qquad (9.38)$$

As can be understood from the structure of (9.38), the only term that remains nonzero is the product of the diagonal term with $X^{ii} \to 1$ for $i = 1$ and $X^{ii} \to 0$ otherwise. Hence, the Capelli determinant is $1 + 1/u$ for all values of K. The antisymmetrizer construction has thus clarified the origin of the common value $1 + 1/u$.

9.4 Quantum determinant of SU(K) Yangian

We now derive the quantum determinant of $\hat{T}_0$ in (9.5) explicitly. Because site-dependent factors in (9.6) commute with each other, the quantum determinant of the product is equal to the product of each quantum determinant. The latter has been calculated as the Capelli determinant:

$$\text{qdet}\left(1 + \frac{\hat{X}_i}{u - \hat{D}_i}\right) = 1 + \frac{1}{u - \hat{D}_i} = \frac{u + 1 - \hat{D}_i}{u - \hat{D}_i}, \qquad (9.39)$$

where $\hat{X}_i$ is the operator-valued matrix whose elements are X_i^{ab} with $1 \le a, b \le K$. Then we obtain

$$\text{qdet}\,\hat{T}_0(u) = \frac{\hat{\Delta}_N(u + 1)}{\hat{\Delta}_N(u)}, \qquad (9.40)$$

where $\hat{\Delta}_N(u) = \prod_i(u - \hat{D}_i)$. Since the quantum determinant is a scalar in $Y(sl_K)$, the result of its action is independent of the physical state chosen. Hence we take the fully polarized state where $K_{ij} = 1$ for any i and j. Then we obtain

$$\hat{D}_i \to -\frac{1}{2}\sum_j{}'(w_{ij} + 1) + N - i. \qquad (9.41)$$

Thus qdet $\hat{T}_0(u)$ is determined once the configuration of the coordinates is fixed. For a periodic lattice, great simplification occurs since summation over

j gives a zero result for w_{ij}. Then the eigenvalues are given by

$$\hat{D}_i \to \frac{1}{2}(N+1) - i, \tag{9.42}$$

which ranges from $(N-1)/2$ for $i = 1$ to $-(N-1)/2$ for $i = N$. Hence we obtain

$$\hat{\Delta}_N(u) = \prod_{i=1}^{N}\left(u - \frac{1}{2}(N-1) + i - 1\right), \tag{9.43}$$

$$\text{qdet}\,\hat{T}_0(u) = \left[u + \frac{1}{2}(N+1)\right] \Big/ \left[u - \frac{1}{2}(N-1)\right]. \tag{9.44}$$

For both periodic and non-periodic lattices, the quantum determinant of $Y(sl_K)$ does not depend on a many-spin state. This is in contrast to that of $Y(gl_K)$, where the quantum determinant does depend on a supermultiplet. We remark additionally that the Yangian commutation relation (9.23) leads to the relation for any element of the monodromy matrix:

$$\left[\text{qdet}\,\hat{T}_0(u), \hat{T}_0^{ab}(v)\right] = 0, \tag{9.45}$$

for any values of the spectral parameters.

9.5 Alternative construction of monodromy matrix

We now discuss another way of deriving conserved quantities and the monodromy matrix using the Lax pair [22, 163]. A merit of the Lax pair formalism is that the explicit form of $\hat{T}_{ab}^{(n)}$ for general n can be derived easily. The monodromy matrix we now derive is written as $T_0(u)$. Later we show the equivalence of $T_0(u)$ to $\hat{T}_0(u)$ given by (9.6). We introduce the Lax operator L by

$$L_{ij} = (1 - \delta_{ij})\theta_{ij}P_{ij}, \tag{9.46}$$

where $\theta_{ij} = z_i/(z_i - z_j)$ is related to $w_{ij} = (z_i + z_j)/(z_i - z_j)$ as

$$w_{ij} = \theta_{ij} - \theta_{ji} = 2\theta_{ij} - 1. \tag{9.47}$$

The generating function of operators $T_{ab}^{(n)}$ is given by

$$T_0(u) = X_0^{ab}T_{ab}(u) = 1 + \sum_{ij} P_{0i}\left(\frac{1}{u - L}\right)_{ij}, \tag{9.48}$$

$$T_{ab}(u) = \delta_{ab} + \sum_{ij} X_i^{ba}\left(\frac{1}{u - L}\right)_{ij} = \sum_{n=0}^{\infty} \frac{1}{u^n}T_{ab}^{(n)}, \tag{9.49}$$

where X_0^{ba} changes state a to b at an auxiliary site 0, and $T_{n=0}^{ab} = \delta_{ab}$. Let us write down lower-order terms of $T_{ab}^{(n)}$. Expansion of (9.49) gives

$$T_{ab}^{(1)} = \sum_i X_i^{ba} = J^{ba}. \tag{9.50}$$

Here we have introduced the notation J^{ba} for a component of the generalized total angular momentum, which is conserved. In the next order we obtain

$$T_{ba}^{(2)} = \sum_{i \neq j} \theta_{ij} \sum_c X_i^{ac} X_j^{cb}. \tag{9.51}$$

Thus both $\hat{T}_1^{ba}$ and $\hat{T}_2^{ba}$ are the same as those given by (9.14) and (9.15).

Let us now prove that the monodromy matrix defined by (9.49) indeed satisfies the Yang–Baxter relation. We follow the method of Bernard *et al.* [22] and work with the coordinate exchange operator K_{ij} with each coordinate z_i being not necessarily on a periodic lattice. For a bosonic wave function Ψ_B, the symmetry results in $K_{ij} P_{ij} \Psi_B = \Psi_B$. However, the wave function becomes no longer symmetric after applying K_{ij} alone. This means that the action of K_{ij} brings the wave function out of the bosonic Fock space. Examples of useful identities within the bosonic Fock space are

$$P_{jl} P_{ij} = K_{ij} K_{jl} = K_{il} K_{ij} = K_{lj} K_{il}, \tag{9.52}$$

where the first equality can be verified by considering a three-particle system.

Let us introduce an operator D_i by

$$D_i = \sum_j{}' \theta_{ij} K_{ij}. \tag{9.53}$$

The commutation rule is given by

$$[D_i, D_j] = (D_i - D_j) K_{ij}, \tag{9.54}$$

which can be derived by direct calculation. Alternatively, it follows from the relation

$$D_i = -\lim_{\lambda \to \infty} z_i \pi_i, \tag{9.55}$$

where π_i has been defined by (7.21), and satisfies $[\pi_i, \pi_j] = 0$. We work with the bosonic wave function from now on. Then we obtain the equivalence

$$D_i^n = \sum_j (L^n)_{ij}, \tag{9.56}$$

which can be verified by direct calculation. We demonstrate the case of $n = 2$ as follows:

$$\theta_{ij}K_{ij}\theta_{il}K_{il} = \theta_{ij}\theta_{jl}K_{ij}K_{il} = \theta_{ij}\theta_{jl}K_{jl}K_{ij} = \theta_{ij}P_{ij}\theta_{jl}P_{jl}, \qquad (9.57)$$

where the first and last expressions correspond to D_i^2 and $(L^2)_{ij}$, respectively, after summation over indices $j(\neq i)$ and $l(\neq j)$. Then the monodromy matrix is recast into the form

$$T_0(u) = 1 + \sum_i P_{0i}(u - D_i)^{-1}. \qquad (9.58)$$

In the Yang–Baxter equation, we encounter the product of permutations, which are transformed as

$$P_{00'}P_{0i}P_{0'j} = K_{ij}P_{0i}P_{0'j}, \quad P_{0'j}P_{0i}P_{00'} = P_{0'j}P_{0i}K_{ij}. \qquad (9.59)$$

Then the Yang–Baxter relation is reduced to the form

$$(u - v + K_{ij})(u - D_i)^{-1}(v - D_j)^{-1} = (v - D_j)^{-1}(u - D_i)^{-1}(u - v + K_{ij}). \qquad (9.60)$$

By taking the inverse of both sides, and multiplying by $u - v + K_{ij}$ from the left and from the right, we obtain

$$(u - v + K_{ij})(v - D_j)(u - D_i) = (u - D_i)(v - D_j)(u - v + K_{ij}). \qquad (9.61)$$

The equality can also be verified by using the commutation rule (9.54). Hence it has been proven that the monodromy matrix in the form of (9.49) indeed satisfies the Yang–Baxter equation in the bosonic Fock space.

We have seen that the generators J^{ab} and Λ^{ab} are equivalently given from a $1/u$ expansion of either $\hat{T}_0(u)$ or $T_0(u)$. It is shown now that higher-level generators are also equivalent in $\hat{T}_0(u)$ and $T_0(u)$ by using Yangian algebra. Namely, the commutation rule between $T_{ab}^{(n)}$ is given with $n \leq m$ by

$$\left[T_{ab}^{(n)}, T_{cd}^{(m)}\right] = \sum_{k=0}^{n}\left(T_{cb}^{(k+m)}T_{ad}^{(n-k-1)} - T_{cb}^{(n-k-1)}T_{ad}^{(k+m)}\right), \qquad (9.62)$$

which is obtained by expansion of $T_{ab}(u)$ in powers of $1/u$ in the Yang–Baxter relation. Putting $n = 1$ and $b = c$, we obtain for $m \geq 2$

$$T_{ad}^{(m+1)} = \sum_b \left[T_{ab}^{(1)}, T_{bd}^{(m)}\right] + T_{ad}^{(m)}, \qquad (9.63)$$

where we have used $T_{ab}^{(0)} = \delta_{ab}$. Therefore, any higher-level generator $T_{ab}^{(n)}$ can be recursively derived from J^{ab} and Λ^{ab} by (9.63). Since the two lowest-order terms are the same for $\hat{T}_0(u)$ and $T_0(u)$, the whole series must also be the same.

Let us turn to the quantum determinant for a non-regular lattice. Since qdet $T_0(u)$ is independent of the spin state, we take the fully polarized state where all sites have spin state 1. The monodromy matrix can be made lower triangular because the fully polarized state is a YHWS. The diagonal elements except for $T_{11}(u)$ are all unity. Then we obtain

$$T_{11}(u) = \text{qdet}\, T_0(u) = 1 + \sum_{i,j=1}^{N} \left(\frac{1}{u - \Theta} \right)_{ij} = \frac{\Delta_N(u+1)}{\Delta_N(u)}. \qquad (9.64)$$

Here Θ is the $N \times N$ matrix with matrix elements θ_{ij}, and $\Delta_N(u)$ is its characteristic polynomial as given by $\Delta_N(u) = \det(u - \Theta)$. We emphasize that the discussion so far *does not* assume a regular periodic lattice. Therefore, $\det(u - \Theta)$ depends on the configuration of the coordinates. The nature of $Y(sl_K)$ appears in the point that $\det(u - \Theta)$ is common to all states, given the coordinate configuration. Provided z_i forms a regular lattice on the ring, the matrix Θ can be diagonalized in terms of momentum eigenstates. Then we obtain

$$\Delta_N(u) = \prod_{i=1}^{N} \left(u - \frac{1}{2}(N-1) + i - 1 \right) = \hat{\Delta}_N(u), \qquad (9.65)$$

where the last quantity has appeared in (9.43). In the regular lattice, the commutation relation with the Hamiltonian is given by

$$[H_{\text{SU}(K)}, L_{ij}] = \sum_l (M_{il} L_{lj} - L_{il} M_{lj}), \qquad (9.66)$$

with

$$M_{ij} = 2(1 - \delta_{ij}) h_{ij} P_{ij} - 2\delta_{ij} \sum_k h_{ik} P_{ik}. \qquad (9.67)$$

The pair of operators L_{ij} and M_{ij} is called the Lax pair.

Because of the property $\sum_i M_{ij} = \sum_j M_{ij} = 0$, we obtain

$$[H_{\text{SU}(K)}, \sum_{ij} L_{ij}] = 0. \qquad (9.68)$$

Namely, $\sum_{ij} L_{ij}$ is conserved. In a similar fashion, it can be shown that

$$[H_{\text{SU}(K)}, \sum_{ij} (L^n)_{ij}] = 0, \qquad (9.69)$$

for any power n. Furthermore, observing that

$$[H_{\mathrm{SU}(K)}, X_i^{ab}] = \sum_j X_j^{ab} M_{ji}, \qquad (9.70)$$

we find that the quantity

$$T_{ab}^{(n)} = \sum_{ij} X_i^{ba}(L^n)_{ij} \qquad (9.71)$$

is also conserved. In this way we have derived the explicit form of higher conserved quantities $\sum_{ij}(L^n)_{ij}$ and $T_{ab}^{(n)}$.

9.6 Drinfeld polynomials

The eigenstates of the $\mathrm{SU}(K)$ spin chain give representations of the Yangian. Each eigenstate constitutes a supermultiplet, which is characterized by either a ribbon diagram or a motif. We now introduce another equivalent description in terms of a set of polynomials $P_i(u)$, where i runs from 1 to $K - 1$. These are called Drinfeld polynomials, and provide information equivalent to that given by motifs or ribbon diagrams [43, 44]. The description in terms of Drinfeld polynomials is algebraically the most systematic, but least pictorial. We provide an intuitive, although less rigorous, route toward Drinfeld polynomials in the following.

All the information about Yangian properties is contained in the monodromy matrix, which can be made lower triangular in the space of YHWS. If one takes an $m \times m$ submatrix from the top left and neglects the rest, the eigenfunctions consist of states belonging to the $\mathrm{SU}(m)$ Yangian $Y(sl_m)$ with $m \leq K$. In this way we can classify the YHWS states according to the chain

$$Y(sl_2) \subset Y(sl_3) \subset \cdots \subset Y(sl_K). \qquad (9.72)$$

The basis set which is consistent with this chain is called the Gelfand–Zetlin basis. The monodromy matrix acting on a YHWS $|w\rangle$ is of the form

$$\hat{T}_0(u)|w\rangle = \begin{pmatrix} \hat{T}_{11}(u) & 0 & 0 & \cdots & 0 \\ * & \hat{T}_{22}(u) & 0 & \cdots & 0 \\ \vdots & \vdots & \vdots & \cdots & \vdots \\ * & * & * & \cdots & \hat{T}_{KK}(u) \end{pmatrix} |w\rangle. \qquad (9.73)$$

The quantum determinant in this representation is simply given by

$$\mathrm{qdet}\,\hat{T}_0(u) = \hat{T}_{11}(u)\hat{T}_{22}(u-1)\cdots\hat{T}_{KK}(u-K+1), \qquad (9.74)$$

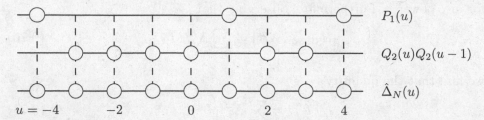

Fig. 9.2. Distribution of zeros shown by circles for the Drinfeld polynomial $P_1(u)$ and $Q_2(u)Q_2(u-1)$. A pair of successive zeros in $Q_2(u)Q_2(u-1)$ represents an SU(2) singlet in the many-spin states, while zeros of $P_1(u)$ specify the set of momenta of spinons.

which takes the value given by (9.44). Although the quantum determinant does not have information on $|w\rangle$, the ratio $\hat{T}_{m+1,m+1}(u)/\hat{T}_{mm}(u)$ does depend on each YHWS.

In the simplest case of $K = 2$, for example, we write

$$\hat{T}_0(u)|w\rangle = \frac{Q_2(u+1)}{Q_2(u)}\begin{pmatrix} P_1(u+1)/P_1(u) & 0 \\ * & 1 \end{pmatrix}|w\rangle, \qquad (9.75)$$

with $P_1(u)$ and $Q_2(u)$ being polynomials of u. Without loss of generality we can choose the coefficient unity for their highest power, which is referred to as monic. Comparing this with the result of the quantum determinant, we obtain

$$\frac{\hat{\Delta}_N(u+1)}{\hat{\Delta}_N(u)} = \frac{P_1(u+1)Q_2(u+1)Q_2(u)}{P_1(u)Q_2(u)Q_2(u-1)}, \qquad (9.76)$$

which leads to $\hat{\Delta}_N(u) = P_1(u)Q_2(u)Q_2(u-1)$. This relation restricts zeros of $Q_2(u)$ so that they should not be adjacent. A pair of zeros in the product $Q_2(u)Q_2(u-1)$ represents a singlet included in the N-body spin state. Figure 9.2 illustrates the situation. The poles of $\hat{T}_{22}(u)$ describe momenta of down spins, of which those in $Q_2(u)$ are part of an SU(2) singlet. On the other hand, the poles of $\hat{T}_{11}(u)$ describe momenta of up spins. Of these poles, those in $Q_2(u)$ are part of an SU(2) singlet, while those in $P_1(u)$ correspond to spinons. The polynomial $P_1(u)$ is an example of a Drinfeld polynomial.

The structure of roots in the monodromy matrix will become more transparent when we proceed to the case of SU(3). We bring the monodromy matrix into the lower triangular form, and represent $\hat{T}_{ii}(u)$ in terms of monic polynomials $P_1(u), P_2(u)$, and $Q_3(u)$ in the following way:

$$\hat{T}_{11}(u)/\hat{T}_{22}(u) = P_1(u+1)/P_1(u), \tag{9.77}$$

$$\hat{T}_{22}(u)/\hat{T}_{33}(u) = P_2(u+1)/P_2(u), \tag{9.78}$$

$$\hat{T}_{33}(u) = Q_3(u+1)/Q_3(u), \tag{9.79}$$

which is a generalization of (9.75). Then the quantum determinant is given by

$$\frac{\hat{\Delta}_N(u+1)}{\hat{\Delta}_N(u)} = \frac{P_1(u+1)P_2(u+1)P_2(u)Q_3(u+1)Q_3(u)Q_3(u-1)}{P_1(u)P_2(u)P_2(u-1)Q_3(u)Q_3(u-1)Q_3(u-2)}, \tag{9.80}$$

which shows that $\hat{\Delta}_N(u)$ is given by the denominator on the RHS.

Now it is easy to generalize to arbitrary K. We define Drinfeld polynomials $P_m(u)$, with $1 \le m \le K-1$, and $Q_K(u)$, which describes the SU(K) singlet, by

$$\hat{T}_{mm}(u)/\hat{T}_{m+1,m+1}(u) = P_m(u+1)/P_m(u), \tag{9.81}$$

$$\hat{T}_{KK}(u) = Q_K(u+1)/Q_K(u). \tag{9.82}$$

The quantity $\hat{\Delta}_N(u)$ appearing in the quantum determinant of $\hat{T}_0(u)$ is given by

$$\begin{aligned}
\hat{\Delta}_N(u) &= P_1(u)P_2(u)P_2(u-1)P_3(u)P_3(u-1)P_3(u-2)\dots \\
&\times P_{K-1}(u)P_{K-1}(u-1)\dots P_{K-1}(u-K+2) \\
&\times Q_K(u)Q_K(u-1)\dots Q_K(u-K+1). \tag{9.83}
\end{aligned}$$

The N successive zeros from $-(N-1)/2$ to $(N-1)/2$ in $\hat{\Delta}_N(u)$ are distributed to those of $P_m(u-l)$ and $Q_K(u-l)$ on the RHS of (9.83). Hence, zeros in $P_m(u)$ should be separated by m at least, and those in $Q_K(u)$ by K at least. The product $Q_K(u)\dots Q_K(u-K+1)$ specifies the location of the SU(K) singlet in the momentum space. Similarly, the product $P_m(u)\dots P_m(u-m+1)$ specifies the location of generalized spinons. Figure 9.3 illustrates the distribution of zeros in the SU(3) case with $N = 9$. The relevant Drinfeld polynomials are given by

$$P_1(u) = (u+2)(u-4), \quad P_2(u) = (u+4)(u-2), \quad Q_3(u) = u+1. \tag{9.84}$$

The same supermultiplets are represented in terms of the motif by $(1)()(11)$ $(1)()$, and in terms of the ribbon diagram of Fig. 9.4.

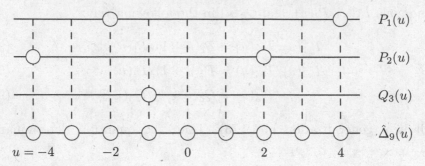

Fig. 9.3. Illustration of roots distribution in SU(3) Drinfeld polynomials with $N = 9$.

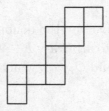

Fig. 9.4. Ribbon diagram corresponding to the set of Drinfeld polynomials of (9.84).

9.7 Extension to supersymmetry

Let us generalize the SUSY t–J model so that the number of bosonic species is K_B, and the number of fermionic species is K_F. Our primary purpose is to show that H_{tJ} commutes with the supersymmetric monodromy matrix, and the conserved quantities are obtained by expansion of the monodromy matrix. For this purpose we work with the following form of the Hamiltonian:

$$\tilde{H}_{tJ} = \sum_{i \neq j} h_{ij} K_{ij}, \qquad (9.85)$$

where K_{ij} is the coordinate exchange operator of any kind of particle. In the original Hamiltonian H_{tJ}, we have the graded permutation operator $\tilde{P}_{ij}$ instead of K_{ij}. We simulate the vacant state by a fermion according to (6.2) in Chapter 6. It is also possible to take the alternative picture where the spin chain corresponds to the high density limit of electrons, i.e., fermions. In the latter case we should replace K_{ij} by $-K_{ij}$ in (9.85) and simulate the vacant state by a boson. The same physical quantities result from either picture. Namely, $Y(sl_{K_\mathrm{B}|K_\mathrm{F}})$ and $Y(sl_{K_\mathrm{F}|K_\mathrm{B}})$ supersymmetries give the same physics.

The operator $\tilde{P}_{ij}$ is chosen so that the combined action on Ψ_{BF}, which is a mixture of fermions and bosons, leads to $\tilde{P}_{ij}K_{ij}\Psi_{\mathrm{BF}} = \Psi_{\mathrm{BF}}$. In accordance with the commutation rule between fermions, we modify the permutation operator to the graded one as follows:

$$\tilde{P}_{ij} = \sum_{\alpha\beta} p(\beta) X_i^{\alpha\beta} X_j^{\beta\alpha}, \qquad (9.86)$$

where $p(\beta) = 1$ if β belongs to the bosonic species, and $p(\beta) = -1$ for fermionic species.

As in the SU(K) chain, we introduce the lattice version of the Cherednik–Dunkl operators by

$$\hat{D}_i = \sum_j{}' \left[\theta(j-i) - \theta_{ij}\right] K_{ij}, \qquad (9.87)$$

which commutes with $\tilde{H}_{tJ}$. Next, we modify the monodromy matrix as

$$\tilde{T}_0(u) = \left(1 + \frac{\tilde{P}_{01}}{u - \hat{D}_1}\right) \cdots \left(1 + \frac{\tilde{P}_{0M}}{u - \hat{D}_N}\right) = \sum_{ab} p(b) X_0^{ab} \tilde{T}_{ab}(u), \quad (9.88)$$

$$\tilde{T}_{ab}(u) = \sum_{c\cdots e} \left(1 + \frac{X_1^{bc}}{u - \hat{D}_1}\right) \cdots \left(1 + \frac{X_M^{ea}}{u - \hat{D}_M}\right). \qquad (9.89)$$

Correspondingly, the R-matrix is modified as

$$\tilde{R}_{ij}(u) = 1 + \tilde{P}_{ij}/u. \qquad (9.90)$$

This modified form also satisfies the Yang–Baxter relation [2]:

$$\tilde{R}_{00'}(u - v)\tilde{T}_0(u)\tilde{T}_{0'}(v) = \tilde{T}_{0'}(v)\tilde{T}_0(u)\tilde{R}_{00'}(u - v). \qquad (9.91)$$

From the form of (9.88), the commuting property follows:

$$[\tilde{H}_{tJ}, \tilde{T}_0(u)] = 0, \qquad (9.92)$$

as in the case of (9.11). Hence, expansion of $\tilde{T}_{ab}(u)$ in powers of $1/u$ gives the set of conserved quantities. The first two are given by

$$\tilde{T}_{ab}^{(1)} = \sum_i X_i^{ba}, \qquad (9.93)$$

$$\tilde{T}_{ab}^{(2)} = \sum_{i\neq j}\sum_c p(c)\theta_{ij} X_i^{bc} X_j^{ca} = \sum_{i\neq j}\sum_c \theta_{ij} X_i^{ba} \tilde{P}_{ij}. \qquad (9.94)$$

By using the relation $w_{ij} = 2\theta_{ij} - 1$, we obtain from (9.94) the conserved quantity

$$\Lambda^{ab} = \sum_{i \neq j} \sum_c p(c) w_{ij} X_i^{bc} X_j^{ca}, \qquad (9.95)$$

which serves as the supersymmetric Yangian generators. Some examples have been given in Section 6.5.1 for the special case of SU(2,1) or $Y(sl_{2|1})$.

10

Uglov's theory

In this chapter, we give an example of how the Yangian theory works in practice in the calculation of physical quantities. Using the Yangian representation theory, calculation of the dynamical correlation function of the Sutherland model with gl_K symmetry can be performed in the same way as (a modified version of) the single-component Sutherland model [189]. Namely, Uglov [189] showed that the spin and the spatial momentum can be unified as a fictitious momentum. This spin–momentum unification is particularly powerful for exact derivation of spin correlation functions [189,197]. Further development for the SU(K) chain has also been achieved [198,199]. Uglov's theory is outlined in this chapter.

In Section 10.1 we introduce as a prelude the symmetric Macdonald polynomials, which include the symmetric Jack polynomials as a special case. Then in Section 10.2, Uglov symmetric polynomials are introduced as another limit of symmetric Macdonald polynomials. We explain in Section 10.3 an isomorphism between the Fock space of the Sutherland model with SU(2) internal symmetry and the space of Laurent symmetric polynomials. The isomorphism by which the Yangian Gelfand–Zetlin basis is mapped onto the Uglov polynomials preserves the inner product. By this isomorphism, density and spin-density operators in the U(2) model find their correspondence in the single-component model. In this way, the calculations of dynamical density and spin-density correlation functions in the U(2) Sutherland model reduce to those of a modified version of the single-component Sutherland model.

10.1 Macdonald symmetric polynomials

We have seen in Section 2.5 that the Jack symmetric polynomials can be defined as the homogeneous symmetric polynomials of $z = (z_1, \ldots, z_N)$

satisfying conditions of triangularity (2.177) and orthogonality (2.196). Macdonald symmetric polynomials P_μ are defined in a similar way, with the following weight function $w(z; q, t)$ for the norm:

$$w(z; q, t) = \prod_{i \neq j} \frac{\left(z_i z_j^{-1}; q\right)_\infty}{\left(t z_i z_j^{-1}; q\right)_\infty}, \tag{10.1}$$

where we have introduced the notation

$$(z; q)_\infty = \prod_{r=0}^{\infty} (1 - zq^r). \tag{10.2}$$

We observe the behavior

$$\frac{(z; q)_\infty}{(q^\lambda z; q)_\infty} = \frac{\prod_{r=0}^{\infty}(1 - zq^r)}{\prod_{r=0}^{\infty}(1 - zq^{r+\lambda})} = \prod_{r=0}^{\lambda-1}(1 - zq^r) \xrightarrow[q \to 1]{} (1 - z)^\lambda. \tag{10.3}$$

Then in the limit of $q \to 1$ with $t = q^\lambda$, the weight function tends to

$$w(z; q, t) = \prod_{i \neq j} \frac{\left(z_i z_j^{-1}; q\right)_\infty}{\left(t z_i z_j^{-1}; q\right)_\infty} \to \prod_{i \neq j}(1 - z_i z_j^{-1})^\lambda. \tag{10.4}$$

With this preparation we define the inner product for two symmetric functions $f(z_1, \ldots, z_N)$ and $g(z_1, \ldots, z_N)$ as

$$\langle f, g \rangle_{N,q,t}^0 = \prod_{i=1}^{N} \oint_{|z_i|=1} \frac{dz_i}{2\pi i z_i} w(z; q, t) f^*(z) g(z). \tag{10.5}$$

The Macdonald symmetric polynomial $P_\mu(z_1, \ldots, z_N; q, t)$ specified by a partition $\mu \in \Lambda_N^+$ is defined as the homogeneous symmetric polynomial that is orthogonal with respect to the inner product (10.5):

$$\langle P_\mu, P_\nu \rangle_{N,q,t}^0 \propto \delta_{\mu\nu}. \tag{10.6}$$

Furthermore, it satisfies the triangularity

$$P_\mu(z_1, \ldots, z_N; q, t) = m_\mu + \sum_{\nu \in \Lambda_N^+, \text{s.t.} \nu < \mu} v_{\mu\nu}(q, t) m_\nu, \tag{10.7}$$

where m_μ is a monomial symmetric polynomial, and the ordering "$<$" has been defined in Section 2.1.2. The triangularity (10.7) can be rewritten in another way in terms of the Schur functions s_μ defined by (2.173). Namely, s_μ is given by the limit $\lambda \to 1$ of symmetric Jack polynomials J_μ which have triangularity property (2.177). Hence, Schur functions also have the triangularity property:

$$s_\mu = m_\mu + \sum_{\nu(<\mu)} v_{\mu\nu} m_\nu, \tag{10.8}$$

from which the triangularity of m_μ

$$m_\mu = s_\mu + \sum_{\nu(<\mu)} \tilde{v}_{\mu\nu} s_\nu \tag{10.9}$$

follows. Owing to (10.9), the triangularity (10.7) is equivalent to

$$P_\mu(z_1, \ldots, z_N; q, t) = s_\mu + \sum_{\nu(<\mu)} \tilde{v}_{\mu\nu}(q, t) s_\nu, \tag{10.10}$$

with the coefficients $\tilde{v}_{\mu\mu}(q, t)$ depending on q and t. The Jack polynomials J_μ are obtained as a degenerate limit of P_μ:

$$\lim_{q \to 1} P_\mu(z_1, \ldots, z_N; q, t = q^\lambda) = J_\mu(z_1, \ldots, z_N). \tag{10.11}$$

Let us quote some basic properties of P_μ without proof. According to [126], the integral norm is given by

$$\langle P_\mu, P_\mu \rangle^0_{N,q,t} = c_N(q, t) \prod_{s \in D(\mu)} \frac{1 - q^{a'(s)} t^{N-l'(s)}}{1 - q^{a'(s)+1} t^{N-l'(s)-1}} \frac{1 - q^{a(s)+1} t^{l(s)}}{1 - q^{a(s)} t^{l(s)+1}}, \tag{10.12}$$

with

$$c_N(q, t) = \langle 1, 1 \rangle^0_{N,q,t} = N! \prod_{1 \le i < j \le N} \frac{(t^{j-i}; q)_\infty (qt^{j-i}; q)_\infty}{(t^{j-i+1}; q)_\infty (qt^{j-i-1}; q)_\infty}. \tag{10.13}$$

The power-sum decomposition formula is given by [117]

$$p_n = \sum_{i=1}^N z_i^n = (1 - q^n) \sum_{\mu, \text{s.t.} |\mu| = n} \frac{\prod_{s \in D(\mu) \setminus (1,1)} t^{l'(s)} - q^{a'(s)}}{\prod_{s \in D(\mu)} 1 - t^{l(s)} q^{a(s)+1}} P_\mu, \tag{10.14}$$

for non-negative integers m. We can confirm that the formulae (10.12) and (10.14), respectively, reduce to (2.191) and (7.310) in the limit $t = q^\lambda, q \to 1$.

The dynamical density correlation function of the single-component Sutherland model can be derived using the formula of the integral norm (2.191) and the power-sum decomposition formula (7.310) of the symmetric Jack polynomials. Uglov has revealed the remarkable fact that eigenfunctions of the U(K) Sutherland model can be mapped onto the Macdonald polynomials with a particular choice of (q, t). Namely, wave functions with both space and spin degrees of freedom are mapped onto functions with only spatial degrees of freedom. Then exact results on the dynamics of the U(K) Sutherland model can be obtained in a way similar to that used in Section 2.7. We note in passing that the Ruijsenaars–Schneider model [152, 153] mentioned in the Introduction has the symmetric Macdonald polynomials as eigenfunctions. Correspondingly, the dynamical density correlation function of the Ruijsenaars–Schneider model has been derived [117] using (10.12) and (10.14).

In the next section, we introduce Uglov polynomials as symmetric polynomials which correspond to the Yangian Gelfand–Zetlin basis of the U(2) Sutherland model.

10.2 Uglov polynomials

Uglov introduced [188] a new kind of symmetric polynomial under the name "gl_N Jack polynomials", which can be obtained as a special case of Macdonald polynomials. Here we call these polynomials "Uglov polynomials" in order to distinguish them from the U(K) Jack polynomials. Let γ be a positive real number and consider the limit

$$q = -p, \quad t = -p^\gamma, \quad p \to 1 \tag{10.15}$$

in (10.5) and (10.7). In this limit, the symmetric Macdonald polynomial reduces to the Uglov polynomial $P^{(\gamma)}$. We set $\gamma = 1 + 2\lambda$ in the following. Then the weight function (10.1) tends to

$$w(z; -p, -p^{2\lambda+1}) \to \prod_{i \neq j}(1 - z_i^2/z_j^2)^\lambda. \tag{10.16}$$

The integral norm for polynomial functions f, g reduces in the Uglov limit to

$$\langle f, g \rangle^0_{N,q,t} \to \{f, g\}_{N,\lambda}. \tag{10.17}$$

Here the inner product on the RHS is given by

$$\{f, g\}_{N,\lambda} = \text{C.T.} \left[\tilde{\Delta} f^*(z) g(z) \right] = \prod_{i=1}^N \oint_{|z_i|=1} \frac{\mathrm{d}z_i}{2\pi \mathrm{i} z_i} \tilde{\Delta} f^*(z) g(z), \tag{10.18}$$

where the symbol C.T. denotes the constant term, and

$$\tilde{\Delta}(z_1, \dots, z_N) = \left(\Delta(z_1^2, \dots, z_N^2) \right)^\lambda \Delta(z_1, \dots, z_N)$$

$$= \prod_{1 \leq i \neq j \leq N} \left(1 - z_i^2/z_j^2\right)^\lambda \left(1 - z_i/z_j\right). \tag{10.19}$$

The U(K) generalization of the Uglov limit is given by [188]

$$q = \omega^K p, \quad t = \omega^K p^{K\lambda+1}, \quad p \to 1, \tag{10.20}$$

where $\omega = \exp(2\pi i/K)$ is the Kth root of unity. The weight function in this case tends to

$$w(z; q, t) \to \prod_{i \neq j} (1 - z_i^K / z_j^K)^\lambda. \tag{10.21}$$

For describing the integral norm of the Uglov polynomials with respect to the inner product (10.18), we introduce the following quantities:

$$c(s) = a'(s) - l'(s), \quad h(s) = a(s) + l(s) + 1, \tag{10.22}$$

which are called the content and the hook length, respectively. Using these, the integral norm of the Uglov polynomial is given by

$$\left\{ P_\mu^{(\gamma)}, P_\mu^{(\gamma)} \right\}_{N,\lambda} = c_N^{(\gamma)} \prod_{s \in D(\mu), \text{s.t.} c(s) = N \bmod 2} \frac{a'(s) + \gamma(N - l'(s))}{a'(s) + 1 + \gamma(N - l'(s) - 1)}$$

$$\times \prod_{s \in D(\mu), \text{s.t.} h(s): \text{even}} \frac{a(s) + \gamma l(s) + 1}{a(s) + \gamma l(s) + \gamma}, \tag{10.23}$$

where

$$c_N^{(\gamma)} = \{1, 1\}_{N,\lambda} = N! \prod_{1 \leq i < j \leq N} c^{(\gamma)}(j - i) \tag{10.24}$$

and

$$c^{(\gamma)}(k) = \begin{cases} \dfrac{\Gamma\left(\gamma(k-1)/2 + 1\right) \Gamma\left(\gamma(k+1)/2\right)}{\Gamma^2\left((\gamma k + 1)/2\right)}, & (k : \text{odd}), \\[4mm] \dfrac{\Gamma\left(\gamma(k-1)/2 + 1/2\right) \Gamma\left(\gamma(k+1)/2 + 1/2\right)}{\Gamma\left(\gamma k/2 + 1\right) \Gamma\left(\gamma k/2\right)}, & (k : \text{even}). \end{cases} \tag{10.25}$$

We refer to the original literature [126, 188] for the proof of the basic properties of the Macdonald and Uglov polynomials.

10.3 Reduction to single-component bosons

The nice original idea of Uglov is to make a one-to-one-correspondence between one-particle states of particles with spin and those of spinless particles. In the case of SU(2), this mapping is realized by doubling the spatial momentum and adding a fictitious momentum $\alpha = 1$ (for ↑) or 2 (for ↓) to distinguish the spin. In this way the odd momentum is assigned exclusively for spin-up particles, and even one for spin-down particles. The one-particle basis for the resultant spinless particles is specified by a momentum $\mu(\in \mathbf{Z})$. The relation

$$\mu_i = 2\kappa_i - \alpha_i + 2 \tag{10.26}$$

gives a one-to-one correspondence between the indices of the two bases (κ, α) and μ. Although Uglov chose the combination $\mu \to -2\kappa + \alpha$ in the original treatment, we follow the choice (10.26) that keeps the sign of momentum.

Figure 10.1 illustrates for $N = 2$ the one-to-one correspondence described in (10.26). In the left column, the states of fermions with spin one-half are shown. The corresponding states of spinless fermions are shown in the middle column. In the right column, the corresponding states in terms of spinless bosons are shown, as will be explained soon. Between the states of spinless fermions and spinless bosons, there is a one-to-one correspondence via the relation

$$\mu_i = \nu_i + N - i, \tag{10.27}$$

where ν_i satisfying $\nu_1 \geq \cdots \geq \nu_N$ is the index of N-particle states of spinless bosons. Then ν is related to κ via

$$\nu_i = 2\kappa_i - \alpha_i + 2 - N + i, \tag{10.28}$$

which results from (10.26) and (10.27). The right column of Fig. 10.1 shows the diagrams describing the bosonic states.

When we take $z^\kappa v_\alpha(\sigma)$ as the one-particle basis of fermions with spin one-half, we obtain the N-particle basis

$$
\begin{aligned}
u_{\kappa,\alpha}(\{Zi\}, \{\sigma_i\}) &= \mathrm{Asym}\left[z_1^{\kappa_1} \cdots z_N^{\kappa_N} v_{\alpha_1}(\sigma_1) \cdots v_{\alpha_N}(\sigma_N) \right] \\
&= \sum_{p \in S_N} \mathrm{sgn}(p) z_{p(1)}^{\kappa_1} \cdots z_{p(N)}^{\kappa_N} v_{\alpha_1}(\sigma_{p(1)}) \cdots v_{\alpha_N}(\sigma_{p(N)}),
\end{aligned}
\tag{10.29}
$$

with $\alpha = (\alpha_1, \ldots, \alpha_N) \in [1, 2]^N$. On the other hand, we can take the Schur functions $s_\nu(z_1, \ldots, z_N)$ as the basis of N-particle symmetric polynomials. The latter basis consists of the set $\mathcal{S}_N$ of symmetric functions and corresponds to the bosonic picture of the particles. Now we introduce an isomorphism $\Omega : \mathcal{F}_{N,\kappa} \to \mathcal{S}_N$ corresponding to the spin–momentum unification together with bosonization by

$$\Omega(u_{\kappa,\alpha}) = s_\nu, \tag{10.30}$$

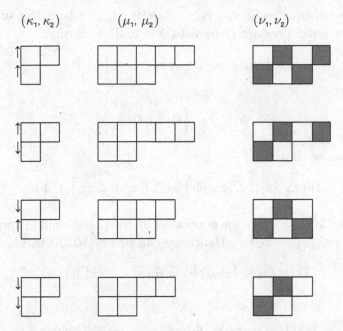

Fig. 10.1. Correspondence between the original momentum of each spin (left) and that in the effective single-component systems (center and right). The middle column is for effective fermions, and the right column for effective bosons with the checkerboard pattern to be discussed in Section 10.5. Note that the number of unshaded squares is three for all four cases in the right column. The number (= 3) gives the momentum of the two-particle system.

where (κ, α) and ν are related via (10.26) and (10.27). An important property of this isomorphism Ω is conservation of the inner product:

$$\Phi, \tilde{\Phi} \in \mathcal{F}_{N,\kappa} \rightarrow \langle \Phi, \tilde{\Phi} \rangle = \left\{ \Omega(\Phi), \Omega(\tilde{\Phi}) \right\}_{N,\lambda}, \qquad (10.31)$$

which is called isometry.

The isometry (10.31) can be proved by using the case of $\lambda = 0$ as reference. First we note that Schur functions s_ν can be written as $a_\mu / a_{\delta(N)}$, where

$$\mu = \nu + \delta(N) = (\nu_1 + N - 1, \nu_2 + N - 2, \ldots, \nu_{N-1} + 1, \nu_N) \qquad (10.32)$$

and a_μ is the Slater determinant,

$$a_\mu = \text{Asym} z_1^{\mu_1} \cdots z_N^{\mu_N} = \sum_{p \in S_N} \text{sgn}(p) z_{p(1)}^{\mu_1} \cdots z_{p(N)}^{\mu_N}. \qquad (10.33)$$

It then follows that

$$\{s_{\tilde{\nu}}, s_\nu\}_{N,0} = \prod_{i=1}^{N} \oint_{|z_i|=1} \frac{\mathrm{d}z_i}{2\pi i z_i} a_{\tilde{\nu}+\delta(N)}^* a_{\nu+\delta(N)} = N! \delta_{\tilde{\nu},\nu}.$$

Similarly, we obtain $\langle u_{\tilde{\kappa},\tilde{\alpha}}, u_{\kappa,\alpha}\rangle_{\lambda=0} = N!\delta_{\tilde{\kappa},\kappa}\delta_{\tilde{\alpha},\alpha}$, where the suffix $\lambda = 0$ indicates the inner product (8.6) with $\lambda = 0$. The relation

$$\langle u_{\tilde{\kappa},\tilde{\alpha}}, u_{\kappa,\alpha}\rangle_{\lambda=0} = \{s_{\tilde{\nu}}, s_{\nu}\}_{N,0} \tag{10.34}$$

leads to

$$\langle \tilde{\Phi}, \Phi\rangle_{\lambda=0} = \Big\{\Omega(\tilde{\Phi}), \Omega(\Phi)\Big\}_{N,0}. \tag{10.35}$$

Next we show that

$$\Omega\Big(\Delta^{\lambda}(z_1,\ldots,z_N)\Phi\Big) = \Delta^{\lambda}(z_1^2,\ldots,z_N^2)\Omega\,(\Phi)\,. \tag{10.36}$$

The function $\Delta^{\lambda}(z_1,\ldots,z_N)$ is generated by the power-sum symmetric polynomials p_r and $(z_1\cdots z_N)^{-1}$. Hence we can prove (10.36) by showing

$$\Omega\,(p_r(z_1,\ldots,z_N)\Phi) = p_r(z_1^2,\ldots,z_N^2)\Omega\,(\Phi) \tag{10.37}$$

and

$$\Omega\,\big((z_1\cdots z_N)^{-1}\Phi\big) = (z_1\cdots z_N)^{-2}\Omega\,(\Phi)\,. \tag{10.38}$$

The relations (10.37) and (10.38) are derived as

$$\Omega\,(p_r(z_1,\ldots,z_N)u_{\kappa,\alpha}) = \Omega\left(\sum_{i=1}^{N} u_{\kappa_1,\ldots,\kappa_i+r,\ldots,\kappa_N,\alpha}\right) = \sum_{i=1}^{N} s_{\nu_1,\ldots,\nu_i+2r,\ldots,\nu_N}$$
$$= p_r\left(z_1^2,\ldots,z_N^2\right)s_{\nu} \tag{10.39}$$

and

$$\Omega\,\big((z_1\cdots z_N)^{-1}u_{\kappa,\alpha}\big) = \Omega\,(u_{\kappa_1-1,\ldots,\kappa_N-1,\alpha}) = s_{\nu_1-2,\ldots,\nu_N-2}$$
$$= (z_1\cdots z_N)^{-2}s_{\nu}, \tag{10.40}$$

respectively.

Now we are ready to prove (10.31). We note that

$$\langle \tilde{\Phi}, \Phi\rangle = \langle \tilde{\Phi}, \Delta^{\lambda}(z_1,\ldots,z_N)\Phi\rangle_{\lambda=0}$$
$$= \Big\{\Omega\left(\tilde{\Phi}\right), \Omega\left(\Delta^{\lambda}(z_1,\ldots,z_N)\Phi\right)\Big\}_{N,0}. \tag{10.41}$$

With use of (10.36), the rightmost side of (10.41) is calculated as

$$\Big\{\Omega\left(\tilde{\Phi}\right), \Delta^{\lambda}(z_1^2,\ldots,z_N^2)\Omega\,(\Phi)\Big\}_{N,0} = \Big\{\Omega\left(\tilde{\Phi}\right), \Omega\,(\Phi)\Big\}_{N,\lambda}. \tag{10.42}$$

Thus we arrive at (10.31).

10.4 From Yangian Gelfand–Zetlin basis to Uglov polynomials

The results in the previous section show that an orthogonal basis of $\mathcal{F}_N$ is mapped onto another orthogonal basis of $\mathcal{S}_N$ via Ω. In Chapter 8, we have found that an orthogonal set of $\mathcal{F}_N$ is given by the Yangian Gelfand–Zetlin basis $\Phi_{\kappa,\alpha}$. Hence, $\{\Omega(\Phi_{\kappa,\alpha})\}$ should yield an orthogonal basis of $\mathcal{S}_N$. In the following, we will show the property

$$\Omega(\Phi_{\kappa,\alpha}) = P_\nu^{(2\lambda+1)}, \text{ or equivalently } \Omega^{-1}(P_\nu^{(2\lambda+1)}) = \Phi_{\kappa,\alpha}. \qquad (10.43)$$

Let us first show the second equality. From (10.10) and (10.30), it follows that

$$\Omega^{-1}(P_\nu^{(2\lambda+1)}) = u_{\kappa,\alpha} + \sum_{\tilde{\nu}<\nu} v_{\nu\tilde{\nu}} u_{\tilde{\kappa},\tilde{\alpha}}. \qquad (10.44)$$

The inequality $\tilde{\nu} < \nu$ means either $\tilde{\kappa} = \kappa$ together with $\tilde{\alpha} < \alpha$, or $\tilde{\kappa} < \kappa$. Here the order of $\alpha, \tilde{\alpha} \in W_\kappa$ satisfying $\sum_i \alpha_i = \sum_i \tilde{\alpha}_i$ has been defined in (8.162).

The orthogonal set of $\mathcal{F}_N$ having the form of the RHS of (10.44) is unique. In the expression

$$\Phi_{\kappa,\alpha} = U(\kappa,\lambda)\chi_\alpha = \sum_{\eta,\ \text{s.t.}\ \eta^+=\kappa} E_\eta R_\eta \chi_\alpha, \qquad (10.45)$$

the monomials $z^{\tilde{\kappa}}$ contained in the summand on the RHS satisfy $\tilde{\kappa} < \kappa$ or $\tilde{\kappa} = \kappa$ owing to the triangularity of non-symmetric Jack polynomials. The monomial z^κ comes only from E_κ. From $R_\kappa^{(\lambda)} = 1$ and the triangularity

$$\chi_\alpha = \varphi_\alpha + \sum_{\tilde{\alpha}<\alpha} c_{\alpha\tilde{\alpha}} \varphi_{\tilde{\alpha}}, \qquad (10.46)$$

we see that (10.45) has the form of the RHS of (10.44). The triangularity (10.46) of χ_α has already been illustrated in (8.95) for two-particle and in (8.131)–(8.134) for three-particle systems. Thus we arrive at (10.43).

As an example, we compare the norm of the Yangian Gelfand–Zetlin basis and Uglov polynomials for two-particle cases. We introduce the notation

$$R_2(\alpha,\beta) \equiv \frac{\left\{ P_{\alpha,\beta}^{(2\lambda+1)} P_{\alpha,\beta}^{(2\lambda+1)} \right\}_{N=2,\lambda}}{\left\{ P_{2\kappa_1-2\kappa_2-1,0}^{(2\lambda+1)} P_{2\kappa_1-2\kappa_2-1,0}^{(2\lambda+1)} \right\}_{N=2,\lambda}}, \qquad (10.47)$$

as the two-particle norm relative to the highest-weight state. With use of (10.23), (10.31), and (10.43), we obtain the following results:

$$R_2(2\kappa_1 - 2\kappa_2, 0) = \frac{\kappa_1 - \kappa_2}{\kappa_1 - \kappa_2 + \lambda}, \tag{10.48}$$

$$R_2(2\kappa_1 - 2\kappa_2 - 1, 1) = \frac{\kappa_1 - \kappa_2 + \lambda}{\kappa_1 - \kappa_2 + 2\lambda}, \tag{10.49}$$

$$R_2(2\kappa_1 - 2\kappa_2, 1) = 1, \tag{10.50}$$

which agree with (8.184)–(8.186) derived by a different method. Thus the correspondence between the Yangian Gelfand–Zetlin basis and the Uglov polynomials is demonstrated via the isometric isomorphim Ω.

10.5 Dynamical correlation functions

The dynamical density–density correlation function and spin–density correlation function are calculated using Uglov polynomials. We assume that N is even and $N/2$ is odd so that the ground state is a non-degenerate singlet state. The wave function Φ_g of the ground state has corresponding bosonic (ν) and fermionic (μ) partitions:

$$\nu = (0, 0, \ldots, 0), \tag{10.51}$$

$$\mu = \nu + \alpha = (N - 1, N - 2, \ldots, 1, 0). \tag{10.52}$$

Then we have κ, which is related to the spatial momentum as

$$\kappa = \frac{1}{2}(\mu + \alpha - 2) = \left(\frac{N}{2} - 1, \frac{N}{2} - 1, \frac{N}{2} - 2, \frac{N}{2} - 2, \ldots, 0, 0\right), \tag{10.53}$$

with $\alpha = (1, 2, 1, 2, \ldots, 1, 2)$. We make a Galilean shift for κ to define

$$k_i \equiv \kappa_i - N/4 + 1/2, \tag{10.54}$$

which has a symmetric distribution ranging from $k_1 = N/4 - 1/2$ to $k_N = -N/4 + 1/2$.

The dynamical density correlation function is written as

$$\frac{\langle g, N | \hat{\rho}(x, t)\hat{\rho}(0, 0) | g, N \rangle}{\langle g, N | g, N \rangle} = \sum_{(\kappa, \alpha)} \frac{|\langle \Phi_{\kappa, \alpha}, \hat{\rho}(0, 0)\Phi_g \rangle|^2}{\langle \Phi_g, \Phi_g \rangle \langle \Phi_{\kappa, \alpha}, \Phi_{\kappa, \alpha} \rangle} e^{-i\omega_\kappa t + P_\kappa x}. \tag{10.55}$$

Here the norm $\langle \cdot, \cdot \rangle$ has been introduced into (8.8). The eigenstates $\Phi_{\kappa, \alpha}$ have been defined in (8.169) with (8.168). The energy ω_κ is given by

$$\omega_\kappa = \left(\frac{2\pi}{L}\right)^2 \sum_{i=1}^{N} \left[k_i + \left(\frac{N+1}{2} - i\right)\lambda\right]^2 - E_{g,N}, \tag{10.56}$$

where k_i was defined by (10.54).

Using (2.307), we obtain the density operator

$$\hat{\rho}(0,0) = \frac{1}{L} \sum_{n>0} [p_n(z_1, \ldots, z_N) + p_{-n}(z_1, \ldots, z_N)], \tag{10.57}$$

and (10.55) is rewritten as

$$\frac{2}{L^2} \sum_{(\kappa,\alpha)} \sum_{n>0} \frac{|\langle \Phi_{\kappa,\alpha}, p_{-n}(z_1, \ldots, z_N) \Phi_g \rangle|^2}{\langle \Phi_g, \Phi_g \rangle \langle \Phi_{\kappa,\alpha}, \Phi_{\kappa,\alpha} \rangle} e^{-i\omega_\kappa t} \cos\left(2\pi n x / L\right). \tag{10.58}$$

Using the isometry (10.31) of the mapping Ω we obtain

$$\langle \Phi_g, \Phi_g \rangle = \{\Omega(\Phi_g), \Omega(\Phi_g)\}_{N,\lambda} = \{1, 1\}_{N,\lambda}, \tag{10.59}$$

where the last equality follows from

$$\Omega(\Phi_g) = P_{0^N}^{(2\lambda+1)} = 1,$$

with the suffix 0^N being the partition $(0, 0, \ldots, 0)$. Similarly, the numerator on the RHS of (10.58) is expressed as

$$\langle \Phi_{\kappa,\alpha}, p_n(z_1, \ldots, z_N) \Phi_g \rangle = \{\Omega(\Phi_{\kappa,\alpha}), \Omega(p_n \Phi_g)\}_{N,\lambda}$$

$$= \left\{P_\nu^{(2\lambda+1)}, p_{2n}\right\}_{N,\lambda}, \tag{10.60}$$

where (10.37) has been used in the second equality.

The power-sum symmetric polynomials are decomposed into a linear combination of Uglov polynomials as

$$p_n = \sum_\nu c_{n,\nu} P_\nu^{(2\lambda+1)}, \tag{10.61}$$

where the expansion coefficients $c_{n,\nu}$ can be obtained by taking the limit (10.15) of the power-sum decomposition formula (10.14) of the symmetric Macdonald polynomials. Consequently, we obtain

$$\frac{\langle g, N | \hat{\rho}(x,t) \hat{\rho}(0,0) | g, N \rangle}{\langle g, N | g, N \rangle} \equiv \langle \hat{\rho}(x,t) \hat{\rho}(0,0) \rangle$$

$$= \frac{2}{L^2} \sum_{n>0} \sum_{|\nu|=2n} c_{2n,\nu}^2 \frac{\left\{P_\nu^{(2\lambda+1)}, P_\nu^{(2\lambda+1)}\right\}_{N,\lambda}}{\{1,1\}_{N,\lambda}} e^{-i\omega_\kappa t} \cos\left(2\pi n x / L\right). \tag{10.62}$$

In (10.62), the factors coming from the norm on the RHS are available from (10.23). The energy ω_κ is given by (10.56), which is determined by ν in terms of (10.28).

Now we turn to the dynamical spin–density correlation function given by

$$\frac{\langle g, N | \hat{S}_z(x,t)\hat{S}_z(0,0) | g, N \rangle}{\langle g, N | g, N \rangle} \equiv \langle \hat{S}_z(x,t)\hat{S}_z(0,0) \rangle. \tag{10.63}$$

Since the system has the global SU(2) symmetry, and the ground state is spin-singlet, x and y components of the correlation functions are the same as (10.63). Hence by using $\hat{S}^- = \hat{S}_x - i\hat{S}_y$, we obtain

$$\langle \hat{S}_z(x,t)\hat{S}_z(0,0) \rangle = \frac{1}{2} \sum_{(\kappa,\alpha)} \frac{\left| \langle \Phi_{\kappa,\alpha}, \hat{S}^-(0,0)\Phi_g \rangle \right|^2}{\langle \Phi_g, \Phi_g \rangle \langle \Phi_{\kappa,\alpha}, \Phi_{\kappa,\alpha} \rangle} \exp(-i\omega_\kappa t + iP_\kappa x),$$

$$\tag{10.64}$$

where $\Phi_{\kappa,\alpha}$ has the total spin

$$S_z^{\text{tot}} = -1. \tag{10.65}$$

For practical calculation, it is convenient to rewrite all quantities and the condition (10.65) in terms of Young diagrams $D(\nu)$ for ν. For this purpose, we paint each cell in $D(\nu)$ black if the content $c(s) = a'(s) - l'(s)$ of the cell s is odd. Otherwise we paint the cell white. Then B_ν and W_ν denote the subset of $D(\nu)$ consisting of black and white cells, respectively. Figure 10.1 shows some examples. In terms of B_ν and W_ν, the momentum P_κ and the excitation energy ω_κ associated with $\Phi_{\kappa,\alpha}$ are written with $\gamma = 1 + 2\lambda$ as [188, 197]

$$P_\kappa = 2\pi |W_\nu| / L, \tag{10.66}$$

$$\omega_\kappa = \left(\frac{2\pi}{L} \right)^2 \left\{ n_w(\nu') - \gamma n_w(\nu) + \frac{|W_\nu|}{2}[(N-1)\gamma + 1] \right\}, \tag{10.67}$$

where $|\cdots|$ is the number of cells belonging to each subset, and we have used the notation

$$n_w(\nu) = \sum_{s \in W_\nu} l'(s), \quad n_w(\nu') = \sum_{s \in W_\nu} a'(s). \tag{10.68}$$

The condition (10.65) is rewritten as

$$|W_\nu| - |B_\nu| = 1. \tag{10.69}$$

Since all factors in the matrix element and norm are expressed in terms of the arm, leg lengths and colengths, the thermodynamic limit can be taken in a way similar to that used in Sections 3.8.2 and 2.7.3. The resultant expression obtained from (10.62) is given by (3.202) [197].

We make the Fourier decomposition

$$\hat{S}^{\pm}(0,0) = \sum_{s\in\mathbf{Z}} \hat{J}_s^{\pm}, \quad \hat{J}_s^{\pm} = \sum_{i=1}^{N} z_i^s \hat{S}_i^{\pm}, \tag{10.70}$$

with $\mathbf{Z}$ being integers. The matrix element in (10.64) is then given by

$$\langle \Phi_{\kappa,\alpha}, \hat{S}^-(0,0)\Phi_{\mathrm{g}} \rangle = \sum_{s\in\mathbf{Z}} \langle \Phi_{\kappa,\alpha}, \hat{J}_s\Phi_{\mathrm{g}} \rangle, \tag{10.71}$$

with $\hat{J}_s = \hat{J}_s^+ + \hat{J}_s^-$. The action of $\hat{J}_s^-$ on $u_{\kappa,\alpha}$ is given by

$$\hat{J}_s^- u_{\kappa,\alpha} = \sum_{i=1}^{N} \delta_{1,\alpha_i} u[\kappa_1, \ldots, \kappa_i + s, \ldots, \kappa_N, \alpha_1, \ldots, \alpha_i + 1, \ldots, \alpha_N], \tag{10.72}$$

where we have used the notation $u[\kappa, \alpha] \equiv u_{\kappa,\alpha}$ on the RHS to magnify the set $[\kappa, \alpha]$. Using (10.72) and the corresponding one with $\hat{J}_s^+$, we obtain

$$\Omega\left(\hat{J}_s^+ u_{\kappa,\alpha}\right) = \sum_{i=1}^{N} s[\nu_1, \ldots, \nu_{i+2s-1}, \ldots, \nu_N] = p_{2s-1}s_\nu, \tag{10.73}$$

with the notation $s[\nu] \equiv s_\nu$ for Schur functions. Thus, (10.71) becomes

$$\left\{ P_\nu^{(2\lambda+1)}, \sum_s p_{2s-1} \right\}'_{N,\lambda} = \sum_s c_{2s-1,\nu} \left\{ P_\nu^{(2\lambda+1)}, P_\nu^{(2\lambda+1)} \right\}'_{N,\lambda}, \tag{10.74}$$

provided the condition (10.65) holds. We obtain

$$\langle \hat{S}_z(x,t)\hat{S}_z(0,0) \rangle$$

$$= \frac{2}{L^2} \sum_{n>0} \sum_\nu{}' c_{2n-1,\nu}^2 \frac{\left\{ P_\nu^{(2\lambda+1)}, P_\nu^{(2\lambda+1)} \right\}_{N,\lambda}}{\{1,1\}_{N,\lambda}} e^{-i\omega_\kappa t} \cos\left(P_\kappa x\right), \tag{10.75}$$

where $\sum_\nu'$ means the restriction to states with (10.65).

In (10.75), ω_κ and P_κ are determined by ν since (κ, α) and ν are related to each other through (10.28). The black-and-white painted diagram represents these relations in the form (10.66) and (10.67). The spin–density correlation function is finally obtained in terms of the partition ν as

$$\langle \hat{S}_z(x,t) \hat{S}_z(0,0) \rangle$$

$$= \frac{2}{L^2} \sum_{n>0} \sum_{\nu}{}' c_{2n-1,\nu}^2 \frac{\left\{ P_\nu^{(2\lambda+1)}, P_\nu^{(2\lambda+1)} \right\}_{N,\lambda}}{\{1,1\}_{N,\lambda}} \bar{\mathrm{e}}^{i\omega_\nu t} \cos\left(2\pi|W_\nu|x/L\right), \quad (10.76)$$

where $\sum_{\nu}'$ is restricted to states with (10.69). The thermodynamic limit of (10.76) was given by (3.205).

Afterword

We have discussed the physical and mathematical aspects of one-dimensional quantum particles with $1/r^2$ interactions. On the physics side, we began with the simplest two-body system, and culminated in the exact dynamics of a supersymmetric electronic model. In this way we have tried to draw a comprehensive picture of one-dimensional quantum systems, taking the canonical systems with $1/r^2$ interactions. On the mathematical side, the most important subjects are first the Jack polynomials, and second the Yangians. These two subjects were originally developed independently, but are actually closely interrelated. Our basic viewpoint is that the $1/r^2$ systems provide the most natural working model to synthesize these two subjects, and to sharpen the concepts further.

There are many topics, however, which we could not discuss in this book. This is firstly because of our insufficient understanding of the topics, and secondly because of our intention of making a reasonably sized book. We shall mention some of the omitted topics, hoping that they will be covered on another occasion.

Examples of other canonical systems, which have escaped our discussion in this book, include the Calogero–Sutherland model with harmonic confinement. The spectrum of the model is equidistant as in the simple harmonic oscillator. Because of this feature the model can be compared with the chiral conformal field theory (CFT) with right-going particles only. We refer to a recent review [88] for the systems. In relation to the CFT, there are active studies to investigate the quasi-particle structure for the general case of the internal symmetry. In particular, special kinds of pseudo-particles have been proposed to account for the partition function [24, 32, 33]. The CFT in an advanced version can describe fractional exclusion statistics as well. Namely, the spinon basis naturally arises by taking the partition function of the so-called Wess–Zumino–Witten model and its variants.

Another topic that deserves more detailed discussion is the case with general $SU(K_B, K_F)$ supersymmetry, where K_B denotes the number of internal degrees of bosons, and K_F of fermions. Especially interesting is the applicability of the idea of fractional exclusion statistics to these systems. We have pointed out in Section 5.3 that the statistical parameters may be different in thermodynamics and dynamics, but more detailed analysis is desirable. On the other hand, the exact thermodynamics for these models show that parafermions are the most natural particles to interpret the entropy. The connection between these partial informations should be made.

The most challenging remaining problem is to reach the dynamical results in the thermodynamic limit without doing complicated calculations for finite-sized systems. We have seen that the dynamical results for finite systems have a rather complicated form which, however, simplifies dramatically upon taking the thermodynamic limit. Thus one naturally expects that there must be an easier way to derive the results, which works only in the thermodynamic limit. It is known that the second quantization approach could meet this expectation in very restricted cases. Although interesting suggestions have been made in the framework of the so-called "collective field theory" [14] or "momentum-space bosonization" [51], the results obtained by these theories need more elaboration.

Although we have discussed only one-dimensional systems in this book, there appears a close connection to a special two-dimensional system, namely the fractional quantum Hall system. The Laughlin wave function, which is of Jastrow type, allows its interpretation in terms of composite fermions [26,92, 123]. The wave functions are very similar to those in the Sutherland model, or the Calogero–Sutherland model with harmonic confinement. This similarity leads to natural identification of statistics obeyed by quasi-electrons and quasi-holes. The excitation energy in the bulk has a finite gap. On the other hand, the edge excitation in a droplet-shaped system is gapless. Thus the dynamics should interpolate between the one-dimensional and two-dimensional limits as the location of excitation is changed from edge to bulk. The dynamical response of the system, which has mainly been obtained numerically, deserves more detailed analysis. The physical interpretation goes fairly parallel to that in the Sutherland model. Recently, it has been pointed out that the Jack polynomials may have a negative number for λ provided this number is rational [56]. Then the fractional quantum Hall systems with wave functions that are more complicated than the Laughlin-type one can be described [27].

Finally we mention a rather different approach to dynamics [97, 171]. One derives matrix elements of the local operators, called the form factor, exploiting as far as possible general properties such as the Yang–Baxter relations, unitarity, crossing relations, and assuming the maximal possible analyticity. Namely, the scattering matrix between elementary excitations is used as input data. In this approach not only the $1/r^2$ models but also the nearest-neighbor interaction models can be dealt with. An interesting aspect is the role of the scattering phase shift in the physical quantities. The role in dynamics is to govern the singularity of the spectral weight of the dynamical correlation function. The relation between the scattering phase shift and the singularity of the spectral weight is clear in the $1/r^2$ t–J model and other $1/r^2$ models. For the Heisenberg model and short-range integrable t–J model, form factors have contributions from various numbers of excited quasi-particles. The explicit calculation of matrix elements has been pursued [93, 96].

Further analysis of these topics, on the basis of the fundamental knowledge provided by this book, should deepen our understanding of strongly correlated low-dimensional quantum systems.

References

[1] I. Affleck. *Phys. Rev. Lett.* **56** (1986) 746.

[2] C. Ahn and W. M. Koo. *Phys. Lett. B* **365** (1996) 105.

[3] P. W. Anderson. *Phys. Rev.* **86** (1952) 694.

[4] P. W. Anderson, B. S. Shastry, and D. Hristpulos. *Phys. Rev. B* **40** (1989) 8939.

[5] M. Arai, M. Fujita, M. Motokawa, J. Akimitsu, and S. M. Bennington. *Phys. Rev. Lett.* **77** (1996) 3649.

[6] M. Arikawa, T. Yamamoto, Y. Saiga, and Y. Kuramoto. *J. Phys. Soc. Jpn.* **68** (1999) 3782.

[7] M. Arikawa, Y. Saiga, T. Yamamoto, and Y. Kuramoto. *Physica B* **281&282** (2000) 823.

[8] M. Arikawa, Y. Saiga, and Y. Kuramoto. *Phys. Rev. Lett.* **86** (2001) 3096.

[9] M. Arikawa, T. Yamamoto, Y. Saiga, and Y. Kuramoto. Unpublished (2002).

[10] M. Arikawa, T. Yamamoto, Y. Saiga, and Y. Kuramoto. *J. Phys. Soc. Jpn.* **73** (2004) 808.

[11] M. Arikawa, T. Yamamoto, Y. Saiga, and Y. Kuramoto. *Nucl. Phys. B* **702** (2004) 380.

[12] M. Arikawa and Y. Saiga. *J. Phys. A: Math. Gen.* **39** (2006) 10603.

[13] M. Arikawa. Unpublished (2006).

[14] H. Awata, Y. Matsuo, and T. Yamamoto. *J. Phys. A: Math. Gen.* **29** (1996) 3089.

[15] T. H. Baker and P. J. Forrester. *Nucl. Phys. B* **492** (1997) 682.

[16] T. H. Baker and P. J. Forrester. *Duke Math. J.* **95** (1998) 1.

[17] T. H. Baker and P. J. Forrester. *Ann. Comb.* **3** (1999) 159.

[18] T. H. Baker, C. F. Dunkl, and P. J. Forrester. In *Calogero–Moser–Sutherland Models*, J. F. van Diejen and L. Vinet (eds). Springer-Verlag, Berlin, 2000.

[19] P.-A. Bares, G. Blatter, and M. Ogata. *Phys. Rev. B* **44** (1991) 130.

[20] A. A. Belavin, A. M. Polyakov, and A. B. Zamolodchikov. *Nucl. Phys. B* **241** (1984) 333.

[21] M. C. Bergere. *J. Math. Phys.* **39** (1998) 30.

[22] D. Bernard, M. Gaudin, F. D. M. Haldane, and V. Pasquier. *J. Phys. A: Math. Gen.* **26** (1993) 5219.

[23] D. Bernard and Y.-S. Wu. In *New Developments of Integrable Systems and Long-Ranged Interaction Models*, M.-L. Ge and Y.-S. Wu (eds). World Scientific, Singapore, 1995.

[24] A. Berkovich and B. M. McCoy. In *Statistical Physics on the Eve of the 21st Century*, M. T. Batchelor and L. T. Wille (eds), p. 240. World Scientific, Singapore, 1999.

[25] B. A. Bernevig, D. Giuliano, and R. B. Laughlin. *Phys. Rev. B* **64** (2001) 024425.

[26] B. A. Bernevig, D. Giuliano, and R. B. Laughlin. *Phys. Rev. B* **65** (2001) 195112.

[27] B. A. Bernevig and F. D. M. Haldane. *Phys. Rev. Lett.* **100** (2008) 246802.

[28] H. Bethe. *Z. Phys.* **71** (1931) 205.

[29] H. W. Blöte, J. L. Cardy, and M. P. Nightingale. *Phys. Rev. Lett.* **56** (1986) 742.

[30] A. H. Bougourzi, M. Couture, and M. Kacir. *Phys. Rev. B* **54** (1996) R12699.

[31] A. H. Bougourzi, M. Karbach, and G. Müller. *Phys. Rev. B* **57** (1998) 11429.

[32] P. Bouwknegt and K. Schoutens. *Nucl. Phys. B* **482** (1996) 345.

[33] P. Bouwknegt and K. Schoutens. *Nucl. Phys. B* **547** (1999) 501.

[34] F. Calogero. *J. Math. Phys.* **10** (1969) 2191, 2197.

[35] F. Calogero. *J. Math. Phys.* **12** (1971) 419.

[36] J. L. Cardy. *Nucl. Phys. B* **270** (1986) 186.

[37] I. V. Cherednik. *Duke Math. J.* **54** (1987) 563.

[38] I. V. Cherednik. *Inv. Math.* **106** (1991) 411.

[39] I. V. Cherednik. *Annals Math.* **141** (1995) 191.

[40] I. V. Cherednik. *IMRN* **10** (1995) 483.

[41] R. Claessen *et al.* *Phys. Rev. Lett.* **88** (2002) 096402.

[42] J. des Cloizeaux and J. J. Pearson. *Phys. Rev.* **128** (1962) 2131.

[43] V. G. Drinfeld. *Quantum Groups, Proceedings of the International Congress of Mathematicians*, Vols 1, 2 (Berkeley, CA, 1986). American Mathematical Society, Providence, RI, 1987, p. 798.

[44] V. G. Drinfeld. *Sov. Math. Dokl.* **36** (1988) 212.

[45] C. F. Dunkl. *Trans. AMS* **311** (1989) 167.

[46] C. F. Dunkl. *Commun. Math. Phys.* **197** (1998) 451.

[47] C. F. Dunkl. *Monatsch. Math.* **120** (1998) 181.

[48] F. J. Dyson. *J. Math. Phys.* **3** (1962) 166.

[49] F. J. Dyson. *Commun. Math. Phys.* **19** (1970) 235.

[50] K. Efetov. *Supersymmetry in Disorder and Chaos*. Cambridge University Press, Cambridge, 2005.

[51] C. Efthimiou and A. LeClair. *Commun. Math. Phys.* **171** (1995) 531.

[52] Y. Endoh, G. Shirane, R. J. Birgeneau, P. M. Richards, and S. L. Holt. *Phys. Rev. Lett.* **32** (1974) 170.

[53] F. H. L. Essler. *Phys. Rev. B* **51** (1995) 13357.

[54] F. H. L. Essler, H. Frahm, F. Göhmann, A. Klümper, and V. E. Korepin. *The One-Dimensional Hubbard Model*. Cambridge University Press, Cambridge, 2005.

[55] L. D. Faddeev and L. A. Takhtajan. *Phys. Lett. A* **85** (1981) 375.

[56] B. Feigin, M. Jimbo, T. Miwa, and E. Mukhin. *IMRN International Mathematics Research Notices*, No. 23 (2002) 1223.

[57] H. Frahm and V. E. Korepin. *Phys. Rev. B* **42** (1990) 10553.

[58] H. Frahm. *J. Phys. A* **26** (1993) L473.

[59] P. J. Forrester and B. Jancovici. *J. Phys. (Paris)* **45** (1984) L583.

[60] P. J. Forrester. *Phys. Lett. A* **196** (1995) 353.

[61] P. J. Forrester. *Mod. Phys. Lett. B* **9** (1995) 359.

[62] P. J. Forrester. *J. Math. Phys.* **36** (1995) 86.

[63] P. J. Forrester, D. S. McAnally, and Y. Nikoyalevsky. *J. Phys. A: Math. Gen.* **34** (2001) 8407.

[64] T. Fukui and N. Kawakami. *Phys. Rev. B* **51** (1995) 5239.

[65] D. M. Gangardt and A. Kamenev. *Nucl. Phys. B* **610** (2001) 578.

[66] F. Gebhard and D. Vollhardt. *Phys. Rev. B* **38** (1988) 6911.

[67] I. M. Gelfand and V. S. Retakh. *Funct. Anal. Appl.* **25** (1991) 91.

[68] G. Gentile. *Nuovo Cimento* **17** (1940) 493.

[69] H. S. Green. *Phys. Rev.* **90** (1953) 270.

[70] M. Greiter and D. Schuricht. *Phys. Rev. B* **71** (2005) 224424.

[71] Z. N. C. Ha and F. D. M. Haldane. *Phys. Rev. B* **46** (1992) 9359.

[72] Z. N. C. Ha and F. D. M. Haldane. *Phys. Rev. Lett.* **73** (1994) 2887; **74** (1995) 3501 (E).

[73] Z. N. C. Ha. *Phys. Rev. Lett.* **73** (1995) 1574; **74** (1995) 620 (E).

[74] Z. N. C. Ha. *Nucl. Phys. B* **435** (1995) 604.

[75] Z. N. C. Ha and F. D. M. Haldane. *Phys. Rev. Lett.* **73** (1994) 2887; **74** (1995) 3501 (E).

[76] F. D. M. Haldane. *J. Phys. C* **14** (1981) 2585.

[77] F. D. M. Haldane. *Phys. Rev. Lett.* **60** (1988) 635.

[78] F. D. M. Haldane. *Phys. Rev. Lett.* **66** (1991) 1529.

[79] F. D. M. Haldane. *Phys. Rev. Lett.* **67** (1991) 937.

[80] F. D. M. Haldane, Z. N. C. Ha, J. C. Talstra, D. Bernard, and V. Pasquier. *Phys. Rev. Lett.* **69** (1992) 2021.

[81] F. D. M. Haldane and M. R. Zirnbauer. *Phys. Rev. Lett.* **71** (1993) 4055.

[82] F. D. M. Haldane. In *Correlation Effects in Low-Dimensional Electron Systems*, A. Okiji and N. Kawakami (eds). Springer-Verlag, Berlin, 1994.

[83] F. D. M. Haldane. *Proceedings of the International Colloquium on Modern Quantum Field Theory II*, S. R. Das, G. Mandal, S. Mukhi, and S. R. Wadia (eds). World Scientific, Singapore, 1995.

[84] P. J. Hanlon, R. P. Stanley, and J. R. Stembridge. *Contemp. Math.* **138** (1992) 151.

[85] K. Hikami and M. Wadati. *J. Phys. Soc. Jpn.* **62** (1993) 469.

[86] K. Hikami and M. Wadati. *Phys. Lett. A* **469** (1993) 263.

[87] K. Hikami. *Nucl. Phys. B* **441** (1995) 530.

[88] K. Hikami and M. Wadati. *J. Math. Phys.* **44** (2003) 3569.

[89] L. Infeld and T. Hull. *Rev. Mod. Phys.* **23** (1951) 21.

[90] V. I. Inozemtsev. *J. Stat. Phys.* **59** (1990) 1143.

[91] V. I. Inozemtsev. *Commun. Math. Phys.* **148** (1992) 359.

[92] J. K. Jain. *Phys. Rev. Lett.* **63** (1989) 199.

[93] M. Jimbo and T. Miwa. *Algebraic Analysis of Solvable Lattice Models*. American Mathematical Society, Providence, RI, 1994.

[94] K. W. J. Kadell. *Compos. Math.* **87** (1993) 5.

[95] S. Kakei. *J. Phys. A: Math. Gen.* **29** (1996) L619.

[96] M. Karbach, D. Biegel, and G. Müller. *Phys. Rev. B* **66** (2002) 054405.

[97] M. Karowski. *Phys. Reports* **49** (1979) 229.

[98] M. Karbach, G. Müller, A. H. Bougourzi, A. Fledderjohann, and K.-H. Mütter. *Phys. Rev. B* **55** (1997) 12510.

[99] Y. Kato and Y. Kuramoto. *Phys. Rev. Lett.* **74** (1995) 1222.

[100] Y. Kato and Y. Kuramoto. *J. Phys. Soc. Jpn.* **65** (1996) 77.

[101] Y. Kato and Y. Kuramoto. *J. Phys. Soc. Jpn.* **65** (1996) 1622.

[102] Y. Kato. *Phys. Rev. Lett.* **78** (1997) 3193.

[103] Y. Kato, T. Yamamoto, and M. Arikawa. *J. Phys. Soc. Jpn.* **66** (1997) 1954.

[104] Y. Kato. *Phys. Rev. Lett.* **81** (1998) 5402.

[105] Y. Kato and T. Yamamoto. *J. Phys. A: Math. Gen.* **31** (1998) 9171.

[106] S. Katsura, T. Horiguchi, and M. Suzuki. *Physica* **46** (1970) 67.

[107] N. Kawakami. *Phys. Rev. B* **46** (1992) 1005.

[108] N. Kawakami. *Phys. Rev. B* **46** (1992) 3191.

[109] N. Kawakami. *J. Phys. Soc. Jpn.* **62** (1993) 2270.

[110] N. Kawakami. *J. Phys. Soc. Jpn.* **62** (1993) 4163.

[111] A. N. Kirillov, A. Kuniba, and T. Nakanishi. *Commun. Math. Phys.* **185** (1997) 441.

[112] N. Kitanine, J. M. Maillet, and V. Terras. *Nucl. Phys. B* **554** (1999) 647.

[113] V. G. Knizhnik and A. B. Zamolodchiko. *Nucl. Phys. B* **247** (1984) 83.

[114] F. Knop and S. Sahi. *Inv. Math.* **128** (1997) 9.

[115] M. Kollar and D. Vollhardt. *Phys. Rev. B* **65** (2002) 155121.

[116] M. Kohgi *et al. Phys. Rev. Lett.* **86** (2001) 2439.

[117] H. Konno. *Nucl. Phys. B* **473** (1996) 579.

[118] V. E. Korepin, N. M. Bogoliubov, and A. G. Izergin. *Quantum Inverse Scattering Method and Correlation Functions.* Cambridge University Press, Cambridge, 1993.

[119] Y. Kuramoto and H. Yokoyama. *Phys. Rev. Lett.* **67** (1991) 1338.

[120] Y. Kuramoto. *Physica B* **186–188** (1993) 831.

[121] Y. Kuramoto and Y. Kato. *J. Phys. Soc. Jpn.* **64** (1995) 4518.

[122] L. D. Landau and E. M. Lifshitz. *Quantum Mechanics – Nonrelativistic Theory.* Pergamon Press, Oxford, 1977.

[123] B. Laughlin *et al.* In *Field Theory for Low-Dimensional Systems*, G. Morandi *et al.* (eds). Springer Series in Solid State Sciences, Vol. 131. Springer, Berlin, 2000.

[124] R. B. Laughlin. In *Proceedings of the Inauguration Conference of Asia-Pacific Center for Theoretical Physics*, Y. M. Cho *et al.* (eds). World Scientific, Singapore, 1998.

[125] F. Lesage, V. Pasquier, and D. Serban. *Nucl. Phys. B* **435** (1995) 585.

[126] I. G. Macdonald. *Symmetric Functions and Hall Polynomials*, 2nd edn. Oxford University Press, Oxford, 1995.

[127] I. G. Macdonald. *Sém. Bouurbaki*, No. 797 (1995) 189.

[128] G. Mahan. *Phys. Rev.* **153** (1967) 882.

[129] M. L. Mehta and G. C. Mehta. *J. Math. Phys.* **16** (1975) 1256.

[130] M. L. Mehta. *Random Matrices*, 2nd edn. Academic Press, San Diego, 1991.

[131] M. Metzner and D. Vollhardt. *Phys. Rev. B* **37** (1988) 7382; **39** (1989) 12339 (E).

[132] J. A. Minahan and A. P. Polychronakos. *Phys. Lett. B* **302** (1993) 265.

[133] J. Minahan and A. P. Polychronakos. *Phys. Rev. B* **50** (1994) 4236.

[134] A. Molev. *Handbook of Algebra*, Vol. 3, M. Hazewinkel (ed.), p. 907. Elsevier, Amsterdam, 2003.

[135] J. Moser. *Adv. Math.* **16** (1975) 197.

[136] G. M. Müller, H. Thomas, H. Beck, and J. C. Bonner. *Phys. Rev. B* **24** (1981) 1429.

[137] M. Mohan and G. Müller. *Phys. Rev. B* **27** (1983) 1776.

[138] E. R. Mucciolo, B. S. Shastry, B. D. Simons, and B. L. Altshuler. *Phys. Rev. B* **49** (1994) 15197.

[139] M. V. N. Murthy and R. Shankar. *Phys. Rev. Lett.* **72** (1994) 3629.

[140] O. Narayan and Y. Kuramoto. *Phys. Rev. B* **73** (2006) 195116.

[141] O. Narayan, Y. Kuramoto, and M. Arikawa. *Phys. Rev. B* **77** (2008) 045114.

[142] C. Nayak and F. Wilczek. *Phys. Rev. Lett.* **73** (1994) 2740.

[143] M. Nazarov and V. Tarasov. *Publ. Res. Inst. Math. Sci.* **30** (1994) 459.

[144] M. Nazarov and V. Tarasov. *J. Reine Angew. Math.* **496** (1998) 181.

[145] T. Niemeijer. *Physica* **36** (1967) 377.

[146] M. A. Olshanetsky and A. M. Perelomov. *Phys. Rep.* **94** (1983) 313.

[147] E. Opdam. *Acta Math.* **175** (1995) 75.

[148] A. Polychronakos. *Phys. Rev. Lett.* **69** (1992) 703.

[149] A. P. Polychronakos. *Phys. Rev. Lett.* **70** (1993) 2329.

[150] A. P. Polychronakos. *Nucl. Phys. B* **419** (1994) 553.

[151] A. P. Polychronakos. In *Generalized Statistics in One Dimension*, Proceedings of the Les Houches Summer School, Session LXIX. Springer, New York, 1998.

[152] S. N. M. Ruijsenaars and H. Schneider. *Ann. Phys.* **170** (1986) 370.

[153] S. N. M. Ruijsenaars. *J. Math. Phys.* **38** (1997) 1069.

[154] S. Sahi. *IMRN* **20** (1996) 997.

[155] Y. Saiga and Y. Kuramoto. *J. Phys. Soc. Jpn.* **68** (1999) 3631.

[156] Y. Saiga and Y. Kuramoto. *J. Phys. Soc. Jpn.* **69** (2000) 3917.

[157] J. Sato, M. Shiroishi, and M. Takahashi. *J. Phys. Soc. Jpn.* **73** (2004) 3008.

[158] K. Schoutens. *Phys. Rev. Lett.* **79** (1997) 2608.

[159] D. Schuricht and M. Greiter. *Europhys. Lett.* **71** (2005) 987.

[160] D. Serban, F. Lesage, and V. Pasquier. *Nucl. Phys. B* **466** (1996) 499.

[161] B. S. Shastry. *Phys. Rev. Lett.* **60** (1988) 639.

[162] B. S. Shastry. *Phys. Rev. Lett.* **69** (1992) 164.

[163] B. S. Shastry and B. Sutherland. *Phys. Rev. Lett.* **70** (1993) 4029.

[164] B. D. Simons, P. A. Lee, and B. L. Altshuler. *Phys. Rev. Lett.* **70** (1993) 4122.

[165] B. D. Simons, P. A. Lee, and B. L. Altshuler. *Phys. Rev. Lett.* **72** (1994) 64.

[166] B. D. Simons, P. A. Lee, and B. L. Altshuler. *Phys. Rev. B* **48** (1993) 11450.

[167] B. D. Simons, P. A. Lee, and B. L. Altshuler. *Nucl. Phys. B* **409** (1993) 487.

[168] J. Sirker and A. Klümper. *Phys. Rev. B* **66** (2002) 245102.

[169] R. P. Stanley. *Adv. Math.* **77** (1989) 76.

[170] M. B. Stone, D. H. Reich, C. Broholm, K. Lefmann, C. Rischel, C. P. Landee, and M. M. Turnbull. *Phys. Rev. Lett.* **91** (2003) 037205.

[171] F. A. Smirnov. *Form Factors in Completely Integrable Field Theories.* World Scientific, Singapore, 1992.

[172] B. Sutherland. *J. Math. Phys.* **12** (1971) 246; 251.

[173] B. Sutherland. *Phys. Rev. A* **4** (1971) 2019.

[174] B. Sutherland. *Phys. Rev. A* **5** (1972) 1372.

[175] B. Sutherland. *Phys. Rev. B* **12** (1975) 3795.

[176] B. Sutherland. *Phys. Rev. B* **38** (1988) 5589.

[177] B. Sutherland and B. S. Shastry. *Phys. Rev. Lett.* **71** (1993) 5.

[178] B. Sutherland. *Beautiful Models.* World Scientific, Singapore, 2004.

[179] M. Takahashi. *Thermodynamics of One-Dimensional Solvable Models.* Cambridge University Press, Cambridge, 1999.

[180] K. Takemura and D. Uglov. *J. Phys. A: Math. Gen.* **30** (1997) 3685.

[181] J. C. Talstra and F. D. M. Haldane. *Phys. Rev. B* **50** (1994) 6889.

[182] J. C. Talstra. PhD thesis, Princeton University, Princeton, NJ, 1995, cond-mat/9509178.

[183] J. C. Talstra and F. D. M. Haldane. *J. Phys. A: Math Gen.* **28** (1995) 2369.

[184] J. C. Talstra and S. P. Strong. *Phys. Rev. B* **56** (1997) 6094.

[185] D. A. Tennant, R. A. Cowley, S. E. Nagler, and A. M. Tsvelik. *Phys. Rev. B* **52** (1995) 13368.

[186] L. S. Tevlin and J. L. Birman. *Phys. Lett. A* **227** (1997) 387.

[187] C. A. Tracy and H. Widom. *J. Stat. Phys.* **92** (1998) 809.

[188] D. Uglov. *Quantum Many-Body Problems and Representation Theory.* MSJ Mem., **Vol. 1, p. 193**. Mathematical Society of Japan, Tokyo, 1998.

[189] D. Uglov. *Commun. Math. Phys.* **191** (1998) 663.

[190] H. Ujino and M. Wadati. *J. Phys. Soc. Jpn.* **65** (1996) 2423.

[191] J. Voit. Rep. *Prog. Phys.* **58** (1995) 977.

[192] E. Witten. *Nucl. Phys. B* **185** (1981) 513.

[193] D. F. Wang, J. T. Liu, and P. Coleman. *Phys. Rev. B* **46** (1992) 6639.

[194] Y.-S. Wu. *Phys. Rev. Lett.* **73** (1994) 922; **74** (1995) 3906 (E).

[195] C. N. Yang and C. P. Yang. *J. Math. Phys.* **10** (1969) 1115.

[196] T. Yamada. *Prog. Theor. Phys.* **41** (1969) 880.

[197] T. Yamamoto and M. Arikawa. *J. Phys. A: Math. Gen.* **32** (1999) 3341.

[198] T. Yamamoto, Y. Saiga, M. Arikawa, and Y. Kuramoto. *Phys. Rev. Lett.* **84** (2000) 1308.

[199] T. Yamamoto, Y. Saiga, M. Arikawa, and Y. Kuramoto. *J. Phys. Soc. Jpn.* **69** (2000) 900.

[200] M. R. Zirnbauer and F. D. M. Haldane. *Phys. Rev. B* **52** (1995) 8729.

Index of symbols

$\langle\cdot,\cdot\rangle_c$	combinatorial inner product of symmetric functions	57
$\langle\cdot,\cdot\rangle_0$	integral inner product of symmetric functions	60
$c(s)$	content $a'(s)-l'(s)$	445
$C(u)$	$T_{21}(u)$ in gl_2 Yangian	402
C.T.	constant term	320
d	particle number density N/L	37
$d(s)$	$(a(s)+1)/\lambda + l(s)+1$	330
d_η	$\prod_{s\in\eta} d(s)$	330
$d'(s)$	$d(s)-1 = (a(s)+1)/\lambda + l(s)$	330
d'_η	$\prod_{s\in\eta} d'(s)$	330
$d''(s)$	$a(s)/\lambda + l(s)+1$	349
d''_κ	$\prod_{s\in D(\eta)} d''(s)$	349
δ	$(N-1, N-2, \ldots, 1, 0)$	339
$\hat{\Delta}_N(u)$	$\prod_i (u-\hat{D}_i)$	430
$\Delta(z)$	Vandermonde determinant $\prod_{i<j}(z_i - z_j)$	59
$\delta_{\uparrow\downarrow}$	$(\delta(N_\uparrow), \delta(N_\downarrow))$	375
$\tilde{\Delta}$	$\left(\Delta(z_1^2, \ldots, z_N^2)\right)^\lambda \Delta(z_1, \ldots, z_N)$	444
$\hat{D}_i$	$\underline{d}_i - z_i/\lambda$	423
$D(\mu)$	set of squares s belonging to Young diagram of partition μ	38
$D(u)$	$T_{22}(u)$ in gl_2 Yangian	402
$\hat{d}_i$	Cherednik–Dunkl operators	314
D_{ij}	$(N/\pi)\sin[\pi(i-j)/N]$	235
$\underline{d}_i$	Cherednik–Dunkl operators	423
$e''(s)$	$a'(s)/\lambda + N - l'(s)$	349
e''_κ	$\prod_{s\in D(\eta)} e''(s)$	349
$e(s)$	$(a'(s)+1)/\lambda + N - l'(s)$	330
$e'(s)$	$(a'(s)+1)/\lambda + N - l'(s) - 1$	330
e'_η	$\prod_{s\in\eta} e'(s)$	330
$E_\eta(z)$	non-symmetric Jack polynomial specified by composition η	319
e_η	$\prod_{s\in\eta} e(s)$	330
e_κ	elementary symmetric function	57

$(r)_\kappa$	generalized shifted factorial defined by (7.174) for partition κ	341
$\tilde{R}_{ij}(u)$	R-matrix for t–J model	439
σ	spin coordinate $(\sigma_1, \ldots, \sigma_N)$	392
$\hat{\sigma}$	an element of symmetric group S_N	336
$\hat{\sigma}^{\mathrm{R}}$	operator defined by (7.220)	348
$J_{\kappa/\mu}$	skew Jack functions defined by (7.329)	368
s_κ	Schur functions	57
S_N	symmetric group of order N	57
$\mathcal{S}_N$	set of symmetric functions	446
Sym	symmetrization operator	110, 347
τ	coordinate of internal symmetry	112
$\hat{T}_{ab}^{(n)}$	coefficient of u^{-n} in $\hat{T}_{ab}(u)$	425
$\hat{\tau}$	$K_{N-1} \cdots K_2 K_1$	324
$\hat{T}_{ab}(u)$	monodromy matrix	423
$\hat{T}_i$	Dunkl operator defined by (7.20)	316
Θ	generating operator $z_N \hat{\tau}$ of non-symmetric Jack polynomials	324
$\theta(x)$	Heaviside step function	63
θ_{ij}	$z_i/(z_i - z_j)$	314
$\tilde{T}_{ab}(u)$	monodromy matrix for t–J model	439
$ul(s)$	upper leg length defined by (7.107)	330
$ul'(s)$	upper leg colength defined by (7.109)	330
$U(\kappa; \lambda)$	operator defined by (8.22)	394
w_{ij}	$(Z_i + Z_j)/(Z_i - Z_j)$	182
ξ_i	$1/(\bar{\eta}_i - \bar{\eta}_{i+1})$	327
$X_i^{\alpha\beta}$	$|\alpha\rangle\langle\beta|$ for site i	221
$X^{\alpha\beta}$	operator defined by (8.45)	399
z	complex coordinate $(z_1, \ldots, z_N)$	392
$\mathbf{Z}$	set of integers	37
$\mathbf{Z}_{\geq 0}$	set of non-negative integers	312
ζ_κ	$1^{l_1} l_1! 2^{l_2} l_2! \cdots$	57

Index

Printed in the United States
By Bookmasters